The designer's guide to
wind loading of building structures

Price lists for all available BRE publications can be obtained from:

Publications Sales Office
Building Research Establishment
Garston, Watford, WD2 7JR
Tel: Garston (0923) 664444 (direct line)
      Watford (0923) 684040 (main switchboard)

The first part in this series is:
Cook N J *The designer's guide to wind loading of building structures —
Part 1: Background, damage survey, wind data and structural classification.*
London, Butterworths, 1985.

Supplements to *The designer's guide to wind loading of building structures* include:

Supplement 1. **Cook N J** *The assessment of design wind speed data: manual
worksheets with ready-reckoner tables.* Garston, BRE, 1985.

Supplement 2. **Cook N J, Smith B W and Huband M V** *BRE program
STRONGBLOW: user's manual.* BRE microcomputer package. Garston, BRE,
1985.

Building Research Establishment Report

The designer's guide to wind loading of building structures

Part 2: static structures

N J Cook, PhD, DSc(Eng), FRMetS, CEng, FIStructE
Building Research Establishment

BUILDING
RESEARCH
ESTABLISHMENT
Department of the Environment

Butterworths
London   Boston   Singapore
Sydney   Toronto   Wellington

 PART OF REED INTERNATIONAL P.L.C.

First published 1990

© **Crown copyright 1990**
Published by permission of the Controller of Her Majesty's
Stationery Office

British Library Cataloguing in Publication Data

Cook, N. J. (Nicholas John)
   The designer's guide to wind loading of building
     structures
   Pt. 2. Static structures
   1. Structures. Wind loads. Effects for design
   I. Title II. Series
   624.1'76

   ISBN 0-408-00871-7

Library of Congress Cataloguing-in-Publication Data

   (Revised for volume 2)
   Cook, N. J. (Nicholas John)
   The designer's guide to wind loading of building
     structures.
   (Building Research Establishment report)
   Includes bibliographies and indexes.
   Contents: pt. 1. Background, damage survey, wind
     data, and structural classification – pt. 2.
   Static structures.
   1. Wind-pressure. 2. Buildings – Aerodynamics.
   3. Structural design. I. Title. II. Series.
   TH891.C66 1985   690'.21   85-16621
   ISBN 0-408-00871-7

Composition by Genesis Typesetting, Borough Green, Kent
Printed and bound in Great Britain by Courier International Ltd,
Tiptree, Essex

Foreword

The first phase of BRE research into wind loading of buildings, begun in the early 1960s and directed by Mr C. W. Newberry, was implemented in design practice by the 1970 and 1972 British Standard code of practice for wind loads, CP3 Chapter V Part 2 and by the 1974 BRE *Wind loading handbook* by C. W. Newberry and K. J. Eaton.

This book, Part 2 of *The designer's guide to wind loading of building structures,* provides the designer with the latest methods and data for static structures. It implements the second phase of BRE research, spanning the last two decades, directed in the first decade by myself, followed by Dr K. J. Eaton and J. R. Mayne, and in the last decade by Dr N. J. Cook.

Its preparation has benefited substantially from the cooperation and help of many members of the international community of wind engineers, made possible by the links forged between national research institutes and universities around the world by the International Association of Wind Engineering and its periodic symposia.

The result is therefore a compilation of the most up-to-date information available worldwide in a form suitable for use by structural engineers and others for the design of building structures.

J. B. Menzies
Assistant Director, Building Research Establishment, 1989

Contents

Chapter 14   Review of codes of practice and other data sources   110

Chapter 15   A fully probabilistic approach to design   131

Chapter 16   Line-like, lattice and plate-like structures   153

Chapter 17   Bluff building structures   235

Chapter 18   Internal pressure   309

Chapter 19   Special considerations   341

Chapter 20   Design loading coefficient data   384

Acknowledgements

While most of the content of this Guide comes from the Building Research Establishment's own research programme, a significant proportion has been obtained from other expert sources, either directly under contract to BRE, from already published sources, or previously unpublished data donated by other research institutes.

Current and previous staff of BRE who have contributed directly or indirectly to Part 2 were: Dr P A Blackmore, A J Butler, Dr K J Eaton, Dr J F Eden, H A Fitzjohn, F J Heppel, J R Mayne, Dr J B Menzies, J Patient, Dr M D A E S Perera, D Redfearn, R G Tull and D S White. Also contributing while attached workers were: Dr E Maruta, Nihon University, Japan and Dr B L Sill, Clemson University, South Carolina, USA.

Principal extra-mural contractors were: British Maritime Technology for lattice truss and tower tests and data for flat roofs with curved eaves; the Flint and Neill Partnership, for the review of codes of practice, the 'Reference face' method for loading of lattice trusses and further calibration of the classification procedure of Chapter 10 in Part 1; Mr R I Harris, Wind Engineering Services, for development of the theory for the time-dependent response of internal pressures; Dr G R Walker, James Cook University of North Queensland, for the discussion on fatigue in cyclone-prone areas; Professor B E Lee, Portsmouth Polytechnic, for the discussion and data for shelter of low-rise buildings; Dr C J Wood, Oxford University, Department of Engineering Science, for data on wall and fence loads, including the effect of shelter, and for data on canopy roof loads, with and without under-canopy blockage.

Access to unpublished data or to source data of published material was kindly provided by: Dr J Blessmann, Universidade Federale do Rio Grande do Sul, Brazil; Dr W A Dalgliesh, National Research Council of Canada; E C English and Dr F H Durgin, Massachusetts Institute of Technology, USA; Dr P N Georgiou, Boundary Layer Wind Tunnel Laboratory, University of Western Ontario, Canada; Dr J D Holmes, Commonwealth Scientific and Industrial Research Organisation, Australia; Dr E Maruta, Nihon University, Japan; Dr J A Peterka, Colorado State University, USA; Dr D Surry, Boundary Layer Wind Tunnel Laboratory, University of Western Ontario, Canada; and other individuals and organisations who have permitted me to reproduce original photographs and data figures, acknowledged in the text.

I hope that all those mentioned above and everyone whose published work has been referenced will accept my thanks for their contribution. Finally, special thanks are extended to those who examined portions of the text and data, suggesting many valuable simplifications, additions and improvements: Dr K J Eaton and Dr J B Menzies, BRE; B W Smith, Flint and Neill Partnership; Dr G R Walker, James Cook University of North Queensland; and most especially to T V Lawson of the University of Bristol, Department of Aerospace Engineering, who undertook to read and comment on the whole of the book.

N J Cook
Building Research Establishment

11 About Part 2: static structures

11.1 Scope

The designer's guide to wind loading of building structures attempts to bring to the disposal of the designer the optimum methods and data for the design of buildings and allied structures to resist wind loads. The Guide is organised into three main Parts and, by making the divisions at natural boundaries within the subject, each Part is individually useful. To assist busy designers, the design data have been collated into separate chapters which can be accessed directly, independently of the chapters giving background, theory and discussion. For most static structures the data chapters of the Guide, Chapter 9 of Part 1 and Chapter 20 in this Part, contain all the instructions and data required to complete an assessment.

The first Part of the Guide, *Part 1: Background, damage survey, wind data and structural classification* divides the problem into three fundamental aspects:

(a) the **wind climate**;
(b) the **atmospheric boundary layer**; and
(c) the **building structure**.

Part 1 covers the first two meteorological aspects completely, giving methods and data which enable the designer to assess those wind characteristics at a particular site necessary for the assessment of any building structure. The third aspect of the building structure is introduced by a procedure to classify the building structure according to its expected response, which acts as a signpost to the second and third Parts. Part 1 also gives background information which is helpful to understanding the problem of wind loading, but is not essential in implementing the assessment procedures.

This Part of the Guide, *Part 2: Static structures*, covers the third aspect when the building structure can be assessed as if it were static. The designer is directed to this Part when the building structure is in one of the following classes:

Class A – small static structures,
Class B – moderate static structures,
Class C – large static structures, or
Class D1 – mildly dynamic structures (for ultimate limit states),

defined by the classification procedure in Chapter 10 of Part 1. In this Part of the Guide the assessment methods for static structures are introduced, their accuracy is assessed and their application in design is described. A format of loading coefficient

1

data is defined and calibrated which is compatible with the equivalent steady gust approach adopted by many codes of practice worldwide. Enough loading coefficient data are provided to enable the wind loading of the most common forms of static building structure to be assessed when used in conjunction with the data on wind characteristics at the site given in Part 1. Advice is given for dealing with the less common forms of static structure by means of wind-tunnel tests.

The third Part of the Guide, *Part 3: Dynamic structures*, covers the third aspect when the building structure must be assessed as being dynamic. The designer is directed to this Part when the building structure is potentially in one of the following classes:

Class D1 – mildly dynamic structures (for serviceability limit states),
Class D2 – fully dynamic structures, or
Class E – aeroelastic structures,

defined by the classification procedure in Chapter 10 of Part 1. In this Part of the Guide the assessment methods for dynamic structures are introduced, their accuracy is assessed and their application in design is described. The alternative approach of experimental testing at model scale is discussed and guidelines are given. The aspects of ultimate limits for safety and serviceability limits for structural distress and the tolerance of human occupants to motion of the structure are covered. The remainder of Part 3 is given over to the assessment of the loading and response of a range of typical dynamic structures, with data and worked examples.

11.2 Correspondence with international codes

In specifying performance standards for buildings, the International Organisation for Standardisation (ISO) has defined a series of technical terms for the physical parameters affecting the building structure [1]. The scope of the Guide is encompassed by the following ISO terms:

Agent – whatever acts on a building or part of a building – in this context, the wind;
Action – the influence of an agent – in this context, normal pressures and shear stresses on the surface of the building structure due to the wind;
Effect – the result of an action – in this context, all the effects of wind pressures acting on the building structure, such as strain, deflection, and damage, including the action effects, below;
Action effect – any force, set of forces or moment, resulting from the effect of actions or combinations of actions exerted on a building – in this context, the wind-induced forces and moments acting on all or part of the building structure;

The definition of action effect is intermediate between action and effect, describing the direct or first effect of the action in the loading chain, here corresponding to the integration of the pressures over the surface of the building structure to give forces and moments.

The way that the scope of the Guide corresponds to these ISO terms is shown in Figure 11.1. The characteristics of the agent wind is covered in Part 1. Wind loading of static structures is covered in Part 2 by the combination of actions and action effects, the structural effects being secondary. Wind loading of dynamic structures

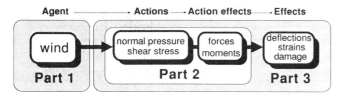

Agent ——————→ Actions ——→ Action effects ——→ Effects

| wind | normal pressure shear stress | forces moments | deflections strains damage |

Part 1   **Part 2**   **Part 3**

Figure 11.1 Wind loading chain

requires the structural effects to be included, and this is covered in Part 3.

ISO recommend a limit-state approach in conjunction with a partial-factor format for codes. This is the aim in the current process towards harmonisation of the European codes. The two concepts of 'limit states' and 'partial factors' are often regarded as being linked, whereas they are quite separate and distinct.

A limit-state approach is a formalised method of identifying the critical states which require analysis. In the context of wind loads, there are two main categories of limit states:

1. Ultimate limit states – the limit states corresponding to the maximum load-carrying capacity or where exceedance would result in structural failure, overturning or collapse.
2. Serviceability limit states – the limit states which are related to the criteria governing normal use, where exceedence would result in unsatisfactory service but not structural failure or collapse.

The first category is mostly concerned with safety and the second with economics of use.

Load-factor and permissible-stress formats are two opposing ways of approaching the principal aim in design, that of ensuring that the design strength or resistance exceeds the design loading by a suitable factor of safety. In the load-factor format, the design loads are multiplied by factors greater than unity to give larger factored loads which must be less than the design strength. In the permissible-stress format, the design strengths are multiplied by factors less than unity to give smaller factored strengths which must remain greater than the design loads. In both these cases, a single safety factor is applied to one side or the other of the load-resistance equation. This factor must cover the uncertainties on both sides of the equation: the uncertainties in load and in resistance, as well as the overall uncertainties of the assessment process. Since these uncertainties are lumped together, it is impossible to decide what part of the global factor covers each individual source of uncertainty.

The partial-factor format is aimed at overcoming the disadvantages of the two older formats, by breaking down the global factor into its component parts. In the simplest form, two factors are used: one increases the design loads to cover the uncertainties in their values; while the other decreases the design strengths to cover their component of uncertainty. The main advantage of this partition is that the partial factor on load can be adjusted for the type of applied load, independently of the partial factor on resistance that can be adjusted for the type of material used. There remains the uncertainty inherent in the design process which depends on how well the mathematical models represent the real structural behaviour. In the two-factor format this additional uncertainty must be absorbed into one or both of the load and resistance factors. An alternative three-factor format has been suggested by Armer and Mayne [2] in which the uncertainty of the design process is treated by a third factor. The value of this factor can then be adjusted as the mathematical models of structural behaviour improve. Further references to the philosophy and implementation of the partial-factor format are given in [3].

The assessment of wind loading in this Guide is directed towards estimation of the design wind loads identified through a limit-state approach. The theory and application of the design loads through partial-factor, load-factor or permissible-stress formats are not directly addressed, since these are subjects in their own right, and are outside the scope of the Guide.

11.3 Philosophy

The philosophy of Part 2 remains that of the Guide as a whole, established in Chapter 4 of Part 1 by the division of the problem into the three fundamental aspects, the **wind climate**, the **atmospheric boundary layer** and the **building structure**. Its implementation is essentially that of finding the optimum compromise between accuracy and complexity through the use of analytical, numerical, semi-empirical and empirical engineering models, as described in Chapter 1 of Part 1. However, in Part 2 the balance between these models is, by necessity, distinctly different than in Part 1. Empiricism in Part 1 is largely confined to the assessment of the characteristics of the wind climate and the effects of topography from meteorological measurements, while the statistical risk and recurrence model is an analytical model and the boundary-layer model is a semi-empirical fit to a numerical model. On the other hand, despite the application of some analytical and numerical models in the assessment process, the loading data presented here in Part 2 are largely empirical and derived directly from measurements. This reflects the lag in the development of practical theories for the complex flows around the range of building shapes encountered. Accordingly, the user will find that some subjects are covered in detail because they have been extensively researched and large amounts of empirical data are available, while other subjects are covered partially or not at all because little or no relevant data exist.

Knowledge and understanding of every subject improves with time. In the case of a young subject like wind engineering, this improvement is rapid. Sometimes it is found that earlier assumptions do not hold and the design advice previously based on these assumptions is not sound and must be modified. In such cases, the flaws in the earlier assumptions are explained and the consequences for design assessments are discussed. In this way the user will be able to recognise and avoid the rare cases of unsound advice in earlier publications. This is not a criticism of earlier work: change is a natural consequence of improved understanding. A key philosophy of this Guide is to encourage the user to develop an understanding of the problem and so avoid pitfalls. The fact that advice has appeared in print does not automatically make it sound. This Guide aims to give the best advice based on current understanding, but this too will be improved upon.

11.4 Sources

The development, testing and application of the engineering models used in this Guide are the result of more than two decades of research in the field now known as wind engineering. This field is so large that it cannot all be covered adequately by any one individual or even any one research organisation. The philosophy of this Guide has been to select and implement the optimum engineering models and data appropriate to each particular problem, irrespective of source. Most of the content of Part 1 came directly from BRE's own research programme, from research in house or from work performed directly under contract, with the remainder coming from already published data. Again, much of the content of Part 2 has come directly from BRE's research programme, but a greater reliance has been placed on externally published data and on unpublished data obtained directly from other expert sources in Australia, Brazil, Canada, Switzerland, West Germany and the USA. All the engineering models and data adopted for use in the Guide have been calibrated and shown to fit observations to the required accuracy.

11.5 Structure

The system which was adopted in Part 1 to number individual chapters, sections, sub-sections, etc., has been continued through this Part, so that the first chapter of this Part is Chapter 11. The system enables accurate cross-referencing within the Guide as a whole. Whenever a concept, equation or parameter is re-introduced in the text after an appreciable interval, the reader is pointed to its original introduction or definition by a cross-reference in the text. Thus §14.1.2.2 refers to the paragraph 'Denmark' in the section on 'Head codes' of Chapter 14 in this Part, whereas §2.2.8.5 refers back to the paragraph 'Induced forces' of Chapter 2 in Part 1. All references to other publications made in this Part are listed at the end of the book and are denoted in the text by reference numbers in square brackets, thus [1] refers to the first reference in the list.

Material that is not suitable for inclusion in the main chapters because it is too detailed or complex, but is nevertheless considered relevant, has been included in a number of appendices. The lettering sequence of the appendices in Part 1 has been retained. Appendix A, Nomenclature, has been amended to include the additional nomenclature of Part 2 and reproduced in full. The first new appendix of this Part is Appendix I. From time to time, newly obtained data which fills gaps in the main Parts, rapid aids to design in the form of simplified 'ready-reckoner' tables or programs for business or personal computer, or material that is considered too specialised for the majority of designers, will be published separately in a series of Supplements. A list of the Supplements currently available appears on page ii, opposite the title page.

11.6 Content

The first half of Part 2, Chapters 11–15, can be regarded as background and theory to the assessment of wind loading, helpful to the understanding of the whole problem but not essential to implementing the design procedures. Chapter 11, 'About Part 2: static structures', summarises the scope, use, structure and content of this Part of the Guide. Chapter 12, 'Assessment methods for static structures', defines the loading coefficients for pressure, shear stress, forces and moments, then introduces the assessment methods for wind loads established over the last several decades. Chapter 13, 'Measurement of loading data for static structures', introduces the methods and equipment by which wind loading data are acquired at full and model scale, then reviews the principles and development of methods for simulating the atmospheric boundary layer, and its effects on structures at model scale in the wind tunnel. Chapter 14, 'Review of codes of practice and other data sources', describes the implementation of wind loading data in codes of practice by reviewing the 'head code' of five nations, two specialised UK codes and several other published sources of design data. Chapter 15, 'A fully probabilistic approach to design', introduces the probability model adopted as the basis for this Part of the Guide which accounts for the joint variability of the loading due to the wind climate, atmospheric boundary layer and structure, in a manner which complies with the 'ideal approach' given in Chapter 4 of Part 1, and then defines the 'pseudo-steady format' used to implement the design data.

The second half of Part 2, Chapters 16–20, discusses the characteristics of the wind loading of a variety of building structures and presents the design loading coefficients for these forms. The variety of building form is divided into the categories line-like, lattice, plate-like and bluff-body models established in Chapter 8 of Part 1. Chapter 16, 'Line-like, lattice and plate-like structures' describes the characteristics of these forms in terms of forces and moments. Chapter 17, 'Bluff building structures', describes the characteristics of the external pressures on bluff-body forms of most conventional buildings, splitting the range by form into curved structures and flat-faced structures, and not forgetting the hybrid forms that occur when the individual forms are combined together. Chapter 18, 'Internal pressure', describes the characteristics of pressures inside bluff-body forms of structure, caused by the external pressure field acting on openings and porosity in the external skin, and modified by the position and porosity of internal divisions. Chapter 19, 'Special considerations' reviews some particular aspects omitted from earlier chapters and other design guides. Some of these aspects are difficult to implement as design rules, like the shelter effects caused by groups of buildings and the negative-shelter effects caused by neighbouring tall buildings. Most have been studied only recently, and the range of data is not sufficient for general guidance. A few are merely conceptual, such as strategies for optimising the design or for reducing wind loads. Several are quite specialised, like variable-geometry and air-supported structures. One is potentially very important, that is, the question of fatigue which played a major part in the failures in Australia due to cyclones Althea and Tracy, and may become more important in temperate regions with the current trend towards steel-framed, steel-sheet clad light industrial buildings. All the design data currently available which meets the requirements of the Guide for scope and accuracy have been collated and are presented in the standard design format in the last chapter, Chapter 20, 'Design loading coefficient data', together with the rules for applying them with the design wind data of Chapter 9 of Part 1.

There are five appendices to Part 2, containing more specialised data or advice. Appendix A, 'Nomenclature', is the first appendix of Part 1 defining the symbols used in the Guide, updated to include the new symbols used in Part 2. The remaining appendices are all new and continue the sequence from Part 1. Appendix I, 'Bibliography of modelling accuracy comparisons for static structures', is a list of references to published studies illustrating how well contemporary model-scale simulation techniques are able to represent the real world. Appendix J, 'Guidelines for *ad-hoc* model-scale tests' gives guidance to assist the designer to commission a wind-tunnel test and obtain the form of data he needs for design with confidence. Appendix K, 'A model code of practice for wind loads' gives the principal data of

Parts 1 and 2 of the Guide, simplified into a code format in the form of a series of BRE Digests which replaced the venerable Digest 119 'The assessment of wind loads' during 1989. The form and content of Appendix K is in line with the draft UK wind loading code BS6399 Pt2 which is due to replace the 1972 wind loading code, CP3 ChV Pt 2[4]. Appendix L, 'Mean overall loading coefficients for cuboidal buildings' presents a self-consistent set of overall force and base moment data discussed in Chapter 17 in more detail than is given in Chapter 20. Appendix M, 'A semi-empirical model for pressures on flat roofs', introduces a prototype mathematical model based on vortex dynamics for predicting the external pressure distribution over flat roofs. This model has the potential for development as a more general model to account for roof pitch and other architectural features such as parapets, as well as for walls of buildings, enabling more accurate computer-based design methods to be devised. At least this is the current hope at BRE. In the meantime, however, the tabulated forms of design data as presented in this Guide must serve.

12 Assessment methods for static structures

12.1 Introduction

A general description of building aerodynamics was given in Chapter 8 of Part 1, where the fluctuating nature of the wind loads, their action on static structures and the assessment methods were introduced. It is appropriate here to reiterate the model behaviour of static structures before embarking on the detail of the assessment methods.

Figure 12.1 represents some key parameters in the loading process in the form of 'traces', named from the use of pen-recorders to draw or trace the fluctuation with time. Trace (a) represents the reference dynamic pressure, q_{ref}, which is the pressure ideally recoverable from the kinetic energy of the wind at the reference location. It is given by the Bernoulli equation, Eqn 2.6 (§2.2.2), which may be redefined here as:

$$q = \tfrac{1}{2} \rho_a V^2 \tag{12.1}$$

in terms of the instantaneous wind speed, V, and the density of air, ρ_a ($= 1.225 \, \text{kg/m}^2$ in the UK, see §6.2.3.4).

The dynamic pressure trace contains fluctuations over a wide range of frequency, corresponding to the complete spectrum of the wind climate and atmospheric boundary layer shown in Figure 4.4 (§4.3). These fluctuations with time are often described as being 'dynamic', and a steady value as 'static', but these terms have been reserved in the Guide to define the dynamic and static classes of structural response. In order to prevent confusion, the other commonly used terms: **steady** for constant with time and **fluctuating** for varying with time, will be used.

Trace (b) represents the variation of surface pressure, p_1, at some point on the windward face of a building. The surface pressure is positive and closely follows the fluctuations of dynamic pressure, so that many individual fluctuations can be recognised in both traces. This corresponds to the real pressures on the west face of the slab block in Figure 8.17. Similarly, trace (c) represents the surface pressure, p_2, at some point on the side face of a building, corresponding to the real pressures in Figure 8.19. Here the surface pressure is negative (suction) and largely follows the inverted shape of the dynamic pressure trace (a). Although some individual features can still be recognised in both traces (c) and (a), there are many more additional features in trace (c) of comparable size or larger that were generated by the flow around the building and are not directly related to the incident turbulence.

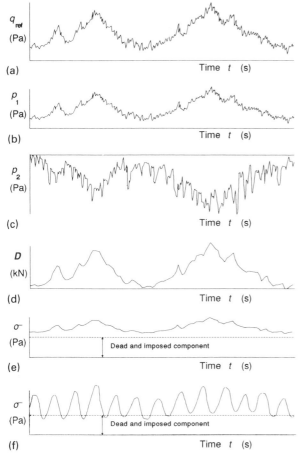

Figure 12.1 Typical wind loading traces: (a) dynamic pressure; (b) pressure on windward face; (c) pressure on side face; (d) drag of building; (e) stress in static building; (f) stress in dynamic building

Trace (d) represents the drag, D, the component of the total force on the building acting in the wind direction (§2.2.8.5). The drag follows the lower-frequency fluctuations of dynamic pressure, but there is less contribution in the high frequencies. This occurs because the smaller eddies in the flow do not act simultaneously over the surface of the building, as described by the aerodynamic admittance (§8.4.1).

Trace (e) represents the stress in a structural member of a static building structure (§10.2) due to the action of the fluctuating drag force, D, and any dead or imposed loads. The contribution due to wind may be a small or a large proportion of the total stress, depending on the weight and form of the structure and its susceptibility to wind, but it follows the fluctuations of drag exactly. This is in essence the definition of the term static used in this Guide (§4.5); namely that the loading effects, the instantaneous stresses, strains and deflections, are all proportional to the instantaneous loads. The principal design parameter for the ultimate limit state of a static structure is the maximum load likely to be experienced in its lifetime. The designer ensures that the design strength of the structure exceeds the stresses produced in the structure by a suitable safety margin to account for the variation of material strengths and for uncertainties in the assessment (§11.2). Consideration of the other loading effects, such as strain or deflection is secondary and is usually limited to checks on the stability of the deflected structure, or to serviceability limits such as interference between components and weathertightness.

Conversely, trace (f) represents the stress in a structural member of a dynamic building structure (§10.2). Although the envelope of the maximum stress generally follows the low-frequency component of the drag trace (d), the stress also fluctuates significantly at the natural frequency of the structure, corresponding to vibration of the structure in its first mode. Here the maximum stress is not directly related to the maximum load, but is also a function of the modal response characteristics of the structure (§8.6.3). This is covered in Part 3 of the Guide.

12.2 Analysis and synthesis

Wind engineering is based, like all other scientific fields, on the two complementary processes of analysis and synthesis. Analysis is the process of resolving a complex phenomenon into its simple components in order to discover the general laws or principles underlying the phenomenon. Synthesis is the process of proceeding from a collection of general laws or principles to deduce their consequences in combination. Each is the complementary or reverse process of the other. In practice, a complex phenomenon is observed, analysed into its general laws, then synthesised into a theory which accounts, not only for the observations, but also for the effects which occur under different conditions. It follows that the analysis process must be reversible to allow the synthesis process to occur.

In the assessment of the wind climate and atmospheric boundary layer in Part 1, it was not necessary to address the specific role of these two processes, but their action is evident. Taking the wind climate as an example, analysis is represented by the alternative methods given in Appendix C for determining the statistical climate parameters from observed data, whereas the corresponding synthesis is represented by the wind climate map and S-Factors in §9.3 for making design predictions. Note that the analysis process here is not perfectly reversable because fitting the model distribution to the data is an averaging process. The actual position of the data values around the fitted model distribution is treated as variance error or 'scatter'. As a consequence, predictions made by the synthesis process produce estimates that lie exactly on the model line and it is not possible to reproduce the original data values.

In the review of assessment methods that follows, it will be seen that some of the analysis methods are not reversible because part of the variation of the data is averaged out or suppressed. Consequently, the synthesis of design methods using these data must use a compensating theory or assumptions to replace this missing information. In general, the early simple analysis methods require more assumptions in the synthesis than contemporary, more complex, analyses. Unfortunately, the range of building types and shapes is so large that insufficient contemporary data exist to cover all the common cases, and it will be necessary to employ the simpler data for some applications.

12.3 Loading coefficients

The loading coefficients are the loading actions and effects reduced to the form of non-dimensional parameters (§2.2.5). They are principally dependent on the shape of the building structure, so are sometimes called 'shape factors'.

12.3.1 Reference wind speed and dynamic pressure

Chapter 4 of Part 1 showed that the problem of assessment is ideally divided at the spectral gap, between the macrometeorological and the micrometeorological peak of the wind spectrum of Figure 4.4, usually at a period of one hour ($T = 1$ h). The fluctuations of the hourly-mean wind speed, $\overline{V}\{z\}$, with time are for periods greater than one hour and characterise the wind climate only, as shown in Figure 4.5(b). The hourly-mean wind speed also varies with height, characteristic of the atmospheric boundary layer over the particular terrain. Accordingly, a reference height, z_{ref}, must be chosen for the reference wind speed, $\overline{V}_{ref}$, relative to the height of the building structure. Thus:

$$\overline{V}_{ref} = \overline{V}\{z = z_{ref}, T = 1\,\text{h}\} \tag{12.2}$$

The reference hourly-mean dynamic pressure, $\overline{q}_{ref}$, corresponding to the reference hourly-mean wind speed, $\overline{V}_{ref}$, is given from 12.1, by:

$$\overline{q}_{ref} = \tfrac{1}{2}\, \rho_a\, \overline{V}_{ref}^2 \tag{12.3}$$

Strictly, $\overline{q}_{ref}$, is only a notional dynamic pressure calculated from the reference wind speed. It is not the hourly-mean dynamic pressure at the reference height, $\overline{q}\{z = z_{ref}\}$, since the latter includes the mean-square terms from the three components of turbulence, given by 2.9 (§2.2.3), and so is typically about 6% larger.

12.3.2 Local and global coefficients

It is useful to be able to distinguish **local** values at a point from **global** values averaged over an area or the whole of the building structure. Accordingly, the symbol c will be used to denote a local coefficient and the symbol C to denote a global coefficient defined by:

$$C = \int c\, \mathrm{d}\phi \tag{12.4}$$

where ϕ is the influence coefficient defining the weighting of the local loading coefficient, c, over the loaded area (see §8.6.2.1). Thus c_p is the pressure coefficient representing the pressure at a single point and C_p is the coefficient representing the pressure averaged over a defined area.

12.3.3 Forms of coefficient

The loading coefficients, C and c, are the actions and action effects non-dimensionalised (§2.2.5) by the reference dynamic pressure. Use of an hourly-mean dynamic pressure removes the contributions of the wind climate, but leaves the contributions from both the atmospheric boundary layer and the structure. This definition is the required form for ideal analysis (see §4.4), giving statistical independence and stationarity of records longer than $T = 1$ h. The values

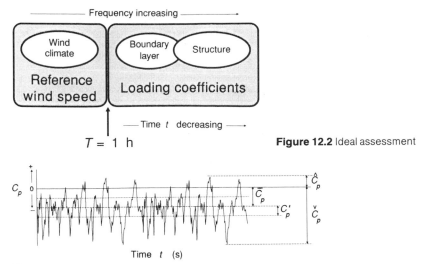

Figure 12.2 Ideal assessment

Figure 12.3 Typical pressure coefficient trace

of loading coefficient are therefore dependent on the site terrain characteristics as well as the shape and size of the building structure. This is indicated in Figure 12.2, using the format of Figure 4.7.

The loading coefficients fluctuate in value with time, and thus possess all the usual mathematical and statistical properties. These are indicated by the standard mathematical notation adopted in Part 1 (Appendix A) applied to the coefficient symbol. A typical trace of pressure coefficient, C_p, is shown in Figure 12.3. Using this as an example:

$\overline{C}_p$ indicates the mean value,

C'_p indicates the rms value,

$\hat{C}_p$ indicates the maximum (most positive) value, and

$\check{C}_p$ indicates the minimum (most negative) value,

as marked on the diagram. In the wind data of Part 1, the only extreme of interest was the maximum value, e.g. gust speed, $\hat{V}$. In the case of the loading actions, however, both maxima (most positive) and minima (most negative) are of interest. The notation for maximum value, ^, requires the complementary notation, ˇ, for minimum value. The maximum value does not need to be positive nor need the minimum be negative in sign, as it is in Figure 12.3, but this will often be the case in practice.

The best form of coefficient for design is one which is the least sensitive to variations in the load duration, the site terrain, etc. and is also simple to apply. A special form of loading coefficient that minimises dependence on the atmospheric boundary layer, is defined in Chapter 15. This is used as the standard for the design data of this Part of the Guide and several recent codes of practice.

12.3.4 Pressure coefficients

Fluctuations of pressure due to wind are very small compared with the absolute atmospheric pressure ($p_{atm} \simeq 100\,kPa$), so that it is always more convenient to measure wind induced pressure, p, as the difference from the ambient static pressure (§2.2.2), than as an absolute value. Accordingly, the pressure coefficient, C_p, is defined by:

$$C_p = p / \bar{q}_{ref} \qquad (12.5)$$

Pressure at any single point acts equally in all directions (Pascal's law §2.2.1), so is a scalar quantity. Pressure acts on a solid surface as a normal stress, so the resulting force takes its direction from the normal to the surface and its value from the integral of the pressure over the area of the surface.

Normal pressure is the dominant wind action for the majority of building structures. In diagrams showing pressure distributions acting on a building it is conventional to indicate positive pressures by arrows acting normally into the surface and negative pressures (suctions) by arrows acting normally out from the surface. Figure 12.4(a) represents the distribution of the local pressure coefficient, c_p, along the middle of a cuboidal building with the wind normal to the front face. As pressure or stress is force per unit area, the global pressure coefficient, C_p, for an area is equivalent to a uniformly distributed load. This is represented in Figure 12.4(b) for the same case as (a), with the roof divided into convenient areas. Because of these properties, the pressure coefficient is independent of any coordinate axis convention, unlike the force and moment coefficients defined below. This makes the pressure coefficient a particularly convenient parameter for describing the wind loading.

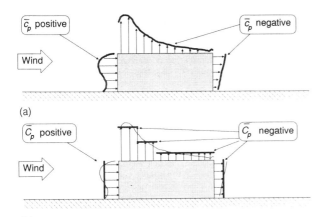

(a)

(b)

Figure 12.4 Representing pressure coefficients: (a) local pressure coefficient; (b) global pressure coefficient

The most positive value of pressure coefficient possible is obtained by bringing the flow to rest and recovering all the kinetic energy, i.e. the ideal dynamic pressure. Hence in smooth uniform flow, the highest possible *mean* pressure coefficient is $\bar{c}_p = 1.0$. In uniform turbulent flow the contributions from the three turbulence components raises the highest possible mean value to:

$$\bar{c}_p = (1.0 + S_u^2 + S_v^2 + S_w^2) \qquad (12.6)$$

from from Eqns 2.9 and 9.30. In the atmospheric boundary layer, the available kinetic energy increases with height above ground, but Eqn 12.6 applies approximately when the reference height is at the top of the building (see §12.4.1.1), giving the maximum $\bar{c}_p \simeq 1.06$ typically. The highest possible instantaneous value occurs when the kinetic energy of the maximum gust is recovered, giving $\hat{c}_p = S_G^2 \approx 2.5$, typically at $z_{ref} = 10\,m$ in open country (Category 2).

The most negative value of pressure coefficient is theoretically unlimited and may reach very high values in the accelerated flow at sharp corners of buildings, particularly along the upwind eaves of low-pitch roofs. Values as low as $\bar{c}_p \simeq -5$ and $\hat{c}_p \approx -15$ have been recorded near the upwind corners of $-15°$ pitch (troughed) roof models.

12.3.5 Shear stress coefficients

The action of shear stress is due to friction between the surface of a building structure and the local wind speed past the surface. The shear stress coefficient is defined by:

$$C_\tau = \tau / \bar{q}_{ref} \tag{12.7}$$

Unlike pressure, shear stress is a vector quantity which acts in a given direction parallel to the surface. The shear stress is represented in diagrams by a half-arrow symbol aligned in the direction of the vector, as indicated in Figure 12.5 for the same example building as Figure 12.4.

On aerodynamically smooth surfaces (§2.2.7.1) the shear stress is generated entirely by the action of viscosity, so that the vector of the local shear stress coefficient, c_τ, acts in the direction of the local wind speed vector. In the example of Figure 12.5(a) the *mean* shear stress acts downwards over the lower two-thirds and upwards over the upper third of the front face, away from the front stagnation point. On the roof, the shear stress acts upwind in the separation bubble at the

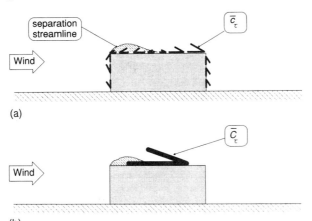

(a)

(b)

Figure 12.5 Representing shear stress coefficients: (a) local shear stress coefficient; (b) global shear stress coefficient

upwind eave and downwind on the remainder of the roof after flow reattachment. The global coefficient, C_τ, for a surface is given by the integral of the local vectors. In most cases of aerodynamically smooth surfaces, the global shear stress will be insignificant compared with the normal pressures because the viscosity of air is so small. Exceptions where shear stress is significant occur when the area of attached flow is particularly large, such as roofs of very long low buildings, large plate-like structures, canopies or dutch barns when the wind is parallel to the surface.

On aerodynamically rough surfaces (see §2.2.7.2) the viscous shear stress is replaced by normal pressures acting on the individual elements of roughness. When the roughness is several orders of magnitude smaller than the surface dimensions, the resulting action is usually represented *as if it were* an equivalent shear stress. When the roughness is random and homogeneous, such as a crushed stone or gravel rendering, equivalence is almost exact and the local coefficient, c_τ, again acts in the direction of the local wind speed vector. When the roughness is not random, but has corrugations aligned in a specific direction, the shear stress will have a strong directional dependence and may not align with the wind speed vector.

Suppose, for example, that the roof of the building shown in Figures 12.4 and 12.5 is clad with a contemporary profiled metal sheet system. With the wind normal to the axis of the corrugations, the surface is aerodynamically rough and gives rise to the situation shown in Figure 12.6. Here the local shear stress coefficient, c_τ, is

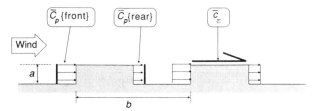

Figure 12.6 Equivalent shear stress

equivalent to the difference between the normal pressure coefficient on the vertical faces of the corrugations, thus:

$$c_\tau = (C_p\{\text{front}\} - C_p\{\text{rear}\})\, a\,/\,b \tag{12.8}$$

where a is the height and b is the horizontal pitch of the corrugations. (The ratio a/b transforms the vertical area of the corrugations to the horizontal area of the surface required for the 'per unit area' component in the definition of stress.) With the wind parallel to the axis of the corrugations, the surface is aerodynamically smooth and the shear stresses revert to the much smaller viscous values. At intermediate angles there will be two components, a very small viscous component parallel to the corrugations and the much larger, usually dominant, normal pressure component from Eqn 12.8. As the size of the surface roughness elements is increased compared with the size of the building structure, this concept of equivalent shear stress becomes less useful. Eventually, it becomes better to assess the loading in terms of the distribution of normal pressures on the individual roughness elements.

12.3.6 Force coefficients

The wind-induced forces are action effects resulting from the normal pressure and shear stress actions integrated over all or part of the building structure. Force coefficient is defined by:

$$C_F = F\,/\,(\bar{q}_{\text{ref}}\, A_{\text{ref}}) \tag{12.9}$$

where A_{ref} is a reference area for the building structure, which is required to make the coefficient non-dimensional (see §2.2.5). Force is a vector quantity so has a direction and a point of application in the three dimensions of space in addition to the magnitude. This requires the coordinate axes convention to be defined.

There are two standard axes conventions:

1 **Wind axes**. Here the horizontal axes are aligned in the in-wind and cross-wind directions. The forces are the in-wind *drag* force, D, and the cross-wind *lift* force, L (§2.2.8.5). The lift force may act in either or both of the two orthogonal crosswind directions.

2 **Body axes**. Here the axes are aligned relative to the building structure. The forces are the x-axis force, F_x, the y-axis force, F_y, and the z-axis force, F_z.

Wind axes are generally used when the interest is in the aerodynamics, since drag can be related to the momentum loss in the wake and lift can be related to vorticity (see §2.2.8.5). Body axes are more often used in the design of building structures because the body forces relate directly to the structural strength requirements.

In the general case, the body axes may be skewed to the wind axes so that they do not coincide, as shown in Figure 12.7. Here both sets of axes have a common origin at the centroid of the building plan at ground level. The x and y body axes are horizontal and the wind direction is skewed by an angle θ from the x-axis. The z-axis is vertical and common to both sets. **Note**: The symbol θ indicates mean wind direction *relative to the structure*, i.e. in body axes, whereas the symbol Θ, introduced in Part 1, indicates mean wind direction *relative to North*.

The corresponding force coefficients are indicated by arrows, with each force defined as positive in the positive direction of the corresponding body axis. The

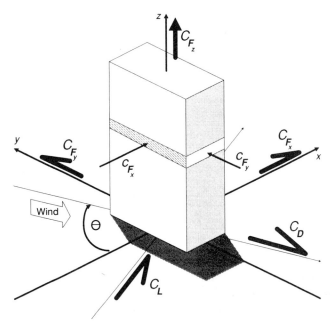

Figure 12.7 Representing force coefficients

global lift C_L, drag C_D, x-force C_{F_x}, and y-force C_{F_y}, coefficients for the whole building are indicated using the shear arrows since they are all base shears, whereas the vertical z-force coefficient C_{F_z}, is indicated by the conventional arrow. (Note that as the y-axis points into the wind for the direction shown, C_{F_y} will have a negative value in this example.)

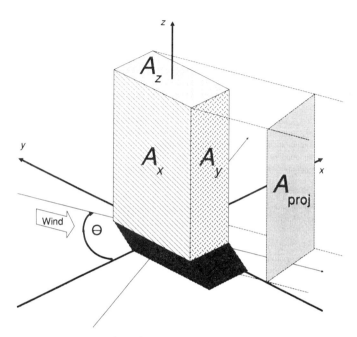

Figure 12.8 Alternative reference areas

Several alternative definitions for the reference area can be used, as indicated in Figure 12.8. The areas of each face of the building, A_x, A_y and A_z, are fixed in value. However, the horizontally projected area of the building in the wind direction, A_{proj}, sometimes called the *shadow area*, changes in value with wind direction. In the example of Figure 12.8, A_{proj} is given by:

$$A_{proj} = A_x \cos\theta + A_y \sin\theta \tag{12.10}$$

The projected area, A_{proj}, is usually adopted for structures of circular or elliptical section (there being no faces), and it is often also used for lattice structures in combination with a factor for porosity (see §8.2.1 and §16.3).

Lift and drag coefficients are defined using a fixed reference area, usually the area of the largest face, A_x, thus:

$$C_D = D / (\bar{q}_{ref} A_x) \tag{12.11}$$

$$C_L = L / (\bar{q}_{ref} A_x) \tag{12.12}$$

although, either A_y or A_{proj} could be used.

The body-axis force coefficient for each axis is usually defined using the area of the corresponding face for that axis as the reference area, thus:

$$C_{F_x} = F_x / (\bar{q}_{ref} A_x) \tag{12.13}$$

$$C_{F_y} = F_y / (\bar{q}_{ref} A_y) \tag{12.14}$$

$$C_{F_z} = F_z / (\bar{q}_{ref} A_z) \tag{12.15}$$

This choice of reference area makes the body-axis force coefficients directly equivalent to the difference in the global pressure coefficients across opposite faces, i.e:

$$C_{F_x} = (C_p\{\text{front}\} - C_p\{\text{rear}\}) \tag{12.16}$$

Either set of wind and body-axis global force coefficients may be determined from the other using the normal trigonometrical relationships and making due allowance for the reference areas used. For example:

$$C_D = C_{F_x} \cos\theta - C_{F_y} \sin\theta\, A_y/A_x \tag{12.17}$$

$$C_L = C_{F_x} \sin\theta + C_{F_y} \cos\theta\, A_y/A_x \tag{12.18}$$

Note that the term A_y/A_x converts the reference area used in Eqn 12.14 to that used in Eqns 12.11 and 12.13.

Local force coefficients are usually used to define the loads acting on thin strips of long structures, as indicated in Figure 12.7, for use with the strip model of §8.2.2. Reducing the strip to an infinitesimal width makes c_{F_x} the coefficient of load per unit length. The global coefficients are obtained by integrating all strips, thus:

$$C_{F_x} = \int_0^1 c_{F_x}\, \mathrm{d}(z/H) \tag{12.19}$$

for the example of Figure 12.7.

12.3.7 Moment coefficients

The wind-induced moments are the action effects resulting from the the normal pressure and shear stress actions multiplied by their moment arms and integrated over all or part of the building surface. Moment coefficient is defined by:

$$C_M = M / (\bar{q}_{ref} A_{ref} l_{ref}) \tag{12.20}$$

where A_{ref} is a reference area and l_{ref} is a reference moment arm. The reference area can be any of the alternative definitions of Figure 12.8. The reference moment arm is often taken as the height, H, of the structure for base moments.

Either wind axes or body axes conventions can be used, but body axes are more often used in design, as illustrated in Figure 12.9. The direction of each moment has been indicated using the conventional *right-hand corkscrew rule*, giving a positive moment in the clockwise direction when viewed upwards along the relevant axis.

Moment coefficients are only required when the force coefficients are defined to act through the fixed origin of the coordinate system as indicated in Figure 12.7. An alternative approach is to define the location of the force vector that gives no net moment. This location is usually called the centre of force or, more often by invoking equivalence to the global pressure coefficients given by Eqn 12.16, as the **centre of pressure**. This is indicated in Figure 12.9 for C_{F_x} as 'CP'. In this example, the force coefficient C_{F_x} acting a distance $z = z_{F_x}$ above the y-axis contributes $C_{F_x} z_{F_x}/H$ to the moment coefficient C_{M_y}. The vertical force coefficient, C_{F_z}, will

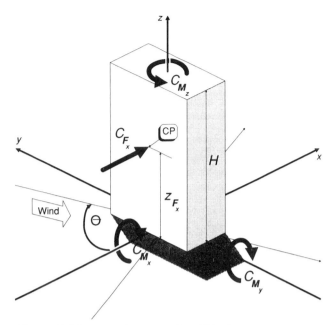

Figure 12.9 Representing moment coefficients

also contribute to C_{M_y} if 'CP' for the roof is not in the centre. This gives the general cases:

$$C_{M_x} = - C_{F_y} z_{F_y}/H + C_{F_z} y_{F_z}/H \qquad (12.21)$$

$$C_{M_y} = + C_{F_x} z_{F_x}/H - C_{F_z} x_{F_z}/H \qquad (12.22)$$

$$C_{M_z} = - C_{F_x} y_{F_x}/H + C_{F_y} x_{F_y}/H \qquad (12.23)$$

when the reference moment arm $l_{\text{ref}} = H$.

12.3.8 Simplified approach of the Guide

Owing to the many possible combinations of axes conventions, reference areas and reference moment arms it is vitally important to be certain of the definitions of force and moment coefficients used in any source of design data. On the other hand, pressure and shear stress coefficients give no such problems. The policy of the Guide will be to favour the use of these direct actions and to express data wherever possible in terms of distributions of pressure coefficients with the position of the corresponding centre of pressure.

12.4 Assessment methods

12.4.1 Quasi-steady method

12.4.1.1 Quasi-steady vector model

This is the earliest model for the action of turbulent wind, devised when only mean values of the loading coefficients were available. It makes the simple, but inaccurate assumption that the building structure responds to the atmospheric turbulence eddies as if they were steady changes of mean wind speed and direction, so that the fluctuations of the loading correspond exactly with the variations of the incident wind. Thus the instantaneous normal pressure, p, is given by:

$$p = \tfrac{1}{2}\,\rho_a\,V^2\{t\}\,\bar{c}_p\{\theta + \alpha,\,\beta\} \qquad (12.24)$$

where V is the magnitude of the instantaneous wind speed vector. The pressure coefficient, $\bar{c}_p$, is the mean value corresponding to the instantaneous azimuth angle, α (the angle in the horizontal plane measured from the mean wind direction, θ) and the elevation angle, β (angle normal to the horizontal plane), of the wind speed vector. The magnitude, V, of the wind speed vector is given by:

$$V^2\{t\} = (U + u)^2 + v^2 + w^2 \qquad (12.25)$$

in terms of the mean wind speed, U, and the turbulence components, u, v, and w (see §2.2.3). The azimuth and elevation angles are given by:

$$\alpha = \arctan(v\,/\,[U + u]\,) \qquad (12.26)$$

$$\beta = \arcsin(w\,/\,\overline{V}) \qquad (12.27)$$

In reality, this model is only likely to be good for eddies that are much larger than the structure.

In the quasi-steady vector model, all variation with time is contained in the wind speed, while the structure is assumed not to contribute any fluctuations. The pressure coefficient is therefore assumed constant with time, so is indicated in Eqn 12.24 as a mean value, but is a function of the instantaneous azimuth and elevation angles. The instantaneous pressure or any other action is assumed to be the same as would occur in a steady wind from the same angle, hence the term 'quasi-steady'. The division of the problem is not at the spectral gap between the wind climate and the boundary layer, but is between the atmospheric boundary layer and the structure, as shown in Figure 4.7(b).

The ideal division of Figures 4.7(f) and 12.2 is restored by dividing both sides of Eqn 2.24 by $\tfrac{1}{2}\,\rho\,\overline{V}^2$, which gives:

$$c\{t,\,\theta + \alpha,\,\beta\} = (V^2\{t\}\,/\,\overline{V}^2)\,\bar{c}\{\theta + \alpha,\,\beta\} \qquad (12.28)$$

for the general case of any loading coefficient. This implies that the quasi-steady vector assumption works and there is no interaction between the atmospheric boundary layer and the structure.

The quasi-steady vector model is a good model for the local actions on lattice plates, the form of Eqn 12.24 being identical to the lattice-plate model, Eqn 8.3 (see §8.4.1). The quasi-steady vector model is also applied to local actions on bluff bodies (see §8.4.2) where the equivalence is not so good. In this case the atmospheric turbulence is distorted by the divergence of the flow [5], and additional fluctuations are introduced by the structure, whereas neither effect is accounted for by the model. The quasi-steady vector model remains a good model for global actions on lattice plates. It is adequate for bluff bodies, provided the effect of the various eddy sizes in the atmospheric turbulence is accounted for by the aerodynamic admittance function, $\chi\{n\}$, because this excludes most of the smaller body-generated fluctuations (see §8.4.1 and §8.4.2.1). This model is also currently the best available for the modal forces of dynamic structures, where the aerodynamic admittance is replaced by the joint acceptance function, $J\{n\}$ (see §8.6.3.2 and Part 3).

12.4.1.2 Linearised model

The full quasi-steady vector model can be simplified by removing the small second-order components. Firstly, the squared terms in Eqn 12.25 contribute only a few percent in typical intensities of turbulence, so may be removed leaving:

$$V^2\{t\} \simeq U^2 + 2Uu \qquad (12.29)$$

Secondly, the variation of $\bar{c}_p\{\theta + \alpha, \beta\}$ with azimuth and elevation angle either side of the mean wind direction can be assumed to be linear for small v and w, giving:

$$\bar{c}\{\theta + \alpha, \beta\} \simeq \bar{c}\{\theta\} + (v/\overline{V}) \, \partial\bar{c}\{\theta\}/\partial\alpha + (w/\overline{V}) \, \partial\bar{c}\{\theta\}/\partial\beta \qquad (12.30)$$

so that Eqn 12.28 now becomes:

$$c\{t\} \simeq (1 + 2u/\overline{V}) \, [\bar{c}\{\theta\} + (v/\overline{V}) \, \partial\bar{c}\{\theta\}/\partial\alpha + (w/\overline{V}) \, \partial\bar{c}\{\theta\}/\partial\beta] \qquad (12.31)$$

Expanding and discarding the small turbulent cross-products leaves:

$$c\{t\} \simeq \bar{c}\{\theta\} + 2u/\overline{V} \, \bar{c}\{\theta\} + (v/\overline{V}) \, \partial\bar{c}\{\theta\}/\partial\alpha + (w/\overline{V}) \, \partial\bar{c}\{\theta\}/\partial\alpha \qquad (12.32)$$

as the expression for the instantaneous loading coefficient. Averaging Eqn 12.32 with time gives the mean value:

$$\bar{c} = \bar{c}\{\theta\} \qquad (12.33)$$

showing that all the turbulence terms are lost. Similarly, the rms value is:

$$c' = [\overline{(c - \bar{c})^2}]^{1/2}$$
$$\simeq [(2[u'/\overline{V}]\bar{c}\{\theta\})^2 + ([v'/\overline{V}] \, \partial\bar{c}\{\theta\}/\partial\alpha)^2 + ([w'/\overline{V}] \, \partial\bar{c}\{\theta\}/\partial\alpha\beta)^2]^{1/2} \qquad (12.34)$$

in which the rms turbulence intensity terms can be replaced by the corresponding turbulence intensity S-factors of Chapter 9 (§9.4.2). Near the ground the intensity of the vertical turbulence component is less than half the in-wind component and the vertical terms in Eqns 12.32 and 12.34 are sometimes also discarded.

12.4.1.3 Equivalent steady gust model

Two further simplifications are often made to produce the equivalent steady gust model, introduced in §8.6.2.3, which forms the basis of many codes of practice.

The first simplification is to discard the cross-wind and vertical turbulence terms in Eqn 12.32, leaving:

$$c \simeq (1 + 2 \, u/\overline{V}) \, \bar{c}\{\theta\} \qquad (12.35)$$

Eqn 12.35 implies that the instantaneous pressure exactly follows changes in dynamic pressure. Hence it is *assumed* that the shapes of the surface pressure traces (b) and (c) in Figure 12.1 scale exactly to the dynamic pressure trace (a). The mean loading coefficient, $\bar{c}$, is still given by Eqn 12.33, but Eqn 12.34 for the rms loading coefficient, $\bar{c}'$, now simplifies to:

$$c' = 2 \, S_u |\bar{c}\{\theta\}| \qquad (12.36)$$

The amount of simplification used to reach this point from the full quasi-steady vector model is quite extensive. Some aspects are oversimplified, which can lead to errors unless the problem is recognised. Discarding the cross-wind and vertical turbulence terms has the effect of modelling the atmospheric turbulence as temporary variations of wind speed, with no changes of azimuth or elevation angle.

So long as the partial derivatives, $\partial \bar{c}/\partial \alpha$ and $\partial \bar{c}/\partial \beta$, are small compared with the mean, $\bar{c}$, this simple model is good. However, when the mean, $\bar{c}$, is zero this model erroneously predicts that the rms is also zero.

A good illustration of this problem is given by the flow around circular cylinders. Figure 12.10(a) shows the distribution of mean pressure coefficient, $\bar{c}_p$, measured by BRE around a full-scale silo. Figure 12.10(b) shows the corresponding measured rms pressure coefficient, c'_p, and also the model prediction from Eqn 12.36.

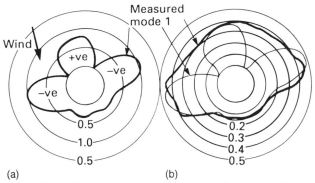

(a)   (b)

Figure 12.10 Pressure around a circular silo: (a) mean c_p; (b) rms c'_p

Although the model is adequate where the mean is large, it underestimates the rms where the mean changes in sign from pressure to suction. Similar comparisons in the wind tunnel[5] and at full scale[6] show that this problem is less serious with sharp-edged structures, because the mean pressure jumps from positive on the windward face to negative on the other faces at the corners. However, these and other studies demonstrate the dependence on frequency, owing to the action of the aerodynamic admittance (see §8.4.2.1). For turbulence eddies larger than the structure the correspondence is good, but with increasing frequency Eqn 12.36 progressively overestimates the rms value as the smaller eddies become uncorrelated over the structure.

Accordingly, the second simplification is to limit the duration of the extreme by adopting the time constant, t, given by Eqn 8.12. This defines the smallest duration load which acts simultaneously over the structure, the 'equivalent steady gust load' (§8.6.2.3) which gives the model its name. In the general case, l, the general size parameter defined in Eqn 10.2 replaces the specific size parameter in Eqn 8.12, giving:

$$t \bar{V} = 4.5 \, l \tag{12.37}$$

The order of the three variables, t, $\bar{V}$ and l, in Eqn 12.37 gives its popular name of the 'TVL formula', initially suggested by T V Lawson[7].

The main strength of the equivalent steady gust model is its simplicity and robustness in estimating the extremes: the maximum loading coefficient, $\hat{c}$, and minimum loading coefficient, $\check{c}$. As the instantaneous pressure is assumed to follow the fluctuations of the dynamic pressure, the extremes are assumed to be caused by the maximum gust. A maximum, $\hat{c}$, is obtained when $\bar{c} > 0$ (positive) and a minimum, $\check{c}$, when $\bar{c} < 0$ (negative), thus:

$$\hat{c} \approx (\hat{q}/\bar{q}) \, \bar{c}\{\theta\} = (\hat{V}/\bar{V})^2 \, \bar{c}\{\theta\} = S_G^2 \, \bar{c}\{\theta\} \tag{12.38}$$

replacing the Gust ratio by the Gust S-factor, S_G, of Chapter 9 (§9.4.3). The deficiency around $\bar{c} = 0$ is not usually serious because extremes tend to be large where the mean is also large. The inability of the model to account for building-generated fluctuations in regions of separated flow is usually more serious and requires a more sophisticated model.

12.4.1.4 Analysis of loading coefficients

In order to adopt one of the quasi-steady models, it is only necessary to determine the mean characteristics of the loading coefficients, as the fluctuating components are assumed to come only from the atmospheric turbulence. The characteristics of the mean loading coefficient, $\bar{c}$, are determined for all possible combinations of azimuth and elevation angles. This is possible in wind tunnels giving uniform incident flow for lattice structures and for structures isolated from the ground. Data for many common structural sections have been published. ESDU's Wind Engineering series of Data Items [8] is a good source of such data. For conventional bluff building structures, it is essential that the atmospheric boundary layer is correctly represented in the wind tunnel (see §2.4.3 and §8.3.2). The variation with azimuth is obtained by measuring all wind directions, but obtaining the variation with elevation angle is impractical owing to the constraint of the ground surface. The turbulence of the atmospheric boundary layer also tends to smooth out the fine detail of the variation with direction by the action of directional smearing (§8.4.2.2).

Figure 12.11 represents the equivalent steady gust model for extreme loads on static structures, where values of the Gust S-factor, S_G, and the mean loading coefficient, $\bar{c}$, are required to implement Eqn 12.38. The analysis of S_G is represented in (a) and requires the wind characteristics only, the structure having no influence. Design values are given in Chapter 9. The analysis of the mean loading coefficient is represented in (b), where all the fluctuations of load, including any building-generated fluctuations, are removed by the averaging process.

12.4.1.5 Synthesis of design values

Design values of load are obtained by reversing the analysis process. To implement the quasi-steady vector model for instantaneous pressure, values of instantaneous wind speed, $V\{t\}$, and mean loading coefficient, $\bar{c}$, are inserted into Eqn 12.24. Similarly, the linearised model is implemented using Eqn 12.31. The linearised model forms the basis of most of the assessment methods in current use for dynamic and aeroelastic structures, as described in Part 3 of the Guide. Elements of the model appear in the tests for galloping and divergence in the form of the loading coefficient derivative terms in Eqns 10.9 and 10.10 of the classification procedure.

For static structures the main design requirement is to estimate the extreme loading, for which the equivalent steady gust model is most often employed. Using pressure as an example, the design peak value, $\hat{p}$, is given by:

$$\hat{p}\{t, \theta\} \simeq \tfrac{1}{2}\, \rho_a\, \bar{V}^2\{\theta\}\, S_G^2\{t, \theta\}\, \bar{c}_p\{\theta\} \tag{12.39}$$

This is represented in Figure 12.11(c). Note that all the design parameters vary with wind direction: the reference mean wind speed from the directional wind climate, the gust S-factor from the directional variation of terrain roughness and the mean loading coefficient from the shape of the structure. On the other hand, all variation

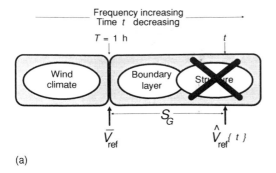

(a)

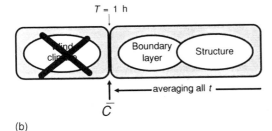

(b)

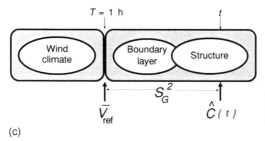

(c)

Figure 12.11 Equivalent steady gust model: (a) analysis of incident flow; (b) analysis of loading coefficient; (c) synthesis of peak coefficients

of load duration is confined to the Gust S-factor, which is the essence of the equivalent steady gust model.

Most wind loading codes worldwide (see Chapter 14), including the current UK Code [4] simplify Eqn 12.39 by lumping the mean wind speed and Gust S-factor together to give a design gust speed, $\hat{V}\{t\}$, thus:

$$\hat{p}\{t, \theta\} \simeq \tfrac{1}{2} \rho_a \, \hat{V}^2\{t, \theta\} \, \bar{c}_p\{\theta\} \tag{12.40}$$

The accuracy of the quasi-steady method, and that of the later methods below, is assessed in Chapter 15.

12.4.2 Peak-factor method

12.4.2.1 Davenport's model

In the early 1960s, working from the statistical foundations of Rice [9], Davenport proposed [10] and subsequently refined [11, 12] the statistical concepts (§8.4.1) that form the basis of most of the assessment methods for dynamic structures. A large part of this work is also of great significance for static structures. Of particular relevance is the concept that the difference of extremes from the mean can be assessed in terms of the rms value by the equation:

$$\hat{c}\{t\} = \bar{c} + g\{t\}\, c' \tag{12.41}$$

for the maximum, and the complementary equation:

$$\check{c}\{t\} = \bar{c} - g\{t\}\, c' \tag{12.42}$$

for the minimum, where the factor, $g\{t\}$, is called the peak factor.

Given that the probability distribution function of the loading is Gaussian or Normal (see Appendix B), Davenport predicted[*] that the peak factor for the average maxima (probability of exceedence $Q = 0.43$) is given by:

$$g = \sqrt{[2\ln(vT)]} + 0.57 / \sqrt{[2\ln(vT)]} \tag{12.43}$$

where T is the observation period for the extremes and v is the rate at which the mean value is crossed. Davenport[11] interprets the mean crossing rate, v, as the frequency at which most of the energy in the spectrum is concentrated. For static structures, this will be near the peak of the spectrum of atmospheric turbulence. For dynamic structures, the spectrum is modified by the structural response, but this makes very little difference to the value of g.

12.4.2.2 General model

Davenport's model predicts an extreme of the shortest duration (more exactly, over the same range of frequency as the rms value) for a Gaussian parent. This is usually appropriate for estimating the peak response of lightly damped dynamic structures, when values of g in the range of 3–5 are common. It will be seen shortly that the probability distribution of the loading in the regions of separated flow around bluff structures is not Gaussian and here Davenport's model might not be appropriate.

The peak-factor concept given by Eqns 12.41 and 12.42 forms a more general method when values of peak factor for a wide range of applications have been determined experimentally. Full-scale studies by the National Research Council of Canada[13,14] on high-rise buildings in Montreal and Toronto showed that Davenport's model worked well in the attached flow regions, but that values of g and c' were higher in separated flow regions, especially in the high suction regions at the windward corners. High values of peak factor imply that the fluctuations contain occasional high peaks or 'spikes', which Dalgliesh[13] found to come in intermittent bursts with $g \simeq 6$ to 7. Later measurements by BRE on the glazing of a high-rise block in Birmingham[15] and on the roof of a domestic house[16] yielded values of $g \simeq 10$ also in intermittent bursts.

For this to be a viable method of assessing extremes from the mean and rms values, the value of the peak factor must be consistent. A calibration of the peak-factor method was made by BRE at model scale as part of a more comprehensive review[17] of assessment methods for static structures (see Chapter 15), using a range of common building forms. Results of this calibration are given in

---

[*] The original equation in reference [10] was later modified to the form of Eqn 12.34 given here, which can be found in references [11,12] and all subsequent implementations of Davenport's approach. The constant of 0.57 is the Euler number.

Table 12.1 for load durations of $t = 1$, 4 and 16 s, where $\bar{g}$ is the value of peak factor averaged for all points on the building and all wind directions, and g' is the standard deviation of the variation of the individual points from this value. In each case the measured rms value of the loading coefficient, $c'\{t\}$, corresponded to the same time duration as the extreme, $\hat{c}\{t\}$. For $t = 16$ s, where the majority of building-generated fluctuations have been averaged out, the value of peak factor remains in the range predicted by Davenport's model. However, values of peak factor increase as t becomes shorter, even though the corresponding rms coefficients, $c'\{t\}$, also increase, indicating a general increase in the 'peakiness' of the coefficient. The variation of $\bar{g}$ with structural form is small compared with the variation with t. Measurements [17] show that departures from the averaged value $\bar{g}$, indicated by g', occur mainly along lines of intermittent flow reattachment where the bursts of 'spikes' discovered at full scale were again found.

Further full-scale measurements on a high-rise building in Tokyo [18], contemporary with the model-scale calibration, confirm the range of values in Table 12.1, but the intermittency found previously [13, 14, 15, 16, 17] was not reported. This is probably because the measurements were confined to the windward and leeward faces for a small range of near normal wind directions and no measurements were made in regions of intermittent flow reattachment.

Table 12.1   Calibration of peak-factor method

Model	$t = 1$ s		$t = 4$ s		$t = 16$ s	
	$\bar{g}$	g'	$\bar{g}$	g'	$\bar{g}$	g'
1 Low-rise building	5.42	1.13	4.74	0.83	4.31	0.70
2 Grandstand	5.22	1.12	4.66	0.82	4.09	0.58
3 Cube	5.82	1.38	5.07	1.15	4.19	0.83
4 Cuboid 3 : 1 : 1	5.78	1.38	4.84	0.91	4.08	0.60
5 Tower 1 : 1 : 3	5.17	1.37	4.51	1.09	3.86	0.62
6 Hipped 45° roof on cuboid 2 : 1 : 1	5.63	1.15	4.82	1.12	4.02	0.57
All six models	5.51	1.29	4.77	1.01	4.09	0.69

12.4.2.3 Analysis of loading coefficients

For the peak-factor method to be applied, it is assumed that values of peak factor, g, are available for the form of the structure: either estimated from Davenport's equation, Eqn 12.43, or from previous direct measurements. It is possible to make measurements of peak factor on a model of the building to be assessed, but this requires measurement of extremes in addition to the mean and rms so that equations 12.41 and 12.42 can be solved. If the facility for direct measurement of extremes is available, then estimation of the peaks by the peak-factor method becomes unnecessary.

The characteristics of the mean loading coefficient, $\bar{c}$, and of the rms loading coefficient, $c'\{t\}$, are determined for all possible wind directions. The frequency range of the rms value should include all periods between $T = 1$ h and the duration, t, of the desired extreme, as indicated in Figure 12.12(a).

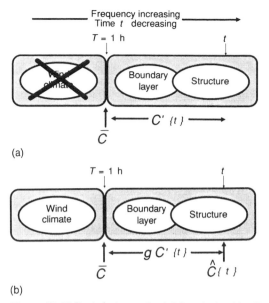

(a)

(b)

Figure 12.12 Peak-factor method: (a) analysis of loading coefficient; (b) synthesis of peak coefficients

12.4.2.4 Synthesis of design values

Design values of load are obtained by substituting the measured values of $\bar{c}$ and c' into Eqns 12.41 and 12.42, and this is represented by Figure 12.12(b). Note that the frequency range of the rms and the duration of the extreme are indicated as the same value of t.

When integrating values of extreme local coefficients, $\hat{c}$ and $\check{c}$, use of the 'TVL formula' of Eqn 12.37 (§12.4.1.3) to obtain the design value of t will ensure that the area of the structure is fully loaded (see §8.6.2.3). But at any point on the structure there remains the problem of deciding which of $\hat{c}$ or $\check{c}$ should contribute to which of the global coefficients $\hat{C}$ and $\check{C}$. With most conventional sharp-edged bluff buildings the choice is obvious: for example, $\hat{c}$ on the windward face and $\check{c}$ on the leeward face both contributing to the global maximum, $\hat{C}$. However, where the loading changes sign on a face (as with most curved structures) or where the influence function for the loading changes sign (as with the Eiffel-type lattice tower example of §8.6.2.1, Figure 8.26(b)), the choice may be more difficult. In such cases the solution is to revert to the principle of the equivalent steady gust model (§12.4.1.3) and take the maximum, $\hat{c}$, when $\bar{c} > 0$ (positive) and the minimum, $\check{c}$, when $\bar{c} < 0$ (negative). This problem is common to all of the other assessment methods which follow.

The peak-factor method can be extended to extremes of arbitrary duration by defining the peak factor as being duration dependent, thus: $g\{t\}$, instead of the rms loading coefficient. Now the rms, c', is not averaged at t, but contains the full spectrum of fluctuations. In this event, the duration dependence of $g\{t\}$ must be pre-determined, either through knowledge of the spectrum or by direct measurement. This approach was used earlier in §9.4.3.1 to determine the gust S-factor, where the term $0.42 \ln(3600/t)$ in Eqn 9.37 represents $g\{t\}$. Substituting $t = 1\,\mathrm{s}$ (about the shortest gust represented) gives a value of $g = 3.4$.

12.4.3 Quantile-level method

12.4.3.1 Lawson's model

The quasi-steady and peak-factor methods are both ways of *estimating* the extreme loading from the time-averaged quantities, mean and rms. This and the remaining methods described here are all techniques of direct measurement. The quantile-level method works by determining the value of pressure coefficient that corresponds to given values of the cumulative distribution function (CDF), P (see Appendix B) of the complete parent data record, in a similar manner to that of parent wind speeds (§5.2.3).

This approach was developed at the University of Bristol by Lawson[19]. By frequent application to design studies, he was able to refine the method to be economical as well as practical. His choice of the design risk was made at about 0.05%: that is, $P_{\hat{c}} = 0.0005$ for the minimum design value and $P_{\check{c}} = 0.9995$ ($Q_{\check{c}} = 0.0005$) for the maximum design value. He explained this choice as follows[19]:

> The short term averaging time used for buildings varies from 1 to 15 s: 0.05% of the time is 1/2000 which is about 2 s in the hour. It would appear, to my mind at least, that we have got our probablilities about correct if, when we measure the value of pressure averaged over about 3 s in a wind which simulates the hour of highest wind speed in the lifetime of the building, we use the value of pressure which is exceeded for only 2 s in that hour.

However, to obtain an accurate direct measurement of this value requires a data length of several hundred hours, which would be impossible at full scale and uneconomic in the wind tunnel, even allowing for the contraction in model time scales (see §13.4.1). But at much higher risks, for example at around 1%, the method is very economic to operate. Lawson[7] devised a method of extrapolating to the design risk from this point. The basis of this extrapolation is discussed in §12.4.3.5, but first it is instructive to consider the methods of analysis and the form of the resulting probability distributions.

12.4.3.2 General model

In essence, Lawson selects the extremes with an average rate of occurrence of about one per hour. In the context of this Guide, where t is the duration of the extreme and T is the standard observation period, the design risks are defined by:

$$P_{\hat{c}} = 1 - (t/T) \qquad (12.44)$$

for the maximum and

$$P_{\check{c}} = t/T \qquad (12.45)$$

for the minimum.

12.4.3.3 Analysis of loading coefficients

Analysis of the loading coefficients by this method is identical to the analysis of the parent wind climate (§5.3.2). Histograms are formed of either the probability density function (PDF), dP/dc, or the cumulative distribution function (CDF), P (see §B.2). The value of the loading coefficient is sampled at equal intervals of time. In the case of the PDF, the span of values is divided into small ranges or 'cells' and the number in each cell is counted. In the case of the CDF, the number of values exceeding each range is counted. In practice, it is usual to obtain the CDF through integration of the PDF by means of Eqn B.2.

12.4.3.4 Form of the probability distribution functions

Fundamental studies by Peterka and Cermak[20] with bluff building models revealed that the observed forms of the PDF fell into two distinct categories. Where the flow was attached to the surface of the building, the distribution functions of local pressure were close to normal or Gaussian (§B.3). This is illustrated in Figure 12.13 by corresponding BRE data[21], where the circle symbols are the data and the curve is the Gaussian model. The distribution functions of atmospheric turbulence are themselves close to Gaussian, so that this result is predicted by the linearised quasi-steady vector model described earlier (§12.4.1.2).

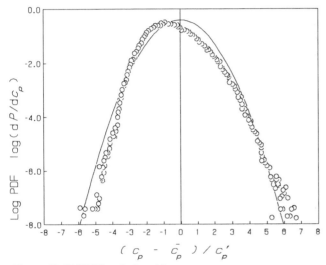

Figure 12.13 PDF for windward face

In regions of separated flow, Peterka and Cermak found that the PDFs diverged from the Gaussian form, particularly in the negative-going tail, becoming exponential in form. This is illustrated in Figure 12.14 by corresponding BRE data[21]. The exponential form plots as a straight line on these axes and it is apparent that both the tails are asymptotic to this form. However, the asymptote of the negative-going tail is the least steep and the Gaussian model progressively underestimates the negative extremes.

The criteria proposed by Peterka and Cermak[20] to distinguish between the two forms was:

1 Gaussian when $\bar{c}_p > -0.1$,
2 Exponential when $\bar{c}_p < -0.25$,

and this is the basis of Lawson's extrapolation procedure[7].

The Gaussian model for the loading coefficient in attached flow is only approximate, even if the atmospheric turbulence was exactly Gaussian and the flow

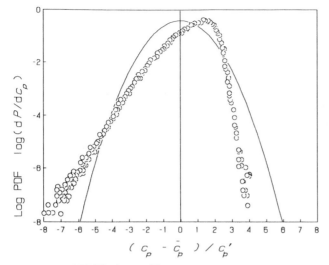

Figure 12.14 PDF for leeward face

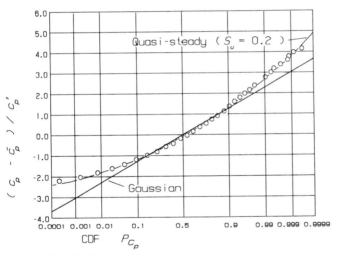

Figure 12.15 CDF on windward face

exactly quasi-steady, owing to the square of the turbulence terms in the full quasi-steady vector model (§12.4.1.1). Holmes[22] has demonstrated that the departure from Gaussian increases with turbulence intensity. His results for a turbulence intensity of 20% ($S_u = 0.2$) are reproduced in Figure 12.15. This shows the CDF of the pressure coefficient on the windward face of a building, normalised to the standard Gaussian measure and plotted on Gaussian axes, which converts the Gaussian CDF to a straight line. It is clear that the measurements diverge from the Gaussian model in the tails, but are a good fit to the full quasi-steady vector model for the corresponding turbulence intensity. This has been confirmed at full scale[18] using rigorous statistical criteria.

12.4.3.5 Extrapolation procedures

Knowledge of the form of the tails of the distribution functions enables reliable extrapolations to be made: that is, by the Gaussian model in attached flow regions and by an exponential model in separated flow regions. Peterka and Cermak's criteria[20] imply a unknown 'grey area' in the range $-0.25 > \bar{c}_p > -0.1$, but Lawson[7] suggests that, as the corresponding extremes in this range are small their effect can be included in the Gaussian model without compromising the design assessment.

Accordingly, Lawson[7] proposes extrapolation to the 0.05% quantile from the 1% quantile and the mean, using the equations:

$$c\{P = 0.0005\} = 1.42\ c\{P = 0.01\} - 0.42\ \bar{c} \qquad \text{and}$$

$$c\{P = 0.9995\} = 1.42\ c\{P = 0.09\} - 0.42\ \bar{c} \qquad \text{for } \bar{c} > -0.25 \qquad (12.46)$$

$$c\{P = 0.0005\} = 1.73\ c\{P = 0.01\} - 0.73\ \bar{c} \qquad \text{and}$$

$$c\{P = 0.9995\} = 1.73\ c\{P = 0.09\} - 0.73\ \bar{c} \qquad \text{for } \bar{c} < -0.25 \qquad (12.47)$$

12.4.3.6 Synthesis of design values

Design values are obtained by taking the values of the measured or extrapolated 0.05% quantiles as representing the design maxima and minima, thus:

$$\hat{c} = c\{P = 0.9995\} \tag{12.48}$$

for the maximum, and

$$\check{c} = c\{P = 0.0005\} \tag{12.49}$$

for the minimum.

The Gaussian standard measure, $(c - \bar{c})\,/\,c'$, of Eqn B.11 to which the data of Figures 12.13 to 12.15 were normalised, is the same as the definition of peak factor, g, of Eqn 12.41, so gives a direct comparison between the two methods. For the windward-face data of Figure 12.15, the extreme of interest is the maximum in the positive-going tail. At the design risk of 0.05%, $P = 0.9995$, the Gaussian model corresponds to $g = 3.5$ which is in the middle of the range predicted by Davenport's model[11, 12] (§12.4.2.1). However, Holmes' data and the quasi-steady theory[22] correspond to $g = 5$, which is in the middle of the range of the calibrated data[17] of Table 12.1. This confirms that Lawson's pragmatic choice[19] of the 0.05% quantile has indeed 'got our probabilities about correct'.

The choice between maximum and mimimum values must again be made when integrating local extreme coefficients. Lawson also comments on this problem[7] as follows:

There are two 0.05% values of pressure at every location (to be more accurate there are the 0.05% and 99.95% values), one from each tail of the distribution, and a *decision* has to be made to accept one or the other. It would be pessimistic to accept the one which gave the greatest overall load. The choice must be made with the flow pattern in mind, and rests upon the answer to the question 'When a sudden gust (increase in wind speed) occurs, does the pressure at the location rise or fall?' If the answer is 'rise', take the more positive value; if 'fall', the more negative.

This is the essence of the equivalent steady gust model (§12.4.1.3) and requires the maximum when $\bar{c} > 0$ (positive) and the minimum when $\bar{c} < 0$ (negative) as established for the peak-factor method (§12.4.2.4).

12.4.4 Extreme-value method

12.4.4.1 Form of the probability distribution functions

It has been established (§12.4.3.4) that the form of the *parent* distribution functions is exponential in the tails for separated flow regions and is approximately Gaussian for attached flow regions. Accordingly, the form of the *extreme* distribution functions can be deduced in exactly the same manner as for the extreme wind climate (§5.3.1). In the case of the wind climate, only the maximum values of wind speed or dynamic pressure were of interest. In this case, however, both the minimum, $\check{c}$, and the maximum, $\hat{c}$, values of the loading coefficient are of interest, otherwise the result is similar.

The argument in §5.3.1.2 for the Fisher–Tippett Type 1 (FT1) distribution[23, 24] given by:

$$P = \exp\left(-e^{-y}\right) \tag{12.50}$$

as the asymptote follows in exactly the same manner for the CDF of the maxima and the minima of extremes drawn from both the exponential and the Gaussian forms of the parent distributions. Now the reduced variate, y, is given for the maximum coefficient, $\hat{c}$, by:

$$y_{\hat{c}} = a\left(\hat{c} - U\right) \tag{12.51}$$

where U is the mode and $1/a$ is the dispersion, as before (§B.5.1). The reduced variate for the minimum coefficient, $\check{c}$, is given by substituting for $\hat{c}$ in the same equation.

12.4.4.2 Single and average extremes

The most likely value for the single extreme in each observation period is the modal value, U, but this is more likely to be exceeded (63%) than not (37%). In early experiments single extremes were often abstracted without regard to the observation period or the risk of exceedence. This is sometimes still practised, often defended on the grounds that rigorous extreme-value analysis is uneconomic, even though the resulting pattern of pressure coefficient contours over a building surface is often 'disorganized'[25]. As the number of observation periods increases, the mean value of the extremes converges onto the FT1 mean ($y = 0.577$) which is less likely to be exceeded (43%) than not (57%). This is the level of the risk against which Davenport's model[11, 12] for the peak factor (§12.4.2.1) is calibrated. The global extreme coefficient obtained by integrating single extreme local coefficients also converges to the FT1 mean with increasing number of measurement points.

Data from a study[26] of assessment methods for extremes have been re-plotted in Figure 12.16 to illustrate the variability inherent in the single extreme values compared with the mean extreme. For 12 locations in separated flow, the range or 'spread' of 50 minima has been plotted against the mean of the minima, thus giving an approximate estimate for the 98% confidence band, and individual estimates with more than 20% variation are common. The skew in the range bars either side of the mean line (marked '+0%') mirrors the skew of the FT1 distribution. It is

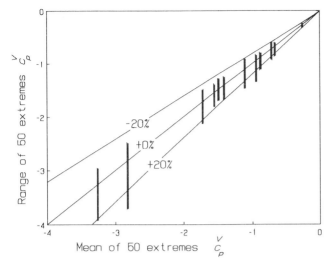

Figure 12.16 Prediction by single extreme

clear that the 'disorganized' nature of pressure coefficient contours drawn from single extreme values is due to the natural variation of the single estimate and is not a property of the flow.

12.4.4.3 Extreme-value analysis

The logical approach to the analysis of extremes is to determine the values of the Fisher–Tippett Type 1 parameters: mode, U and dispersion, $1/a$. The established methods of extreme-value analysis were described in §5.3.2 for estimating the wind climate parameters. As there is only one maximum and one minimum in each observation period, it would be quite uneconomic to collect enough data to form the probability density functions. Instead, all the methods estimate the cumulative distribution functions directly by means of the order statistics[24].

Figure 12.17 shows Gumbel's method of extreme-value analysis (§5.3.2.1) applied to minima of three durations, $t = 1, 4$ and $16\,s$, over 16 observation periods of $T = 1\,h$ from a wind tunnel model. The application of Gumbel's method to the loading coefficient extremes is exactly as described in Appendix C for the wind climate. Unbiased estimates of mode, U, and dispersion, $1/a$, are obtained using Lieblein's 'best linear unbiased estimators'[27] (§C.3.3). Three datum lines are marked on the diagram:

'$y = 0$' – marks the mode, with a risk of exceedence of 63%;
'$y = 0.58$'– marks the mean, with a risk of exceedence of 43%;
'$y = 1.4$' – marks a value with a risk of exceedence of 22%.

The last value is the design value predicted by the fully probabilistic design method described in Chapter 15. These datum lines represent the typical range of design values and demonstrate that, unlike the case of the wind climate, extrapolation past the range of the data is not required.

Unfortunately, the use of 16 hours of data is quite impractical at full scale and uneconomic at model scale, and measures equivalent to the extrapolation procedures for the quantile-level approach (§12.4.3.5) need to be adopted. The

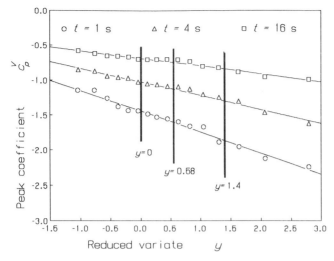

Figure 12.17 Gumbel plot of minimum coefficient

rate of acquisition of extremes can be increased sixfold by reducing the observation period to $T = 600\,\text{s}$ (10 minutes), the upper limit of the spectral gap indicated in Figure 4.4, without invalidating the ideal division of the problem (§4.3, §12.3.1). Acquisition of 16 extremes then requires under 3 hours of data, or 96 seconds in the wind tunnel with the typical 1/100 contraction of time scale (see §13.4.1). This brings the economy of the Gumbel method of extreme-value analysis to a par with the quantile-level method using Lawson's extrapolation procedure (§12.4.3.5).

Correcting the 10 minute observation period results to the 1 hour standard is done by the method established in §5.3.1.4 for the wind climate. The CDF is related to other periods within the spectral gap by Eqn 5.16 which, for transforming from 10 minute to 1 hour, becomes:

$$P\{T = 3600\,\text{s}\} = P^6\{T = 600\,\text{s}\} \tag{12.52}$$

The Fisher–Tippett formulae, Eqns 5.12 and 5.13, give the exact relationships:

$$1/a\{T = 3600\,\text{s}\} = 1/a\{T = 600\,\text{s}\} \tag{12.53}$$

$$U\{T = 3600\,\text{s}\} = U\{T = 600\,\text{s}\} + \ln 6/a\{T = 600\,\text{s}\} \tag{12.54}$$

The dispersion is unchanged, but the mode is increased. This is equivalent to shifting the points on Figure 12.17 by $y = \ln 6 = 1.8$ to the left, so that the range of the data is only just exceeded by the range of design values. No extrapolation is required for the mode or the mean, while the extrapolation required for $y = 1.4$ is minimal. Figure 12.18 shows some typical results[28] of applying this transformation to data measured for observation periods of 10 minutes and 1 minute, compared with measurements for the full 1 hour period. Agreement with the data transformed from 10 minutes is very good. However the 1 minute period is shorter than the minimum ideal value and lies near the middle of the micrometeorological peak. The agreement of the 1 minute data is not good, but it is significant that the error is conservative, overestimating by between 10% and 20%.

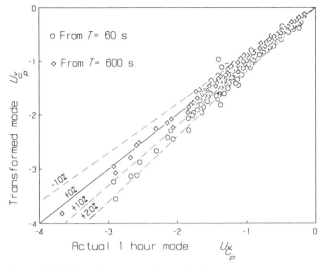

Figure 12.18 Transformation of mode to $T = 1\,\text{h}$

Use of the Gumbel method with $T = 600\,\text{s}$, followed by the transformation to 1 hour, has been the standard method of analysis at BRE for the past decade. Much of the data in the later chapters were assessed by this method.

The variability of the FT1 parameters is expressed by the characteristic product, $\Pi = aU$ (see §5.13.1.2). For the UK wind climate Π was about 10 for wind speed and 5 for dynamic pressure (§5.3.4.1). Typical data for the pressure coefficient on a range of building shapes[29] is plotted in Figure 12.19 which shows that the characteristic product lies in the range $5 < \Pi < 20$. This will be relevant to the operation of the fully-probabilistic design method described in Chapter 15 which, in its simplest form, predicts that the design value corresponds to the reduced variate of $y = 1.4$.

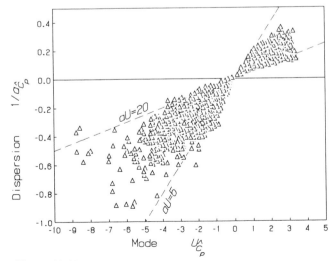

Figure 12.19 Range of Fisher–Tippett parameters

12.4.4.4 Prediction from the parent

Estimation of maximum wind speeds from the parent distributions was discussed in §5.3.1.1. The case for loading coefficients is similar, except that minima must also be considered. The exact CDF of maxima for a given observation period, T, is the CDF of the parent raised to the power of the number of *independent* values, N, in the observation period. Thus, following Eqn 5.11:

$$P_{\hat{c}} = P_c^N \tag{12.55}$$

For the minima in the other 'tail' of the parent CDF, the corresponding equation is:

$$P_{\check{c}} = (1 - P_c)^N \tag{12.56}$$

Davenport[11] equates the number of independent values, N, to the number of mean crossings in the period:

$$N = \nu\, T \tag{12.57}$$

where ν is the mean crossing rate (§12.4.2.1). For extreme values of duration t, the number independent values cannot exceed the limit given by:

$$N < T/t \tag{12.58}$$

and use of this value in Eqns 12.55 and 12.56 should yield a conservative result. An example [30] of using this approach is given in Figure 12.20 using the standard Gumbel format, where the line is the FT1 model calculated from the parent and the data points are directly measured extremes.

The desire to predict the extremes from the parent rather than measure them directly comes from the need to be economic. A great deal of valuable information is lost when only the extremes are abstracted. On the other hand, a great deal of extra data analysis and computation is done when all the parent data are used. This approach is still largely experimental and few design data have been acquired in this manner.

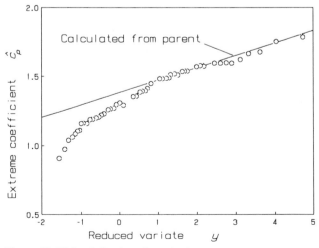

Figure 12.20 Prediction from the patent

12.4.4.5 Prediction from mth highest extremes

Peterka[27] recently proposed a middle course which makes use of the mth highest extremes, and so is a very similar approach to the storm analysis of the wind climate (§5.3.2.3, §C.5). Interest in the properties of mth highest extremes is not new because they are relevant to the theory of level crossings and other attributes which support Davenports' statistical concepts[10, 11, 12]. They are also of direct use in predicting the failure of glazing and other materials that are sensitive to accumulated damage (see §19.9). In the case of the wind climate, independent maxima can be easily abstracted because individual storms are readily identified. The definition of loading coefficient results in fluctuations which are consistent in their properties with time, i.e. they are *statistically stationary*. Abstraction of most, even if not all, of the independent extremes is not straightforward because recognisable independent periods like 'storms' do not occur.

A procedure to abstract independent peaks[14], called 'spikes', which was used in the full-scale studies by the National Research Council of Canada, was defined by:

1 A spike value must exceed a threshold value at least at least twice the rms from the mean. Both mean and rms are measured over the entire sampling interval.
2 A spike must be separated from adjacent spikes by returns towards the mean that extend at least the threshold value from the spike value.

Only large extremes are selected by the first criterion. By requiring that consecutive spikes are separated, the second criterion ensures statistical independence. Unfortunately, these criteria call for considerable analysis effort. This is practical for the relatively short record lengths obtained at full scale, which are usually assessed some time after the event. In the wind tunnel, however, assessment is usually performed in 'real time', i.e. as fast as the data are acquired, and this procedure would not be economic.

Peterka proposes[31] a much simpler and faster procedure: to sample the data without testing for independence, but to retain only the m highest values. A value of m is selected which is large enough to include much more data than the single extreme, but is not so large that the extremes are significantly correlated or insufficiently converged towards the FT1 asymptote. By examining a range of values Peterka selected $m = 100$ as being the optimum value. Analysis then proceeds by the standard Gumbel method to obtain the CDF of the mth highest extremes, then this is transformed to the CDF of the maxima through Eqn 5.11 (Eqn 12.55). In this method only a single 1 hour observation period is necessary, giving a threefold increase in speed over the standard BRE method using 16 10 minute periods (§12.4.4.3).

Peterka also tested the effect of increasing the number of observation periods from $n = 1$ to 5, and found that $n = 2$ gave a significant increase in consistency of prediction over $n = 1$, but that higher values of n gave increasingly smaller improvements. This method with $m = 100$ and $n = 2$ is compared with the standard Gumbel method with 100 extremes ($m = 1$, $n = 100$) in Figure 12.21, which shows the range of results obtained[31] for the design value at $y = 1.4$ (see Figure 12.17) over 50 trials. Just as the independent storm analysis makes more efficient use of the wind climate data than annual extremes, so Peterka's proposal makes more efficient use of loading coefficient data.

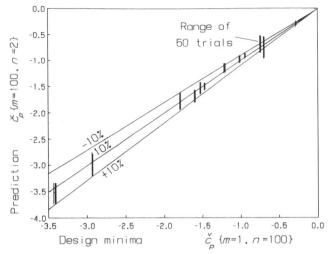

Figure 12.21 Prediction from *m*th highest extremes

12.4.4.6 Synthesis of design values

Design values are obtained by taking the loading coefficient corresponding to a design value of the reduced variate, y, either by reading directly from the Gumbel plot of Figure 12.17 or, more usually, by subsituting for y in the Fisher–Tippett Type I equation (Eqn 12.51). Thus:

$$\hat{c} = U_{\hat{c}} + y / a_{\hat{c}} \qquad (12.59)$$

for the maximum, $\hat{c}$. The value for the minimum, $\check{c}$, is given by substituting the corresponding values of U and $1/a$ in the same equation. Eqn 12.59 assumes that the mode and dispersion estimates have been measured over, or corrected to, the standard 1 hour observation period. A more general equation, combining Eqns 12.53 and 12.54 with Eqn 12.59 is:

$$\hat{c} = U_{\hat{c}}\{T\} + [y + \ln (3600 / T)] / a_{\hat{c}}\{T\} \qquad (12.60)$$

where T is the observation period used (in seconds).

The choice of design value for the reduced variate, y, was touched on in §12.4.4.3. Davenport's model for the peak-factor method (§12.4.2.1) is calibrated to the mean, $y = 0.58$. The fully probabilistic design method of Chapter 15 predicts $y = 1.4$ as the design value. The calibration of the general peak-factor model given by Table 12.1 (§12.4.2.2) and Lawson's 0.05% design quantile (§12.4.3.5) correspond approximately to $y = 1.4$.

12.5 Dynamic amplification factor

The deflection of a static structure is proportional to the loading as in Figure 12.1(e). As the structure becomes more dynamic, it oscillates at its natural freqency as in Figure 12.1(f). Energy is stored in the oscillation of the structure and the deflection is no longer related directly to the loading trace. The classification procedure of Chapter 10 defined a sub-class of dynamic structures as D1 – mildly dynamic. The philosophy behind the term mildly dynamic is that the amount of energy stored is a small part of the whole and that it is possible to calculate the deflection as if the structure were static and then multiply this deflection by a dynamic amplification factor, γ_{dyn} (§10.8.2.1), to obtain the deflection of the mildly dynamic structure. Thus the dynamic amplification factor is a function of the structural response. Exactly the same result is obtained when the factor is applied to the quasi-steady loads to give the *equivalent* load for a static structure.

Eqn 10.11 for the dynamic amplification factor, γ_{dyn}, may now be derived using the preceding static assessment methods. The value of γ_{dyn}, is defined as the ratio of the actual peak deflection of the structure to the quasi-steady peak deflection:

$$\gamma_{dyn} = \hat{x}_{D1} / \hat{x}_A = \hat{c}_{x_{D1}} / \hat{c}_{x_A} \tag{12.61}$$

where $\hat{x}$ is peak deflection and the subscripts D1 and A denote Class D1 – mildly dynamic and Class A – static response, respectively. This is also the ratio of the peak coefficients of deflection, $\hat{c}_x$, by the definition of the standard form of coefficient (§12.3.3).

Consider $\hat{c}_{x_A}$, the quasi-steady peak coefficient, first. Quasi-steady response is the datum for the classification procedure and, although this may be given by any of the quasi-steady methods in §12.4.1, the equivalent steady gust model of Eqn 12.38 (§12.4.1.3) is the most convenient, giving:

$$\hat{c}_{x_A} = S_{G_A}^2 \bar{c}_x \tag{12.62}$$

Now consider $\hat{c}_{x_{D1}}$, the actual peak coefficient for the mildly dynamic structure. The mean and the fluctuating components must be separated, to allow the enhancement of the fluctuating component by the response of the structure to be included. For this, the peak-factor model of Eqn 12.41 is the most convenient, giving:

$$\hat{c}_{x_{D1}} = \bar{c}_x + g_{D1} \, c'_{x_{D1}} \tag{12.63}$$

where g_{D1} is the peak factor for the mildly dynamic response. An equation for γ_{dyn} can now be written using these two models, by substituting Eqns 12.62 and 12.63 into Eqn 12.61, thus:

$$\gamma_{dyn} = (\bar{c}_x + g_{D1} \, c'_{x_{D1}}) / S_G^2 \, \bar{c}_{x_A}$$
$$= (1 + g_{D1} \, c'_{x_{D1}} / \bar{c}_x) / S_{G_A}^2 \tag{12.64}$$

At this stage the dynamic component is confined to the term $g_{D1} \, c'_{x_{D1}} / \bar{c}_x$.

The peak factor is given in §12.4.2.1 by Eqn 12.43, where it was noted that its value did not change significantly with the degree of structural response to atmospheric turbulence. We may therefore assume $g_A = g_{D1} = g$, a general value for all structures, without a significant loss of accuracy. Equating the value of the quasi-steady peak coefficient, $\hat{c}_{x_A}$, given by the equivalent steady gust model, Eqn 12.38, to that given by the peak-factor model, Eqn 12.41, gives:

$$g \, c'_{x_A} / \bar{c}_x = S_{G_A}^2 - 1 \tag{12.65}$$

which is similar to the previous term in Eqn 12.64, except the quasi-steady rms coefficient replaces the dynamic rms coefficient.

The structural reponse parameter, R, given by the classification procedure is defined as the ratio of the actual mean square displacement to the quasi-steady mean square displacement (§10.3), that is:

$$R = x_{D1}'^2 / x_A'^2 = c_{x_{D1}}'^2 / c_{x_A}'^2 \tag{12.66}$$

and combining this with Eqn 12.65 gives the required term in Eqn 12.64:

$$g\, c'_{x_{D1}} / \bar{c}_x = (S_G^2 - 1)\, \boldsymbol{R}^{1/2} \tag{12.67}$$

Finally, substituting Eqn 12.67 into Eqn 12.64 gives the required equation for γ_{dyn}:

$$\gamma_{dyn} = [1 + (S_{G_A}^2 - 1)\, \boldsymbol{R}^{1/2}] / S_{G_A}^2 \tag{12.68}$$

given in Chapter 10 as Eqn 10.11.

12.6 Concluding comments

The design values of loading actions required for static structures are the maximum (largest positive) and minimum (largest negative) expected in the lifetime of the structure. Data to give the design wind speed characteristics of any required risk are given in Chapter 9. It is usual in design to use a design wind speed of the same risk as the required design loading (see Chapter 14). In this case, the synthesis of design values by the various assessment methods for static structures is the answer to the question:

what is the value of the loading coefficient that results in a design load of the desired design risk, given a wind speed of the same design risk?

Each of the various codification methods discussed in Chapter 14 are attempts to answer this question. All, bar one, are approximate methods or contain pragmatic assumptions. The one exception uses the fully-probabilistic design method described in Chapter 15 which, in simplified form, is the basis for the design data of this Guide.

13 Measurement of loading data for static structures

13.1 Introduction

This chapter does not attempt to teach the designer measurement and modelling techniques. Such expertise is available from the specialist wind engineering consultant. Rather, the chapter aims to convey sufficient information to allow the designer to assess the value of design data available to him and to decide whether *ad-hoc* tests are necessary. In most cases, the design data given in the Guide, in the relevant code of practice or elsewhere will be sufficient. But when *ad-hoc* tests are required, this chapter will assist the designer to choose between the various test options, to commission the test and to assure the quality of the result. Accordingly, the mechanics of measurement of loading coefficients at full and model scale are reviewed in the following sections. More emphasis is placed on modelling technique because this is the area in which the designer is most likely to become involved and to be required to make choices in an unfamiliar and sometimes confusing field.

As only current methods which give the accuracy required for design are reviewed, it is appropriate first to consider the historical development of the subject, expanding on the brief review in Chapter 2.

13.2 Historical development

13.2.1 Full-scale tests

The earliest measurements were made exclusively at full scale and some of the more significant experiments are listed in Table 13.1. Initially the aim was to obtain reliable values directly for use in design. In 1884, reporting his experiments made for the design of the Forth Rail Bridge and with the Tay Bridge disaster still of major concern, Baker[32] wrote that the design

> necessarily involved many matters of pure conjecture, which rendered it impossible to state with precision what factor of safety would belong to the Forth Bridge. The same remark of course applies even now with equal force to every other bridge, because there exists a lamentable lack of data respecting the actual pressure of the wind on large structures. Mr Fowler and I have spared no pains during the past two years to contribute something to the general fund of information; and other engineers, doubtless are experimenting – *for experiments, and not speculations, are wanted.*

Table 13.1   Some early full-scale experiments

1884 Baker	**Forces on flat plates at site of Forth Rail Bridge** [32]. Equipment: one 300 square foot board and two 1.5 square foot boards mounted on spring balances. Measurements: found forces were not steady and that smaller areas were loaded more by gusts than larger areas.
1894 Irminger	**Forces on a cylindrical gas holder in Copenhagen** [33]. Equipment: floating cylindrical gas holder was tethered against the wind by a rope. Measurements: tension in the rope gave the drag of the whole gas holder. The difference in the internal gas pressure from still-air conditions gave the lift force on the roof.
1900 Eiffel	**Displacement of the Eiffel Tower in Paris** [34]. Equipment: vertically-mounted telescope on the ground aimed at a target at the top of the tower. Cup-and windmill-anemometers at various levels on the tower. Measurements: deflection of the tower with simultaneous reference wind speed and direction.
1930 Dryden and Hill	**Pressure distribution around circular chimney** [35]. Equipment: 200 ft high circular chimney. Manometers. Measurements: circumferential pressure distribution 41 ft from top where the diameter was 11.4 ft.
1933 Bailey	**Pressures on railway-car shed** [36]. Equipment: 100 ft long by 42 ft wide shed, 23 ft high to eaves, 1:2 duopitch roof. Multi-tube manometer and camera. Reference Dynes anemometer on a 40 ft pole, 35 ft upwind of shed. Measurements: pressure coefficient distribution at 4 tappings in each slope of the roof and at single tappings at 17 ft above ground in the walls at the centre of the shed.
1938 Rathbun	**Pressures and deflection of the Empire State Building** [37]. Equipment: the Empire State Building, one anemometer on a mast at the top at 1263 ft above ground, multi-tube manometers with cameras, 22 extensometers, collimator and target, and a plumb-bob extending from the 86th to the 6th floor damped in an oil bath. Measurements: simultaneous pressures around the building on three floors, stress in columns and deflection between 86th and 6th floors.

Early measurement equipment was fairly crude, but ingenuity of approach was certainly not lacking as when Irminger [33] measured the wind loads on a large cylindrical gas holder. As the gas holder was of the floating type, he was able to measure the overall drag by tethering the gas holder against the wind and measuring the tension in the tether. Gas pressure in this type of gas holder is maintained by the weight of the holder, so Irminger was also able to measure the overall uplift on the roof of the gas holder from the difference in gas pressure from still-air conditions.

Later experiments [34, 35, 36, 37] were progressively better equipped and measurements became more extensive and more accurate. By the 1930s experiments were beginning to be made at model scale in wind tunnels, and the rôle of full-scale tests had changed to include acquiring 'benchmark' data to verify developing theories and models. For its time Rathbun's experiment [37] on the

Empire State Building was extremely ambitious and successful. This is despite the fact that the data are 'chequered by anomalous values'[38], a problem that still occurs today! Nowadays the problem is one of maintaining sensitive electronics in field conditions, whereas Rathbun's problem was the simpler, but no less trying, one of keeping manometers air- and water-tight. Nevertheless, Davenport[38] has shown that a great deal of useful information can be gleaned from these old data after re-analysis using modern techniques and re-assessment in the light of current knowledge. Figure 13.1 shows Rathbun's data from the plumb-bob, reworked by Davenport[38] to show the mean deflection in each axis (scaled against the square of the wind speed) in terms of wind direction. These mean deflection effects are proportional to the loading actions. Comparing the form of Figure 13.1 with the modern data of Figures 17.23 and 17.24 (§17.3.1.3), shows that it is consistent even to showing the small negative lobe for the cross-wind (east–west) action effect at small wind angles.

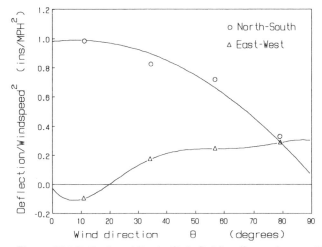

Figure 13.1 Deflection of Empire State Building (from reference 38)

The peak of full-scale experimentation occurred in the decade 1970–80. A survey[39] in 1974 showed 103 current or recently completed experiments on a wide range of structures, and these are summarised in Table 13.2. Since this time the rate of full-scale experiments has declined, but the range and accuracy of the data obtained from them have continuously improved. The need for full-scale data always remains because new theories or analysis techniques must be verified by full-scale measurements using the same contemporary analysis techniques. This case has been put many times, for example[38]:

> There is a continuing need to maintain an adequate source of full-scale data so that new theories and modelling procedure can be tested. Such data should preferably contain results from experiments on several structures so as to reduce the uncertainty. Conclusions reached from an ensemble of full-scale experiments are likely to be of significantly greater value than those reached from any individual experiment. The sharing of full-scale data should be encouraged.

Table 13.2   Full-scale experiments recently completed or in progress in 1974

(a) Tall buildings

Country	Steel	R/Concrete	Composite	Other
Australia	1	3		
Brazil		1		
Canada	21			
Czechoslovakia	1	3		
France	1			
Hong Kong	1			
Japan	10		2	
Netherlands	1			
New Zealand			1	
South Africa	2	1		
UK	1	2		
USA	7	1		2
W. Germany		4		

(b) Low buildings (< 10 m high)

Country	Types
Phillipines	7 single family dwellings.
UK	2-storey brick housing in Aylesbury, Buckinghamshire.
	Reinforced concrete flats in Southampton.
USA	Housing in Montana.

(c) Towers

Poland	Steel lighting towers, Krakow.
UK	61 m high lattice lighting tower.
	330 m high concrete and steel television tower, Emley moor.
	23 m high lattice microwave tower, Scotland.
	41 m high lattice microwave tower, Scotland.
USA	33 m high lattice mast in Hawaii.
W. Germany	220 m high concrete and steel television tower, Hernisgrinde.
	272 m high concrete and steel television tower, Hamburg.
	294 m high concrete and steel television tower, Munich.
	139 m high concrete television tower.

(d) Bridges

Japan	Cable-stay box, suspension pipe truss, Arakawa Konpira.
USA	Golden Gate suspension bridge, San Francisco.
	Cable-stayed box-girder, Sitka Harbour, Alaska.

(e) Cooling towers

Canada	114 m high, 75 m diameter natural tower, Muskingum River.
UK	Several Central Electricity Generating Board towers.
USA	139 m high, 97 m diameter natural draft tower, Martin's Creek, PA.
W. Germany	112 m high, 75 m diameter natural draft tower, Scholven.
	105 m high, 71 m diameter natural draft tower, Weisweiler.
	115 m high, 85 m diameter natural draft tower.

Table 13.2 (Continued)

Country	Types
(f) Chimneys	
Canada	194 m high reinforced concrete stack, Nanticoke.
Poland	225 m high reinforced concrete stack, Krakow.
W. Germany	Six stacks of various types.
	200 m high concrete stack, Irsching.
Japan	120 m high steel stack, Hiroshima.
(g) Other structures	
Japan	78 m high gantry crane, Nagasaki.
Sweden	136 m high portal crane, Malmo.
USA	Cable roof, Deerfield.
	Flag pole, Chicago.

13.2.2 Model-scale tests

In 1893, just before his full-scale experiment with the gas holder, Irminger constructed what was probably the world's first wind tunnel. This'was a 115 mm by 230 mm section duct, 1 m in length, let into the base of a 30 m high chimney. The updraft of the chimney drew air through the tunnel at speeds between 7½ and 15 m/s, controlled by a simple sliding shutter. Experiments in this tunnel were naturally crude, for Irminger and his associates were feeling their way in a totally new field, but many previously undiscovered effects were observed for the first time. Model-scale experiments became common from about 1930 after the development of the wind tunnel as a routine tool in aircraft design. Five of the first extensive series of test programmes [40, 41, 42, 43, 44, 45, 46] are listed in Table 13.3.

In 1927, during his eightieth year [47], Irminger began a comprehensive series of tests on building shapes in collaboration with Nøkkentved which took a decade to complete [40, 45]. Unfortunately all these data were collected in smooth uniform flow and it is now known from the later work of Jensen [48], described in §2.4.3, that results from bluff building models without a good simulation of the atmospheric boundary layer are very different from what actually occurs in nature (§8.3.2). Over the same period, Flachsbart made an extensive series of tests on lattice sections and frames [41, 42, 43] and later, with Winter, on complete lattice structures [44], again in smooth uniform flow. This time the results remain useful today, owing to the way that the lattice model depends only on the local wind vector (§8.3.1 and §8.4.1). Quasi-steady theory (§12.4.1) enables the effects of the atmospheric boundary layer to be assessed through the action of influence (§8.6.2.1) and admittance (§8.4.2.1) functions.

The effect of the thin boundary layer on the floor and walls of the wind tunnels confused early investigators. As the floor boundary layer was a different thickness in each wind tunnel, it was found that the same building model tested in each wind

Table 13.3 Some early model-scale experiments

1930 Irminger and Nøkkentved	**Wind pressures on wide range of building shapes** [40]. Equipment: 303 × 302 mm wind tunnel, producing smooth uniform flow at about 20 m/s. Single inclined-tube manometer. Models comprising plates, spheres, cones, cubes and building shapes. Measurements: mean pressure coefficient distributions.
1934 Flachsbart	**Wind forces on single lattice girder frames** [41, 42, 43]. Equipment: Göttingen (aeronautical) wind tunnel, giving smooth uniform flow, with mean force balance. Models of structural sections including flat, square, triangular and circular bars. Models of single lattice frames. Measurements: inwind and cross-wind mean force coefficients in wind and body axes, including variation with proportions, solidity, wind angle and Reynolds number.
1935 Flachsbart and Winter	**Wind forces on spatial lattice girder frames** [44]. Equipment: Göttingen (aeronautical) wind tunnel, giving smooth uniform flow, with mean force balance. Models of lattice frame sections in pairs and forming square sections, and models of complete masts. Measurements: inwind and cross-wind mean force coefficients in wind and body axes, including variation with proportions, solidity, and wind angle.
1936 Irminger and Nøkkentved	**Wind pressures on wide range of building shapes** [45]. Equipment: 303 × 302 mm wind tunnel, producing smooth uniform flow at about 20 m/s. Single inclined-tube manometer. Models comprising open and enclosed building shapes, solid and perforated screens. Measurements: mean pressure coefficient distributions. Comparisons between models with ground-plane and with reflected image.
1943 Bailey and Vincent	**Wind pressures on buildings including adjacent buildings** [46]. Equipment: 3 ft square wind tunnel operating between 10 and 14 m/s, *with wall boundary layer*. Models of cuboids, duopitch buildings and the Empire State building at 1/240 scale. Multi-tube manometer. Measurements: mean pressure coefficient distributions including effect of roof pitch, spacing between pairs of buildings of similar and different sizes.

tunnel would give different results. Attempts were made to isolate the building model from the floor boundary layer by suspending it in the middle of the wind tunnel like an aircraft model. Without the restraint of the ground, air flowed under the model building which was clearly not correct. Potential flow theory indicated that a second model forming a mirror-image of the main model would produce symmetrical flow either side of a streamline at the missing ground plane, but when this was tried strong vortex shedding occurred caused by interaction between the two shear layers (§2.2.10.4) from the main model and its 'reflection'. Finally, vortex shedding was suppressed by adding a ground plate downwind of the model only. This was known as the 'image' technique of modelling.

Unique among the early data is the work of Bailey and Vincent reported in 1943 [46]. By this time, much more was known about the structure of the atmospheric boundary layer. Bailey and Vincent compared this with the boundary layer on the wall of their wind tunnel and found a reasonably good match to the mean wind speed profile at a scale factor of 1/240. The data they obtained compares well with modern data although only mean pressures were measured. Figure 13.2 shows how the pressure coefficient on the front face of a typical house varies with the separation distance of an identical house directly upwind. The form of these data is identical with the modern data of Figure 19.1 for a pair of walls (§19.2.2.1).

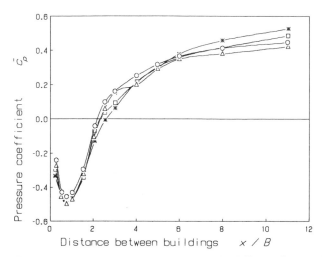

Figure 13.2 Loading of downwind building of pair (from reference 46)

Inappropriate model techniques using smooth uniform flow with floor-mounted or suspended image models continued to be used through the 1950s [49] and into the middle of the 1960s [50]. Most of these data found their way into codes of practice and many remain in force to this day. For example, the use of parapets to control the regions of high local suctions at the periphery of flat roofs is based on tests in 1964 using smooth uniform flow [51], and the findings were not checked in proper atmospheric boundary-layer simulations until 1982 [52].

Following Jensen's work [48], efforts were made to understand the interaction between the boundary layer and bluff structures by modelling parts of the atmospheric boundary-layer structure in isolation from the rest. For example, Baines [53] showed the importance of the profile of mean velocity and demonstrated how simulating this alone could reproduce the main features of the mean flow. Based on this work, a number of 'profile only' simulations was spawned which could be used in aeronautical wind tunnels (see §13.5.3.1), which led to a new generation of model studies [54]. Conversely, Vickery [55] showed that turbulence alone could change the mean as well as the fluctuating forces on a bluff body. Later work in the same vein on flat plates [56] was directed towards investigating this effect and verifying theories for the aerodynamic admittance approach (§8.4.1).

The state of the field at the first international conference in 1963 was mentioned in Chapter 2. Model techniques reported varied from uniform flow image modelling [57], profile only modelling [53] and Jensen's full boundary-layer simulation from five years earlier, reiterated by Franck[58]. The paper by Colin and d'Havé[59], comparing the image models and uniform flow with floor-mounted models and wind speed profile, is also worthy of note since this greatly assisted in eradicating the inappropriate image modelling method.

In the remainder of this chapter modern measurement and modelling techniques are reviewed. Where obsolete or inappropriate techniques are mentioned, this is done to assist the designer to recognise and so avoid them.

13.3 Measurement techniques

13.3.1 Wind speed and turbulence

13.3.1.1 Full-scale techniques

This subject was introduced in Chapter 7 for the purpose of calibration of the atmospheric boundary-layer characteristics at a site. Such data are also required for reference and scaling purposes when measurements are made on structures at full scale. A range of instruments are suitable for full-scale use, criteria being size and resistance to exposure to the weather.

Cup anemometers give the magnitude of the horizontal wind vector and weathercock vanes give the corresponding direction. The Meteorological Office standard cup and vane will measure speeds from about 1 m/s to the maximum recorded and resolve gusts down to about 1 s duration, and several of these were shown in Figure 7.15. Better gust response is obtained with special lightweight cups and vanes, such as the Porton type also shown in Figure 7.15, which will resolve gusts to about 0.1 s but may not withstand the very high speeds found in cyclones. Most types give electrical signals that are usually fed to recorders which produce traces on a paper chart, but may also be analysed directly.

Better frequency response and resolution of the wind speed vector into orthogonal axes are required when turbulence characteristics are to be acquired. Propellor anemometers of the Gill[60,61] type, shown in Figure 7.16, are suitable for this purpose and will resolve frequencies up to about 2 Hz at 10 m/s. The directional response of these anemometers is not quite cosinusoidal[60,61,62] so requires corrections[63,64] which are usually routinely applied. The frequency response improves with increasing wind speed, so the alignment shown in Figure 7.16 is not the best because the vertical propellor sees a component which is zero on average. Response is improved by tilting the array so that all three propellors point 45° to the mean wind vector, then the data are transformed back to the wind axes during the analysis process. Even higher frequencies, up to 30 Hz, may be resolved using sonic anemometers. These measure the time delay in ultra-sonic signals transmitted between three sets of detectors in an orthogonal array. The difference in the time delay between both directions along the same path is proportional to the wind speed. The average delay for both directions is a function of air temperature, so the sonic anemometer will also measure turbulent heat fluxes. Both types give electrical signals proportional to the instantaneous wind vector along each orthogonal axis. The difference in response between these two types of anemometer is reflected in their respective costs.

Most of the model-scale techniques listed below may also be used at full scale, although the instruments may not be rugged enough for long-term exposure and will require frequent calibration. Wind speed data may also be obtained from dynamic pressure measurements (§13.3.2.1).

13.3.1.2 Model-scale techniques

Measurements of wind speed and turbulence characteristics through simulations of the atmospheric boundary layer and around models are usually made using hot-wire anemometers. These work on the principle that the heat lost by a wire kept at a constant high temperature is a function of the wind speed normal to the wire. Figure 13.3 shows a common form of probe, with a 0.005 mm diameter tungsten wire welded between two fine prongs about 3 mm apart. The wire is

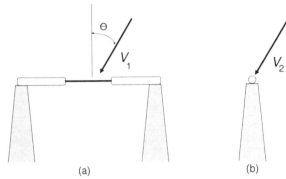

Figure 13.3 Hot-wire probe: (a) yaw; (b) pitch

heavily gold plated at either end to give a low electrical resistance which is insensitive to the flow, and ensures that the effect of any flow disturbance caused by the prongs is not measured. The active part of the wire is the 1.2 mm wide central region. A voltage is applied to the wire by electronics which heats it to a constant temperature of about 200°C. Wind flow past the wire tends to cool it and the electronics constantly adjusts the voltage to keep the wire temperature constant. The output signal, E, is given by the King's law equation:

$$E^2 = k_A + k_B V^{1/2} \tag{13.1}$$

where k_A and k_B are calibration constants, and V is the magnitude of the wind speed vector normal to the wire, This equation is very non-linear. In the high intensities of turbulence close to the ground, it is essential that the response is made proportional to V **before** any analysis is attempted. This can be done mathematically, as the first step in a digital analysis, but is more commonly done by a special circuit or 'lineariser' in the anemometer electronics. The signal is also sensitive to changes in ambient temperature which must either be kept constant or monitored and corrections applied. As the wire responds to the normal vector, ideally there is a cosine response, $V = V_1 \cos\theta$, to the yaw angle, θ, in Figure 13.3(a), but a uniform response, $V = V_2$, to the pitch angle in (b). In practice, the cosine response in yaw is good over the range $-70° < \theta < 70°$ but is affected by the body of the probe at large pitch angles.

A typical single-wire probe is shown in Figure 13.4(a). The frequency response is very high, from 0 to 100 kHz typically, so the hot-wire is the principal instrument for measuring turbulence. In moderate intensities of turbulence ($< 20\%$) the linearised output signal, A, of a single wire aligned normal to the mean flow gives the mean and the inwind turbulence component, thus:

$$A = U + u \tag{13.2}$$

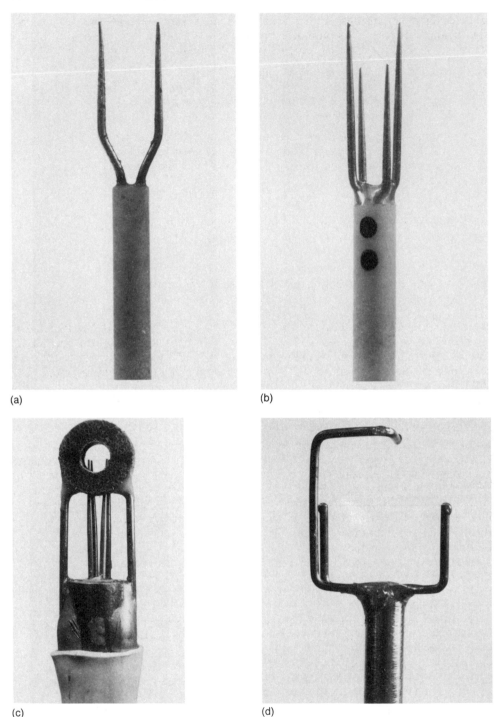

(a)   (b)

(c)   (d)

Figure 13.4 Typical hot-wire probes: (a) single-wire probe; (b) X-probe; (c) McGill probe;
(d) pulsed-wire probe

but at higher intensities the squared turbulence terms u^2, v^2 and w^2 become significant. Because of its omnidirectional response to pitch, a single wire aligned vertically is useful for measurements of wind speed around buildings near to the ground, but gives only the magnitude of the horizontal wind speed vector (Eqn 12.25 without the w^2 term). The signal is rectified, so that if the flow reverses in direction the signal stays positive in value. In high intensities of turbulence measurements of mean speed are overestimated, the rms is underestimated, but peak values are likely to be good.

Two components of turbulence can be measured by a pair of wires each aligned at 45° to the flow and normal to each other, forming an X-shape. A typical 'X-probe' is shown in Figure 13.4(b). In this case the two linearised output signals, A and B are:

$$A = U + u + v \quad \text{and} \quad B = U + u - v \tag{13.3}$$

when the cos45° and sin45° terms for the inclination of the wires are included in the calibration. Adding the two signals gives the mean and in-wind components:

$$U + u = (A + B)/2 \tag{13.4}$$

and subtracting the signals gives the cross-wind component:

$$v = (A - B)/2 \tag{13.5}$$

Recalling that a single wire is good for the range $-70° < \theta < 70°$, the range common to both wires is reduced to $-25° < \theta < 25°$, which limits the turbulence intensity for accurate measurements to about 10% ($S_u = 0.1$). For Eqns 13.4 and 13.5 to work it is important that the probe is accurately aligned with the mean flow direction, otherwise corrections for the actual wire angles are required[65]. Also particular care is needed to set equal gains for both wires of an X-probe, especially when measuring the Reynolds stress $-uw$ (§7.1.2).

In very turbulent flows, when the flow may reverse in direction, the conventional hot-wire probe is not accurate and special probes are required. The McGill probe shown in Figure 13.4(c) was developed[66,67] to make measurements in low-speed reversing flows. The probe comprises a parallel pair of hot wires aligned across the centre of a hole though a disc-shaped shield. The proportions of the shield are specially chosen so that the wind speed through the hole is equal to the incident wind speed vector along the axis of the hole. Two wires are required to detect the direction of the flow through the hole. One wire is always in the hot wake of the other, so gives a smaller signal. An electronic circuit chooses between the two signals, passing the larger and blocking the smaller. One of the signals, if passed to the output is passed without change, whereas the other, if passed, is inverted. Thus flow in one direction gives a positive signal and in the other direction gives a negative signal.

A completely different approach is adopted in the probe of the pulsed-wire anemometer[68] in Figure 13.4(d). Here a sudden pulse of electricity down the centre wire heats the adjacent air which advects with the flow. The other pair of wires normal to the pulsed wire are fast-acting thermometers and detect the pulse of warm air as it passes. The value of the wind speed component along the single axis normal to all three wires is given by the time delay and the sign by whichever of the two detector wires senses the pulse. Unlike the conventional hot-wire anemometer which gives a continuous analogue signal, the pulsed-wire anemometer is a digital instrument, passing individual values at rates up to about 50 Hz.

This has implications for analysis: mean, rms and peak measurements are straightforward, but computing spectra to frequencies above 50 Hz requires some statistical sleight-of-hand.

Another anemometery technique using lasers has recently become practical. The optical alignment of early laser anemometers was so critical that they were mounted on very heavy optical benches. In altering the measurement position it was often easier to move the wind tunnel than to move the anemometer! However, the use of fibre-optic cables to transmit the laser light has enabled small probes, like that in Figure 13.5(a), to be developed. Two beams of coherent light are projected by a lens to intersect about 100 mm from the tip of the probe, Figure 13.5(b). Interference fringes form in a small diamond-shaped region where the beams intersect (but are much closer spaced than indicated in Figure 13.5(c)), and any small particle of dust carried through the flow is periodically illuminated. The lens focusses the reflected light from the particles on a detector. The frequency of this illumination is proportional to the velocity vector in the plane of the beams. In the standard form the fringes are stationary, so particles moving in either direction give the same result and it is not possible to distinguish the sign of the wind speed. However, the frequency of one beam can be changed slightly by passing it through a Bragg cell, making the interference fringes move rapidly across the sensing region. Zero wind speed returns a signal at the moving fringe frequency, positive and negative wind speeds increase or decrease this frequency, respectively. More than one directional component can be simultaneously measured at the same location by using additional beams of differently coloured or polarised light. With two colours and polarisation, all three turbulence components can be measured. In practice it is necessary to 'seed' the flow with micron-sized particles and to filter the data to remove signals from large dust particles that do not move at the wind speed.

13.3.2 Static, total and surface pressures

13.3.2.1 Static and total pressures

At full scale, the reference static and total pressures may be obtained from a pitot-static probe, Figure 2.3 (§2.2.2), mounted on a vane which keeps it pointing into the wind. The reference dynamic pressure and hence the wind speed can then be obtained from their difference (Eqn 2.6). The Dines anemometer operates on this principle. When the wind speed is acquired by anemometry (§13.3.1.1) the dynamic and total pressures can be deduced provided the static pressure is known. Other forms of static pressure probe which do not need to be turned into the wind are more reliable at full scale, these include tappings in the ground surface and special probes that make use of directional symmetry, such as the NBS probe shown in Figure 13.6(a). Sometimes it is impossible to find a site unaffected by surrounding buildings or other obstructions. The static pressure is the datum from which the pressure coefficient is measured (Eqn 12.5). Another datum such as the internal pressure of the building can be substituted, but then the value of the internal presure from static is unknown. Internal pressures are the subject of Chapter 18.

The reference wind speed in the wind tunnel is often acquired from the dynamic pressure, particularly when surface pressures are being measured, only in this case the pitot-static probe is held fixed relative to the wind tunnel, pointing directly into the wind. A problem with fixed pitot-static probes is that the high intensities of

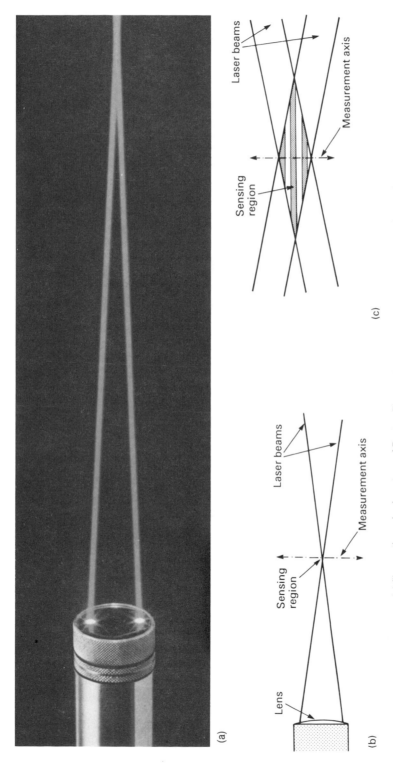

Figure 13.5 Laser anemometer: (a) fibre optic probe (courtesy of Dantec Electronics Ltd.); (b) light paths; (c) sensing region

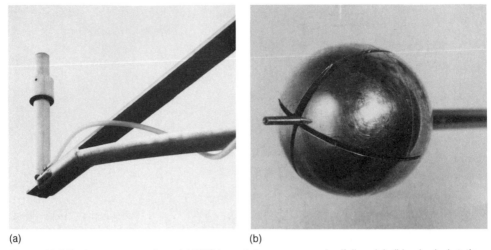

(a)   (b)

Figure 13.6 Static pressure probes: (a) NBS type static pressure probe (full scale); (b) spherical static probe (model scale)

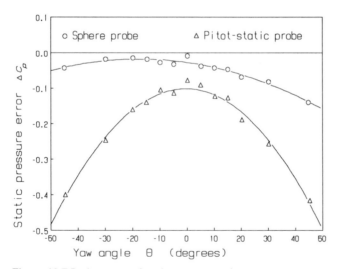

Figure 13.7 Performance of static pressure probes

turbulence cause significant changes in the instantaneous wind direction (§12.4.1.1) giving components of flow normal to the probe axis. Any deviation of the flow from along the axis gives an error in static pressure in the negative direction, and an offset of $c_p = 0.06$ at 20% turbulence intensity ($S_u = 0.2$) is typical [69]. The sphere probe shown in Figure 13.6(b) was developed from a probe for two-dimensional flows [70] to be suitable for use in three-dimensional flows and to include a total pressure tapping [71]. The response of this probe to yaw in a simulation of the atmospheric boundary layer is compared with that of a standard pitot-static tube in Figure 13.7. It is seen that the sphere probe is far less sensitive to yaw than the

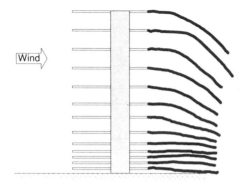

Figure 13.8 Total pressure rake

pitot-static probe, and that the pitot-static probe consistently indicates a lower static pressure. Away from the influence of any structures, the static pressure in a wind tunnel is constant through the simulated atmospheric boundary layer in the horizontal and vertical cross-wind directions, but decreases in the in-wind direction. (This gradient of static pressure balances the drag of the rough ground surface.) The vertical profile of dynamic pressure is often obtained from an array of total pressure tubes called a 'rake', as shown in Figure 13.8, referenced against a single static probe **in the same cross-section.**

13.3.2.2 Surface pressures

The normal surface pressure is the dominant wind action for the majority of structures (§12.3.4). Measurements can be made by transducers mounted directly at the surface, as in Figure 13.9(a), or by ducting the pressure through tubes from holes in the surface known as 'tappings', as in Figure 13.9(b). Both types are used at full and model-scale, although the most common choices are as shown in Figure 13.9. In the first case the transducer must replace part of the surface of the structure without affecting the flow, so it should be flush with the surface. This can be a problem if the surface is curved or is corrugated, when a flat region may need to be formed around the transducer, or tappings used instead. Similarly, tappings must be flush and normal to the surface, as shown by the examples (i)–(iii). Protruding tappings, as in example (iv), disturb the flow and should be avoided. Tappings that are not normal to the surface, example (v), will be affected by the local flow direction. These are sometimes required when access inside a model is restricted, particularly near corners, but are better replaced by tappings of form (i).

13.3.2.3 Manometers

The simplest form of manometer is the U-tube manometer, comprising a U-shaped glass tube containing a quantity of water or another suitable liquid. One end of the tube, the passive or reference limb, is left open to the ambient static atmospheric pressure, or is connected to the static pressure tapping of a pitot-static probe or in the wall of a wind tunnel to provide the datum reference pressure (§12.3.4). When a pressure is applied to the other, active end of the tube, the liquid is displaced until the pressure is balanced. The difference in height of the liquid in each limb of the tube is a direct measure of the pressure from the hydrostatic equation, Eqn 2.3 (§2.2.1). When both limbs of the tube have the same bore, the displacement from the initial level is the same in each limb, but in opposite directions, each half the

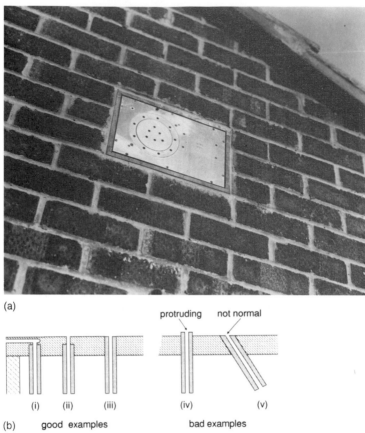

(a)

(b)   good examples   bad examples

Figure 13.9 Measuring surface pressures: (a) surface-mounted transducer at full scale; (b) pressure tappings at model scale

total displacement. It is therefore quicker to read the displacement of one side only against a fixed scale then to double the value, although only half the resolution accuracy is obtained.

Most of the full resolution accuracy is restored by making the diameter of the passive limb very much bigger than the active limb so that, on the principle of the hydraulic ram, most of the displacement occurs in the active limb. The sensitivity may be increased by inclining the active limb from the vertical, so that a given vertical displacement causes a greater measurable displacement along the tube. Taking both effects together, the total vertical displacement, z, in terms of the measured displacement, s, along the active limb is given by:

$$z = s \cos\beta \, (1 + [d_{\text{active}} / d_{\text{passive}}]^2) \tag{13.6}$$

where β is the angle of declination of the active limb from vertical and d is the diameter of the active and passive limbs as indicated by the subscripts. There is a practical maximum limit to the declination angle because the liquid meniscus becomes increasingly sensitive to variation of tube straightness and bore diameter. In most commercial inclined-tube manometers the fixed scale is calibrated directly in units of pressure, accounting for both Eqn 13.6 and Eqn 2.3.

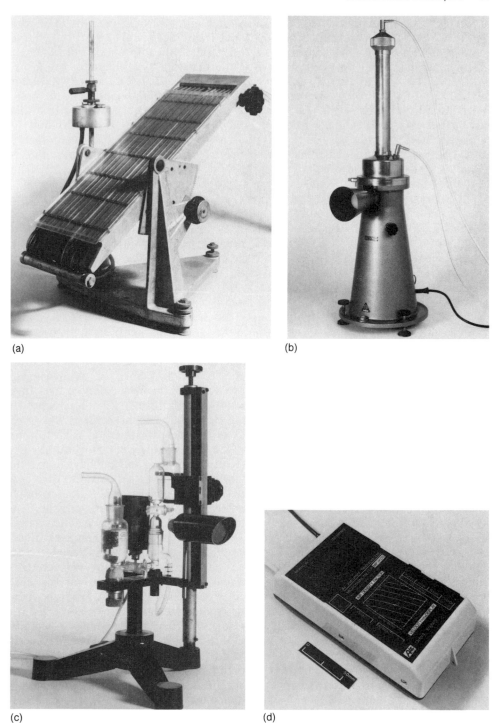

(a)

(b)

(c)

(d)

Figure 13.10 Types of manometer: (a) multi-tube manometer; (b) Betz manometer; (c) Chattock manometer; (d) electronic manometer

Multi-tube manometers of the form shown in Figure 13.10(a) were the main tool of the early experimenters at full and model-scale. They are formed by a large number of individual active tubes, mounted parallel to each other on a flat scale plate, connected to a single common passive limb with a large diameter reservoir. The tubes are selected for straightness and constant bore and are usually held flat against the scale plate by many small leaf springs. The declination angle of the scale plate is adjustable from vertical to nearly horizontal. Conventionally, the first and last tube at each end of the plate are connected in common with the passive limb to reference static pressure and so display the reference level. The end positions are used because the wide scale lines on the plate must be horizontal and this is achieved by adjusting leveling screws until both tubes read the same value. The remaining tubes are connected to the various pressures to be measured. The level in the common reference reservoir may go up or down, depending on the proportion of positive and negative pressures applied to the tubes. This is corrected by raising or lowering the reservoir to bring the reference level onto a convenient scale line as 'zero'. In this way the total displacement is always read so that the relative diameters of tubes and reservoir are irrelevant and the sensitivity depends only on the cosine of the declination angle. With water as the fluid and at the maximum practical declination of about 75°, reading the scale to 1 mm allows pressure to be resolved to about 2.5 Pa. At full scale the response of a multi-tube manometer is usually fast enough to resolve gusts of a few seconds duration, but at model-scale this is equivalent to several minutes. Use of the multi-tube manometer at model scale is therefore confined to the measurement of mean pressures only.

Sensitivity may also be increased by more accurate measurement of the fluid displacement. There are two main approaches to this problem: one by measuring the displacement from zero, analogous to the action of a spring balance; the other by measuring the action required to restore the displacement to zero, analogous to a beam balance. The first type, the 'Betz' or projection manometer is shown in Figure 13.10(b). This consists of a large bore U-tube with unequal height limbs (more a sort of J-tube) containing water, with a glass float in the taller limb. When a suction is applied to the tall limb, or a positive pressure to the short limb, the water level in the taller limb rises, lifting the float. Hanging beneath the float is a very finely graduated glass scale which is illuminated by a small bulb. A much magnified image of the scale is focussed onto a screen which enables the displacement to be read to about 0.1 mm, corresponding to about 1 Pa. The second type, the 'Chattock' manometer, shown in Figure 13.10(c), has an extra kink in the U-tube (giving it a shape more like a W) in which a quantity of oil is trapped. The magnified image of one oil–water mensicus projected onto a screen is used to indicate 'balance'. The height of the water reservoir on one side can be raised or lowered, while a tap in the other limb can allow or prevent the fluid from moving. Initially, with both limbs of the manometer connected to ambient static pressure and the tap open, the reservoir position is adjusted to align the meniscus exactly with a pointer on the screen. The tap is then closed and the pressure to be measured is applied to the manometer. The reservoir is raised by approximately the expected pressure and balance is tested by slowly opening the tap. If the meniscus rises the reservoir must be raised further, or lowered if the meniscus falls, until balance is achieved. The amount by which the reservoir was raised represents the pressure directly and this is measured by a vernier scale to an accuracy of about 0.025 mm, corresponding to about 0.25 Pa. In practice, the Betz manometer with its direct display of pressure, is commonly used to measure the reference dynamic pressure in wind tunnels. The Chattock manometer, with its ingenious but cumbersome balancing procedure, is suitable only for measuring steady pressures and its rôle is usually confined to calibrating transducers.

Recent improvements in transducers (see §13.3.2.4, below) and microcomputers have enabled electronic manometers to be made which have stable calibration characteristics and display the pressure directly in engineering units. Other advantages of a microcomputer is the ability to linearise the transducer response, make unit conversions or to transform dynamic pressure to wind speed. Figure 13.10(d) shows an electronic manometer with a digital display and selectable pressure ranges of 199.9 Pa, 1999 Pa and 3 kPa, which can be read in steps of 0.1 Pa, 1 Pa and 10 Pa respectively, and an additional wind speed range of 70 m/s in steps of 0.1 m/s.

13.3.2.4 Transducers

Pressure transducers are devices which generate a voltage signal in response to applied pressure. Ideally the response should be linear in amplitude and constant with frequency. When BRE embarked on its programme of wind pressure measurements on full-scale buildings during the 1960s, there were no suitable transducers in existence so it was necessary to develop one specially for the purpose [72]. More modern transducers for full- and model-scale use are shown in Figure 13.11. In essence, the active pressure is applied to one side of a diaphragm and a reference pressure is applied to the other side. Changes in the applied pressure cause the diaphragm to deflect and this is detected electrically, using capacitive or inductive methods or by strain gauges. Transducers differ in size and frequency response, those for use at model-scale being smaller and able to resolve higher frequencies.

At full scale the pressure transducer is often mounted with the diaphragm flush with the building surface, but with tubing connecting the back of the diaphragm to the reference pressure. As incident wind conditions at full scale change continuously with the changing weather, simultaneous measurements must be made at all locations, requiring an individual transducer at each location. Much care must be taken in designing the reference pressure tubing, especially when many transducers are connected to a common point as usual at full scale [73,74].

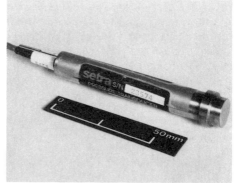

(a)   (b)

Figure 13.11 Typical flush diaphragm transducers: (a) for measurements at full scale; (b) for measurements at model scale

Deflection of the diaphragm of one transducer moves the air in the reference tubing which, in turn, deflects the diaphragms of other transducers connected in common, giving a spurious signal called 'cross-talk'. This effect is additional to the response of the tubing described below, but is a less serious problem with modern transducers which have much stiffer diaphragms than the original BRE transducer.

At model scale the transducer is generally too large to mount flush with the model surface, unless the average pressure over a large area is required, so the active side is usually connected by tubing to a pressure tapping (§13.3.2.2). Incident wind conditions are kept constant in the wind tunnel and may be reproduced at will. It is therefore not necessary to make measurements simultaneously at all locations. Instead measurements can be made sequentially using a pressure-scanning switch to connect the tappings in turn to a single transducer as shown by Figure 13.12(a). The transducer from Figure 13.11(b) may be seen protruding from one end a 'Scanivalve' pressure-scanning switch, through an annular connector from which the pressure tubes pass to tappings on the model. The valve switches between 48 connections, driven by a solenoid and controlled and monitored through electrical connections at the other end.

13.3.2.5 Tubing response

The length of tubing that connects each tapping to the transducer introduces distortions into the pressure traces which are a function of the tubing length and diameter, the path through any pressure-scanning switch, and the internal volume of the transducer. The length and diameter of the tubing connecting the back of the transducer to the reference pressure also affects the trace, depending on the stiffness of the transducer diaphragm, but this is usually of secondary importance. The distortions are a function of frequency, so are most conveniently described by the standard frequency response function approach (as used earlier in §8.4).

There are two principal components: organ-pipe resonance of the tubing in the middle range of frequency and Helmholtz resonance of the internal volume of the transducer at the high end of the range. With the Scanivalve/transducer combination shown in Figure 13.12(a), the Helmholtz resonance is more than critically damped and sets an effective upper limit to the range of about 500 Hz [75]. The organ-pipe resonance selectively amplifies frequencies at the harmonics appropriate to its length. These distortions affect all measurements of fluctuating pressure components, including the peak values required for static structures [76]. A theory was developed in 1965 which enables the response characteristics of any given tubing system to be calculated [77].

Only those analyses made in the frequency domain: spectra, cross-spectra, etc., can be corrected retrospectively. In general it is necessary to correct the pressure traces before analysis. Four main options are available:

1 Use very short tubing so that the fundamental organ-pipe frequency is well above the range of interest.
2 Acquire the distorted signal, but correct the distortion before analysis.
3 Insert an electronic circuit which corrects the distortion between the transducer and the acquisition/analysis system.
4 Modify the characteristics of the tubing to remove the distortion.

The first option is only practical when the distance between tappings is small, or when a separate transducer is used for each tapping.

For the second option, the frequency response function or *transfer function* of the tubing is first determined by calculation [76] or experiment [77]. The 'raw' fluctuating pressure signal, $p_{\text{raw}}\{t\}$, acquired through the tubing is Fourier transformed into the frequency domain, $p_{\text{raw}}\{n\}$. Dividing by the transfer function gives the undistorted pressure, $p\{n\}$, and Fourier transforming back into the time domain gives the required pressure signal, $p\{t\}$. This is called the inverse transfer

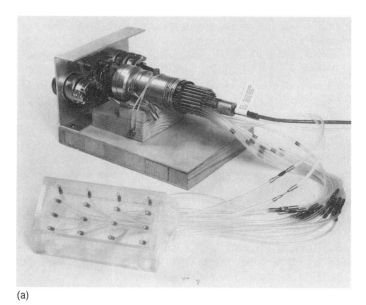

(a)

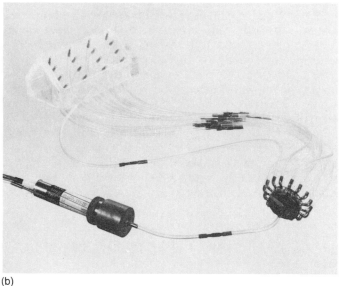

(b)

Figure 13.12 Typical hardware for surface pressure acquisition: (a) 'Scanivalve' pressure scanning switch, pressure tubes and model; (b) 16-tube pneumatic averager, transducer, pressure tubes and model

function method and its application is illustrated in Figure 13.13 by data reproduced from reference [78]. In each of the three cases shown, the 'A signal' is the undistorted signal obtained from a flush transducer, while the 'B signal' is the signal through 10 foot (3 m) long tubing with various stages of correction. The top traces show that the uncorrected signal is delayed in time (representing the transit time through the tubing), that fluctuations are less sharp (loss of high frequencies) and that the peaks are reduced in value. The last is most serious, representing a considerable underestimate in peak loading, and occurs because the first resonance of the very long tubing is well inside the range of interest and is strongly damped. With shorter tubing, particularly when the duration of a peak pressure coincides with the period of a harmonic, the peaks can be exaggerated. The middle traces of Figure 13.13 show the result of applying the full inverse transfer function method, correcting the 'B signal' so that it is almost indistinguishable from the undistorted signal (except at the very beginning and end of the trace, which is a feature of the fast Fourier transform algorithm). Sometimes when devising correction procedures, the phase may be neglected in favour of optimising only the amplitude response, but the bottom traces show that some peaks are reduced while others are increased. Clearly, it is important to correct both amplitude and phase. The major disadvantage of this method is that the mathematical manipulation involved: two Fourier transforms and a division, require a digital data processor and are a considerable analysis overhead.

The third option, that of an electronic correction circuit based on active filters, is sometimes applied. The main aim here is to generate a transfer function with exactly the inverse characteristics of the tubing in amplitude and phase. Special analogue circuits have been developed which linearise the fundamental and first two harmonics [78, 79]. If digital methods are used, the problem reverts to the inverse transfer function method above.

A number of techniques have been developed for the fourth option of modifying the tubing characteristics, all based on including a restriction in the tube to increase the damping. Some are purely empirical approaches that require experimental calibration: a length of yarn may be inserted in the tubing [80], or else a length of metal tube may be bent to form a restriction [78]. In the latter case, the tube can be encased in resin after the bend has been adjusted to give the optimum effect, so that the calibration does not change. More successful are the methods in which a length of very small-bore tube is inserted, as the optimal length, bore and position of the insert can be calculated using the Bergh–Tidjeman theory [77], or its development for this purpose by Gumley [81, 82], although it is prudent to verify the calibration experimentally [75, 83]. The aim of the method is to obtain a measured pressure signal, $p_M\{t\}$, at the transducer end of the tubing which is identical to the surface pressure, $p\{t\}$, at the tapping end of the tubing, but which may be delayed in time by an amount Δt corresponding to the transit time down the tubing, thus:

$$p_M\{t\} = p\{t + \Delta t\} \tag{13.7}$$

Consider both pressure signals in terms of their Fourier components in magnitude, $A\{n\}$, and phase, $\phi\{n\}$:

$$p\{t\} = \int A\{n\} \cos(2\pi n t + \phi\{n\}) \, dn \tag{13.8}$$

giving:

$$p\{t\} = \int A\{n\} \cos(2\pi n t) \, dn \tag{13.9}$$

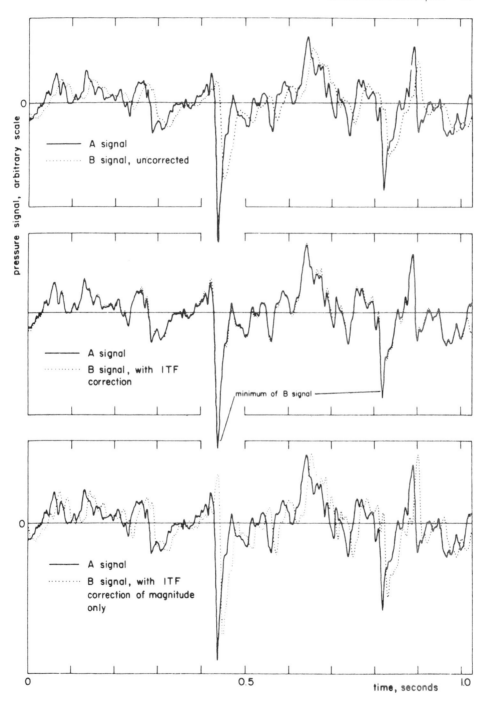

Figure 13.13 Correction of tubing response by ITF method (from reference 78)

$$p_M\{t\} = \int A\{n\} \cos(2 \pi n [t + \Delta t]) \, dn \tag{13.10}$$

Substituting Eqns 13.9 and 13.10 into Eqn 13.7 shows that the amplitude of the transfer function must be unity through the required frequency range to keep the Fourier components, $A\{n\}$, the same, but that the phase of the transfer function is given by:

$$\phi\{n\} = 2 \pi n \, \Delta t \tag{13.11}$$

that is, changing linearly with frequency.

The effectiveness of the method is demonstrated by Figure 13.14, reproduced from reference [78]. Figure 13.14(a) shows how the fundamental organ-pipe resonance of a 2 ft (0.6 m) long tube reduces in amplitude as the restriction is increased. The optimum response is within 5% of unity in the range 0–100 Hz. Figure 13.14(b) compares the phase response of the optimal restriction to the original tube response, showing that the restrictor gives the required linear change with frequency. Finally, Figure 13.14(c) compares the undistorted trace from a flush transducer, the 'A signal', with the signal through the tubing with the optimal restrictor, the 'B signal'. There is the expected time delay between the two traces, but otherwise they are practically identical, including the peak values. Restrictors may be seen in the tubing shown in Figure 13.11. These were formed by collapsing a 10 mm long section of brass tube onto a 0.33 mm diameter hard steel wire, then withdrawing the wire. This simple method of manufacture, originally devised at the University of Western Ontario, is now standard at BRE and elsewhere owing to the consistency of characteristics obtained [75].

13.3.3 Forces and moments

13.3.3.1 Integration of surface pressures

On most bluff bodies where the contribution of normal pressures greatly exceeds that of the shear stresses, the wind-induced forces and moments may be obtained from the normal pressures by integration over the relevant part of the building structure (§12.3.6). Tappings must be spaced sufficiently close together that each measured pressure is representative of an area around the tapping, called the 'tributary area', and that all the tributary areas combine to represent the whole surface. At full scale, the instantaneous loads are obtained by algebraically summing the product of the tributary areas and the corresponding instantaneous pressures measured simultaneously at all points.

This is not possible at model scale when a single transducer and scanning switch is used to measure pressures sequentially, but the mean forces and moments can be obtained by integration of the mean pressures. When two transducers are used, rms values of the forces and moments can be obtained by integration of the variances and covariances of the individual tapping pressures [84]:

$$F' = \sum_{i=1}^{N} \sum_{j=1}^{N} (\overline{p_i p_j} A_i A_j)^{1/2} \tag{13.12}$$

where i, j indicate the individual tappings of each pair, A_i, A_j are the corresponding tributary areas and N is the number of tappings. Given the mean and rms values of

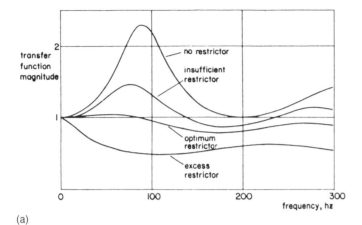

(a)

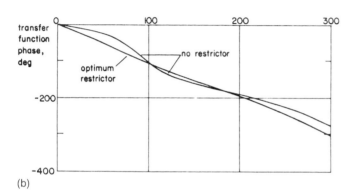

(b)

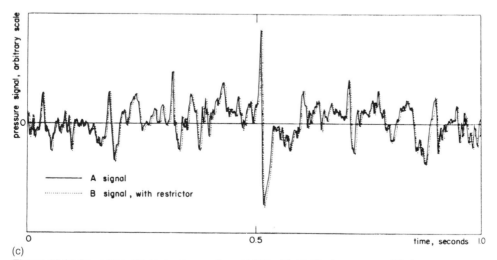

(c)

Figure 13.14 Correction of tubing response by restrictor: (a) amplitude response; (b) phase response; (c) typical pressure traces (from reference 78)

the forces and moments, peak values can be estimated by the peak-factor method (§12.4.2). The principal disadvantage of this approach is that the analysis time is proportional to the *square* of the number of tappings, since the covariance of each pair must be measured.

In 1978 Surry and Stathopoulos [85] proposed that if a number of pressure tubes from surface tappings were connected to a common point, the pressure at that point would be the instantaneous average of the individual tapping pressures. Justification of the approach, based on dimensional analysis, required that tube lengths should be sufficiently long and the differences between the individual tapping pressures sufficiently small to maintain laminar flow in the tubes. (When the tubes are fitted with restrictors laminar flow is maintained in all practical cases.) This case was supported by experimental measurements. More recently, Gumley's development [81,82] of the Bergh–Tidjeman tubing response theory gave this approach a sound theoretical basis.

The method is now known as 'pneumatic averaging'. Figure 13.12(b) shows typical hardware: a number of identical tubes (with restrictors) are connected to a common point in the star-shaped manifold, and a single tube (also with a restrictor) connects this point directly to a transducer. A number of averaging manifolds can be connected to the transducer by a pressure-scanning switch (§13.3.2.4). Note that on the model shown, 16 tappings are equally spaced on a four-by-four rectangular grid, so that the tributary area for each tapping is equal and the measured pressure represents the load on the building face as a uniformly distributed load. In this case the measured pressure, $p_M\{t\}$, is given by:

$$p_M\{t\} = \Sigma p\{t\} / N \tag{13.13}$$

where N is the number of tappings averaged, and the load on the face, $F\{t\}$, is:

$$F\{t\} = p_M\{t\} A \tag{13.14}$$

where A is the face area. By altering the number and spacing of the tappings, so changing the relative tributary areas, the measured pressure can be weighted to give other action effects, such as moments. Figure 13.15 shows typical tapping arrangements for obtaining the shear force and moment at the base of a tower. In (a) 16 tappings are equally spaced, so that the contribution from each tapping to the average pressure is equally weighted and the mean force on the face is obtained through Eqns 13.13 and 13.14. In (b) the tappings are not equally spaced, but are

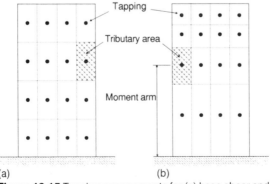

(a)   (b)

Figure 13.15 Tapping arrangements for (a) base shear and (b) base moment

arranged so that the product of each tributary area and the corresponding moment arm to the base are equal, so that each tapping is weighted to contribute equally to the base moment. The measured pressure is equivalent to the uniformly-distributed load that gives the same base moment, so that:

$$M\{t\} = p_M\{t\}\,A\,H\,/\,2 \tag{13.15}$$

For the base shear or moment of the whole building a second set of tappings is required on the opposite face, connected to a second transducer, and the difference between the two transducer signals is measured. The principle can be extended to accommodate most practical influence functions, where all tappings in positive lobes are connected in common to one transducer, all tappings in negative lobes to a second transducer, and the difference of the signals is measured. A good example of this approach is the estimation of bending moment in the 'knee' joint of a portal frame by weighted pneumatic averaging in the plane of the frame described in references [86] and [87].

The accuracy of the method relies on having sufficient closely spaced tappings to resolve strong gradients of pressure and small turbulent eddies, the two main sources of error. The latter is avoided by setting an upper frequency limit to the data, using the 'TVL-formula' (§12.4.1.3 and §15.3.5). The eddies measured by each tapping are assumed to be correlated over the whole tributary area, whereas this is the case only for larger eddies. Figure 13.16 compares the spectra of base bending moment for the case illustrated by Figure 13.15(b) with that obtained directly from a high-frequency range balance [88]. The match is excellent below

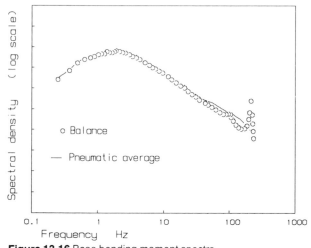

Figure 13.16 Base bending moment spectra

about 50 Hz but it is clear that the pneumatic average tends to overestimate at higher frequencies. (The sharp peak in the balance spectrum is the dynamic response of the model, which is discussed in §13.3.3.2 below.) More recent developments of the method include the use of a porous polythene skin to give a very fine grid of tappings, where the weighting is achieved by varying the active area [89]. As the method gives the equivalent average pressure, the sensitivity does not depend on the size of the loaded area, so is particularly useful for small areas such as cladding panels where direct measurement is impractical.

13.3.3.2 Direct measurement

The alternative to integration of pressures is direct measurement using some form of balance or dynamometer. The first balances designed for aeronautical use were intricate arrangements of platforms suspended on pivots which separated the forces and moments into orthogonal axes, each measured by an individual beam balance. These balances were generally fixed in the roof of the working area, from which the aircraft model would hang from struts or wires 'flying' inverted, and measured only steady values. Replacing the beam balances with force transducers enabled fluctuating forces and moments to be measured. Balances are still made to this design for use with building models [90], but are mounted on the wind-tunnel floor instead of the roof. These fixed balances work in wind axes, whereas body axes are more relevant to building studies (§12.3.6).

A high frequency range is required in order to study peak loads for static structures and load spectra for dynamic structures, and this demands that the balance be stiff. The majority of force transducers are based on strain gauges, so that this stiffness acts against sensitivity. Tchanz's approach to the problem was to make the balance and model very light [91,92] by replacing the balance platform by a single L-shaped girder. Models for this balance are made using polystyrene foam, allowing measurements to about 300 Hz. The balance is sufficiently compact to be installed in a small turntable, allowing it to be rotated with the model, so giving body-axis forces and moments. With this balance, like all based on strain gauges, the load range is small and significant overloads will damage the transducers. To increase the range of the balance, the transducers must be replaced with others which are stiffer and so less sensitive. Typically, strain-gauge based balances have a range of three orders of magnitude, so that a 10 N range transducer will be able to resolve about 0.01 N.

Transducers based on piezo-electric crystals are many orders of magnitude stiffer than strain-gauge transducers and offer a major advantage in range and frequency response. As they give a small electric charge instead of a voltage, they require special charge amplifiers to give usable signals. Unfortunately these amplifiers allow the charge to leak slowly away so that steady values decay and only fluctuations above a certain frequency are measurable. A balance [88] designed at BRE which uses piezo-electric crystals is shown in Figure 13.17. It consists of four three-component piezo-electric transducers clamped between two thick circular

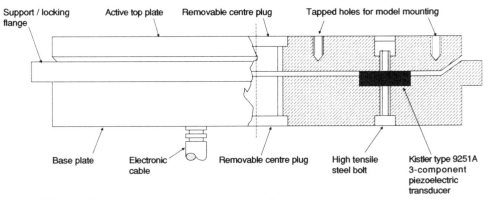

Figure 13.17 BRE 6-component high-frequency range balance

steel plates by a high-tensile bolt through each transducer. The range of each transducer is 50 kN in compression only, so the bolts are pre-tensioned to 25 kN to bring the transducers to the centre of their range. A flange on the bottom plate is used to clamp the balance firmly to a massive (0.5 tonne) concrete block on rubber antivibration mountings which isolates the balance from external vibrations and prevents the heavy top plate from acting as an accelerometer. Owing to its circular form, the whole balance can be rotated with the model to change the wind direction, giving the forces and moments in body axes. Alternatively, the balance can remain fixed, but the model mounted at various wind directions on the active top plate of the balance, giving the forces and moments in wind axes. The removable central plugs in both plates allows pressure tubing to pass through the balance (the stiffness of the tubing being insignificant in comparison with the balance), and this is how the data of Figure 13.16 were obtained.

The signals from the four three-component transducers are resolved into the forces and moments in the three orthogonal directions by the control electronics. The gain of the amplifiers is adjustable in steps to change the range from ± 5 kN to ± 1 N and at this most sensitive range the balance will resolve 0.001 N. Now one advantage is apparent: the balance cannot be broken accidentally but the sensitivity is ten times greater than a comparable strain-gauge balance. Despite the massiveness the first natural frequency of the balance is 750 Hz for the horizontal forces and 1200 Hz for the horizontal moments. Hence the next advantage: there is no need to keep the models light. The model mass may be 12 kg before the first natural frequency is reduced by 70% to about 500 Hz; also provision is made in the control electronics to remove the model weight from the output signal by grounding the charge, allowing the most sensitive range (1 N) to be used with any weight of model. In use, however, the frequency response is set by the model which is always more flexible than the balance. This balance was used for the comparison with pneumatic averaging in Figure 13.16 and the peak in the balance data at 300 Hz corresponds to the lowest natural frequency of the aluminium model. To be able to employ the full frequency range of the balance, a number of cylindrical cores have been made using carbon fibre, with a first natural frequency exceeding 1 kHz, around which building models are constructed.

The decay time constant, corresponding to the lowest measurable frequency, is a function of the impedance of the amplifiers in the balance electronics, and this can be set to three values 'long', 'medium' and 'short'. At the 'medium' time constant the balance gives all fluctuations above 0.16 Hz which is generally sufficient for the measurement of spectra. At the 'long' time constant the balance gives all fluctuations above 0.000 16 Hz corresponding to a period of 6250 s, and this is more than sufficient to obtain reliable mean values [88, 93].

Instead of a general purpose balance, another approach is to install transducers in the model supports or to mount the model directly on dynamometers [94, 95, 96]. This is usually the only practical approach for full-scale buildings and structures. In BRE experiments, for example, strain gauges were installed on the steel reinforcement at the base of the Post Office Tower [97], a reinforced-concrete communication tower in London, while for a two storey building at Aylesbury dynamometers were mounted between pile-caps and the building to give overall loads, and also between the roof and the walls to give roof loads [98]. Also at full scale it becomes practicable to support small elements such as cladding panels directly on transducers [99], where at model-scale the forces would be too small to measure.

13.4 Full-scale tests

13.4.1 Rôle of full-scale tests

So far in this chapter both full-scale and model-scale techniques have been described, although the reader should have discerned a distinct emphasis on model-scale. This is because full-scale tests are of very limited direct use to the designer, although the designer is often indirectly influenced by their contribution to codes and design regulations. The rôle of full-scale tests may be divided into three main aspects:

(a) tests on unique designs;
(b) tests on mass-produced designs; and
(c) tests on components.

The first rôle is the prerogative of the professional wind engineer, who uses the data as 'benchmarks' for model/full scale comparisons to verify model accuracy (§13.5.5) and to 'calibrate' codes of practice and building regulations. This kind of work is usually only undertaken by national research laboratories, such as the UK Building Research Establishment, the US National Bureau of Standards, the National Research Council of Canada and the Commonwealth Scientific and Industrial Research Organisation in Australia. Tests on a built structure are of no direct use to the designer of that structure other than to prove his design. While they may be of use for future similar structures, these will be the concern of other designers and, more significantly, other clients. Nevertheless, the range of full-scale data is always increasing and it is possible that useful design information may be gleaned from a past experiment. The best source of such data is the *Journal of Wind Engineering and Industrial Aerodynamics* and the proceedings of the first five IAWE conferences [100, 101, 102, 103, 104] (the sixth and later conference papers are published in the journal).

The second rôle is where full-scale tests become useful to designers in R and D departments of large companies producing similar structures in volume. Suitable structures will usually be small, such as steel-framed industrial buildings, prefabricated garages and sheds, mobile homes and caravans, lattice masts and hoardings. Larger structures constructed in volume by public authorities may also be tested by the corresponding central research laboratory, such as hyperboloid-shell cooling towers and lattice power-transmission pylons. These tests are invariably carried out over many years to satisfy long-term development plans.

The third rôle is where full-scale tests become useful in the shorter term to companies making structural components in volume. These may be lightweight cladding for walls or roofs, roofing tiles and slates, purlins and trusses, external insulation systems, glazing items from single components to complete units, roller-shutter doors, ventilators and fire vents, solar panels and satellite transmission receiving dishes. This approach may useful when the structural component is particularly difficult to justify by calculation as, for example, with loose-laid external roof insulation panels (see §18.8) where full-scale tests have contributed directly to design rules as well as extending knowledge of the fundamental aerodynamic loading mechanism.

13.4.2 Commissioning full-scale tests

Any full-scale test will be both expensive and time consuming if it is to be accurate and useful, so should never be attempted without first having taken professional advice. Ask the following four questions:

1 Is the acquisition of test data necessary to the design?
2 Is it essential to work at full scale?
3 Will the resulting data be worth the cost of acquisition?
4 Will the data be obtained within the required time period?

and proceed only if the answer to all four questions is yes.

Expertise in full-scale experimentation is the result of much time and money, mistakes made and lessons learned. It will always be better to obtain the assistance of an experienced test laboratory than to embark alone on a full-scale test programme, otherwise the first several years are likely to be spent making the same mistakes and learning the same lessons. Even with experienced assistance, success within a realistic time period is not guaranteed. Murphy's law as pertaining to full-scale experiments is *acquisition equipment which works perfectly in calm weather always fails as soon as a strong wind occurs*.

13.5 Model-scale tests

13.5.1 Principles of model-scale testing

In Part 1 of the Guide (§2.4.4) the process of modelling was likened to making 'three wishes', in that scale factors may be chosen for each of the three primary dimensions of mass, length and time. The term 'full scale' really only refers to the length scale factor, $\mathscr{L} = 1$, as the wind speed of the tests is unlikely to be near the design value (§2.4.1), so the **design full-scale** conditions to be modelled will be called **prototype** conditions here.

The aim of modelling is to obtain dynamic similarity, which is the ideal state when the ratios of all the actions in the model are identical to the prototype. This occurs when all the relevant non-dimensional parameters (§2.4.2) have equal values in the prototype and model. Unfortunately, it has already been demonstrated (§2.4.4) that it is impossible to obtain similarity of all the non-dimensional parameters simultaneously except when the three primary scale factors are all unity. In practice the problem becomes one of deciding which of the various non-dimensional parameters need to be matched to obtain the required accuracy, and this can be determined only by experiment.

The relevant non-dimensional parameters have traditionally been named after the researchers who established their importance: Reynolds number, Re, was introduced in §2.2.6 and then was used to deduce the characteristics of boundary layers and vortices. The single most important parameter when modelling structures on the Earth's surface was shown by Jensen[48] to be the ratio of structure's height, H, to the aerodynamic roughness of the ground, z_o (the log-law parameter (§7.2.1.3.2)), hence it is only right that this is now recognised as the Jensen number[105], Je $= H/z_o$.

Jensen's work required such a fundamental change to modelling practice that is it natural that his work would be critically reviewed before adoption. One early review[106], while admitting the importance of the boundary layer, sought to imply that Jensen's scaling laws were incomplete because the depth of the boundary layer, z_g, was not included as a parameter, and that some Reynolds number dependence could be found with sharp-edged models in uniform flow. It was proposed that z_g be included by splitting H/z_o into z_g/H, an 'immersion ratio', and z_o/z_g, a 'relative roughness', then that the latter should be related to the power-law exponent, α. Dependence on the immersion ratio was demonstrated when the building occupied a significant proportion of the boundary layer, but its effect is confined to very tall buildings, whereas buildings in the surface layer are independent of boundary-layer height (see §7.2.1.3).

During the next decade, modelling procedures were developed which used boundary-layer depth and power-law exponent as the principal parameters, particularly in the USA [107] where most of the work concerned tall buildings. Special boundary-layer wind tunnels were built to allow the new simulation procedures to be exploited. The work done in the 1960s led to a generation of design codes of practice which included the effects of the atmospheric boundary layer, but were based on the power-law model (see Chapter 14). However, with the development of boundary-layer theory, the log-law model gradually gained favour over the power-law, particularly in Europe where concern was for squatter buildings, eventually leading to the Deaves and Harris model (§7.2.1.3.3) on which most of the present design guidance is based. Recent reviews of the similarity requirements [108, 109] and text books [110, 111, 112] agree that the log-law model is best, but that the power-law model remains a useful empirical approximation. Jensen number is now accepted as the principal scaling parameter for model simulations, with the boundary-layer depth becoming significant only for the tallest structures.

Non-dimensional parameters for which similarity is **always** required for static and dynamic structures are:

Jensen number: $Je = H / z_o$

This is the fundamental scaling parameter established by Jensen [48, 105] which relates the length scale factors of the structure and the atmospheric boundary-layer simulation. When a self-consistent 'full-depth' simulation is achieved (§13.5.2) Je similarity is sufficient. But when a 'part-depth' simulation is used and in 'artificial' simulations it is necessary to match the turbulence explicitly (usually through the u-component spectrum (see §13.5.3)) and to consider the relative height of atmospheric boundary layer and structure, z_g/H, the immersion ratio. A mismatch of Je by a factor of two to three is acceptable [113] (see §13.5.5).

Strouhal number: $St = n D / V$

Strouhal number was derived to define the characteristic frequency, n_s, of vortex shedding, but the Guide extends this restricted definition and uses St as the general reduced frequency for any frequency-dependent parameter. The reciprocal of St is known as the reduced velocity, $V / n D$. As the reciprocal of frequency is time, similarity of St is needed to match the duration of gust loads between model and prototype. Essentially, St is the dimensional statement that velocity is length/time and is therefore required in **all** models in which time or frequency dependence is represented.

A non-dimensional parameter for which similarity is **sometimes** required for static and dynamic structures is:

Reynolds number: $Re = \rho_a V D / \mu$

Reynolds number was discussed in §2.2.6–§2.2.9. It describes the relative importance of fluid inertia and viscosity. In the majority of wind engineering applications Re is very large and fluid inertia forces dominate, leading to complex turbulent flows around bluff bodies (§2.2.10). The exception is the region close to the surface of structures (§2.2.7) where viscosity is significant and controls flow separation and re-attachment (§2.2.10). Curved structures are senstive to Re, particularly through the 'critical range' around $Re_{D(crit)} \simeq 2 \times 10^5$. On the other hand, sharp-edged structures are insensitive to Re in boundary-layer flows. At length scale factors smaller than about 1:10, Re similarity cannot be achieved unless the air is compressed or replaced by another fluid. However, comparisons with full scale (§13.5.5) show that with sharp-edged structures it is sufficient to ensure the Re is large enough for inertia forces to dominate, and that with curved structures for Re to be in the subcritical or supercritical range as appropriate.

A non-dimensional parameter for which similarity is **not** required for static structures, but is **always** required for dynamic structures is:

Scruton number: $Sc = 4 \pi m \zeta / (\rho_a B^2)$

Full dynamic similarity of dynamic structures requires that density number, ρ_s/ρ_a, elasticity number, $E/\rho_s V^2$, and structural damping ratio, ζ, be matched. However Scruton showed that matching Sc (which he called the 'mass-damping parameter' and is derived from the other three parameters) was sufficient[114]. This is discussed in Part 3.

A non-dimensional parameter for which similarity is **not** required for static structures, but is **sometimes** required for dynamic structures is:

Froude number: $Fr = V / \sqrt{(g D)}$

This describes the relative importance of inertia and gravity forces. It is sometimes alternatively expressed as the gravity number, gD/V^2. Similarity is essential when gravity is a significant component of the stiffness of the structure, e.g. suspension bridges, but is not required for static structures. One way of matching Fr is to increase gravity by adding mechanical stiffness.

Non-dimensional parameters for which similarity is usually not required for static or dynamic structures are:

Rossby number: $Ro = V / D \Omega$

This relates angular velocities to the rotation of the Earth and is required when representing large-scale atmospheric motions involving rotations, including tornadoes (§5.1.2.3) and the Ekman spiral in the upper regions of the atmospheric boundary layer (§7.1.4). It is required when simulating large-scale atmospheric motions when similarity is achieved by rotating the whole wind tunnel.

Richardson number: $Ri = g (\partial T/\partial z + \lambda_A) / [T(\partial u/\partial z)^2]$

Richardson number describes the stability of the atmosphere, which is neutral (adiabatic) when $Ri = 0$. Stability in the wind tunnel is neutral unless modified by heating or cooling the ground surface or the air. The atmosphere is usually assumed to be neutrally stable in strong winds (§7.1.1), so that $Ri = 0$ automatically and the need to make a specific match is avoided.

Perhaps the most complete modern review of the principles of model-scale testing is given by *Wind tunnel modeling for civil engineering applications*[114], which is the proceedings of an international workshop on this subject held in 1982, and contains many of the individual references quoted in this chapter.

13.5.2 Boundary-layer wind tunnels

Most aeronautical wind tunnels are too short to develop good boundary-layer simulations. Some were used to develop artificial techniques of reproducing boundary-layer characteristics, usually just the profile of mean wind speed (§13.5.3.1). In others, previously unused parts of the return flow circuit of other tunnels were found to be suitable for growing a boundary-layer simulation. In the main, however, new wind tunnels were designed specifically for the purpose [107, 115, 116, 117, 118, 119, 120, 121, 122, 123, 124, 125, 126]. Figure 13.18 shows the two main configurations: (a) closed return, where the flow recirculates;

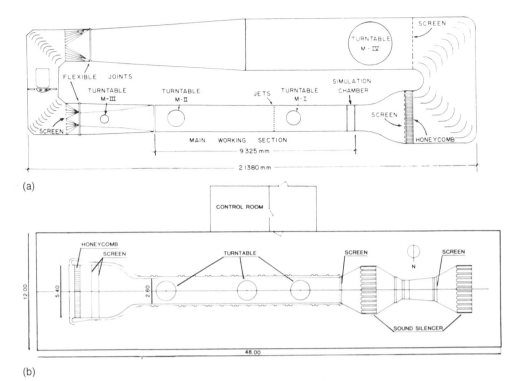

(a)

(b)

Figure 13.18 Two typical boundary-layer wind tunnels: (a) closed return tunnel (Universidade Federal do Rio Grande do Sul, Brazil from reference 120); (b) open return tunnel (Danish Maritime Institute from reference 126)

and (b) open return, where the flow is drawn in at one end and is discharged at the other. Each system has advantages and disadvantages: the closed circuit requires less power, the working area is near atmospheric pressure and it retains tracer smoke, model sand or snow particles and other pollutants, but the flow gradually heats up and it is more expensive to build; the open circuit is cheaper to build and maintains constant flow temperature by dissipating heat to the laboratory, but the working section is usually above or below atmospheric pressure requiring elaborate seals around doors and turntables to prevent leaks and the laboratory soon becomes polluted by tracer smoke or gases.

Certain generic features are common to most wind tunnels. All will have one or more fans to drive the air through the tunnel. Early tunnels were always driven by axial propellors, but centrifugal fans, which are more tolerant of turbulent flow, are now becoming popular. Wind speed is usually controlled by varying the speed of the fan, but some axial fans rotate at constant speed and the pitch angle of the blades is varied instead. With open return tunnels the fan is often downstream of the working section so that turbulence from the fan does not contaminate the flow. Centrifugal fans do not give the flow the rotation or 'swirl' characteristic of axial fans. Sometimes these are placed upstream of the working section to give a 'blower tunnel', which has the advantages that the working section near the open end of the tunnel, where acoustic noise is least (see below) and where the static pressure is close to atmospheric. One or more turntables in the working section floor on which the models are mounted enables the incident wind direction to be changed. A 'contraction', a reduction in cross-sectional area which improves the uniformity of the flow, is often placed immediately upstream of the working section.

The main characteristic of the boundary-layer wind tunnel is the length of the working section in which the atmospheric simulation is grown. In some tunnels the cross-sectional area can be varied by changing the height of the roof to control the gradient of static pressure [107, 115, 126]. Some specialised tunnels have provision for heating or cooling the air and the ground surface to simulate non-adiabatic conditions [107, 121]. Modern tunnels often incorporate a special area at the upstream end of the working section called a 'flow processing section' where turbulence grids, walls and other elements of simulation hardware can be easily inserted [116, 117, 120, 126].

Figure 13.19 shows the BRE boundary-layer wind tunnel of the Structural Design Division, which is an open-return tunnel with a centrifugal fan downstream of the working section. When first built in 1973 the working section was 8 m in length [117], but this was later extended to 14 m. Models are mounted only in the

Figure 13.19 BRE boundary-layer wind tunnel

last 3 m (the glazed area), the remainder being used for growing the simulation. The cross-section is 2 m wide by 1 m high. The roof height is adjustable only in the model area in order to mitigate model blockage (§13.5.4.1). The floor of the model area is removable: either of two 1.75 m diameter remotely controlled turntables can be inserted, allowing for fast exchange of models; or else a 0.5 tonne concrete block may be installed on anti-vibration mountings, on which the BRE dynamic force balance of Figure 13.17 (§13.3.3.2) or dynamic/aeroelastic models can be mounted. The wind speed may be set at any value between 0 and about 20 m/s and maintained to within 1%.

Acoustic noise in the wind tunnels can contaminate measurements of pressure, depending on the volume of the noise and the acoustic properties of the tunnel. The fan is a common source of noise which is produced when fan blades pass stator blades, fixed struts or each other. Large numbers of blades in the fan may be used to bring the frequencies above the range of the measurements. Another common source of noise is resonance of panels in the tunnel walls, often excited by motor or fan vibrations, but this may be suppressed by stiffening panels to raise their frequency or adding mass to lower their frequency, or by adding damping. One source of acoustic noise that is always present is the turbulence of the atmospheric boundary layer simulation, although this is generally small because acoustic-flow interaction is governed by the Mach number and this is kept small ($M < 0.1$). Nevertheless, resonance of the wind tunnel selectively amplifies frequencies corresponding to dimensions of the wind tunnel, in a similar fashion to 'organ pipe' resonance of pressure tubing (§13.3.2.5). Pressure tubing resonates in ¼-wave fundamental mode, that is, a standing wave with a node (maximum noise level) at the closed end and an antinode (zero noise level) at the open end at a frequency corresponding to a wavelength four times the tubing length. The closed-return wind tunnel of Figure 13.18(a) will have a ½-wave fundamental mode in each limb, node at either closed end and antinode in the middle, at a frequency corresponding to a wavelength twice the limb length. The open-return tunnel of Figure 13.18(b) will also have a ½-wave fundamental mode, but with an antinode at each open end and a node in the middle. The BRE tunnel behaves like a pressure tube, in ¼-wave fundamental mode with an antinode at the inlet and a node at the fan. Further harmonics are probable in all cases.

Acoustic resonance is reduced by a contraction which decorrelates the phase between the acoustic and velocity components of the standing wave. The contamination of pressure signals may be further reduced by selecting antinodes for the model working areas and avoiding the nodes, but this is not always possible in practice. Electronic filtering of the signal to remove these components is not a satisfactory solution as the flow-induced fluctuations in the same frequency range are also lost. However, if the acoustic noise signal is separately acquired by a static pressure probe in the same cross-section as the model, it can be subtracted from the data signal electronically or during the analysis stage.

13.5.3 Simulation of the atmospheric boundary-layer

13.5.3.1 Historical development

The initial response to Jensen's model law [48] was twofold, as indicated in Figure 13.20 for about 1965. Long boundary-layer wind tunnels began to be constructed [107, 115] in which boundary layers could be grown naturally on the

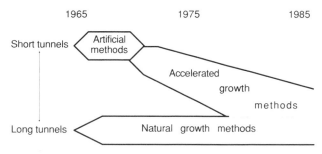

Figure 13.20 Evolution of simulation methods

working section floor that had been covered with roughness elements scaled to the terrain. It was quickly confirmed[115] that this gave an accurate representation of the atmospheric boundary layer, as predicted by Jensen. Natural growth of the boundary layer is relatively slow, depending on the terrain roughness, as shown by Figure 13.21. Even in these long tunnels, the largest scale factors obtained naturally were about 1:500, so that some additional artificial thickening of the boundary layer was necessary to achieve larger scales. (Early work assumed that the atmospheric boundary-layer depth in *design* conditions was about 500 m, the value measured in moderately strong winds. Estimates of depth have regularly been revised upwards in value, and now exceed 2000 m in design conditions (see Chapter 9).)

In many research laboratories the resources to construct long tunnels were not available and attempts were made to reproduce the some of the boundary-layer characteristics by completely artificial methods. It had already been established that shear flows could be generated by a grid of differentially spaced rods[127] or a curved gauze screen[128], and these techniques were used to reproduce the profile of mean wind speed using the power-law model. These artificial methods, described by Lawson[129], are called 'velocity profile only' simulations because only the mean velocity profile is represented. As these methods were unable to reproduce the turbulence characteristics, the mean wind speed profiles rapidly degenerate downstream, particularly when disturbed by the presence of a building model. While the major effects of mean wind profile described in §8.3.2 are reproduced adequately on single building models, the mean effects on groups of models are poorly represented and transient gust effects are not represented at all. Consequently, these methods are no longer regarded as suitable for serious design applications. Figure 13.20 indicates their use ending in about 1970, although they are still employed for demonstration and teaching purposes.

The use of artificial methods declined when methods were found of accelerating the natural growth of boundary layers without significantly changing their characteristics. The very short (1:1 length:height) aeronautical working sections were unsuitable for these methods, so they were often implemented in the longer (> 4:1) return flow sections. First methods were polarised into two distinct approaches. One approach attempted to represent the full depth of the atmospheric boundary layer, and was used for tall building studies and diffusion or effluent dispersal problems over long distances at scales around 1:500. Notable among these 'full-depth' methods is the method developed over many years at the UK Central Electricity Research Laboratories[130, 131, 132, 133] whose character

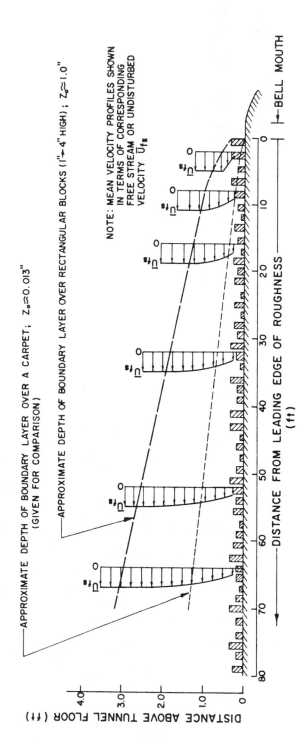

Figure 13.21 Natural growth of atmospheric boundary-layer simulation (from reference 115)

has been investigated in great detail. The other approach sought to represent only the region near the ground surface at the larger scale factors required for low-rise building studies. This is exemplified by the method devised by Cook while at Bristol University and developed further at BRE [134]. These are only two out of many methods, too many to review individually here, and instead the reader is directed to references [135, 136, 137, 138].

Development of simulation methods has continued to increase their accuracy. The trend towards their use in longer wind tunnels has tended to blur the distinction between completely natural and accelerated growth methods, as indicated in Figure 13.20. From the three elements common to modern simulation methods, the generic name 'roughness, barrier and mixing-device methods' has been coined [136, 138].

13.5.3.2 Roughness, barrier and mixing-device methods

A typical arrangement of simulation hardware is shown in Figure 13.22, which corresponds to a suburban simulation at about 1:250 by the method of Cook [134, 136, 138]. The rôle of the roughness is the same as in a naturally grown layer; it represents the roughness of the prototype full-scale ground surface. The roughness is the most important component in that it establishes the values of the three logarithmic-law parameters, z_o, u_* and d (§7.2.1.3.2). The barrier and mixing-device are the 'artificial' part of the simulation. The barrier gives an initial ground-level momentum deficit and depth to the boundary layer which is mixed into the developing simulation by the mixing-device. The flow is tricked by the barrier into believing the fetch of roughness to be longer, and by the mixing-device that the barrier is not there at all! In the ideal case, the flow in the test area near the downwind end of the roughness should have the characteristics of a boundary layer grown over a much longer fetch of the same roughness, without any additional characteristics imposed by the barrier or mixing-device.

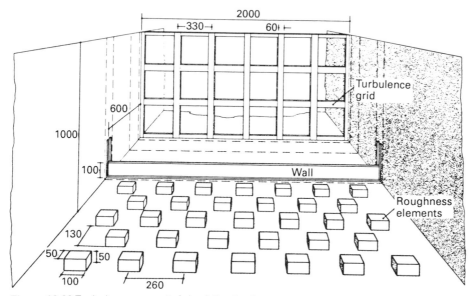

Figure 13.22 Typical arrangement of simulation hardware

Barrier and mixing-device vary in form between methods and there is often interaction between them. The mixing-device shown in Figure 13.22 is a plane grid which gives uniform turbulence, as in this case only the surface region is represented in the simulation[134]. CERL's full-depth method uses Counihan's[131] elliptic wedges, shown in Figure 13.23, to give turbulence that decreases in intensity towards the top of the simulation. The barrier may be the

Figure 13.23 Hardware for full-depth simulation: back—elliptic wedge vorticity generators, 800 mm high; middle—elliptic-wedge vorticity generators, 400 mm high; front—castellated wall

plain wall shown in Figure 13.22, as in the original Bristol method[134], but mixing is improved by perforations, castellations[131] (as in Figure 13.23) or triangular teeth[136] at the top of the wall. Standen[139] devised an array of tapered spires that combines the functions of barrier and mixing-device. Instead of a solid wall some methods form the barrier by jets of air blowing upwind[140, 141], where the momentum of the jets gives the momentum deficit, or cross-wind[142], where the jets form a fluid wall. In general, deeper boundary layers and hence larger length scale factors are obtained by lengthing the roughness fetch or by raising the barrier height. The maximum height the barrier can be before it imposes its own characteristics on the flow depends on the size and the fetch of the roughness, so that larger length scale factors can be obtained for the rougher urban simulations than for smoother rural simulations in the same wind tunnel.

Data from two 'families' of boundary layers are now used to illustrate the scope and accuracy of typical simulations. Figure 13.24 shows the vertical profiles of mean velocity obtained in boundary layers grown over a surface roughness of: (a) gravel, representing rural terrain; and (b) cuboidal blocks in a staggered array at a plan-area density $a = 0.15$ (§9.2.1.2), representing urban terrain, both taken from reference [138]. In every case, the surface region is fitted to the log-law model (§7.2.1.3.2) and the corresponding values of aerodynamic roughness, z_o, and zero-plane displacement, d, are given. (Note that each profile is separated by an

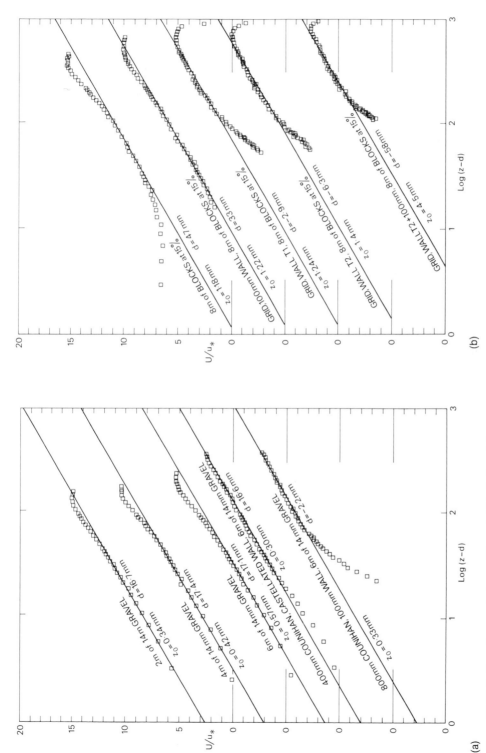

Figure 13.24 Profiles of mean wind speed through (a) typical rural simulations and (b) typical urban simulations

offset on the wind speed axis, and that the axes are transposed relative to the standard graphical format of Chapters 7 and 9.)

The top three profiles of (a) show the profiles obtained over the gravel roughness when the boundary layer is allowed to grow naturally over fetchs of 2 m, 4 m and 6 m, and these compare well with the characteristic shape in Figure 7.8. The final 6 m fetch gives a boundary-layer depth of $z_g = 190$ mm, corresponding to a development length of 30 heights (comparable with Figure 13.21) but giving a length scale factor of only about 1:10 000. The bottom two profiles show the result of deepening the boundary layer over the 6 m-long fetch using Counihan's method [131] (the vorticity generators and barrier of Figure 13.23), in development lengths of 15 and 7.5 heights and giving length scales of 1:5000 and 1:2500, respectively. In every case except the last, the log-law parameters are approximately constant, indicating that the gravel roughness controls the flow. The zero-plane displacement, d, measured from the tunnel floor approximately equals the gravel thickness as expected. However, the log-law parameters for the last profile are different: in particular d is apparently negative, and this is a characteristic indicator that the barrier is too high for this fetch of roughness.

Similarly, the top profile of (b) shows the profile obtained when the boundary layer is allowed to develop naturally over an 8 m fetch of the 'blocks' roughness. The other profiles correspond to the simulation hardware of Figure 13.22 with increasing barrier height. The bottom of each profile extends into the interfacial layer between the blocks (§7.1.3), where the wind speed is slower behind each block and faster between the blocks. As before, the zero-plane displacement becomes negative when the barrier is too high, but with the rougher surface the maximum useful barrier height is increased. The boundary layer becomes so thick that it merges with the thinner boundary layer growing downwards from the roof, displacing the outer 'velocity-deficit' region of the boundary layer (§7.2.1.3.2) and extending the log-law region towards the roof. This is the essence of 'part-depth' simulations, that the top of the prototype boundary layer is excluded, but the surface layer is represented at a larger scale.

The data of Figure 13.24 form 'families' of profiles that develop gradually from 'natural' to 'accelerated-growth' 'full-depth' simulations in the case of (a), and from 'natural' to 'accelerated-growth' 'part-depth' simulations in the case of (b). This confirms the view expressed earlier, that these distinctions are blurred in modern simulation methods (Figure 13.20). However, in the case of 'natural' growth the turbulence characteristics are certain also to be correctly represented, but with 'accelerated-growth' methods the turbulence characteristics should be independently checked. Figure 13.25 shows the spectrum of the inwind turbulence component, u, at a height of $z = 300$ mm above the ground using the 'blocks' roughness with: (a) no barrier; and (b) the highest barrier, corresponding to the first and last profiles of Figure 13.24(b). The prototype atmospheric turbulence spectrum of Figures 7.13 and 9.24 (the solid curve) has been fitted to the data points, and in both cases the match is excellent. The spectrum of (b) is shifted to lower frequencies than (a), indicating a larger value for the integral length parameter, $\Lambda_{x,u}$, and hence a larger linear scale factor.

The linear scale of the simulation is determined by comparison with the prototype full-scale data of Chapter 9. It is effectively set by the two length parameters describing the mean wind speed and turbulence characteristics: aerodynamic roughness, z_o (§9.2.1.1); and inwind integral length parameter, $\Lambda_{x,u}$. The match can be done by trial-and-error, but is more conveniently done by

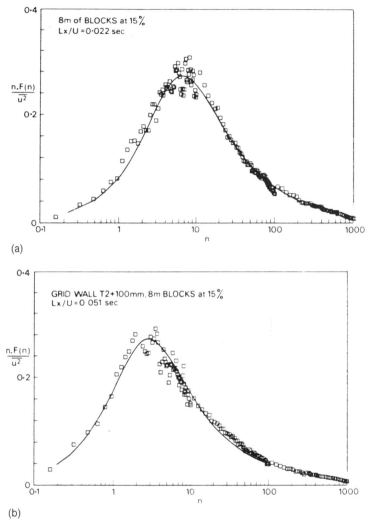

(a)

(b)

Figure 13.25 Spectra of inward turbulence component for urban simulations: (a) no barrier; (b) highest barrier

substituting the measured values transformed to full scale $z_o / \mathscr{L}$ and $\Lambda_{x,u} / \mathscr{L}$ into Eqn 9.44 and solving [143] for the linear scale factor, $\mathscr{L}$. For the two cases of Figure 13.25 this process gives $\mathscr{L} = 750$ for (a) and $\mathscr{L} = 300$ for (b), showing that the boundary-layer growth was accelerated by a factor of 2.5.

Given an estimate for the scale factor from the mean wind speed and inwind turbulence spectrum, the remaining turbulence characteristics should be checked. Figure 13.26 shows the vertical profiles of the three turbulence components and the Reynolds stress (normalised against the friction velocity, u_* (§7.3.1.2)). The prototype atmospheric profiles corresponding to the indicated linear scale factors have been fitted to the data (the solid curves). In all cases the match is very good. The highest barrier produces slightly excessive vertical turbulence (w-component)

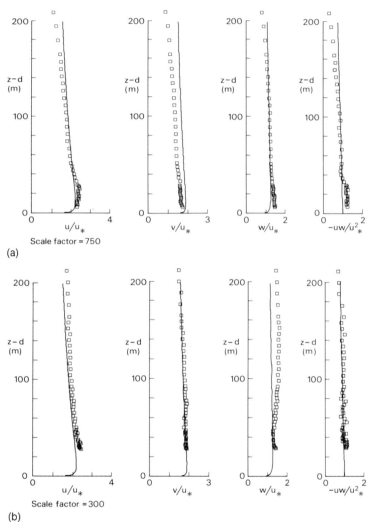

Figure 13.26 Profiles of turbulence intensity and Reynolds stress for urban simulations: (a) no barrier; (b) highest barrier

and this is another characteristic indicator that the barrier is too high. In a 'part-depth' simulation the height range over which the match should be good depends on the height of interest, but is typically at least ¾ of the working section height. When the wind tunnel is too short to achieve the desired linear scale, either of two courses of action are possible:

1 The linear scale of the building model must be reduced to match the reduced linear scale of the simulation.
2 The linear scale of the simulation must be increased by artificial means, provided the consequent reduction in simulation accuracy is acceptable.

One way of accelerating the boundary-layer growth further is to distort the surface roughness. Tieleman *et al.* [135] suggest using roughness that is initially too large, but tapers gradually to the correct size near to the model. Tests at BRE have confirmed that this is effective and also allows a higher barrier to be used, resulting in acceptable simulations at linear scale factors as high as 1/100. When larger scales of turbulence are required, Bienkiewicz *et al.* [144] describe a method of generating large eddies using oscillating aerofoils, but this approach is experimental and is better suited to research than to design applications at present.

13.5.3.3 Simulation of a particular site

13.5.3.3.1 Proximity modelling. The random or regular arrays of roughness elements used in the simulation methods produce the correct general flow characteristics approaching a site. However, conditions at a particular site are more strongly influenced by the local terrain than by the terrain far upwind (see Figure 9.9). In an urban area the site may be directly affected by neighbouring buildings. At some distance from the site, it is necessary to change from a general simulation to a detailed representation of the site, usually called a 'proximity model'. The construction of detailed proximity models is time consuming and expensive, particularly for urban areas. If too small an area is represented, the results will not reflect site conditions accurately. If too large an area is represented, in order to fit all the proximity model on the turntable the linear scale factor may become too small. It is therefore necessary to have a rational approach to this problem which will be considered here in two stages: local topography and local terrain roughness.

13.5.3.3.2 Local topography.   When the site is influenced by topographic features it will be necessary to include these in the model. But if the topography is extensive, the resulting scale factor may be too small to represent the building structure. In this case the problem may be approached in two stages:

(a) a model of the topography is prepared at a suitable scale and the wind conditions at the site are measured (see §7.5.2); and
(b) a model of the building is prepared at a larger scale and the measured wind conditions are reproduced to this scale in the wind tunnel.

A good example of this process is given by the design study for the Hongkong and Shanghai Bank in Hong Kong[145] performed by the University of Western Ontario[146]. Figure 13.27(a) shows wind speeds being measured over a 1:2500 model of Hong Kong and Figure 13.27(b) shows the 1:500 model of the Bank building in the proximity model. The topographic model data are also applicable to other developments in the vicinity, so have also been used for the nearby Exchange Square project[147].

13.5.3.3.3 Local terrain roughness.   The extent of the detailed proximity model is often determined by cost or time considerations alone, but consideration of the effect of adjacent roughness enables a rational approach to be adopted[148]. The area of surface roughness that affects the flow at various heights was determined by Pasquill[149] who derived a table of dimensions for this area which he called the 'roughness footprint'. If one were to insist that every part of the roughness footprint that affects the height of the model was modelled in detail, then the radius of the proximity model would be about ten building heights (4 m in Figure 13.27(b)). Fortunately this is not necessary, since where the footprint contains a large number of similar elements, i.e. buildings, the individual effect of a single

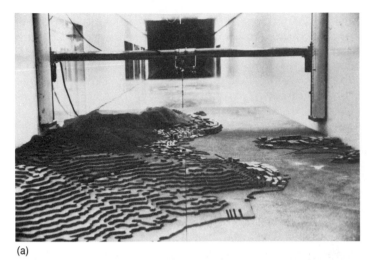

(a)

(b)

Figure 13.27 Two-stage test for topographic effects: (a) topographic model at 1:2500 scale; (b) building and proximity models at 1:500 scale. (Building = 1 Queen's Road Central, Hong Kong; testing laboratory = University of Western Ontario; consulting engineers = Ove Arup & Partners; client = Hong Kong & Shanghai Banking Corporation)

element is indistinguishable from the combined effect of the others and here a form of general roughness should be used. Only when the footprint contains a few elements can the individual effect of a single element be recognised and here a detailed model is required. Successive studies at Bristol with arrays of cubes [150] and tower blocks [151] suggest that five rows of similarly sized elements are sufficient to mask the individual effects of the next upwind row. Accordingly, it is suggested [148] that the minimum extent of the detailed proximity model in urban areas should be five blocks or streets. This approach breaks down outside the detailed proximity model when one individual element is sufficiently different from the rest of the elements in the footprint, i.e. a high-rise tower, open space or dominant street line in an area of low-rise housing. In this case it is sufficient to make local modifications to the general surface roughness by inserting a model of that element in the appropriate position. This is illustrated in Figure 13.28 which shows the instrumented building next to London Bridge on the south bank of the River Thames in the foreground and the detailed proximity model for a radius of 260 m on the turntable. In the background, the large tower of Guy's Hospital is reproduced in the general roughness.

Figure 13.28 Dominant building model upwind of detailed proximity model. (Building = 1 London Bridge, London; testing laboratory = Building Research Establishment; consulting engineers = Mott, Hay and Anderson; client = St Martins Property Group)

13.5.4 Modelling the structure

13.5.4.1 Wind-tunnel blockage

When a building is modelled in a boundary-layer wind tunnel by the methods just described (§13.5.3), the roughened floor of the wind tunnel represents the ground plane of the terrain. Unfortunately, the side walls and roof of the wind tunnel are solid boundaries that have no full-scale counterpart. The instantaneous flow around a cuboidal building in full scale is represented in plan by Figure 13.29(a). When this is represented in a wind tunnel, the streamlines are confined by the walls, as shown in Figure 13.29(b). The flow accelerates in the constricted region between the model and the walls, increasing the loading of the building. This effect, called 'blockage', depends on the relative cross-sectional areas of the model and tunnel, A_{model}/A_{tunnel}, called the 'blockage ratio'. The effect cannot be removed by just removing the walls in the test area, i.e. in an 'open-jet' wind tunnel, because then the effect reverses, so that the streamlines are displaced too far and the loading is decreased.

Two examples of this effect on bluff bodies in smooth uniform flow are shown in Figure 13.30: (a) shows the radial pressure distribution on the front and rear faces of a circular disc [152]; and (b) shows the pressure distribution on the front and rear

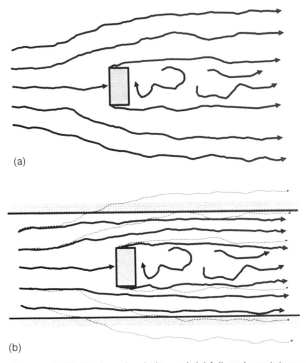

(a)

(b)

Figure 13.29 Blockage in wind tunnel: (a) full-scale prototype; (b) model scale in closed-section wind tunnel

faces of a long flat plate [153], expressed as pressure coefficients referred to the the incident (unblocked) flow conditions. McKeon and Melbourne [152] remark on their data for the disk, (a):

> It is immediately apparent that most of the wall constraint effect is on the downstream face exposed to the wake pressure and that it is uniform across this face. On the upstream face even for 20% blockage there is no apparent effect on the pressures in the centre of the plate *but a distortion of the pressure distribution does occur for the higher blockages towards the edge of the plate.*

The final caveat in italics will be seen, below, to be significant. For their flat plate data, (b), Ranga Raju and Vijya Singh [153] have no doubts:

> This is clear proof that blockage corrections cannot be looked at as a velocity increment, since, if this were the case, the pressure distribution on both sides of the plate ought to have been affected by the blockage.

We will examine these assertions shortly.

Several methods for correcting the mean drag for this effect have been proposed. One of the earliest by Maskell [154], intended to correct the mean drag of stalled wings, was later extended by Cowdrey [155] to apply to solid bluff bodies. This **does** treat the blockage effect as an increase in wind speed in the flow outside the wake of the body, in compensation through the continuity equation, Eqn 2.4 (§2.2.2), for the air displaced by the model and for the loss of momentum in the wake. The

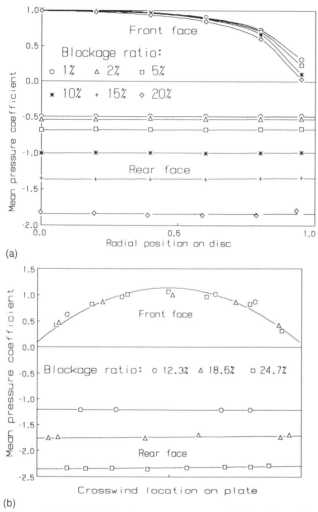

(a)

(b)

Figure 13.30 Effect of blockage on pressures on (a) disc (from reference 152) and (b) flat plate (from reference 153)

blockage effect is modelled as a simple local increase in velocity, giving a consequent increase in dynamic pressure, Δq, of:

$$\Delta q/q = \varepsilon \, C_D \, A_{\text{model}}/A_{\text{tunnel}} \tag{13.16}$$

where ε is a coefficient dependent on the model shape, determined experimentally. The measured mean drag can then be corrected by the factor $1/(1 + \Delta q/q)$. Maskell's theory is linearised in a similar manner to the linearised quasi-steady vector model of §12.4.1.2, so that there are no squared terms in Eqn 13.16. This limits the method to small blockage ratios, and it becomes less reliable above about 10% blockage. Complex shapes of model make determination of ε difficult and approximate, but the method has the advantage that it can be used to correct existing data. Another method developed by NASA[156] overcomes these problems by utilising the reaction of the blockage effects on the static pressure distribution on the wind-tunnel walls. This 'wall signature' must be directly measured during the model test, so the method cannot be used to correct existing data. Burton [157] has demonstrated that correction procedures for **mean** forces by this method can be made almost automatically and suggests it is applicable to blockage ratios as high as 60%.

So, if correction methods based on a simple velocity increment work for mean forces, why is the effect on the mean pressures in Figure 13.30 not similar on the front and rear faces? The answer is that not only is the flow momentum conserved through the continuity equation, Eqn 2.4 (§2.2.2), but there is also a tendency to conserve energy through the Bernoulli equation. The flow attempts to keep the total pressure, p_T, constant and the increase in the dynamic pressure is accompanied by corresponding decrease in static pressure, p_S. The zero pressure

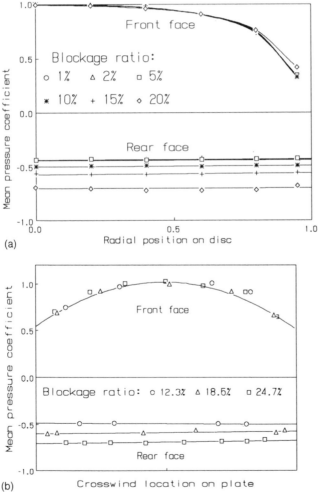

(a)

(b)

Figure 13.31 Effect of blockage on pressures on (a) disc referred to model location and (b) flat plate referred to model location

axes in Figure 13.30 move downwards with increasing blockage, but values near C_p = 1 remain nearly constant. Figure 13.31 shows the pressure data referenced against the local dynamic and static pressures estimated by Maskell's method as implemented by ESDU [158]. In (a) the data for the disc at blockage ratios less than 10% collapse together, **even on the front face**, but the method progressively underestimates the correction (overestimates the loading) at blockages above 10%. In (b) the data for the plate show the same trend. In dealing with loads, Maskell [154] had no need to address the question of static pressure, since its effect integrates to zero over the whole model. However the static pressure variation, which is the basis of the 'wall signature' method, is clearly required when considering surface pressures (Eqn 12.5, §12.3.4). The previous assertions [152, 153] are wrong because they neglect the static pressure variation, and it is clear that blockage can be modelled as a local increase in velocity combined with a local decrease in static pressure.

All the work described above was performed in uniform flow with models in the centre of the working section. The velocity profile of the atmospheric boundary layer affects the action of blockage. Work at Bristol [159] indicates that models on the ground are less susceptible to blockage and that 10% blockage may be acceptable without correction, although most workers regard 5% as the 'safe' limit. In any case, the correction methods apply only to the **mean** pressures and forces, while it is the **extremes** that are required in the design of static structures. The effects of blockage are therefore best minimised by using small models to keep the blockage ratio low. When the model size becomes too big, there are two other options. Some wind tunnels, including the BRE tunnel in Figure 13.19, are fitted with moveable roofs so that the area of the working section can be increased locally to offset the model blockage. There is some argument over the degree of compensation required. The gradient of static pressure on the roof can be monitored using a manometer and the roof adjusted to give a constant (isobaric) value, but this overcompensates. Lawson [7] suggests that the best compromise is to adjust the roof so that it follows the mean streamline expected at roof height, as determined by smaller-scale model test, but even this is a constraint to the turbulence. A more recent suggestion by Parkinson [122] is to use a working section with slatted walls, about half solid and half open, to give characteristics midway between an open-jet and a conventional closed wind tunnel. This, he demonstrates, is 'tolerant' to blockage ratios up to about 25%.

A new wind tunnel is being built at BRE during 1989 which will employ Parkinson's suggestion. In the meantime, the solution adopted at BRE is to keep the blockage ratio below about 8% and to obtain the reference values of dynamic and static pressure **in the same cross-section as the model**. Thus the effects of blockage are minimised as shown between Figures 13.30 and 13.31, except that the reference pressures are measured and not estimated. Using this strategy in combination with the moveable roof, Hunt [160] has demonstrated that changing between normal blockage (flat roof) and an isobaric roof (overcorrection) has a negligable effect on the results up to 8% blockage.

13.5.4.2 Lattice structures

Owing to the small size of the individual members, it is usually impossible to reproduce complete models of lattice structures at a scale compatible with atmospheric boundary-layer simulations. However, this is not a problem, because it has already been established that the lattice model (§8.2.1) responds to the atmospheric boundary layer in a quasi-steady manner (§8.3.1, §8.4.1). This enables the loading to be deduced from the mean loading coefficients of the individual members by summation over the whole structure. The mean loading coefficients are determined in smooth uniform flow or in uniform flow with small-scale turbulence. This approach is valid as long as the solidity ratio, s, is small so that the components act independently. Solidity ratio is defined by:

$$s \{\theta\} = \text{total area of all individual members} / \text{envelope area} \qquad (13.17)$$

where both areas are the areas projected horizontally along the mean wind angle, θ, and the envelope area is A_{proj} of Figure 12.8.

Typical structures for which this approach is valid are indicated in Figure 13.32. The individual members are likely to be line-like, so will be modelled as described in §13.5.4.3. Decorrelation of the gust effect over large lattices is accounted for by an admittance function (§8.4.1) or the equivalent steady gust model (§8.6.2.3, §12.4.1.3).

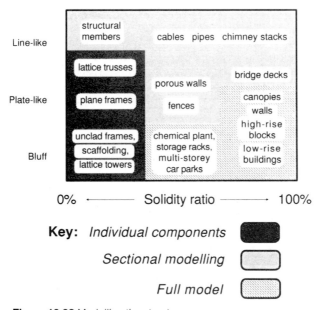

Figure 13.32 Modelling the structure

With multiple lattices, the summation of individual member loads will progressively overestimate the total loading with increasing solidity ratio, as downwind members become shielded in the wakes of upwind members. When the lattice is also line-like, as with square- or triangular-section masts, trusses or booms, this may be modelled by reproducing a short section as shown in Figure 13.33. This shows a section of lattice mast supported between the tunnel walls. A turbulence grid upstream gives small-scale turbulence to represent the high-frequency end of the atmospheric turbulence spectrum which affects the wakes of the members and hence the degree of shielding. When the structure is three-dimensional, a full model is appropriate as in Figure 13.34 which shows a 1:10

Figure 13.34 Full model of lattice mast and ancillaries (courtesy of British Maritime Technology Ltd)

Figure 13.33 Sectional model of lattice mast (courtesy of British Maritime Technology Ltd)

linear scale lattice communication tower complete with dish antennae and corresponding ancillaries. This model was used to prove the provisions of the British Standard on lattice towers and masts, BS 8100, and was tested in smooth and turbulent uniform flow and in the boundary-layer simulation[*] shown. Owing to the large number of individual members it is not practical to measure pressures, so that sectional and full models are invariably mounted on force transducers.

13.5.4.3 Line-like structures

Line-like structures are long slender structures that tend to follow strip theory (§8.2.2), so can be represented in the wind tunnel by sectional modelling. Owing to their slenderness, many of these structures will be dynamic or even aeroelastic. Static models are used to determine the steady aerodynamic loading coefficients and the response is calculated from quasi-steady theory, but this subject is covered in Part 3. Some typical structures are indicated in Figure 13.32. Plate-like structures, such as suspension bridge decks, are treated in the same manner as the lattice sectional models described above. With bluff cylinders where vortex shedding occurs (§2.2.10.4), it is necessary for the modelled section to be longer than the correlation length of the vortex shedding (see Part 3).

Full models are often used when the structure is a vertical cantilever and the effects of the profiles of mean wind speed and turbulence are desired to be modelled rather than calculated. Figure 13.35(a) shows a full model of the CN Tower, Toronto, in the boundary-layer wind tunnel of the University of Western Ontario where the base forces and moments are being measured to ensure overall stability of the structure. Figure 13.35(b) shows a sectional model of the restaurant and microwave gallery where the surface pressures are being measured for the design of the toroidal cladding of the microwave gallery. Aeroelastic models of the full tower and of the cylindrical antenna section at the top were also made (see Part 3), illustrating the occasional need for a 'family' of models to encompass all the design aspects.

13.5.4.4 Plate-like structures

Sectional models are used for very long plate-like structures such as suspension bridge decks or long boundary walls. Three-dimensional structures such as canopy roofs, dutch barns, grandstands, road signs and hoardings will generally require a full model. Unless the 'plate' is very thick, when it may be bluff rather than plate-like, there is usually some difficulty in installing pressure tappings and tubing within the scale thickness and in finding a path out of the model for the tubing. Figure 13.36 shows one solution for a canopy roof, where pneumatic averaging manifolds (§13.3.3.1) have been installed within the canopy thickness and that the reduced number of 'averaged' pressure tubes lead out of the model through the support legs. When the 'plate' is thin, for example a steel-framed grandstand roof sheeted on one side of the frame, it may be necessary to measure the pressure on each side separately, with the back of the tappings and the tubing exposed to the wind flow.

[*] It is evident that, although the tower is fully immersed in the simulation, the linear scale factors of tower and simulation are mismatched by at least a factor of ten. The requirement here was only to reproduce the mean velocity profile and hence the mean loads, but this approach would not be adequate for a solid three-dimensional structure. Gust loading may then be deduced by applying the TVL-formula of §12.4.1.3.

(a)

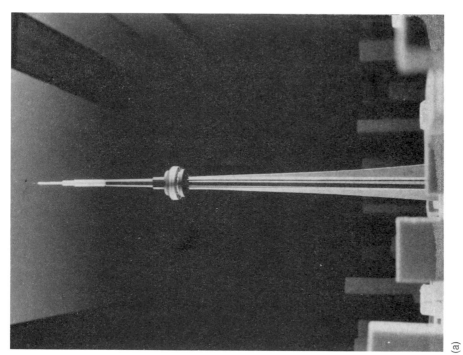

(b)

Figure 13.35 (a) Full and (b) Sectional models of the CN Tower, Toronto. Testing laboratory = Boundary-Layer Wind Tunnel Laboratory, University of Western Ontario; consulting engineers = Roger Nicolet and Associates; client = CN Tower Limited

Figure 13.36 Pneumatic averaging inside canopy roof (courtesy of the University of Oxford)

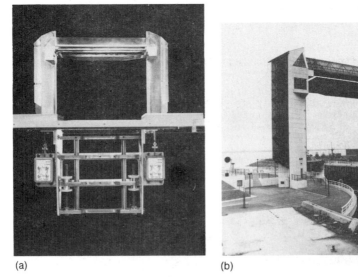

(a)   (b)

Figure 13.37 River Hull tidal surge barrier: (a) model barrier with force transducers; (b) barrier at full scale. Testing laboratory = Building Research Establishment; consulting engineers = Sir M McDonald and Partners; client = Yorkshire Water Authority

Where overall forces are required it is easier to support the 'plate' on dynamometers. Figure 13.37(a) shows this solution applied to the turn-over gate of the River Hull Tidal Surge Barrier at 1:200 scale. The barrier gate is supported on three force transducers underneath the model by rods passing through the towers, enabling the vertical (lift) force and the two horizontal moments (pitch and roll) to be measured. The design problem here was not the gate or the main supports, which were designed to withstand a large head of water, but the secondary guide mechanism that rotates the gate to a vertical position as it drops, for which the moment about the main supports (pitch) due to wind was critical. Accordingly, the transducers are mounted on a frame which can be moved vertically and the gate rotated to represent the action of the full-scale barrier, shown in Figure 13.37(b).

13.5.4.5 Bluff structures

13.5.4.5.1 Sharp-edged structures.   Most buildings are of this form, which is the easiest of all to model. The external shape determines the aerodynamic characteristics and this should be represented with sufficient detail to satisfy the measurement requirements. The example of Figure 13.28 is reasonably typical. This building has an almost smooth facade which is represented as a hollow shell, in which there is plenty of room inside for the insertion pressure tappings and tubing. This is even easier to install in squat structures, such as the model of the Princess of Wales Conservatory, Kew, Figure 13.38(a) which contained 412 tappings. Half of the tappings at a time were connected through optimised tubing to five Scanivalve manifolds, Figure 13.38(b). These were sequentially connected through a single Scanivalve pressure-scanning switch to a transducer beneath the model (Figure 13.12(a)). In slender tower blocks the tubing may be too short for this procedure, in which case one wall can be made removable and the Scanivalve is installed inside the model.

Flow separation from sharp-edged models occurs at the sharp corners, giving the required insensitivity to Reynolds number (§13.5.1). The need to model surface

(a)   (b)

Figure 13.38 Princess of Wales Conservatory, Kew. Testing laboratory = Building Research Establishment; consulting engineers = Property Services Agency; client = Royal Botanic Gardens, Kew

texture depends on the required measurement detail. Models for structural loadings are usually made smooth, even when there are small mullions or recessed glazing panels, etc. For detailed cladding loading close to surface features, or where these features are of a significant size in proportion to the building, they must be represented. For example, the structural steelwork of the Princess of Wales Conservatory is exposed and is reproduced in the model, Figure 13.38(a).

13.5.4.5.2 Curved structures. The position of flow separation on curved structures is controlled by the way the boundary layer develops on the curved surface and is dependent on the Reynolds number, Re (§2.2.6, §2.2.7). The example of the flow around circular cylinders was described in Part 1 (§2.2.10.2). It was noted that there are two principal flow regimes 'subcritical' and 'supercritical' where the loading coefficients are relatively constant, separated by a 'transcritical' region where the loading coefficients are affected by Re, incident turbulence and the surface roughness of the structure. In the subcritical regime at low Re the drag coefficient is high, $\overline{C}_D \simeq 1.2$, while in the supercritical regime at high Re it can fall to less than half, $\overline{C}_D \simeq 0.5$ (Figure 2.21), but is strongly dependent on the surface roughness (see §16.2.2.1). This general behaviour extends to all structures with curved surfaces, barrel vaults, domes, cooling towers, etc.

Because of this dependence on Re, it is never possible to obtain exact similarity between the prototype and model (§2.4.4, §13.5.1). In order to get an acceptable match it is necessary to ensure that the model is in the same regime as the prototype. Except where the cross-section is very small, i.e. cables, lamp-posts, circular lattice elements, the prototype structure is likley to be supercritical while the reduced linear and velocity scales are likely to make the model subcritical. Fortunately, the critical Re is reduced by increasing surface roughness, as shown in

Figure 13.39 Al Shaheed Monument and Museum, Baghdad. Testing laboratory = Building Research Establishment; consulting engineers = Ove Arup and Partners

Figure 2.23, so that by artificially roughnening the model[161] it is possible to mimic higher Re by factors of up to about 6[162]. Figure 13.39 shows a model of the Al Shaheed Monument and Museum, Baghdad, in the BRE boundary-layer wind tunnel. The outer convex faces have been roughened by a coating of sand, but a small area around each pressure tapping has been left clear to avoid localised influence. (Note the reference pitot-static tubes in the same cross-section as the model, as recommended earlier (§13.5.4.1).) This process is not as simple as it may look, because the value of the drag coefficient in the supercritical regime is also affected by the roughness, and some care must be taken in interpreting the results.

13.5.5 Review of accuracy

13.5.5.1 General

As the majority of design wind load data currently available comes from wind-tunnel tests, the designer needs to know that the models accurately represent full-scale characteristics. Verification is an integral part of the development process of any modelling technique. The 'benchmarks' for comparison are the small amounts of good-quality full-scale data from tests on actual structures in the natural wind. Additionally, the reliability of standard modelling techniques may be assessed from cross-comparisons at model scale. A large number of research papers have been published on this subject. A bibliography of papers relevant to contemporary techniques has been assembled and is reproduced as Appendix I.

13.5.5.2 Full-scale to model-scale comparisons

The principal method of verification of modelling methodology and technique is by comparison with full-scale 'benchmarks'. One of the best set of full-scale data for high-rise buildings comes from the National Research Council of Canada's (NRCC) experiments between 1973 and 1980 on Commerce Court, a tall building

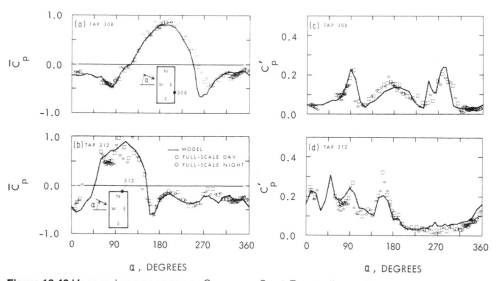

Figure 13.40 Mean and rms pressures on Commerce Court, Toronto (from references 14 and 164)

in central Toronto which is strongly affected by other close tall neighbours. This building was modelled on several occasions in the NRCC wind tunnel [14, 163, 164] using a boundary-layer simulation and proximity model similar to those in Figures 13.27(a) and 13.28. In Figure 13.40 the mean and rms pressure coefficient for two typical positions on the walls are compared for all wind directions. Similarly, peak pressure coefficients at several locations are assessed in Figure 13.41, in terms of the probability density of the peak factor (§12.4.2). Finally, the base shear and moment coefficients are compared in Figure 13.42. Each figure shows an excellent correspondence between model and full scale, verifying the atmospheric boundary layer, proximity and building models used. Dalgliesh [164] concludes that the

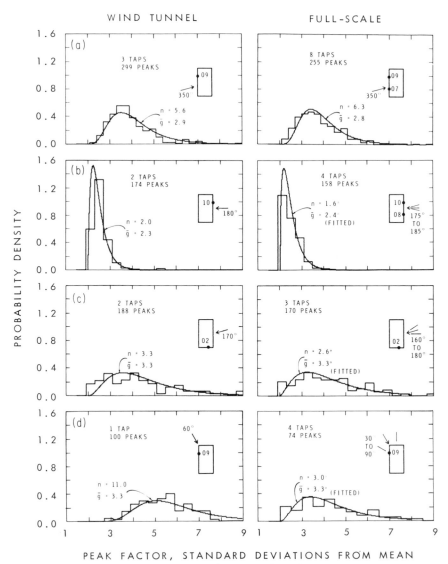

Figure 13.41 Peak factors for pressures on Commerce Court, Toronto (from references 14 and 164)

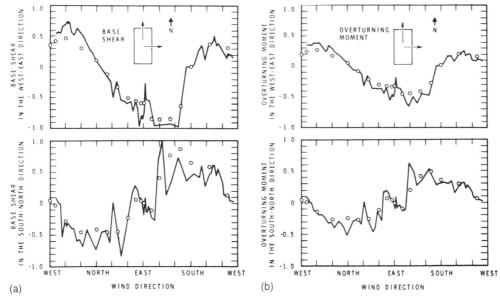

Figure 13.42 (a) Base shear forces and (b) base moments on Commerce Court, Toronto (from reference 163)

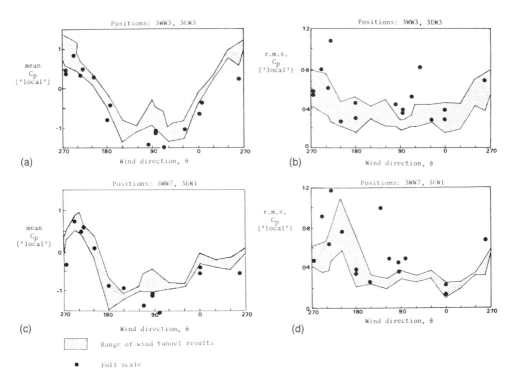

Figure 13.43 Mean and rms pressures on Aylesbury House (from reference 166): (a) mean pressure, positions 3WW3, 3EW3; (b) rms pressure, positions 3WW3, 3EW3; (c) mean pressure, positions 3WW7, 3EW1; (d) rms pressure, positions 3WW7, 3EW1

discrepancies are within the range of experimental error expected for the full-scale data, about 10%–15%.

Similar comparisons have been made on low-rise buildings, many using the full-scale data obtained between 1972 and 1975 from the BRE experimental house at Aylesbury [16, 98] and Marshall's data for a mobile home reported in 1977 [165]. Mean and rms pressures at several locations on the Aylesbury house are shown in Figure 13.43, reproduced from Holmes' review [166]. The model-scale data come from nine different tests in four different wind tunnels at several scales and with differing degrees of proximity modelling [167]. The range of these data is represented by a band which indicates the limits to the typical variation of modelling accuracy. The largest scale factor used was 1:25, which is difficult to reproduce accurately [168] (§13.5.3.2). The discrepancies are clearly greater than for the high-rise Commerce Court, particularly in the rms values. However these still remain within the range of experimental error for the full-scale data. The reason for the larger full-scale errors is that the Aylesbury experiment was conducted over 20 years ago. The data acquisition equipment was less reliable and the analysis techniques used were cruder than their modern equivalents. Even so, the match is reasonably good.

13.5.5.3 Model-scale to model-scale comparisons

Comparisons between model tests can be useful in two main ways:

1 Parametric studies show the sensitivity of the models to variations in the incident wind conditions, building geometry and the match between the linear scale factors of the building model and wind simulation.
2 Tests with the same model in different wind tunnels quantify the reliability of the modelling process, enabling rational values of safety factor to be set.

An example of the first process is given by the work of Hunt [160] in which he tested four sizes of cube in two different boundary-layer simulations, and came to the following main conclusions:

1 Pressures on the windward face are strongly dependent on the incident velocity profile.
2 Pressures on the other faces depend on the intensity and integral length parameters of the incident turbulence.
3 The desired incident wind conditions are more difficult to achieve when the linear scale factor is large (as already noted above).
4 Blockage ratios up to 8% change the design loadings by less than 2% when the reference static and dynamic pressures are measured in the same cross-section as the building model (see also §13.5.4.1).
5 Mismatching the linear scales of the building and the atmospheric boundary-layer simulation by a factor of 2 changes the loading on the windward face by between 5% and 10% and in the high local suction regions by between 20% and 30%, underestimating the loads when the building model is too large (the most common case in practice) and overestimating when the building model is too small.

This last conclusion has been confirmed by other tests [168] and is the basis of the scaling rules for *ad-hoc* tests given several codes of practice (and in Appendix G).

The bibliography of Appendix I also contains research papers and discussions on the effects of immersion ratio, H/z_g. These arguments are sometimes used as evidence for 'full-depth' and against 'part-depth' simulation methods. However the effects are largely academic in that they are only significant at immersion ratios of 0.5 or greater, which are far in excess of practical building heights. Assertions that 'part-depth' methods are only 'partial' representations, in the sense that something important is missing, are not supported by the experimental evidence. Nothing important is missing from the lower region of the atmospheric boundary layer represented by 'part-depth' methods when the mean wind and turbulence characteristics have been properly matched (§13.5.3.2), except for the upper region of the Ekman layer (§7.1.4) which is too remote from the ground to effect most structures and is not correctly represented in 'full-depth' simulations owing to the lack of Coriolis parameter. The assertion is most often used to defend the use of shallow boundary layers, mismatched in scale to the building model, which seriously underestimates the loading. Indeed it is clear that properly scaled models in both types of simulation (small models in 'full-depth' and large models in 'part-depth' simulations) give closely similar results. In addition, however, Holmes[169] notes that the larger linear scales offered by the 'part-depth' methods:

> enable much more detailed modelling of the building geometry to be achieved, and alleviate to some extent the anxiety about possible Reynolds number effects. Also the effective frequency response is considerably greater at the larger scales.

Holmes also notes that the common practice of measuring the reference dynamic pressure at gradient height in 'full-depth' simulations leads to errors that are avoided when a reference height near the building height is used as in Figure 13.39.

Examples of the second process are given by two international comparative experiments: the Commonwealth Advisory Aeronautical Research Council (CAARC) standard tall building experiment, reported in 1980[170], and the IAWE Aylesbury Comparative Experiment (ACE), currently in progress. The

Table 13.4   IAWE Aylesbury comparative experiment. Regressions between data from four wind tunnels [171, 172]

| | | Mean pressures | | Rms pressures |
		Slope	Intercept	Slope
Run code	Tunnel			
A11	1 CSIRO	1.02	0.06	1.26
	2 Surrey	0.93	−0.03	0.83
	3 Madrid	1.28	−0.07	1.15
	4 UWO	0.77	0.04	0.75
A31B	1 CSIRO	1.16	−0.05	1.07
	2 Surrey	1.04	0.08	1.04
	3 Madrid	0.90	−0.07	1.02
	4 UWO	0.89	0.03	0.86
A32/A7	1 CSIRO	1.04	0.13	1.29
	2 Surrey	0.98	−0.03	0.78
	3 Madrid	1.08	−0.08	1.02
	4 UWO	0.89	−0.01	0.90
A35F	1 CSIRO	1.09	−0.03	1.22
	2 Surrey	0.96	0.01	0.94
	3 Madrid	1.07	−0.03	1.03
	4 UWO	0.88	0.04	0.81

CAARC experiment was principally concerned with modelling the dynamic response of a tall cuboidal building, but included comparisons of pressure measurements. Figure 13.44 compares pressure spectra from four tests, showing a good match between the three tests in scaled atmospheric boundary-layer simulations, but a poor match with the single test (marked 'NPL') in a grid-generated artificial simulation (§13.5.3.1).

IAWE–ACE is concerned with modelling selected sets of full-scale data from the BRE house at Aylesbury. Four 1:100 scale models of the house are being circulated for testing in over 20 wind tunnels around the world. When complete in 1989–90, this will provide an enormously valuable data set addressing the difficulties of low-rise modelling at large-scale factors, but some comparisons have already been made of the early data returns[171,172]. Figure 13.45 shows the regressions obtained for the mean pressures at the individual tappings for one data run from four tunnels (three independent pairs). Correlation is very high, giving regression coefficients between 0.96 and 0.98, but there is significant variation in the regression intercepts and slopes. Regression parameters for the individual tunnels

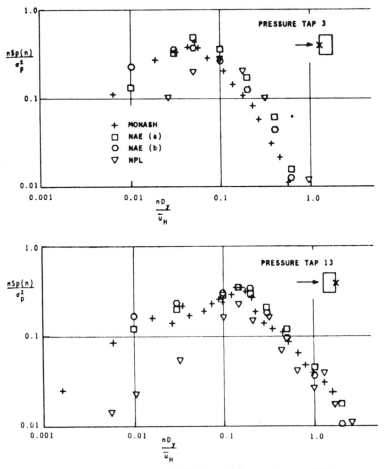

Figure 13.44 Pressure spectra on CAARC model (from reference 170)

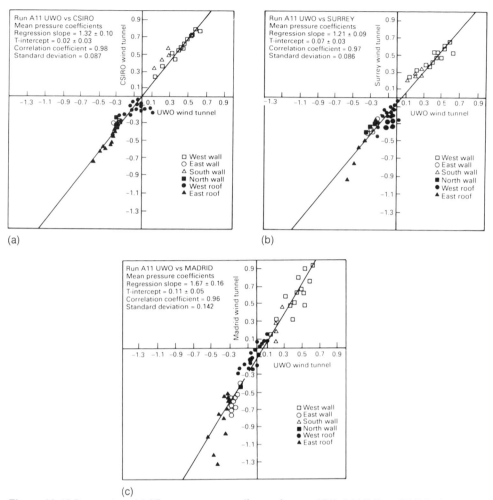

Figure 13.45 Regression of ACE mean pressures (from reference 171): (a) UWO vs CSIRO; (b) UWO vs Surrey; (c) UWO vs Madrid

regressed against the ensemble average are given in Table 13.4 for four data runs. The slope and intercept of the mean pressures indicate differences in the reference dynamic and static pressures, respectively. Differences between the slopes of the rms pressures and the mean pressures indicate variation in the turbulence characteristics. With a few exceptions, most variations are less than 10%, but it is still a significant proportion of the total variation indicated by the full-scale to model comparisons. The size of these systematic reference errors was a surprise, since it had been widely expected that modelling errors would be random and unattributable. Several possible causes for these errors have already been noted: the tendency for standard static pressure probes to give a negative error in turbulent flow (§13.3.2.1); variation in blockage ratio between model and reference locations (§13.5.4.1); and errors from referencing at gradient height rather than near to the building height [169]. These causes of error can all be eliminated and it is clear that **more care must be taken** in acquiring the reference dynamic and static pressures. This is unlikely to be the only lesson learned from the ACE experiment.

13.5.5.4 Conclusions

From the available evidence it is generally accepted that modern modelling techniques give good estimates of design loads, with accuracy ranging typically from 10% for high-rise buildings in urban areas to 20% for low-rise buildings in open country. The paths to further improvements in technique and procedures can be identified [109] and are being persued. A number of specific conclusions can be drawn:

1 Urban sites are easier to model than open country and larger scale factors can be achieved.
2 Proximity modelling of neighbouring buildings gives good results for urban sites, especially for high-rise buildings.
3 Low-rise buildings are more dependent on local ground roughness details than high-rise buildings.
4 Low-rise buildings depend only on the characteristics of the surface region of the atmospheric boundary layer, scaling principally on Jensen number (§13.5.1), and the immersion ratio is not significant.
5 Immersion ratio becomes significant only for high-rise buildings that occupy a significant proportion of the ABL (i.e. dependent on wind speed, §9.4.1).
6 For low-rise buildings part-depth simulation methods are as accurate as full-depth methods, with the additional benefits of more modelling detail, higher Re and better frequency response.
7 Reference values of static and dynamic pressure should be measured in the same cross-section as the building model to minimise blockage effects, and near the height of the building in preference to gradient height to minimise errors in ABL modelling.
8 The linear scales of the building model and atmospheric boundary-layer simulation should be matched to better than a factor of 2.
9 If all the recommendations in these conclusions are met, repeatability should be better than the typical range 10%–20%.

13.5.6 Commissioning model-scale tests

Model-scale tests are of much greater direct use to the designer than full-scale tests (§13.4.1), as they are the only practical way of determining wind loads before construction. Using the model-scale techniques described in this chapter, the designer can obtain loading data for structural design calculations of static structures and of static components (i.e. cladding) of dynamic structures. Modelling of dynamic and aeroelastic structures is covered in Part 3 of the Guide.

Ad-hoc model tests are necessary only when there are no reliable loading data available for the external shape of the structure. The main purpose of the Guide and all wind loading codes of practice is to present such data for the most common building shapes in a form suitable for the majority of static structures. Hence *necessity* for model tests will be the exception rather then the rule. Innovations in design do commonly lead to unusual forms for which no data are directly available, as for the Princess of Wales Conservatory shown in Figure 13.38 or the Al Shaheed

Monument and Museum shown in Figure 13.39. However, for both these cases and many others, conservative estimates of wind loading sufficient for design purposes can be deduced from this Guide by extrapolating from the more common shapes.

In many cases, the main justification for model tests is the precision and detail of the resulting data which allows the position and size of the structural components to be optimised. The degree of optimisation depends on how well and how early the tests are integrated into the design process. At worst, when the design is already complete, tests only serve to verify the adequacy of the design. At best, when the tests are made before the external shape is fixed, the data can be used to modify the shape to reduce the wind loading. This requires a degree of cooperation between the architect, the structural engineer and the consultant wind engineer which, though rare, is happily becoming more common. Through the more typical chain of design responsibility, the structural engineer is commonly presented with a external shape fixed by the architect for which he must design the supporting structure. Here the model test is still of great value: the fixed external shape determines the wind loading characteristics so no optimisation of loading is possible, but the loading data may be used to optimise the structural components. In addition to the main structural aspects, the loading data may also be useful to the services engineer to give service loads for ventilation and heating plant, emergency smoke extraction vents, etc. Measurements of wind speeds around the structure will determine the environmental conditions for pedestrian safety and comfort and for the effect on neighbouring buildings. Model tests can also predict the movement of exhaust gases from flues, vents, etc., and their impact on the vicinity. Thus model-scale tests are appropriate when the value to the designer of the detailed data to the designer justifies the cost, even when adequate codified data are available.

The modelling measurement procedures reviewed in this chapter represent the 'analysis' part of the process. The final step is the 'synthesis' of the data into design values. Codes of practice employ one or more of the methods reviewed in Chapter 12 to implement the data available at the time of drafting, as described in the next chapter, Chapter 14 – 'Review of codes of practice and other data sources'. The design data for static structures presented in this part of the Guide have been synthesised by a method developed at BRE, which uses data analysed by the extreme-value method of §12.4.4. This accounts for the joint risks of the wind climate, atmospheric boundary layer and structure and is described in Chapter 15 – 'A fully-probabilistic approach to design'. Its implementation has been reduced to a very simple form compatible with the format of current codes of practice.

The method is also directly applicable to *ad-hoc* model-scale tests, and is standard at BRE and other wind tunnel laboratories in the UK and overseas. Figure 13.46 shows an example of the resulting design pressures obtained from the model of the Al Shaheed Monument and Museum for the wind direction of Figure 13.39, plotted as isopleth contours over the surface and viewed in elevation from four orthogonal directions. Although this synthesis method is considered to be currently the best, the other simpler approaches of Chapter 12 remain adequate. Next best is the quantile-level method of §12.4.3, and Figure 13.47 shows the worst cladding pressures for the Exchange Square, Hong Kong, for all wind directions derived by this approach [147]. In this case the isopleth contours are drawn on the surface of each of the twin towers shown split at one corner and 'unwrapped'. Note that, although the two towers are antisymmetric, this is not reflected in the surface pressure distributions because of shielding by the neighbouring Connaught Tower [147]. This isopleth contour form of presentation is strongly favoured as it

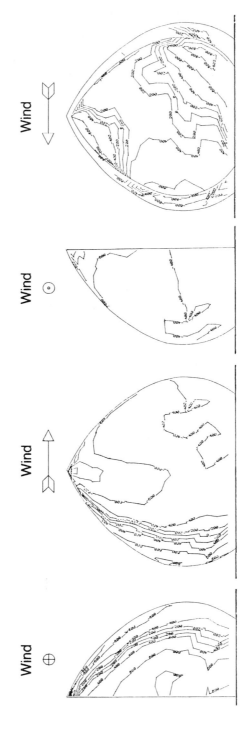

Figure 13.46 Al Shaheed Monument and Museum, Baghdad, design pressures for azimuth 270°. Testing laboratory = Building Research Establishment; Consulting engineers = Ove Arup and Partners

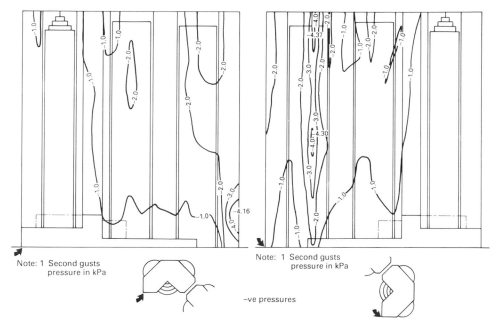

Figure 13.47 Exchange Square, Hong Kong, worst cladding pressures (all azimuths) (from reference 147). Testing laboratory = University of Bristol; consulting engineers: Ove Arup and Partners

gives an immediate and useful impression of the complete loading distribution. A tabular form is usually better for application in structural calculations, especially when using computer-aided design methods.

The necessary provisions for wind tunnel tests listed in Appendix G of Part 1 will ensure that the data are of the required accuracy to be used in place of a Code, and mirror the formal requirements of the UK Code, CP3 ChV Pt2, after the 1986 amendments. Fuller guidance, which includes advice on defining the test program, tendering, selecting a contractor, defining the data format and quality assurance, is given in Appendix J – 'Guidelines for *ad-hoc* model-scale tests'.

14 Review of codes of practice and other data sources

14.1 Codes of practice

14.1.1 Introduction

The typical designer's first, and often only, experience of wind loading comes in the application of the current code of practice. There is some variation in the form of the code between countries, but all use one or more of the design methods in Chapter 12. In order to implement any code, an adequate archive of compatible data must exist to service it, acquired as described in Chapter 13. Also, many codes remain unchanged for more than a decade, so that the methods and data in current use may typically be several decades old. The majority of current codes are still based on the simpler quasi-steady methods. While several current codes contain some extreme-value data, there has not been sufficient data until recently to construct a working code using extreme-value methods. The legal status of codes differs between countries. In the UK, for example, the codes are produced by the British Standards Institution, an independent body, and are not mandatory. However, the UK wind loading code CP3 ChV Pt2 1972 [4] is invoked as one of the means of meeting the mandatory provisions of the UK Building Regulations [173] in Approved Document A.1/2 [174].

This chapter reviews the way codes synthesise design values using five examples that were in force or in the process of drafting in 1988. These are all national 'head' codes, i.e. the principal code intended to cover the majority of typical buildings. Each code is reviewed individually in terms of the method of synthesis, the scope covered, the presentation format and the ease of use. The principal characteristics of these codes are summarised in Table 14.1. These data have been taken from a review by the Flint and Neill Partnership, a firm of consultant structural engineers with expertise in wind engineering, under contract to BRE as part of the preparation for revision of the UK code.

Most of the UK structural or materials codes direct the user to the wind loading head code CP3 ChV Pt2 1972 [4] for the loading data to implement their design methods. However, some individual forms of building or component have particular attributes that make the head code inappropriate, so must supply their own wind loading data, and several of the specialised codes required to cover these cases are also reviewed. In the last section of this chapter other data sources that are sometimes used in place of, or to augment the data from, codes of practice are reviewed.

Table 14.1 Principal characteristics of five codes of practice for static structures

General:

Country:	UK	Denmark	USA	Canada	Switzerland
Year:	1972 (+1986)	1982	1982	1985	1988
Code:	CP3 ChV pt2	DS 410	ANSI A58.1		SIA 160
Scope:	Buildings	Structures	Buildings	Buildings	Structures
Exclusions:	Dynamic or unusual shape	Taller than 200 m	Unusual shape or response	No guidance	(Taller than 100 m implied)
Dynamics:	Excluded	Appendix	Appendix	Appendix	Excluded
Loading model:	Quasi-steady	Quasi-steady	Quasi-steady + **extreme**	Quasi-steady + **extreme**	**Extreme**
Simple method:	None	None	None	Includes most buildings	None

Wind climate and atmospheric boundary-layer data:

Basic wind:	Gust speed map	10 min mean speed value	Fastest mile speed map	Hourly-mean pressure values	Gust dynamic pressure zones
Special areas:	None	None	Mountain zones	None	Mountain zones
External data:	No guidance	No guidance	Permitted	No guidance	No guidance
Basic risk:	1 in 50 years	1 in 50 years	1 in 50 years	Variable	1 in 50 years
Direction:	from 1986	No guidance	No guidance	No guidance	No guidance
Terrain types:	4	3	4	3	3
Terrain fetch:	Minimum value	Change model	Not considered	Not considered	Change model
Profile model:	Power-law on speed	Log-law on speed	Power-law on speed	Power-law on pressure	Log-law on pressure
Load duration:	3 classes by size	Fixed	Gust factor	Gust factor	Fixed
Topography:	Guide model from 1986	Take height from base	No guidance	Use exposed terrain type	Take height from base of hill

Loading coefficient data:

Form of data:	Mean	Mean	Mean + extreme	Mean + extreme	Extreme
Building walls:	Yes	Yes	Yes	By structure	By roof type
Boundary walls:	No	Yes	Yes	Yes	Yes
Flat plates:	No	Yes	Yes	Yes	Yes
Flat roof:	Yes	Yes	Yes	Yes	Yes
Monopitch roof:	Yes	Yes	Yes	Yes	Yes
Duopitch roof:	Yes	Yes	Yes	Yes	Yes
Multi-span roof:	Yes	Yes	No	No	Yes
Hipped roof:	Use duopitch	No	No	No	No
Mansard roof:	No	No	No	No	Yes
Curved:	No	Yes	Yes	Yes	Yes
Open canopies:	Yes	Yes	Yes	No	Yes
Parapets:	No	No	No	No	No
Eave overhangs:	As wall	Yes	Yes	No	No
Cylinders:	Yes	Yes	Yes	Yes	Yes
Lattice frames:	Yes	Yes	Yes	Yes	Yes
Lattice towers:	No (BS8100)	Yes	Yes	Yes	Yes
Structural sections	Yes	Yes	No	Yes	Yes
Wires, cables:	Yes	Yes	Yes	Yes	Yes
External data:	Wind tunnel tests permitted: requirements given	'Determined by experiment' if not given or if H > 200 m	Wind tunnel tests permitted: requirements given	Wind tunnel tests permitted: references given	Wind tunnel tests permitted: no guidance given
Load factors:	Refer to structural codes	Limit state method Factors given	Permissible stress and strength methods	Permissible stress and limit state methods	Factor for combined wind and snow

14.1.2 Head codes

14.1.2.1 United Kingdom

14.1.2.1.1 Scope of code. Until 1986, CP3 ChV Pt2 1972[4] covered all buildings, structures and components thereof within the UK. However, with the increasing use of specialised codes, structures other than buildings were removed from the scope by the 1986 amendments. Also specifically excluded from the scope are:

(a) buildings that are of unusual geometric shape, i.e. out of the range of loading coefficient data given;
(b) buildings that have unusual site locations, i.e. out of the range of meteorological data given; and
(c) buildings susceptible to dynamic excitation, i.e. Class D – dynamic structures for which the equivalent steady gust method is invalid.

The 1986 amendments gave references to design methods for dynamic structures. These and the other excluded buildings may be assessed by 'experimental methods', which include wind tunnel tests that meet the necessary provisions. Among the more common excluded structures are bridges and lattice towers, often covered in other overseas head codes (including the Swiss code, below), which have individual specialised codes in the UK.

14.1.2.1.2 Method and models. As one of the earliest of wind loading head codes, CP3 ChV Pt2 1972 employs the simplest of the quasi-steady methods, the equivalent steady gust model (§12.4.1.3). This assumes that all the fluctuations of load correspond to fluctuations of the incident wind speed in the wind direction, so that the extreme load is coincident with the peak gust. The peak pressure is given by Eqn 12.39 or Eqn 12.40 depending on whether the reference wind speed is an hourly-mean or a gust speed, respectively.

CP3 ChV Pt2 1972 uses the maximum gust at 10 m above open country with an annual risk of exceedence $Q = 0.02$ (50 year return period) as the reference. This is quoted as having a 3 s duration, but current knowledge suggests that it is nearer 1 s in duration. The value of the reference gust speed varies with geographical location and includes the effects of site altitude. The directional characteristics were unknown in 1972, so code required that the wind was 'assumed to blow from any horizontal direction'. Directional characteristics were added as a discretionary option by the 1986 amendments. Adjustment for risk of exceedence and exposure period is made using the V-model (§5.3.1.3), although the standard risk is almost invariably used (values less than the standard risk and exposure period are not permitted by the UK Building Regulations). The minimum exposure period for temporary structures is 2 years and no seasonal variation is given.

Adjustment for terrain roughness is made using four roughness categories, without regard for the length of the upwind fetch (but corresponding to values typical of the UK). Adjustment for height above ground is made by the power-law model (§7.2.1.3.1). Adjustment for topography was initially either by a fixed (inadequate) factor for 'exposed hills' or by an empirical procedure for cliffs and escarpments. This was replaced in 1986 by a procedure based on the data in Chapter 9 of this Guide. Decorrelation of gusts over large areas is accounted for by using the TVL-formula, Eqn 12.37, to define three load durations: 3 s (nearer 1 s), 5 s and 15 s, for cladding, small buildings and large buildings, respectively.

Loading coefficients are all mean values, as required by the equivalent steady gust model, however the majority were derived from early wind tunnel tests in smooth uniform flow. Although the accuracy of many of these data is doubtful, comparisons with modern data show that the values are almost always overestimates. The model data was augmented by some full-scale data obtained by BRE through the 1960s, particularly in the regions of high suction near the upwind corners of side walls and around the periphery of low-pitched roofs. (The sole exception is the canopy roof data in Table 13 which was revised in the 1986 amendments using peak pressure coefficient data, using the same process as the Swiss code §14.1.2.4 and Chapter 15.)

14.1.2.1.3 Format. The UK wind loading head code CP3 ChV Pt2 1972 [4] defines the assessment method and supplies the wind climate, atmospheric boundary layer and structural loading coefficient data to support it, in a single comprehensive document. (In addition, guidance is given on the deposition of ice on lattice masts and cables because this affects the wind loading, by increasing the effective size of the elements, in addition to the obvious effect of increasing the dead weight.)

The equivalent steady gust equation, Eqn 12.40, is implemented in three discrete steps: design wind speed, design dynamic pressure and design loading. The design wind speed[*], V_S, is obtained by multiplying the reference gust speed by a number of adjustment factors (the original S-factors from which the format of Chapter 9 was developed).

$$V_S = V S_1 S_2 S_3 S_4 \tag{14.1}$$

The reference gust speed, V, is presented as isopleths on a map of the UK. The rôle of each S-factor is as follows:

- S_1 – 'Topography factor' is equivalent to the topography factor, S_L (§9.4.1.7), except that it applies to gust speeds rather than the hourly-mean, hence:

$$S_1 = S_L S_G / G \tag{14.2}$$

  i.e. normalised by the ratio of the Gust Factor of the site to that for flat terrain. This factor is unity when the site is not affected by topography, and this is defined as 'where the average slope of the ground does not exceed 0.05 within a kilometre radius of the site'. The procedure for calculating S_1 adopted in the 1986 amendments follows the method and data of §9.4.1.7 exactly, but with the additional normalisation of Eqn 14.2.

- S_2 – 'ground roughness, building size and height above ground factor', as its name implies, is a 'jack of all trades' factor that accounts for all of the remaining atmospheric boundary-layer characteristics. Thus:

$$S_2 = S_E S_Z G / G_B \tag{14.3}$$

  Note that the Fetch Factor, S_X, does not appear in Eqn 14.3 because the effect of fetch is not recognised by the code. Values are given in a table for three load durations depending on building/element size (in $\simeq 5\%$ steps), four categories of terrain roughness (in $\simeq 10\%$ steps) and for heights above ground between 3 m and 200 m ($\simeq 5\%$ steps near the ground, falling to $\simeq 1\%$ steps at 200 m).

---

[*] The symbol convention of each code of practice is used only in this chapter, so does not appear in Appendix A Nomenclature.

- S_3 – 'statistical factor' is analogous to the Statistical Factor, S_T (§9.3.2.1), except that it is based on the V-model instead of the q-model (§5.3.1.3). S_3 is more onerous than S_T for exposure periods longer than the standard 50 years and less onerous for periods shorter than 50 years. Values are abstracted from a graph to a resolution of about 1%.

- S_4 – 'directional factor', introduced by the 1986 amendments is equivalent to the Directional Factor, S_Θ (§9.3.2.3.1). However, S_4 is normalised to unity in the worst wind direction ($\Theta = 240°T$) as the reference gust speed, V, corresponds to this wind direction. Hence:

$$S_4 = S_\Theta / 1.05 \tag{14.4}$$

Values are given in a table for 30° increments of wind direction. Use of this factor is optional. As $S_4 \leqslant 1$, conservative estimates are obtained if the factor is not used.

The design dynamic pressure, q, corresponding to V_S is calculated from:

$$q = k V_S^2 \tag{14.5}$$

where $k = ½ \rho$ ($= 0.613$ in kg/m^3), i.e. the Bernoulli equation, Eqn 2.6 or 12.1.

Loading coefficients for external surfaces of buildings are defined as overall force coefficients, C_f, external pressure coefficients, C_{pe}, and internal pressure coefficients, C_{pi}. Force and external pressure coefficients are given in the form of tables for common building shapes. The steps in value through the tables are fairly coarse, but this is offset by interpolation. Internal pressure coefficients are given separately in an appendix for typical buildings, depending on the relative permeability of the building faces.

The design load, F, is given by the product of the design dynamic pressure, the loading coefficient for the building shape and the loaded area, thus:

$$F = C_f q A_e \tag{14.6}$$

from the force coefficients, where A_e is the projected area (A_{proj} in Figure 12.8) (there are separate rules for friction drag which account approximately for the area of building surface swept by the wind); or across a building face by:

$$F = (C_{pe} - C_{pi}) q A \tag{14.7}$$

from the pressure coefficients, where A is the area of the face.

The method and models of UK code have been described in some detail to act as a standard for comparison with the other five codes. The following sections will concentrate on the differences, particularly the improvements, from this standard.

14.1.2.2 Denmark

14.1.2.2.1 Scope of code.   The Danish Standard DS 410 [175, 176] is a limit state code that covers all types of actions on structures. The wind loading provisions cover all static or dynamic structures below 200 m in height. Only the provision for static structures will be reviewed here.

14.1.2.2.2 Method and models.   Like the UK code, the Danish code also uses the equivalent steady gust model, but from a 10-minute mean reference wind speed. The annual risk of exceedence is again $Q = 0.02$ (50 year return period), but adjustment to other risks of exceedence or exposure period are made by the q-model (§5.3.1.3) used in this Guide. The minimum permitted exposure period is 1 year and no seasonal or directional variation is given.

Adjustment for terrain roughness is made using three categories, with empirical rules on the extent of fetch required, dependent on height above ground. Adjustment for height above ground is made using the logarithmic law for mean wind speed with an additional turbulence component to give a short duration design gust speed. For sites close to a change of roughness, an empirical interpolation rule is given between the values for each roughness category. The duration of the design gust is quoted as 'a few seconds'. No adjustment is made for decorrelation of gusts over large areas. Topography is accounted for by taking the height above ground from the foot of any escarpment or hill.

The loading coefficients are called 'shape factors', and are mean values as required by the equivalent steady gust model. The code requires that the shape factors are 'determined by tests', warning that 'it should be considered whether the model test is adequately relevant in the actual situation', but giving no specific advice equivalent to the provisions in the 1986 amendments to the UK code. However, example shape factor data are given in an appendix, where the range of structural form covered is slightly greater than that in the UK code. Although the source is not specifically acknowledged, these data are most likely to have been derived from the work of Jensen (§2.4.3, §13.5.1) and if so are mean values derived from models in correctly scaled atmospheric boundary-layer simulations.

14.1.2.2.3 Format. The Danish wind loading provisions are part of a loading head code that also includes dead and imposed loads [175, 176]. The method for static structures is given in the body of the code, with the detailed data and a method for dynamic structures in appendices.

The method is implemented in the same three steps as the UK code, but in an different format. The design wind speed is given by the equation:

$$v = v_b\, k_t\, (\ln[z\,/\,z_o] + 1.3) \tag{14.8}$$

where:

v_b is the reference 10 minute mean wind speed (usually 27 m/s)
k_t is a 'terrain factor', equivalent to the Exposure Factor S_E
z is height above ground, z, and
z_o is the aerodynamic roughness parameter, z_o.

Values of z_o and k_t are tabulated for the three terrain categories. The additional constant 1.3 term in Eqn 14.8 is an equivalent term to the turbulence component of the Gust Factor equation, Eqn 9.37, using a simplified model for the turbulence intensity:

$$S_u = I_u = 1\,/\,\ln[z\,/\,z_o] \tag{14.9}$$

commonly used in codes.

The design dynamic pressure, q, corresponding to v is calculated from:

$$q = \tfrac{1}{2}\,\rho\,v^2 \tag{14.10}$$

i.e. the standard form of Eqns 12.1 and 14.5, but using a higher value of air density ($\rho = 1.28$ kg/m³) than for the UK. Values are given for $z \leqslant 25$ m for the three terrain classes in a graph.

The loading coefficients are given in terms of overall force coefficients, C, external pressure coefficients, c, internal pressure coefficients, c_i, and tangential (shear stress) coefficients, c_t. Values are given for individual shapes, on diagrams of the building shape as in Figure 12.4(b), and by graphs.

The design wind load, w, is given by:

$$w = q\,c\,A \tag{14.11}$$

which, apart from the different symbol notation, is identical to the UK code, Eqn 14.6.

14.1.2.3 USA and Canada

These two codes, the American National Standard ANSI A58.1–1982[177] and the National Building Code of Canada 1985[178] (NBCC) with Supplement[179], are reviewed together because they have strong similarities. The codes differ in units: British Imperial units are still used in the USA, whereas Canada uses the International System of Units (SI) as do the other four codes reviewed and the Guide itself. Other differences are noted in the review.

14.1.2.3.1 Scope of codes.   The ANSI code covers dead, live, wind, snow, rain and earthquake loads on all static or dynamic buildings and structures. Excluded from the ANSI wind loading provisions are structures 'having unusual geometric shapes, response characteristics, or site locations'. The NBCC covers all aspects of Static and Dynamic buildings, including fire, ventilation, heating and plumbing as well as the loading aspects aspects covered by the ANSI code.

14.1.2.3.2 Method and models.   Both codes use a mixture of the equivalent steady gust model and an extreme-value approach for static structures. However, they extend the equivalent steady gust model to apply to dynamic structures by defining a 'gust response factor for flexible buildings' (ANSI) or 'gust effect factor' (NBCC) which accounts for the dynamic effects of the building. The ANSI code gives no guidance for distinguishing between static and dynamic buildings, whereas the NBCC requires buildings higher than 120 m or with proportions height:width greater than 4 to be treated as dynamic. As before, only the static method will be reviewed here.

The reference wind speeds differ. The ANSI code uses the 'fastest-mile' wind speed as reference, which corresponds to the mean values over periods from about 1 minute at 60 mph, down to 30 s at 120 mph. The NBCC uses the hourly-mean dynamic pressure as reference, although this has been derived in most cases from 'fastest-mile' speeds, and gives values for three annual risks, $Q = 0.1, 0.02$ and 0.01. The ANSI code has an 'importance factor' that adjusts the reference wind speed to annual risks of $Q = 0.04, 0.033$ and 0.01. Although not specifically stated, the derivation of the meteorological data implies the use of the V-model for risk in both codes. Neither code gives directional or seasonal variation.

Adjustment for terrain roughness in the ANSI code is made using four categories, each requiring a stated minimum fetch and maximum building height. The NBCC uses three roughness categories, each with a minimum fetch, but with no reference to the building height. In both cases, adjustment for height above ground is made using the power-law model. Neither code has any adjustment for topography.

Loading coefficients are given as mean and as peak values, the latter being disguised [180, 181] by the presentation format as described below. Decorrelation of loading over large areas is accounted for by presenting peak loading coefficients for a range of areas, measured by pneumatic-averaging techniques (§13.3.3.1). In both codes, the range of building form is less than that in the UK code, although the most common shapes are adequately covered.

14.1.2.3.3 Format. The format of the ANSI code and the NBCC is required to cope with the inclusion of both mean and peak loading coefficients. This is achieved in two steps: reference dynamic pressure and design loading. This differs from the UK approach in that the dynamic pressure is a mean value and not a site gust value. Most of the factors for the atmospheric boundary layer, including the adjustment for gusts, are applied with the loading coefficients in the final step.

In the ANSI code the reference 'fastest-mile' wind speed, V, with an annual risk of $Q = 0.02$ (50 year return period) is presented as isopleths on a map of the USA. This encompasses the range of extreme-wind climates from depressions in the north, through thunderstorms in middle latitudes, to hurricanes in the southern coastal areas. Transition through these areas is assumed to be smooth. Some special regions area marked in the mountains prone to katabatic winds (§5.1.3.2) where values are set by the 'authority having juristication'. The ANSI code includes the variation of mean wind speed with height in the first step, so that the reference dynamic pressure, q_z, is height dependent:

$$q_z = 0.002\,56\ K_z\,(I\,V)^2 \tag{14.12}$$

i.e. the Bernoulli equation, Eqn 12.1, where the constant corresponds to $\frac{1}{2}\rho$ and the unit conversion is in British Imperial units. The other parameters are as follows:

- K_z – 'velocity pressure exposure factor' adjusts for height above ground and terrain roughness category, so is analogous to $(S_E\,S_Z)^2$, but uses the power-law model and applies to 'fastest mile' values. Values are tabulated for heights above ground between zero and 500 feet or may be calculated from a power-law equation.

- I – 'importance factor' adjusts the risk of exceedence according to the importance of the building, so is equivalent to S_T^2. Values are tabulated according to structural category, giving $I > 1$ for 'essential facilities' and $I < 1$ for structures that represent 'a low hazard to life.' Values in hurricane-prone areas are increased to allow for the higher dispersion of the extreme-value distribution (see §6.2.2.2).

The NBCC gives the reference hourly-mean dynamic pressure, q, with annual risks of $Q = 0.1$, 0.033 and 0.01 (10, 30 and 100 year return periods) for 643 individual locations across Canada. Adjustment for risk is made by selecting the appropriate value according to the rule: $Q = 0.1$ for cladding and deflection or vibration of structural members (i.e. serviceability limit state); $Q = 0.033$ for strength of structural members (i.e. ultimate limit state) **except** for 'post-disaster buildings' for which $Q = 0.01$.

Both codes obtain the design loading by applying factors to the reference dynamic pressure and mean pressure coefficient to give a design pressure, p, thus:

$$p = q_z\ G_h\ C_p \qquad \text{(ANSI)} \tag{14.13}$$

$$p = q\ C_e\ C_g\ C_p \qquad \text{(NBCC)} \tag{14.14}$$

The rôle of the factors are as follows:

- C_e – 'exposure factor' of the NBCC is equivalent to K_z of the ANSI code, but applies to hourly-mean values.
- G_h – 'gust response factor' (ANSI) and C_g – 'gust effect factor' (NBCC) convert the dynamic pressure from reference values to gust values, so are equivalent to the square of the Gust Factor, S_G^2, of Chapter 9. For dynamic structures, these factors include the dynamic amplification of the structure, equivalent to $S_G^2 \gamma_{dyn}$, and values are obtained by computation or experiment. For static structures, the ANSI code tabulates the values for heights above ground from zero to 500 feet, whereas the NBCC gives two default values: $C_g = 2.0$ for structural members and $C_g = 2.5$ for cladding. (Owing to the difference in reference wind speeds, the values of G_h are smaller than C_g.)

The reason for this choice of format is its ability to mix data derived from mean and peak loading coefficients. The key to this is Eqn 12.38, which is reproduced here as:

$$\hat{c} = S_G^2 \, \bar{c}$$
$$= G_h \, C_p \quad \text{in the ANSI code and}$$
$$= C_g \, C_p \quad \text{in the NBCC} \tag{14.16}$$

i.e. each product is a peak pressure coefficient in disguise.

For common shapes of static structures, values of the products $G_h \, C_p$ and $C_g \, C_p$ are given directly. Values differ between the codes owing to the difference between G_h and C_g from the different reference wind speeds. However, as the gust effect factor, G_h and C_g, are dependent on the terrain roughness, the values of $G_h \, C_p$ and $C_g \, C_p$ are effectively locked to the terrain roughness of the original measurements. These measurements were made in a correctly scaled open-country boundary layer and the design values correspond to a peak loading coefficient with a 20% risk of exceedence [180, 181].

The loading coefficients are also given as mean values, C_p and C_p for use with separate values of G_h or C_g in Eqns 14.13 and 14.14, using the equivalent steady gust model.

14.1.2.4 Switzerland

14.1.2.4.1 Scope of code. The Swiss code SIA 160 [182] is the most recently revised of the five codes reviewed here. It is a limit state code which covers all actions on building structures, including soil, snow, wind, indirect (due to creep, subsidence and temperature), pedestrian, road traffic, rail traffic, live and earthquake loads. The wind loading provisions cover all static structures, and extend to cover dynamic structures less than 100 m in height. Dynamic structures taller than 100 m are specifically excluded.

14.1.2.4.2 Method and models. SIA 160 is the first code to adopt an extreme-value approach exclusively to derive the loading coefficients. Davenport's peak factor model (§12.4.2.1) is also used to assess the decorrelation of the peak loads over large areas. However, both these approaches are transparent to the user because the data are disguised in a quasi-steady format identical to the equivalent steady gust method [183].

SIA 160 uses a 3 second duration gust dynamic pressure as its reference, corresponding to an annual risk of exceedence of $Q = 0.02$ (50 year return).

Adjustment to other risks for serviceability limit states is made using the q-model (§5.3.1.3). Directional and seasonal variation is not given. Three categories of terrain roughness are used: lakeside, open country and urban. Adjustment for height above ground is made using the log-law model. An empirical interpolation rule accounts for the fetch of roughness. Minor topography is accounted for by taking the height from the base of the topography. (The geographical zones of higher basic dynamic pressure account for the major topography.)   A classification procedure is given for assessing the response of buildings up to 100 m in height to distinguish between static and dynamic buildings in this range. While the format of the code permits the assessment of dynamic structures, the user must refer outside the code for a value of the 'dynamic amplification factor'. For buildings taller than 100 m, the code recommends a separate detailed analysis.

Loading coefficients are derived exclusively from peak values, from an extensive series of wind tunnel tests undertaken for this specific purpose, but have been disguised as quasi-steady values by the presentation format described below. The wind-tunnel tests [183] were made by the Swiss Federal Institute of Technology using the best of the simulation and measurement techniques described in Chapter 13. Decorrelation of loading over large areas is accounted for directly in the pressure coefficient values, from pneumatically averaged data (§13.3.3.1), but for the overall force coefficients by a factor derived from the peak-factor model (§12.4.2.1). The range of building forms covered is comparable with the UK code [4], but the data is given in much more detail.

14.1.2.4.3 Format.

The Swiss wind loading provisions are part of a loading head code that includes all actions relevant in design of buildings. The method and the data to implement it are given together in the body of the code.

The format of SIA 160 is quasi-steady, despite the exclusively extreme-value derivation of the loading data, and is implemented as design pressures (actions) or loads (action effects) in a single step. For external surface pressures, q_e:

$$q_e = C_{q_e} C_H q \tag{14.17}$$

The equation for internal pressures is identical except for the subscript 'i' in place of 'e'. For overall normal forces, Q_j:

$$Q_j = C_j C_R C_G C_H q A \tag{14.18}$$

where $j = 1$, 2 or 3 represents inwind, cross-wind and vertical body axes (x, y and z in §12.3.6). The equation for tangential (friction) forces is identical except for the subscript 't' in place of 'j'. The parameters of Eqns 14.17 and 14.18 are as follows:

C_{q_e} and C_{q_i} – external and internal pressure coefficients for the surface area in question. Three types of value are given for the external coefficient in a table for each building form:

C_{q_e} – the average over the whole surface area;

$C_{q_e}^*$ – the average over a local area marked within the general surface area, each for 15° increments of wind direction; and

$\hat{C}_{q_e}$ – the maximum value found within the local area in any possible wind direction. The surface areas are marked on key plan and elevation diagrams for each building form in a similar manner as the UK code [4].

C_j ($j = 1$, 2, 3) and C_t – the normal force coefficients for the three body axes and the tangential force coefficient in the wind direction. Values are given in the same table as the pressure coefficients.

C_H – a 'height coefficient' which adjusts the gust dynamic pressure for height above ground and terrain roughness category. Values are plotted against building height for each of the three categories. Adjustment for fetch is made by interpolating between the categories.

C_R – a 'reduction coefficient' which accounts for the decorrelation of gusts over the whole building, so applies only to the force coefficients. Values are plotted against building height for a range of building proportions. The product $C_H C_R$ is directly equivalent to S_2^2 of the UK code [4].

C_G – a 'dynamic amplification coefficient' which accounts for the resonant component of dynamic buildings and is exactly the same as the dynamic amplification factor, γ_{dyn}, of Chapter 10 and §12.5. For static buildings $C_G = 1$, but for dynamic buildings the value must be calculated from external sources.

q – the reference gust dynamic pressure. Owing to the mountainous nature of the terrain, katabatic 'Foehn' winds (§5.1.3.2) are very significant in some of the Alpine valleys. The country is divided into three zones of exposure, each with a corresponding basic 3 second duration gust dynamic pressure. The lowest value applies to the majority of the country, while the higher two values apply to valleys susceptible to Foehn winds.

The loading coefficients C_q and C_i obtained from the wind tunnel tests are not mean values. They are defined [183] as the peak loading with the design risk of exceedence of $Q = 0.02$ divided by the peak dynamic pressure with the same design risk, thus:

$$C_q = \hat{p}\{Q = 0.02\} / \hat{q}\{Q = 0.02\} \tag{14.19}$$

The peak design loading, $\hat{q}\{Q = 0.02\}$, assessed by extreme-value analysis (§12.4.4), includes the interaction of the building model with the atmospheric boundary-layer model. The peak reference dynamic pressure, $\hat{q}\{Q = 0.02\}$ is representative of the incident flow. Synthesis of design values by Eqn 14.17 reverses this analysis process exactly. This retains the **format** of the equivalent steady gust model without requiring the **assumption** of quasi-steady flow. The values of C_q are close to the values of the mean pressure coefficient, and hence close to the values in the UK code CP3 ChV Pt2 [4], but their differences indicate departures from the quasi-steady model correctly quantified by the extreme-value method of §12.4.4. The comparison with the extreme-value approach of USA and Canada is:

$$\begin{aligned} C_q &= (G_h \, C_p) \, / \, G_h \quad \text{in the ANSI code and} \\ &= (C_g \, C_p) \, / \, C_g \quad \text{in the NBCC;} \end{aligned} \tag{14.20}$$

i.e. each peak coefficient product divided by its corresponding gust effect factor. Removal of the gust effect factor 'unlocks' the loading coefficients from the terrain roughness of the original measurements. The philosophy of this approach has been described by Hertig [183] and is discussed further in Chapter 15, where it is compared with the approach adopted in this Guide.

14.1.2.5 Commentary on the five head codes reviewed

The five head codes reviewed above represent successive stages in the very significant improvements in model and data accuracy developed over the last several decades. They also illustrate that these improvements do not make the codes any more complex to use. In fact the format of all five codes is essentially identical and the improvements are transparent to the user. The apparent differences between the five codes are largely superficial. They are largely due to the way that the various effects are partitioned between the reference values, factors and coefficients. These are chiefly: whether the reference is a mean or a gust value; whether the factors act on the wind speed ('S' factors) or on the dynamic pressure ('C' factors); whether the effects are treated separately by individual factors or together by combined factors; and whether the data are presented in tables or graphs. The Danish code shuns the factor approach entirely, directly implementing its provisions by means of an equation.

CP3 ChV Pt2 1972 [4] is one of the earliest codes to implement the quasi-steady method by the equivalent steady gust model (§12.4.1.3). It was also the first to assess the meteorological data on a probabilistic basis, using the V-model (§5.3.1.3). The fact that this code has survived unchanged for so long indicates that it does work reasonably well. Knowledge has improved greatly since this code came into force and it is now recognised to contain a number of compensating errors and to be unreasonably conservative in some areas. For example, the pressure coefficients for low-pitched roofs are based on typical domestic house shapes, for which they work well, but they result in excessively high predictions of uplift on the modern large plan-area, low-rise industrial buildings that were not envisaged in 1972. Much of the loading coefficient data come from early wind tunnel tests that represented full scale conditions very poorly. Detailed discussion of current loading coefficient data is given in Chapters 16 and 17.

The improvements in the Danish code [175, 176], in comparison with the UK code [4], are:

(a) recurrence model for extreme winds – q-model instead of V-model.
(b) wind profile model – log-law instead of power-law.
(c) changes of terrain roughness – empirical rules added.
(d) loading coefficients – from wind-tunnel tests in properly scaled simulations of the atmospheric boundary layer.

The main improvement in the codes of the USA and Canada is the introduction of extreme-value data in the derivation of loading coefficients. There being insufficient new data to replace the mean data entirely, the format was modified to accommodate both forms, but the choice of format locks the data to the terrain roughness of the measurements. As open country was chosen as the reference terrain, conservative loading estimates are obtained in urban areas. Another improvement is the use of pneumatically averaged data to define the decorrelation of loading over large areas in place of the assumptions contained in the TVL-formula (§12.4.1.3). However, both codes retain the empirical power-law as the wind speed profile model.

The Swiss code incorporates all the improvements in models and measurement technique that are currently available, except that it too retains the empirical power-law model and the adjustment for fetch remains empirical. Most remarkable is the manner in which sufficient extreme-value data were obtained by an intensive series of carefully planned tests, limited in range to that needed to replace all the mean loading coefficients in the previous revision of the code. In its definition of the loading coefficient, Eqn 14.19, the Swiss code approach is the one most like the approach adopted in this Guide and described in Chapter 15. Similar approaches have been used at CSIRO for the revision of the Australian code and at BRE for this Guide and for the replacement of the UK code by BS 6399 Part 2, as discussed in Chapter 15.

14.1.3 Specialised UK codes

14.1.3.1 Introduction

The rôle of the national head codes is to give the loading data, dependent on the external shape of the building but independent of the structural form. These data are normally invoked by the relevant material or structural design codes in one of three ways:

1 Direct – by citing the head code as the definitive data source – as in the UK code of practice for farm buildings [184], BS 5502, which requires that the wind loads should be 'as defined in, and calculated in accordance with, CP3: Chapter V: Part 2: 1972'.
2 Simplified – that is, data from the head code are retabulated or replotted to give only those data relevant for the application, possibly including some simplifying assumptions – as in the UK code of practice for glazing of buildings [185], BS 6262, which cites CP3 ChV Pt2 as the definitive data source, but also gives tables of simplified data that can be used for buildings less than 10 m in height.
3 Incorporated – that is, *ad-hoc* design rules are given that have been calibrated using data from the head code – as, for example, the limiting slenderness ratio of 1:4.5 for masonry chimneys set in the UK Building Regulations [174] by consideration of the range of wind loading relative to the stability of chimneys.

There will always be cases where the data of the head code is not suitable through some peculiarity of the loading model or the building form which is not covered by the scope of the code, when a specialised code is required. These cases are illustrated by the following two UK examples.

14.1.3.2 Slating and tiling

14.1.3.2.1 Scope of code.   The UK code of practice for slating and tiling [186], BS 5534, covers the design and application of slates, tiles, sarking, battens and their fixings. The design aspects include materials requirements, weather resistance, structural stability, durability, thermal insulation, condensation and drainage. The wind loading provisions form part of the structural stability requirements.

14.1.3.2.2 Method and models.   Current UK practice for slated and tiled roofs is shown in Figure 14.1. This is to lay an impermeable underlay of 'sarking' or board on top of the rafters which is retained by the horizontal tile battens. The slates are always nailed to the battens, whereas tiles are generally retained by deadweight and prevented from sliding down the roof by the 'nib' of the tile. Tiles should always be clipped to the battens around the perimeter of the roof and, depending on the exposure, additionally clipped at intervals of several courses. This form of construction gives two surfaces, the outer porous slate or tile surface and the inner impermeable underlay. Between the two surfaces is a void called the 'batten space' which runs horizontally under each course.

BS 5534 recognises three principal components of wind loading on slated or tiled roofs:

(a) from the difference between the normal pressure acting on the external and internal surfaces of the roof;
(b) from the flow of wind over the steps in the outer surface formed by the overlapping tile and slate courses; and

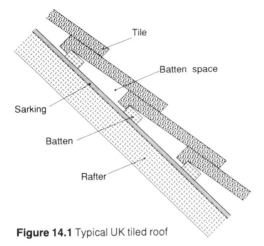

Figure 14.1 Typical UK tiled roof

(c) from the flow of wind through the outer slated or tiled surface in and out of the batten space, driven by differences in the external pressure field.

The first is the normal loading predicted by the head code and is always by far the largest component. If the slates or tiles were laid on battens fixed directly to the rafters, without any underlay, they would have to resist this loading directly; whereupon the other components would be negligible and the head code would be sufficient. However, as modern roofs are constructed with an impermeable underlay between the battens and rafters, the external pressure leaks through the relatively permeable slated or tiled outer surface, so that this load is taken principally by the underlay. This effect has been confirmed by tests[187] at full scale and is the model used in the code for underlay loads.

The second component is much smaller, but gives the principal load on thick tiles when the first component is taken by the underlay. Wind blowing up the courses of tiles produces a distribution of pressure over each course, shown in Figure 14.2(a),

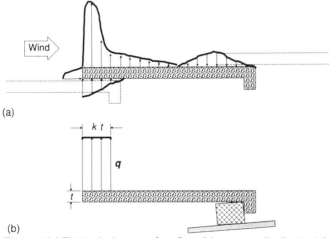

(a)

(b)

Figure 14.2 Tile loads due to surface flow: (a) pressure distribution (after reference 188); (b) code model of loading

which tends to both lift and rotate each tile. This effect has been quantified by tests on full-scale tiles in the wind tunnel [187, 188], and these data form the basis of the code model. The wind speed or dynamic pressure immediately above the tiles is a required parameter. The local dynamic pressure tangential to the surface, q_t, is estimated from the normal surface pressure field using the Bernoulli equation, Eqn 2.6, thus:

$$q_t = p_T - p_S = (1 - C_{p_e}) q_{ref} \tag{14.21}$$

When the Bernoulli equation was derived in §2.2.2, it was noted that its applicability required four assumptions, of which three do not hold. Fortunately, the error from each is conservative in this context: i.e. dynamic pressure is overestimated from the static pressure field (but static pressure would be underestimated from the dynamic pressure). This is the model used in the code for slating and tiling loads.

The third component is smaller still, but will give a significant proportion of the load as the slates and tiles become very thin and the second component reduces towards zero. In regions of the roof where the external normal pressure is highest, air flows between the slates and tiles into the batten space, then out again in regions where the external pressure is low. In regions of outflow, generally the periphery area of the roof, the slates and tiles tend to be lifted. This effect is neglected by the code, but is discussed in Chapter 18 (§18.8.2).

14.1.3.2.3 Format.   BS 5534 divides the loading into two parts: the underlay loads and the slate or tile loads. The starting point for both is the design dynamic pressure, q, and the pressure coefficients for the external and internal surfaces, C_{p_e} and C_{p_i}, appropriate to the building shape and roof pitch taken directly from the head code [4], CP3 ChV Pt2.

The underlay loads are given as a pressure difference across the underlay, q_u, by:

$$q_u = q (C_{p_e} - C_{p_i}) \tag{14.22}$$

which is the same loading predicted by CP3 ChV Pt2 through Eqn 14.7. Slating or tiling laid without underlay or with sealed lap joints are treated by the code as impermeable coverings, and are subject to the same loading as underlays.

The loading on slates and tiles fitted with an impermeable underlay is represented as the local tangential dynamic pressure, q_t from Eqn 14.21, acting on a strip of width kt across the toe of the slate or tile, as shown in Figure 14.2(b); where t is the thickness of the tile and k is a 'step height factor' depending on the form of the tiles. This is a drastic simplification of the actual loading, Figure 14.2(a), designed to represent the most critical loading. The observed mode of failure is lifting of the tile at the toe, followed by rotation around the batten. The simplified model gives a good estimate of the moment around the batten.

14.1.3.2.4 Commentary.   This is a case where the loading model offered by the head code (peak normal pressures) is not representative of the actual loading, due to the peculiarity of the form of construction. Thus special wind loading provisions are essential in the slating and tiling code. Other similar cases where the outer surface is porous and is underlaid by a second impermeable surface include loose-laid insulation boards and paving stones on flat roofs, gravel on flat roofs (where the problem is scour) and decorative cladding systems over impermeable walls. These are discussed in Chapter 18.

14.1.3.3 Lattice towers and masts

14.1.3.3.1 Scope of code. The UK code of practice for lattice towers and masts [189], BS 8100 Part 1, is a limit state code which covers the dead, wind and ice loads for the design of all lattice towers and masts on land. Offshore-mounted lattice towers are specifically excluded. It also covers the dynamic response of free-standing towers, but the response of guyed masts is specifically excluded. Unlike the head code [4] CP3 ChV Pt2, which is restricted to the UK in scope by the meteorological data, the lattice towers and masts code BS 8100 is intended for worldwide application, having been written so as to allow the introduction of external meteorological data.

14.1.3.3.2 Method and models. The wind loading provisions of BS 8100 implement the lattice-plate model (§8.2.1) through the equivalent steady gust model (§12.4.1.3) for static and mildly dynamic (§10.8.2.1) towers and the spectral admittance model (see Part 3) for fully dynamic (§10.8.2.2) towers. Additionally, the line-like and bluff-body models are used where necessary for ancillaries such as cylindrical and dish antennae. BS 8100 was nearing completion when Part 1 of this Guide was published and the opportunity was taken to include the Guide data where the framework of BS 8100 Part 1 permitted. The method and models used in code are fully described in the accompanying Part 2 [190], 'Guide to the background and use of Part 1'.

BS 8100 makes use of the wind climate data for the UK given in Part 1 of this Guide: including the map of basic hourly-mean wind speed, Figure 9.5; the Altitude Factor, S_A; and the Directional Factor, S_Θ. Adjustment for risk of exceedence and other loading uncertainties is provided by the partial factor for wind, together with partial factors on dead and ice loads, and a factor for quality of design and construction acting on the strength. These factors are set by reliability considerations, including the economic consequences of failure and the risk to life.

Adjustment for terrain roughness is made using five categories, without allowance for roughness fetch. Adjustment for height above ground is made using the power-law model. Many communication towers are constructed on the summits of hills and escarpments, making topography an important consideration. Adjustment for topography is made using a different empirical model from the head code [4] and this Guide, but which predicts very similar results [190]. Decorrelation of loading over large towers is accounted for by a gust factor which accounts for the typical response of mildly dynamic towers in addition to the correlation of gusts in the wind, combining the function of the Gust Factor, S_G (§9.4.3), and the dynamic amplification factor, γ_{dyn} (§10.8.2, §12.5).

Loading coefficients are all mean values as required by the equivalent steady gust and spectral admittance models. They are derived principally from wind-tunnel data obtained in uniform smooth or turbulent flow, as required for the lattice-plate model (§8.2.1) where the fluctuations of load are assumed to correspond exactly with gusts in the atmospheric boundary layer. Values are given for individual flat-sided and circular members, for complete panels of square and triangular towers, for single frames, and for linear or discrete ancillaries.

14.1.3.3.3 Format. Owing to the way that BS 8100 integrates the complete design process for dead, wind and ice loads, the format of the code is complex. This complexity is indicated by the flow diagram of Figure 14.3 in which the 'flow'

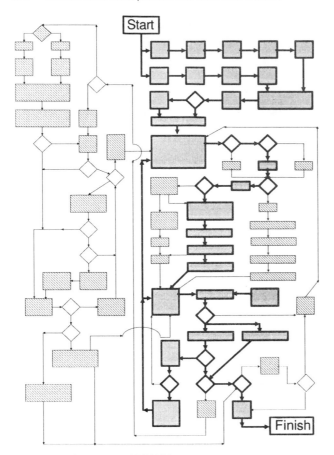

Figure 14.3 Procedure of BS8100

direction is indicated by arrows, each rectangular box represents an individual step in the design process and each diamond-shaped box represents a 'branch' where a choice of path must be made. The thicker boxes and arrows indicate the typical flow for normal designs. These are confined to the right-hand side of the chart which corresponds to the strength considerations for the ultimate limit state, whereas the left-hand side corresponds to the displacement and response considerations for the serviceability limit state. Accordingly, a detailed description of the format is not attempted here. Instead the reader is referred to Part 2 of BS 8100 [190], 'Guide to the background and use of Part 1', which is a commentary on the code with worked examples of the calculations.

14.1.3.3.4 Commentary.   The wind loading provisions of BS 8100 use the same lattice-plate loading model as the head code and employ much of the same data. However the wind loading provisions are integrated with the dead and ice load provisions to produce a procedure that embraces all the design aspects. The flow diagram, Figure 14.3, representing the procedure is made up of many simple steps and the apparent complexity is due only to their number and interaction required to support the comprehensive scope of the code.

This is a case where relevant data have been abstracted from the head code to meet this specific application. However, the opportunity was taken to reassess the head code data and to augment them with new data from wind-tunnel tests commissioned for this purpose [191, 192] and other published sources [193, 194]. Advantage was taken of the symmetry offered by typical square and triangular towers to offer overall values in place of values for individual members. Figure 14.4 shows the provisions for bare square towers compared with the available experimental data.

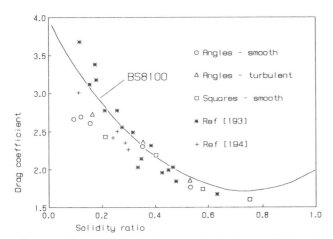

Figure 14.4 Drag coefficient for square lattice towers of flat members

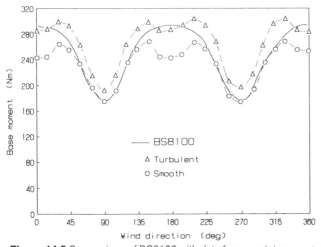

Figure 14.5 Comparison of BS8100 with data from model square tower with eight dishes

Operation of the resistance clauses in the draft code was checked by comparing the code predictions for a number of tower designs with typical configurations of ancillaries against the experimental data from full models. Figure 14.5 gives the comparison for the model tower shown in Figure 13.34, a square tower with a central ladder, cylindrical microwave feeders and eight microwave

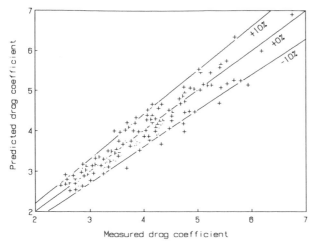

Figure 14.6 Drag of square and triangular lattice towers from BS8100

dishes [195, 196]. BS 8100 is probably among the most thoroughly calibrated codes in current use and predicts the mean drag typically to within ±10%, as demonstrated by Figure 14.6. This was possible only because of the range of form of lattice towers is quite restricted, and because the lattice-plate model is simple (§8.2.1) and particularly amenable to codification.

14.2 Other data sources

14.2.1 ESDU data items

The Engineering Sciences Data Unit (ESDU) was originally formed in 1940 within the Royal Aeronautical Society to provide accurate engineering data to the aircraft industry. Since then, ESDU has expanded its range of data to cover many engineering applications and it is now an independent body. Data from many published and unpublished sources are collated by the ESDU staff and published as data items. This work is guided and monitored by panels of experts provided by industrial companies, government research laboratories and universities on a voluntary basis. Great care is taken to ensure the accuracy of the data provided, principally by comparison of data from several independent sources. Data items on wind began to appear from about 1970, monitored by the fluid mechanics, external flow panel. In 1979 the wind data was separated from the more general fluid mechanics data and is now published in the wind engineering series of data items, supervised by a specialist wind engineering panel.

The wind engineering series of data items is currently divided into four volumes:

1 *Wind Speeds and Turbulence*
2 *Mean Loads on Structures*
3 *Dynamic response*
4 *Natural Vibration Parameters of Structures.*

Volume 1 covers the same ground as Chapter 9 in Part 1 of this Guide, using the same models and data sources, but giving additional detail. (ESDU now market

Table 14.2 Index to ESDU wind engineering sub-series, volume 2 – mean loads on structures (as at December 1988)

Item	Topic
71012	Fluid forces on non-streamline bodies – background notes and description of the flow phenomena.
80025	Mean forces, pressures and flow field velocities for circular cylindrical structures: single cylinder in two-dimensional flow.
81017	Mean forces, pressures and moments for circular cylindrical structures: finite-length cylinders in uniform shear flow.
79026	Mean fluid forces and moments on cylindrical structures: polygonal sections with rounded corners including elliptical shapes.
84015	Cylinder groups: mean forces on pairs of long circular cylinders.
70015	Fluid forces and moments on flat plates.
71016	Fluid forces, pressures and moments on rectangular blocks.
80003	Mean fluid forces and moments on rectangular prisms: surface-mounted structures in turbulent shear flow.
80024	Blockage corrections for bluff bodies in confined flows.
82031	Paraboidal antennas: wind loading. Part 1: mean forces and moments.
82020	Paraboidal antennas: wind loading. Part 2: surface pressure distribution.
81027	Lattice structures. Part 1: mean fluid forces on single and multiple plane frames.
81028	Lattice structures. Part 2: mean fluid forces on tower-like space frames.
82007	Structural members: mean fluid forces on members of various cross sections.

Supplement 2 of this Guide – the BRE wind speed program STRONGBLOW – under licence, but with small modifications to make the data exactly compatible with the corresponding ESDU data items.) Volume 2 gives loading coefficient data as discussed below. Volumes 3 and 4 give data for the dynamic response of structures and are discussed in Part 3.

The loading coefficient data currently offered by ESDU in Volume 2 are confined to mean values, so can only be applied to static structures through the quasi-steady method (§12.4.1). A list of the data items available at the time of writing is given in Table 14.2. The scope of the available data far exceeds that given by the codes, requiring more effort and understanding on the part of the designer to implement them. The ESDU data items include mean loading coefficients for circular or prismatic cylinders, flat plates, individual structural members of lattices and lattice frames, so are of particular value for lattice structures (towers, frames and arrays) and line-like structures (stacks, conveyors and pipelines) where the quasi-steady method gives good results. The range of typical building shapes covered is confined to cuboidal shapes representing flat-roofed buildings. ESDU maintains a policy of continuously upgrading the quality and range of their data and expects to publish new data for typical building shapes, including peak values, based on the same recent sources of data used for this part of the Guide.

ESDU data item volumes may be obtained individually or by subscription to the wind engineering series. By using the subscription method the designer ensures that the data items are kept up to date with the regular amendments and additions.

Details of this service and a complete index of all data items can be obtained from:
ESDU International plc.,
27 Corsham Street,
London, N1 6UA, UK.

14.2.2 Product design manuals, journals and text books

Finally, there are a number of other sources of loading coefficient data available to the designer. Many product associations or individual manufacturers publish design manuals for their products that contain loading data specific to the use of that product. These manuals usually integrate the wind loading data from the head code with the design strength or performance data from the relevant material or structural code, and present the result directly in terms of the required sizes or other design parameters of the product. Thus by using these manuals, the designer avoids making any wind loading calculations at all. However, the designer usually pays a price for this simplification, in terms of an additional hidden safety factor. For example, a glazing manual could present tables of glass thickness and maximum window size for various geographical regions and heights of building over a range of shapes. Each step in the tables must not allow underdesign, so must cope for the most onerous case, i.e. the strongest wind speed in the region, the tallest building and the worst loading coefficient. Then for the application that is just within the bottom of the range of each parameter, the manual will predict the next thickest glass for the largest window. In all, this gives five probable sources of additional safety factor. The use of glazing as the example was purely arbitrary, as the same approach is used for window frames, doors, cladding, slating and tiling and many other standard building components.

Useful loading data may also be obtained from papers in technical journals such as the *Journal of Wind Engineering and Industrial Aerodynamics*. Few papers in journals address the problem of design specifically. The majority are concerned with investigating particular aspects of the aerodynamics with a view to understanding the flow processes involved, so may not be directly applicable to design. If such data are used directly, the designer cannot have the same confidence in the result that is given by the collated and validated data given by codes, ESDU or this Guide. However, for unusual shapes of building this may be the only alternative to wind tunnel testing.

Most text books do not aspire to be design manuals. Their function is to teach the reader the relevant aerodynamic theory and its application in the design process. The data given in these books are intended to be illustrative rather than comprehensive, nevertheless much useful data can be found. This Guide, however, attempts to be the inverse of a text book: its purpose is to give reliable and comprehensive data that are directly useful for design; whereas the theory is intended to be illustrative and to foster an understanding so that the data are used correctly.

15 A fully probabilistic approach to design

15.1 Re-statement of the problem

The division of the problem at the 'ideal' position in the Spectral Gap between the wind climate and the atmospheric boundary-layer, as described in Chapter 4, allows separate and statistically independent analysis of the incident wind conditions, in terms of $\overline{V}$ or $\overline{q}$, and the standard form of loading coefficient, C or c (§12.3). However, by permitting variability of the loading coefficients, synthesis requires an answer to the question posed in §12.6: **what is the value of the loading coefficient that results in a design load of the desired design risk, given a wind speed of the same risk?**

All the assessment methods reviewed in Chapter 12 contain inherent simplifications or assumptions so that their answers to this question are approximations to an unknown degree. The requirement is for a method that takes the extreme-value parameters from the analysis of the wind climate and the peak loading coefficients and synthesises the corresponding extreme-value parameters of the peak design loading, as indicated in Figure 15.1, without imposing any further assumptions or approximations. The exact form of the peak loading (maximum or minimum peak values of pressure, force or moment) is irrelevant to this ideal synthesis, so is represented by the symbol $\hat{X}$ in the diagram and equations of this chapter only.

An approach which accounts for the variabilities of the wind climate and the loading coefficients was developed by the author and J R Mayne at BRE between 1978 and 1982[21,29,197,198,199], greatly assisted by external discussions and contributions[198,200,201,202]. Although complex in its full form, the method reduces to a simplified design method that directly answers the question posed above. This simplified method has been extensively adopted[26,76,183,203]: it is presently the standard method of assessment for wind tunnel data by BRE[203] and CSIRO[76] for the revision of the UK and Australian codes (the major sources of data for this Guide); and it is the basis for the Swiss code SIA 160 (§14.1.4.2) as described by Hertig[183]. The development of the method from the original first-order concept to the final simplified design method is outlined below. Further detail may be found in the original references given above.

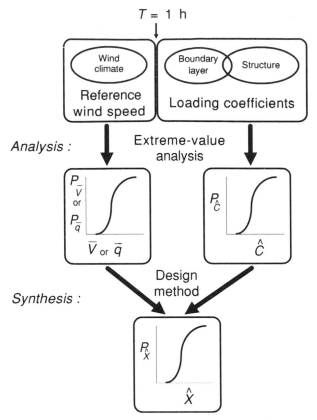

Figure 15.1 Ideal assessment by extreme-value analysis

15.2 A fully probabilistic design method

15.2.1 First-order method

The method was originally developed in terms of wind speed and loading coefficient, the traditional design parameters in the UK. Once the advantage of the q-model over the V-model (§5.3.1.3) had been appreciated, the method was converted to loading coefficient and dynamic pressure. The starting point is the Fisher–Tippett parameters: mode, U and dispersion, $1/a$, for the annual-maximum hourly-mean wind speed, $\overline{V}$ and the peak loading coefficient, $\hat{C}$ or $\hat{c}$. (The method makes no distinction between global or local coefficients, maxima or minima. For convenience, $\hat{c}$ is used in the equations throughout this chapter.)

The statistical independence gained by the ideal division of the problem in Chapter 4 allows the joint probability density function[*], $d^2P / d\overline{V}\,d\hat{c}$, to be obtained from Eqn B.7 as the product of the individual probability densities:

$$d^2P / d\overline{V}\,d\hat{c} = dP / d\overline{V} \times dP / d\hat{c} \tag{15.1}$$

---

[*]Note that the PDF is denoted here by the differential of the CDF (Eqn B.3), to avoid using the usual symbol, p, which is reserved for pressure.

As both the individual probability densities are forced to the FT1 form (§B.5.1) by the extreme-value analysis, the resulting joint PDF is always of the form shown in Figure 15.2, plotted in isometric projection on the V–c plane. The detailed shape of this 'probability mountain' will vary with climate, wind direction, building shape and location on the building, depending on the the relative values of the characteristic products $\Pi_{\overline{V}}$ and $\Pi_{\hat{c}}$.

Any value of peak load, $\hat{X}$, is given in terms of $\overline{V}$ and $\hat{c}$ by:

$$\hat{X} = \tfrac{1}{2}\,\rho\,\overline{V}^2\,\hat{c} \tag{15.2}$$

and lines of constant $\hat{X}$ are drawn on the V–c plane of Figure 15.2. These form a family of curves indicating that any given peak load can be caused by a range of values of $\overline{V}$ and $\hat{c}$: large $\overline{V}$ with small $\hat{c}$; small $\overline{V}$ with large $\hat{c}$; or intermediate combinations. The probability density of a given peak load is the integral of the joint PDF along the corresponding line, as indicated by the shaded region in Figure 15.2:

$$dP\,/\,d\hat{X} = \tfrac{1}{2}\,\rho \int (d^2P\,/\,d\overline{V}\,d\hat{c})\,dl \tag{15.3}$$

where dl is an line element along a contour $\hat{X}$ = constant. This integration, known 'convolution', is difficult to perform analytically but is amenable to numerical methods. The cumulative distribution function, $P_{\hat{X}}$, is the form required for design, and this is obtained from Eqn 15.3 by the simple integration:

$$P_{\hat{X}} = \int (dP\,/\,d\hat{X})\,d\hat{X} \tag{15.4}$$

which represents the volume under the 'probability mountain' up to the $\hat{X}$ contour.

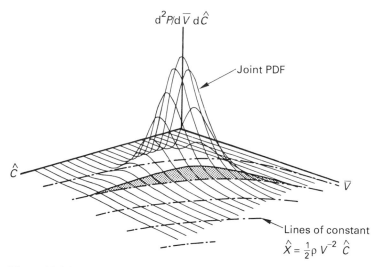

Figure 15.2 Joint probability density function of wind speed and loading coefficient

15.2.2 Full method

In the first-order method only the combination of the annual-maximum hourly-mean wind speed with the maximum loading coefficient within that hour was considered. An important assumption implicit in this treatment is that the maximum load experienced by a structure will occur in the strongest hour of wind. The statistical variation of the loading coefficient admits the possibility that a sufficiently large peak loading coefficient can occur in the **second-strongest** hour of wind to produce a loading in excess of that which occurred in the strongest hour. Similarly, but with decreasing probability, the maximum load could occur during the the third-, fourth-, to Mth strongest hour of wind. Clearly, the first order method requires second- and higher-order corrections to avoid underestimation.

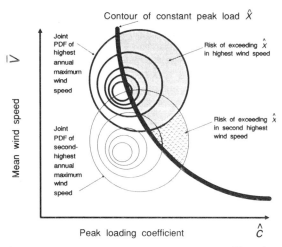

Figure 15.3 First- and second-order joint PDF of $\overline{V}$ and $\hat{c}$

The main problem of introducing corrections is illustrated in Figure 15.3, where the joint PDF of the peak loading coefficient with the highest and second-highest annual maximum hourly-mean wind speeds are each represented as contours on the V–c plane. The risk of a given loading being exceeded in the highest hour of wind, Q_1, is the volume under the first joint PDF outside the $\hat{X}$ contour, i.e. under the shaded area. Hence the corresponding CDF, $P_1 = 1 - Q_1$, is the same as the first-order method. Similarly, the risk of exceeding the same loading in the second-highest hour of wind, Q_2, is the volume under the second joint PDF. Summing these risks for all the Mth highest wind speeds up to $M = 8766$ would converge to the convolution of the extreme loading coefficient and the **parent** wind speed [201]. In order to obtain the risk of **annual extreme** loading, the Mth highest wind speed contributes to the risk **only** when the resultant loading exceeds the contributions from **all** $M - 1$ higher speeds.

This is a process of **conditional** probabilities which is most easily implemented by Monte-Carlo simulation techniques (§5.3.1.6). Simulated values of the M highest extreme wind speeds in a year, are each combined with a simulated peak loading coefficient, and only the highest value of $\hat{X}$ resulting from Eqn 15.2 is retained. Each value represents one annual maximum, and the model is run for a large number of model years to build up the CDF.

The loading is made non-dimensional, $\hat{x}$, when it is normalised by the modal values of wind speed and loading coefficient:

$$\hat{x} = \hat{X} / (\tfrac{1}{2} \rho \, U_{\overline{V}^2} \, U_{\hat{c}}) \tag{15.5}$$

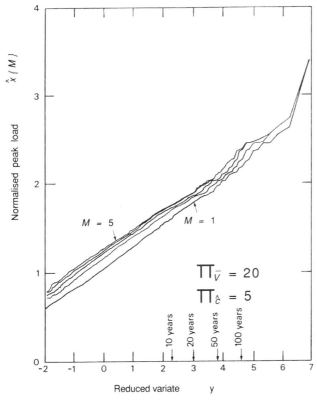

Figure 15.4 Typical results of 1st–5th-order Monte-Carlo models

and the model must be run for the expected range of the characteristic products for wind speed, $\Pi_{\bar{V}} = a_{\bar{V}} U_{\bar{V}}$, and peak loading coefficient, $\Pi_{\hat{c}} = a_{\hat{c}} U_{\hat{c}}$.

Figure 15.4 shows typical results for the first five orders of correction, plotted on standard FT1 axes (Gumbel plot, §C.3.2). Note that each successive order of correction, M, has a smaller effect but consistently increases the mode (value at $y = 0$) while decreasing the dispersion (slope). Limiting forms of the CDF can be deduced by allowing $\bar{V}$ or $\hat{c}$ to become constant [198]:

$$\hat{x} \to 1 + y_{\hat{c}} / \Pi_{\hat{c}} \qquad \text{as } \bar{V} \to \text{constant}, \tag{15.6}$$

but in reducing the dispersion of $\bar{V}$, each hour of wind becomes much like any other hour, hence $\bar{V}_1 \to \bar{V}_2 \to \bar{V}_3 \ldots \to \bar{V}_M$ where $M \to 8766$. This is equivalent to running M trials of the first order, hence:

$$P_{\hat{X}} \to P_{\hat{c}}{}^M = \exp(-e^{-(y_{\hat{c}} + \ln M)} \qquad \text{as } \bar{V} \to \text{constant}, \tag{15.7}$$

i.e. the FT1 form of the loading coefficient is retained, giving a straight line on the FT1 axes. On the other hand:

$$\hat{x} \to (1 + y_{\bar{V}} / \Pi_{\bar{V}})^2 \qquad \text{as } \hat{c} \to \text{constant} \tag{15.8}$$

but with constant $\hat{c}$ the peak loading must occur in the strongest hour of wind, reverting to the first-order model. This time, the limiting CDF appears as a parabola on FT1 axes.

15.2.3 Refinement and verification of full method

The form of the CDF, $P_{\hat{x}}$, computed in the $V{-}c$ plane by fifth-order Monte Carlo model is consistently transitional between an straight line and a parabola when plotted on FT1 axes, as in Figure 15.4. This behaviour is forced by the FT1 models for $\overline{V}$ and $\hat{c}$ and the squared wind speed term in Eqns 15.2 and 15.5. When the statistical criteria applied to the wind speed and dynamic pressure to justify the q-model (§5.3.1.3) are applied to the peak loading, the same result is obtained and a FT1 distribution is expected. Figure 15.4 is almost a straight line, so the data could be linearised and forced to a FT1 fit. However, this proves to be unnecessary when the V-model is replaced by the q-model.

With the q-model, computations are done in the $q{-}c$ plane. The form of the PDFs in Figures 15.2 and 15.3 are unchanged, but the $\overline{V}$ axis becomes the $\overline{q}$ axis and the contours of constant $\hat{X}$ are given by:

$$\hat{X} = \overline{q}\,\hat{c} \tag{15.9}$$

The peak loading is now non-dimensionalised by:

$$\hat{x} = \hat{X} / (U_{\overline{q}} U_{\hat{c}}) \tag{15.10}$$

The limit as $\overline{q} \rightarrow$ constant, Eqn 15.7, is unchanged, but the other limit becomes:

$$\hat{x} \rightarrow 1 + y_{\overline{q}} / \Pi_{\overline{q}} \text{ as } \hat{c} \rightarrow \text{constant} \tag{15.11}$$

The form of the CDF computed in the $q{-}c$ plane by fifth-order Monte Carlo model now remains consistently close to a straight line when plotted on FT1 axes, as expected, allowing values for the FT1 parameters to be fitted to the peak loading. Fitting the non-dimensional peak loading, $\hat{x}$, to the form:

$$\hat{x} = A + B\,y_{\hat{x}} \tag{15.12}$$

gives estimates for the mode, $U_{\hat{x}}$, and dispersion, $1 / a_{\hat{x}}$, of the peak loading:

$$U_{\hat{x}} = A\,U_{\overline{q}} U_{\hat{c}} \tag{15.13}$$

$$1/a_{\hat{x}} = B\,U_{\overline{q}} U_{\hat{c}} \tag{15.14}$$

Results from the fifth-order method[199] are given for a range of $\Pi_{\overline{q}}$ and $\Pi_{\hat{c}}$ in Table 15.1(a). The values in italics correspond to the limiting forms of Eqns 15.7 and 15.11. Gumley and Wood[202] extended the Monte-Carlo model to the 100th order, recomputing over a wider range of characteristic product, and their results are give in Table 15.1(b). With the higher orders of correction, these results exhibit consistently higher modes (A) and lower dispersions (B), continuing the trend in Figure 15.4.

Independent verification of the full method was provided by Harris[201] who succeeded in solving a multi-order convolution of the **conditional** joint PDFs of peak loading coefficient with Mth highest dynamic pressures – a most formidable task! His results are given in Table 15.1(c), together with the number of orders of correction necessary to achieve the given resolution. These indicate that more than five orders of correction are required, but that 100 orders are nearly always sufficient. Harris' work is a rigorous theoretical verification of the full method. He also provides an independent comment on the method, as follows[201]:

> The Cook-Mayne method is attractive for the following reasons: (i) it is logical, in that quantities which vary statistically are treated statistically; (ii) it does not

involve any unrealistic causal relationship between wind force and micro-meteorological fluctuations in the incident wind (as, for instance, does the old-fashioned code approach for leeward surfaces); (iii) the partition of the fluctuations into micro- and macro-meteorological time scales is in accordance with the known physics of strong winds; (iv) the method is based on hourly mean wind data and is thus uniform with modern methods developed for more complex structures; and (v) it uses only data which are measurable either as full-scale meteorological data or in a proper boundary-layer wind-tunnel simulation.

The computations, having been performed once, do not need to be performed again. The full method may be implemented by from the FT1 parameters of $\bar{q}$ and $\hat{c}$ by looking up values of A and B from Table 15.1, interpolating as necessary. The design value of load is then given through Eqns 15.13 and 15.14 and the FT1 equations, Eqns B.18 and B.19, as:

$$\hat{X}^* = U_{\bar{q}} U_{\hat{c}} \{A + B (- \ln(- \ln[1 - Q_{\hat{X}}]))\} \tag{15.15}$$

where $Q_{\hat{X}} = 1 - P_{\hat{X}}$ is the design annual risk. Here the symbol * is introduced to denote the value of the peak loading has been derived through the convolution of the full method.

15.2.4 Simplified method

A major disadvantage of the full method is that it requires access to the FT1 parameters of both dynamic pressure and peak loading coefficient. The first is not a great problem for the UK, since a standard value of $\Pi_{\bar{q}} = 5$ was adopted for the design data in Chapter 9 (§9.3.2.1.2). It may be a problem, however, in other wind climates where the data are less well defined. Adequate peak loading coefficient data can only be obtained by *ad-hoc* design studies or by direct access to research data from boundary-layer wind tunnels, which could not be justified for the majority of typical buildings. A simplification of the full method to make it compatible with the single design loading coefficient value format of codes would be desirable.

The first step in simplification is to test the sensitivity of the full method to variations in the parameters. We still do not have a direct answer to the question: **what is the value of the loading coefficient that results in a design load of the desired design risk, given a wind speed of the same risk?**

Equation 15.15 gives the design peak load. The FT1 q-model gives the corresponding design dynamic pressure of the same risk:

$$\bar{q} = U_{\bar{q}} - \ln(-\ln[1 - Q_{\bar{q}}]) / a_{\bar{q}} \tag{15.16}$$

Using these values, Eqn 15.9 may be solved for the corresponding value of peak loading coefficient, $\hat{c}^*$, which represents the required design value of loading coefficient in Eqn 12.59:

$$\hat{c}^*\{Q\} = \hat{X}^*\{Q\} / \bar{q}\{Q\} \tag{15.17}$$

Again the symbol * is used to denote that the peak loading coefficient has been derived through the convolution of the full method. These values of $\hat{c}^*$ may be examined in terms of their reduced variate, $y_{\hat{c}^*}$, by solving the FT1 equation, Eqn 12.59.

$$y_{\hat{c}^*} = a_{\hat{c}} (\hat{c}^* - U_{\hat{c}}) \tag{15.18}$$

Table 15.1   Fisher–Tippett type 1 parameters for non-dimensional peak loading

(a) Cook and Mayne [199], 5th-order Monte Carlo model, $\hat{x} = A + B y_{\hat{x}}$

$\Pi_{\bar{q}} =$		$\Pi_{\hat{e}} =$ 5	10	15	20	∞
$\Pi_{\bar{q}} = \infty$	A	2.8156	1.9076	1.6052	1.4539	1.0000
	B	0.2000	0.1000	0.6667	0.0500	0.0000
10	A	1.2439	1.1057	1.0627	1.0442	1.0000
	B	0.2145	0.1369	0.1177	0.1099	0.1000
7.5	A	1.2227	1.0957	1.0559	1.0398	1.0000
	B	0.2340	0.1647	0.1479	0.1416	0.1333
5	A	1.1888	1.0795	1.0493	1.0341	1.0000
	B	0.2839	0.2282	0.2139	0.2102	0.2000
2.5	A	1.1395	1.0626	1.0391	1.0261	1.0000
	B	0.4877	0.4311	0.4215	0.4177	0.4000

(b) Gumley and Wood [202], 100th-order Monte Carlo Model, $\hat{x} = A + B y_{\hat{x}}$

$\Pi_{\bar{q}} =$		$\Pi_{\hat{e}} =$ 2	5	7	10	12	15	18	20	30	10^4
$\Pi_{\bar{q}} = 10^4$	A	3.295	1.920	1.657	1.460	1.384	1.306	1.254	1.229	1.153	1.000
	B	0.502	0.202	0.145	0.101	0.083	0.067	0.055	0.050	0.034	0.000
8760	A	3.300	1.924	1.658	1.459	1.383	1.306	1.255	1.230	1.153	1.000
	B	0.505	0.201	0.142	0.101	0.084	0.067	0.056	0.050	0.034	0.000
15	A	2.493	1.484	1.305	1.178	1.133	1.094	1.071	1.060	1.033	1.001
	B	0.404	0.177	0.136	0.102	0.094	0.083	0.078	0.076	0.072	0.067
10	A	2.174	1.355	1.220	1.127	1.096	1.068	1.054	1.046	1.027	1.002
	B	0.413	0.195	0.159	0.131	0.121	0.116	0.113	0.109	0.103	0.102
9	A	2.093	1.324	1.120	1.112	1.089	1.065	1.051	1.044	1.026	1.001
	B	0.426	0.208	0.166	0.141	0.131	0.125	0.122	0.120	0.115	0.112
7.5	A	1.953	1.277	1.168	1.104	1.081	1.059	1.047	1.039	1.021	1.000
	B	0.452	0.222	0.183	0.164	0.158	0.151	0.145	0.142	0.139	0.137
6	A	1.790	1.232	1.146	1.090	1.070	1.054	1.040	1.037	1.021	1.000
	B	0.481	0.263	0.219	0.196	0.192	0.185	0.179	0.179	0.175	0.169
5	A	1.675	1.200	1.126	1.082	1.065	1.048	1.042	1.039	1.021	1.001
	B	0.535	0.300	0.254	0.235	0.227	0.218	0.215	0.213	0.210	0.202
3.5	A	1.501	1.157	1.107	1.070	1.053	1.040	1.026	1.027	1.023	1.002
	B	0.647	0.391	0.352	0.330	0.316	0.305	0.299	0.302	0.296	0.288
2.5	A	1.384	1.130	1.085	1.059	1.054	1.038	1.032	1.030	1.019	1.000
	B	0.802	0.521	0.463	0.441	0.432	0.427	0.427	0.421	0.419	0.401
1	A	1.211	1.100	1.064	1.048	1.052	1.034	1.033	1.032	1.015	0.992
	B	1.667	1.196	1.121	1.085	1.072	1.058	1.053	1.043	1.029	1.008

(c) Harris [201], multi-order analytic solution, $\hat{x} = A + B y_{\hat{x}}$

$\Pi_{\bar{q}} =$		$\Pi_{\hat{e}} =$ 5	10	15	20
$\Pi_{\bar{q}} = 10$	A	1.373	1.130	1.070	1.047
	B	0.176	0.124	0.112	0.107
	(M)	(219)	(80)	(40)	(25)
7.5	A	1.282	1.104	1.059	1.040
	B	0.206	0.157	0.146	0.141
	(M)	(78)	(34)	(20)	(15)
5	A	1.200	1.081	1.049	1.035
	B	0.271	0.225	0.214	0.209
	(M)	(25)	(15)	(12)	(10)
2.5	A	1.127	1.061	1.040	1.030
	B	0.478	0.432	0.419	0.414
	(M)	(11)	(9)	(9)	(9)

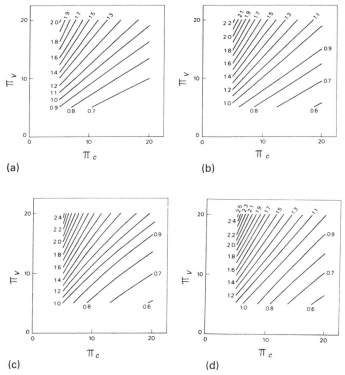

Figure 15.5 Design values of $y_{\hat{c}}$ by 5th-order V-model: (a) $y_{\hat{c}}$ {$R = 10y$}; (b) $y_{\hat{c}}$ {$R = 20y$}; (c) $y_{\hat{c}}$ {$R = 50y$}; (d) $y_{\hat{c}}$ {$R = 100y$}

Figure 15.5 shows the variation of $y_{\hat{c}}$. over the observed range of $\Pi_{\bar{V}}$ and $\Pi_{\hat{c}}$ for four values of return period, $R = 1/Q$. Each graph in Figure 15.5 has contours radiating as almost straight lines from the origin, indicating that the ratio of the characteristic products of $\bar{V}$ and $\hat{c}$ is the principal controlling parameter. Each graph is very similar, indicating that there is only a weak dependence on design risk. These results are almost identical when the q-model is used.

Unfortunately, the use of a range of values for $y_{\hat{c}}$. prevents the use of the code approach of a single design loading coefficient. However, the range of $y_{\hat{c}}$. is not large. It is therefore possible to choose of a single design value somewhere in the middle of this range which is rarely exceeded in practice. The relevant ranges for $\Pi_{\bar{V}}$ and $\Pi_{\hat{c}}$ were estimated by a survey of the UK meteorological data and from BRE's large pool of loading coefficient data [198]. The resulting joint probability histogram is shown in Figure 15.6. The use of $\Pi_{\bar{V}} = 10$ (or $\Pi_{\bar{q}} = 5$) as standard in the UK (§9.3.2.1.2) further limits the effective range of $y_{\hat{c}}$ to give a maximum value of about 1.4. The contour $y_{\hat{c}}. = 1.4$ is marked on Figure 15.6 to show that only 1% of cases (shaded region) would be underestimated in practice. The use of this value overestimates for the typical (modal) case, giving a safety factor of 1.04.

The simplified method is essentially the adoption of a standard design value of $\hat{c}^*$ corresponding to $y_{\hat{c}}. = 1.4$ in Eqn 12.59, giving:

$$\hat{c}^* = U_{\hat{c}} + 1.4 / a_{\hat{c}} \tag{15.19}$$

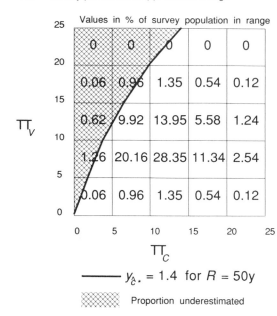

$y_{\hat{c}\cdot} = 1.4$ for $R = 50y$

Proportion underestimated

Figure 15.6 Joint probability of wind speed and loading coefficients from survey

This has now been widely adopted and is often called a "Cook–Mayne coefficient" to distinguish it from coefficients derived from single extremes (§12.4.4.2) or by the peak factor method (§12.4.2). As the reduced variate y is an alternative form of expressing the CDF, P (Eqns 5.12 and 12.50), and hence the design risk, Q, or return period, R, we have a simple answer to the earlier question, namely:

for $y_{\hat{c}\cdot} = 1.4$, $P_{\hat{c}\cdot} = 0.78$, $Q_{\hat{c}\cdot} = 0.22$ and $R_{\hat{c}\cdot} = 4.5\,\mathrm{h}$.

The only assumption contained in the simplification is that the observed ranges of $\Pi_{\bar{V}}$ and $\Pi_{\hat{c}}$ are adequately contained by the design value $y_{\hat{c}\cdot} = 1.4$. Although this assumption was only tested for UK use, it is also valid for world-wide use provided the characteristic product for the annual maximum wind speed or dynamic pressure is not significantly greater.

15.2.5 Calibration of earlier approaches

15.2.5.1 Preamble

The major advantage of the fully probabilistic design method is that it gives an absolute result that does not rely on any assumptions or calibration. It can therefore be used to calibrate the other, earlier approaches. In 1982, a calibration of the the quasi-steady and peak-factor approaches, using linear regression analysis on model-scale data collected in the BRE boundary-layer wind tunnel for this Guide, was reported by the author [17]. This was quickly followed by a calibration of the quantile-level method by Everett and Lawson [204]. The principal results of these calibrations are summarised below.

15.2.5.2 Quasi-steady approach

The equivalent steady gust model is the commonest quasi-steady approach used by codes of practice. It estimates the peak loading from the mean loading by Eqn

12.38, **assuming** that the fluctuations of load are entirely due to gusts in the incident wind. The calibration used data from six wind-tunnel models of different shape, from pressure tappings at locations distributed over all faces of the models.

Table 15.2   Calibration of the equivalent steady gust model

	1 s duration			16 s duration		
Model description	$\hat{c}_p^*/\bar{c}_p$	S_G^2	ε	$\hat{c}_p^*/\bar{c}_p$	S_G^2	ε
1 UWO low-rise model	2.56	2.88	0.34	1.75	1.94	0.14
2 Grandstand	3.04	2.46	0.41	1.83	1.75	0.15
3 Cube	3.67	2.79	0.88	2.13	1.92	0.38
4 Cuboid 3:1:1	3.72	2.79	0.94	2.13	1.92	0.37
5 Tower 1:1:3	2.60	2.36	0.60	1.71	1.71	0.24
6 Hipped roof	3.74	2.74	0.65	2.15	1.92	0.24

The principal results[17] for the six models are summarised in Table 15.2 for loading durations of 1 s and 16 s, with the peak coefficients as the dependent variables and the mean coefficient as the independent variable of the regression. For both durations the observed regression slope for the peaks, $\hat{c}_p^* / \bar{c}_p$ (or $\check{c}_p^* / \bar{c}_p$), and the expected slope, S_G^2, are given together with the rms error, ε. Notice that the observed slope is usually (but not always) greater than the expected slope, indicating a component of the fluctuations additional to the incident turbulence. This is examined in more detail in §15.3.2, below. Of more concern is the high values of the rms error, ε, common to all the data.

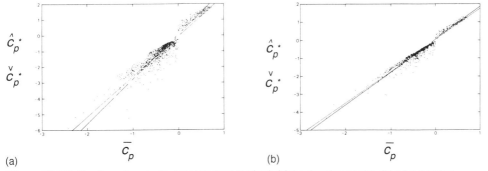

(a)   (b)

Figure 15.7 Calibration of equivalent steady gust method: (a) 1 s duration peaks; (b) 16 s duration peaks

One of the six models: the University of Western Ontario (UWO) low-rise building model[17], is used here to illustrate a typical calibration. Figures 15.7(a) and (b) show the maximum and minimum peak coefficients, $\hat{c}_p^*$ and $\check{c}_p^*$, for durations of 1 s and 16 s, respectively, plotted against the corresponding mean pressure coefficient, $\bar{c}_p$. This is commonly called a 'scatter plot' or 'shotgun diagram' because it forms a random cloud of points when there is poor correlation.

For a perfect regression, if Eqn 12.38 held exactly, the points would all lie along a straight line through the origin with a slope of S_G^2. Linear regression analysis gives two estimates for this line in terms of its slope and intercept: one for the peak coefficient as the dependent variable and the mean coefficient as the independent variable; and *vice-versa* for the other. Both these lines are drawn on the figure.

Inspection of Figure 15.7(a) for the short 1 s duration loading shows a step change between $\hat{c}_p^*$ and $\check{c}_p^*$ either side of zero (where the model is expected to be deficient, §12.4.1.3). This indicates that the regressions of maxima and minima would be better assessed separately. Even so, the scatter would remain very large. The most outlying values of $\check{c}_p^*$ are in excess of a value of 2 away from the regression lines. These correspond to locations around the periphery of the roof in the separated flow regions where the fluctuations from the incident turbulence have been augmented by building-generated turbulence. Inspection of Figure 15.7(b) for the longer 16 s duration loading shows the same trends, but reduced in value, indicating that most of the additional building-generated fluctuations are of short duration and so are more relevant to cladding loads than to structural loads.

15.2.5.3 Peak-factor approach

The peak-factor method estimates the peak loading by Eqn 12.41 as the sum of the mean loading and a proportion of the rms loading given by the peak factor, g. The value of the peak factor is usually estimated using Davenport's model (§12.4.2.1), but linear regression analysis enables the value of g to be calibrated directly. The results from the previous six models [17] were discussed in §12.4.2.2 and summarised in Table 12.1.

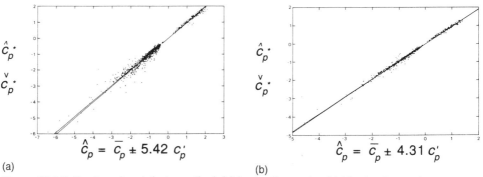

(a)   (b)

Figure 15.8 Calibration of peak-factor method: (a) 1 s duration peaks; (b) 16 s duration peaks

The regression analysis for the UWO model is illustrated in Figure 15.8, showing that the best fit values obtained for the peak factor were $g\{t = 1\,\text{s}\} = 5.24$ and $g\{t = 16\,\text{s}\} = 4.31$. Inspection of Figure 15.8(a) for the short 1 s duration loading shows no step between $\hat{c}_p^*$ and $\check{c}_p^*$ either side of zero and the scatter is reduced in comparison with Figure 15.7(a).

There are still outlying values of $\check{c}_{p}^*$, indicating locations where the value of g is higher, but these are now all less than a value of 1 away from the regression lines. However their locations do not correspond with the outliers of the equivalent steady gust method. On Figure 15.9 the local value of g required for exact

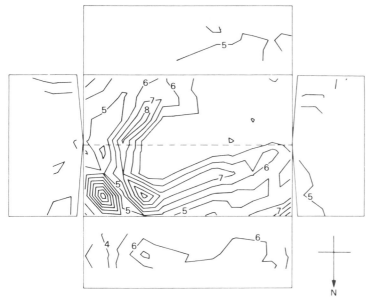

N

Figure 15.9 Equivalent peak factors for 1 s duration pressures

equivalence is shown plotted as contours on the faces of the building. The outliers correspond to highest values which appear principally on the roof in a 'V'-shape along the flow re-attachment lines behind the conical 'delta-wing' vortices (see Figure 8.12, §8.3.2.2.2) and also on the re-attachment position on the long windward side. The regions of high local suction at the periphery of the roof, where the equivalent steady gust model was poor, are adequately assessed by the peak-factor method. Cook[17] postulated a mechanism for this occurrence, as follows:

Separated-flow regions are characterised by very high values of rms coefficient c_p', whereas attached-flow regions give much lower values. The boundaries between these regions are unsteady and move back and forth across the models, creating an intermittency in the flow regime. Here the peak values from the separated flow will dominate those from the attached flow, but the rms value c_p' will find a transitional level, and thus the gust factor is elevated in value. The consequence to design is less serious than with the quasi-steady approach, since the regions of highest design loading are adequately assessed in value, and it is only their spatial extent that is underestimated.

15.2.5.4 Quantile-level approach

The previous two calibrations were simple to perform because it is quick and simple to collect the necessary mean and rms values while performing extreme-value analysis (§12.4.4.3), so this can be done routinely. Calibration of the quantile-level method requires acquiring the distribution function (CDF or PDF) of the parent (§12.4.3.3) simultaneously with the extreme-value analysis, effectively doubling the effort. Accordingly, the calibration of the quantile-level approach against the

Cook–Mayne method was specially performed by Everett and Lawson [204], using data from just four locations on a model of a high-rise building. The four locations were carefully chosen as follows:

(a) high local suction region near the upwind corner of the side face;
(b) attached flow region near edge of windward face;
(c) attached flow region near centre of windward face; and
(d) wake region near centre of leeward face;

to cover the range of typical flow phenomena.

Although the number of points were limited, Everett and Lawson repeated the calibration ten times at each point to quantify the variability of the method. The results of this calibration are summarised in Figure 15.10, which compares the CDF of the ten quantile-level estimates of the design pressure (solid curve) with the corresponding single estimates by the Cook–Mayne method of Eqn 15.19 (dashed lines). The correspondence between the two methods is remarkably good, with equivalence on average in the attached flow region and slight overestimation by the quantile-level method in the separated flow regions. The spread of the quantile-level CDFs is small, indicating that the variability of the method is also small.

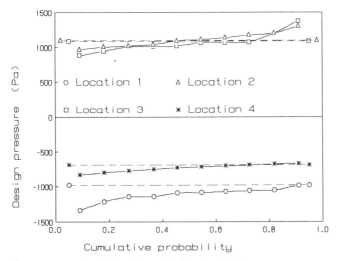

Figure 15.10 Calibration of quantile-level method

15.2.5.5 Conclusion

It is clear that the three approaches are closely equivalent to the Cook–Mayne method **on average**. The systematic and random departures from equivalence have important consequences for safety when these methods are employed in design assessments, particularly in the case of the equivalent steady gust method which forms the basis of most current codes of practice. The Cook–Mayne method is the only currently available method which requires no assumptions and provides design values with a given risk of exceedence.

15.3 The pseudo-steady format

15.3.1 Introduction

The majority of the design data in this Guide have been assessed through the Cook–Mayne method, Eqn 15.19. However, the range of available data is not sufficient to replace all the quasi-steady data in codes of practice. The ideal division of the problem, introduced in Chapter 4 and used in Chapter 12 to define the loading coefficients, and the Cook-Mayne method result in a mean dynamic pressure and peak loading coefficient. This is incompatible with the commonest quasi-steady code approach of peak dynamic pressure and mean loading coefficient. Quasi-steady theory predicts that values of peak coefficients are larger than the corresponding mean coefficients by a factor of S_G^2, typically about 2.5, (Eqn 12.38) so depend on both the terrain roughness and the duration of the peak loading. For example, Figure 15.11 shows values of maximum and minimum peak coefficients for pressures of duration $t = 1\,\mathrm{s}$ and $t = 16\,\mathrm{s}$, using the Cook–Mayne equation, Eqn 15.19, on BRE measurements from a circular silo in full scale. This is the same example used to illustrate $\bar{c}$ and c' in Figure 12.10, but the values are plotted on conventional cartesian instead of polar axes.

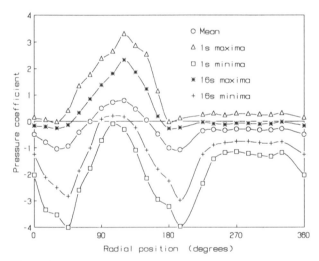

Figure 15.11 Cook–Mayne coefficients for silo

It would be impractical to give different values of peak loading coefficient for all terrain roughness categories and load durations in addition to the many required building shapes. The Canadian and US codes attempt to resolve the incompatibility problem by disguising the peak coefficients as discussed in Chapter 14 (§14.1.2.3.3). However, as these disguised peak coefficients still depend on the terrain roughness and the duration of the peak loading, the values are locked to the unique terrain and load duration for which they were derived: open country and about $t = 1\,\mathrm{s}$, respectively, in the case of the Canadian and US codes.

Although the standard form of peak loading coefficient, defined in §12.3.3, is necessary for ideal division of the problem for analysis (see Chapter 4), it is not essential to maintain this standard in the subsequent design synthesis. The best form of coefficient for design is one which is the least sensitive to variation and is also simple to apply. It would be very useful to adopt a format for design synthesis which enables the quasi-steady mean coefficient data and peak coefficient data assessed by Cook–Mayne, peak-factor or quantile-level methods to be mixed, which also unlocks the data from the restriction of roughness categories (if this is possible), and yet does not violate the requirements of ideal assessment.

15.3.2 Definition of pseudo-steady loading coefficients

The quasi-steady assumption allows the peak value to be estimated from the mean using Eqn 12.38:

$$\hat{c} \simeq (\hat{q} / \overline{q}) \, \overline{c} = (\hat{V} / \overline{V})^2 \, \overline{c} = S_G^2 \, \overline{c} \tag{15.20}$$

where the standard symbol '$\simeq$' indicates an approximation. Consider the following equation of similar form:

$$\hat{c} \equiv (\hat{q} / \overline{q}) \, \tilde{c} = (\hat{V} / \overline{V})^2 \, \tilde{c} = S^2 G \, \tilde{c} \tag{15.21}$$

where the standard symbol '$\equiv$' indicates exact equivalence. For this equation to be always true, the new coefficient, $\tilde{c}$, is forced to take the value the mean would have to be for the quasi-steady assumption to be exact, hence the new '~' (tilde) symbol in place of the mean bar '—'. In practice, this new coefficient will be somewhere near the mean in value, differing most for small load durations and converging onto the mean value at durations of one hour, thus:

$$\tilde{c} \to \overline{c} \text{ as } t \to 3600 \, \text{s} \tag{15.22}$$

that is, $\tilde{c}$ is a weak function of averaging time, t. This new coefficient, $\tilde{c}$, may therefore be called a 'pseudo-steady', i.e. **false**-steady, coefficient. Rearranging Eqn 15.21 gives the definition:

$$\tilde{c} \equiv (\overline{q} / \hat{q}) \, \hat{c} = (\overline{V} / \hat{V})^2 \, \hat{c} = \hat{c} / S_G^2 \tag{15.23}$$

This transformation from $\hat{c}$ to $\tilde{c}$ is an example of a common process called 'normalisation', whereby some non-dimensional parameters are replaced by others of the same dimensions (i.e. exchanging one velocity parameter for another) so as to reduce the variability of the result. In effect, we now have:

$$\hat{X} = \tfrac{1}{2} \rho \, \hat{V}^2 \, \tilde{c} = \hat{q} \, \tilde{c} \tag{15.24}$$

in place of the standard 'ideal' definitions of Eqns 15.3 and 15.9. This does not alter any of the information contained in the peak coefficient; neither does it violate the ideal division of the problem since both the original peak coefficient, $\hat{c}$, and gust factor, S_G, are assessed after the ideal division has been applied to the data. The change from $\hat{c}$ to $\tilde{c}$ is merely a convenient linear transformation which gives a gust reference wind speed instead of a hourly-mean reference wind speed.

15.3.3 Comparison with the equivalent steady gust model

Figure 15.12 shows the data for the example silo in this pseudo-steady format, comprising four curves of pseudo-steady coefficients, representing maximum and minimum values for two load durations, and a single curve of the mean value. The data are essentially identical to that in Figure 15.11, but have been re-normalised by Eqn 15.22. Now it can be seen that the values for the two load durations $t = 1 \, \text{s}$

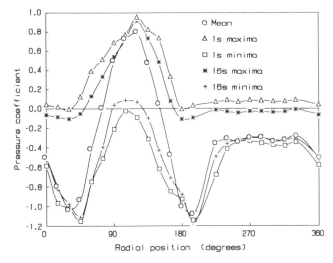

Figure 15.12 Pseudo-steady coefficients for silo

and $t = 16\,\text{s}$ are closely similar. The maximum values are closely similar to the mean when the mean is positive, and the minimum values are closely similar to the mean where the mean is negative, as expected. Near a mean of zero, where the quasi-steady method is deficient, the maxima and minima are non-zero, and have similar absolute values, $|\tilde{c}|$:

The ratio $\tilde{c} / \bar{c}$ gives a direct comparison with quasi-steady response. Values less than unity indicate that not all the fluctuations in the wind result in fluctuations of loading, and this would be expected in the attached flow regions on the windward faces of buildings for short load durations as a consequence of the aerodynamic admittance (§8.4.1). Values greater than unity indicate the contribution of additional building-generated fluctuations, and this would be expected in separated flow regions. Regression analysis between the pseudo-steady coefficient, $\tilde{c}$, and the mean, $\bar{c}$, gives essentially the same result as the calibration in §15.2.5.1 (as in Figure 15.7, for example) but scaled by the factor $1/S_G^2$. Outlying points will be correctly assessed in the same relative positions on the graph. However, as $\tilde{c}$ must be non-zero when $\bar{c} = 0$, regressions for maxima and minima are better performed separately (§15.2.5.1).

Linear regression analysis between the pseudo-steady and mean coefficients has been performed on all the data collected by BRE for this Guide. This pool of data is very much more extensive than for the first calibration [17], comprising values for more than 50 000 locations on a wide variety of shapes of structure. The maxima and minima were analysed separately to give regressions of the form:

$$\tilde{c} = A\,\bar{c} + B \pm C \tag{15.25}$$

where A is the slope and B is the intercept of the regression line, and C is the standard deviation of the scatter about the regression line.

The principal results are summarised in Figures 15.13 to 15.15 for all the data in total and for the data subdivided into data from roofs and data from walls. The number of data values used in each case is given in brackets, showing that the amount of 'roof' data exceeds 'wall' data by almost an order of magnitude and so

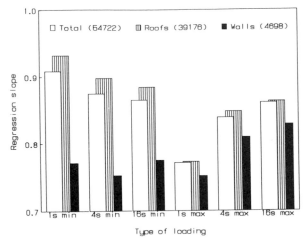

Figure 15.13 Quasi-steady to pseudo-steady regression slope

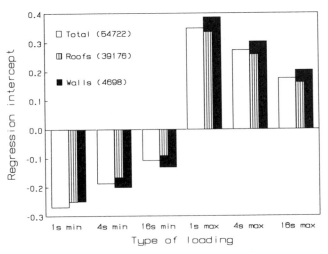

Figure 15.14 Quasi-steady to pseudo-steady regression intercept

dominates the 'total'. In the previous calibration [17], analysis of the maxima and minima together forced the regression line very close to the origin. Releasing this constraint allows the intercepts to take positive values for maxima and negative values for minima, resolving the previous anomaly (§15.2.5.1). This is accompanied by a compensating reduction in the regression slope, giving values that are now less than unity on average.

Some distinct trends are evident:

1 The regression slopes for all maxima increase with increasing load duration in the range $1\,\mathrm{s} < t < 16\,\mathrm{s}$. Maxima occur in regions of attached flow on windward walls and windward facing steep-pitched roofs, where a quasi-steady aerodynamic admittance function (§8.4.2.1) similar to Figure 8.18 would be expected. The regression slope is analogous to the integral of the admittance function up to the

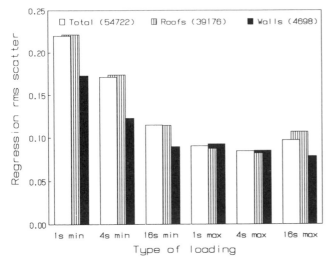

Figure 15.15 Quasi-steady to pseudo-steady regression scatter

critical frequency corresponding to the load duration. As shown by Figure 8.28, this is less than unity for short durations (high frequencies), converging towards unity at long durations (low frequencies).

2 The regression slope for minima show different trends with load duration. Minima occur in regions of separated flow on low-pitched roofs, side and leeward walls. For roofs (which also dominate the total) the regression slopes are high, but **decrease** with increasing load duration. This is due to the additional contribution from building-generated turbulence which spans the same frequency range, as shown in Figure 4.6. Although this is not accounted for by quasi-steady theory, a large proportion of this additional contribution can still be normalised against the mean value [17]. At $t = 1$ s most of the contribution is included, but is excluded by increasing load duration to $t = 16$ s. Increasing the load duration further into the range of the atmospheric boundary-layer turbulence would be expected to reach a minimum, rising again to unity at $t = 3600$ s. The implication is that the separated-flow regions on low-pitched roofs are dominated by coherent effects, the most likely candidate being the 'delta-wing' vortices of Figure 8.12. On the other hand, minima on walls tend to occur in the wake region, where the building-generated turbulence is much weaker and less coherent. For walls the regression slope is much lower and the variation with load duration appears to have caught the expected minimum.

3 The trends with the intercept are much more consistent, with similar values for roofs, walls and in total. The intercept is always positive for maxima and negative for minima. The trend is always for the intercept to reduce in magnitude, converging towards zero, with increasing load duration. This indicates that the non-quasi-steady component decreases as the building-generated turbulence is excluded.

4 The rms scatter also shows different trends. For the minima, the scatter is greater for roofs than for walls, and reduces with increasing load duration in both cases. The scatter for the maxima remains similar for roofs and walls, and fairly constant with load duration. This appears to be intermediate between the trends

for the regression slope and intercept, which is expected since the scatter includes all departures from the average.

This regression analysis is useful in two main ways. Firstly it provides data on the reliability of the quasi-steady equivalent steady gust approach that can be used to set rational values of safety factor for wind loads in codes of practice for actions. Secondly it opens the possibility of using Eqn 15.25 to 'correct' values of mean coefficient, $\bar{c}$, into the pseudo-steady format, $\tilde{c}$. The values of slope, A, and intercept, B, presented in Figures 15.13 and 15.14 allow $\tilde{c}$ to be estimated: but the rms scatter, C, which indicates the standard error of this estimate is very large, so the process is not accurate. Clearly, if there were an accurate relationship quasi-steady theory would always be sufficient, so the Cook–Mayne assessment method and the resulting pseudo-steady coefficients would not be needed. However, this form of correction can be used when only mean data are available. The corrections are made on the basis of the population of data used, which gives the following restrictions:

1 The data are corrected to the average, so that systematic variation caused by other parameters, such as building shape, wind direction or roof pitch, which have all been lumped together in the regression are excluded
2 The corrections, being empirical, are valid only over the range of value observed in the calibration.

The first restriction can be overcome by further subdividing the data by building shape, or value of each parameter and repeating the analysis. The second restriction is less important as the data are so extensive as to cover the likely range of value.

Figure 15.16 shows the results of applying Eqn 15.25 to the mean pressures on the silo in Figure 15.12, compared with measured pseudo-steady values. The values of slope, A, and intercept, B, were taken from the 'total' data to predict the 1 s duration maximum and minimum pressures. The prediction is good in the highly-loaded areas, the maxima in the positive pressure lobe at the front

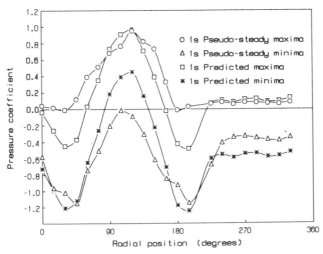

Figure 15.16 Pseudo-steady and corrected mean coefficients for silo

stagnation point and the minima in the suction lobes at either side, but the minima in the wake region at the rear are overestimated. Correspondence between the minima in the positive pressure lobe and the maxima in the suction lobes, i.e. the lowest loading, is poor and the values are underestimated.

The design data presented in this Guide will be in the form of pseudo-steady coefficients wherever possible, so that correction will not be needed.

15.3.4 Advantages of the pseudo-steady format

In summary, the principal advantages of the pseudo-steady format are:

1 Quasi-steady components are removed from the data, minimising the variability with load duration and terrain roughness through the action of the Gust Factor, S_G^2.
2 Such variability that remains will be due to the effect of other parameters, including building form, shape, wind direction, roof pitch, Reynolds number, Jensen number and the other non-dimensional parameters discussed in §13.5.1, and this can now be examined in isolation.
3 Pseudo-steady data may be applied to terrain categories other than that of the original measurements, since the strong quasi-steady component is included in the design gust wind speed, $\hat{V}$, (see §15.3.5 below) provided that any additional effect of Jensen number (§13.5.1) is negligible or can be accounted for.
4 Pseudo-steady data for one load duration may be used when the residual effect of load duration is small (as in Figure 15.12).
5 Pseudo-steady data and mean value data may be mixed together in codes of practice which use a format based on the equivalent steady gust model, thus allowing old data to be progressively revised. Mean values can be replaced directly, one value of $\tilde{c}$ for one value of $\bar{c}$, where the loading is large and second order effects are small. Otherwise, typically where the mean coefficient is near zero, a maximum and a minimum value of $\tilde{c}$ is required.
6 Adoption of pseudo-steady data in such codes is 'transparent' to the user, and there is no need to alter the format of the code or to alter working procedures in any way.

15.3.5 Implementation for design in this Guide

Pseudo-steady coefficient data are implemented in the equivalent steady gust format, in exactly the same manner as described in §12.4.1.5 except that the pseudo-steady coefficient, $\tilde{c}$, replaces the mean. Using the earlier example, Eqn 12.40 for the peak design pressure becomes:

$$\hat{p}\{t, \theta\} \equiv \tfrac{1}{2} \rho \, \hat{V}^2\{t, \theta\} \, \tilde{c}_p\{\theta\} \tag{15.26}$$

with the approximation now replaced by an equivalence. This is the format used in the Guide.

As the maximum peak, $\hat{c}$, can also be replaced by the minimum peak, $\check{c}$, the pseudo-steady coefficient has two possible values. For local values, the coefficient with the larger magnitude is generally the more relevant, except when the response of the structure is non-linear. For example, cladding fixings may be more susceptible to suction than to positive loading. The problem of deciding whether the maximum or minimum values should contribute to a global value, as discussed for the peak coefficients in §12.4.2.4, still remains relevant. The former choice of maximum where $\bar{c} > 0$ and minimum where $\bar{c} < 0$ is still generally appropriate, and usually corresponds to the pseudo-steady coefficient of larger magnitude. Difficulties can arise with load combinations – wind loads and dead loads, or wind loads and snow loads, for example – where a smaller magnitude positive wind load may be more critical than much larger negative loads. In this case the combination that causes the most onerous loading is required. Other difficulties may arise with structural forms, such as arches, that are more susceptible to small asymmetric loads than large uniform loads. Accordingly, every attempt has been made to provide pseudo-steady coefficients both maxima and minima, as in Figure 15.12. Loading coefficient data for various forms of structure are given in the remaining chapters.

In a full implementation, the design gust speed in Eqn 15.26 will be estimated from the design data in Chapter 9 for various gust durations corresponding to the required load durations. This process is reviewed in §9.6 with worked examples. Alternatively, the ready-reckoner tables of Supplement 1: The assessment of design wind speed data: manual worksheets with ready-reckoner tables, or the computer program of Supplement 2: BRE program STRONGBLOW can be used, the latter giving the most precise result for the least effort. (See inside the front flyleaf for details of the supplements to this Guide.)

The load duration, t, in Eqn 15.26 relevant to static structures and their components is adequately assessed by the simple TVL-formula (§8.6.2.3, §12.4.1.3) of Eqn 12.37, reproduced here as:

$$t = 4.5\,l/V \tag{15.27}$$

The length parameter, l, represents the size of the loaded area under consideration. Although l may be obtained by integration of the load influence function (§8.6.2.1–2) through Eqn 8.9, this complication is rarely necessary and the simple representation by Eqn 10.2, here:

$$l = (b^2 + h^2)^{1/2} \tag{15.28}$$

as the diagonal of the loaded area is usually sufficient.

A simpler alternative to the full implementation is given here in Appendix K, A model code of practice for wind loads on buildings, in which the design wind speed data of Chapter 9, the structural classification method of Chapter 10 and the loading coefficient data of the following chapters are further simplified to the form proposed for the new UK code of practice BS6399 Part 2. Inevitably, this simplification results in a loss of detail and an increase in conservatism. The choice between simplicity and precision must be made by the designer.

16 Line-like, lattice and plate-like structures

16.1 Introduction

16.1.1 Form

Owing to the dependence of the loading coefficients on the form of the structure and the vast variety of form found in practice, the range of practical structures is divided between Chapters 16 and 17 in accordance with the loading models described in Chapter 8 of Part 1. Inevitably some unusual forms are bound to be omitted, both here and also in the later design data given in Chapter 20. This is usually because reliable data just do not exist. However, it is hoped that most forms of structure will be covered either directly or by extrapolation from other similar forms. The forms are divided into the main classes as follows:

1 **Line-like structures**: including tall chimneys and stacks; flagpoles and masts; elevated cables, pipelines, conveyors and gantries; long-span bridges; columns, beams and other structural sections.
2 **Lattice structures**: including lattice towers and trusses; electricity and lighting pylons; cranes; unclad building frames; falsework and scaffolding.
3 **Plate-like structures**: including boundary walls and fences; signboards and hoardings; freestanding canopy roofs; bare building façades during renovation.
4 **Bluff structures**: including the majority of typical building shapes; detached and terraced housing; high-rise tower and slab blocks; low-rise factories and warehouses; cylindrical tanks and silos; barrel-vaults and domes.

Loading of structures in classes 1–3 are discussed in this chapter. The whole of Chapter 17 is devoted to structures in class 4 which includes the majority of building structures. Design values for all four classes of form have been collated and are presented in Chapter 20. Before embarking on discussion of these individual cases, it is helpful to consider several aspects of more universal relevance.

16.1.2 Slenderness ratio

Slenderness ratio is a term used here to describe the proportions of the elevation of building structures in the plane normal to the wind direction, i.e. the proportions of A_{proj} in Figure 12.8. Figure 16.1(a) illustrates the extremes of slenderness for rectangular flat plates normal to the wind and free of the ground surface. A square plate is the least slender, with a slenderness ratio $L/B = 1$, and this has a drag coefficient of $C_D = 1.2$. At the other extreme with very slender plates, when L/B is very large, the drag coefficient rises to 2.0. For structures in free air, it does not

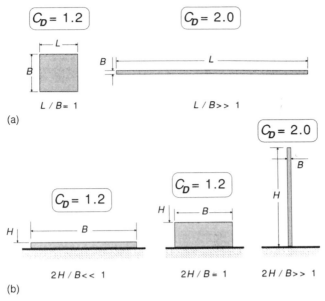

Figure 16.1 Slenderness ratio for flat plates: (a) free plates normal to wind; (b) surface-mounted plates normal to wind

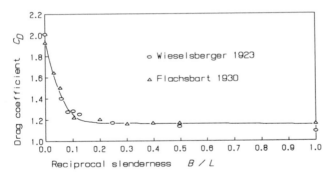

Figure 16.2 Effect of slenderness on drag of flat plates (from reference 43)

matter whether they are aligned horizontally, vertically, or any angle in between. Variation of the drag coefficient between these extremes is plotted in Figure 16.2 against B/L, the reciprocal of the slenderness ratio, as reported in 1930 by Flachsbart [43]. This shows that the drag coefficient remains constant at the low extreme of $C_D = 1.2$ for plates with a slenderness ratio less than about 6 ($B/L > 0.16$), but rises quickly towards the other extreme of $C_D = 2.0$ as they tend towards becoming two-dimensional or infinitely long. This effect is also shared by circular cylinders at subcritical Reynolds numbers, but in this case the drag coefficient changes from $C_D = 1.2$ for very long cylinders to $C_D = 0.5$ (the supercritical value) for short cylinders. Clearly, there must be some fundamental difference between the aerodynamic behaviour at each extreme to produce such a large difference in drag.

For surface-mounted rectangular plates, as shown in Figure 16.1(b), the situation is at first sight similar. Owing to the apparent symmetry afforded by the ground plane, the slenderness ratio $2H/B$ is equivalent by potential flow theory to a free plate twice as high, i.e. with the ground plane at the centreline (§13.2.2). Thus for a plate with a slenderness ratio $H/B = 0.5$ the drag coefficient is again $C_D = 1.2$, and for very tall slender plates the drag coefficient again rises to $C_D = 2.0$ as H/B tends to infinity. However, for long low plates with H/B very small one might reasonably expect the drag coefficient to rise again: but it remains constant at $C_D = 1.2$. The physical difference between these two cases is that the ground plane is attached to the short side of the tall slender plate and to the long side of the long low plate. The same effect occurs with cylinders at subcritical Reynolds numbers: a tall slender cylinder sticking out of the ground plane again has a high drag coefficient of $C_D = 1.2$, while a long hemicylinder sitting on the ground has a low drag coefficient of $C_D = 0.5$.

With the tall plate or cylinder, the shear layers separating from either side are able to interact and so produce periodic vortex shedding (§2.2.10.4), but this is not possible for the long low plate or barrel-vault-like hemicylinder as there is only one shear layer. Thus the orientation of surface-mounted structures is important. Tall structures have a slenderness ratio greater then unity and a higher drag than long low structures which have a slenderness ratio less than unity.

As vortex shedding is essentially a **fluctuating** effect, producing a cyclic lift coefficient (Eqn 2.29), how can this have such a large effect on the **mean** drag? Inserting a long plate behind a cylinder parallel to the flow prevents the shear layers from interacting and suppresses vortex shedding. Roshko [205] measured the mean pressure coefficient around the cylinder and along the centreline of the wake and his results for the wake are reproduced as Figure 16.3. Although the alternate

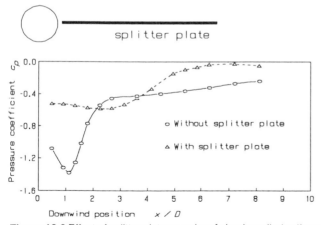

Figure 16.3 Effect of splitter plate on wake of circular cylinder (from reference 205)

vortices produce fluctuating lift of opposite sign and so no mean lift, as each vortex forms behind the cylinder the low pressure in the vortex acts on the rear of the cylinder, always increasing the drag. The pressure distribution on the windward face is not significantly changed. Accordingly, investigations of this effect often concentrate on changes in the pressure on the rear centreline, called the 'base pressure'. Thus long structures in free air such as bridge decks, cables, structural elements, pipelines and conveyors, or very tall structures protruding from the ground such as stacks, lamp-posts or flagpoles behave in a different way from long low surface-mounted structures such as boundary walls and terraced houses. Only the former can shed vortices periodically and can be considered to be line-like.

16.1.3 Fineness ratio

Fineness ratio is a term used to describe the proportions of the plan of building structures relative to the wind direction, and is given by the ratio of the depth in the wind direction, D, to the breadth normal to the wind direction, B. It defines the 'bluffness' (§2.2.10) of the structure, with low values for bluff bodies and high values for streamlined bodies. For example, a circular cylinder, which has a fineness ratio $D/B = 1$, is only one member of a family of elliptical cylinders ranging from the extremes of a flat plate aligned normal to the wind ($D/B = 0$) to a flat plate aligned parallel to the wind ($D/B = \infty$).

The variation of drag coefficient of two-dimensional elliptical cylinders in the middle of this range is shown in Figure 16.4 [206]. As expected, the normal pressure drag decreases with increasing fineness as the elliptical cylinder becomes less bluff

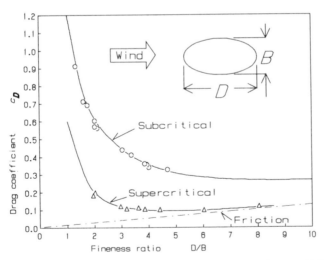

Figure 16.4 Drag of elliptical cylinders (after Hoerner, reference 206)

and more streamlined. However, as the ellipse becomes longer in the wind direction, the area swept by the wind becomes greater and the friction drag increases as shown by the dashed line. Thus there exists an optimum fineness ratio at which the normal pressure and friction components are equal and the total drag is at a minimum, and this is about $D/B = 9$ for subcritical Reynolds numbers and $D/B = 5$ for supercritical.

Recalling the similar variation with slenderness ratio for both circular cylinders and flat plates normal to the wind, it might be expected that cuboidal structures would exhibit a similar decrease in drag with increasing fineness. This is certainly true for fineness ratios greater than unity, but for a long time it was believed that

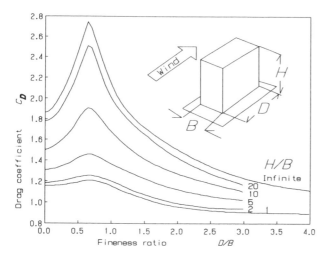

Figure 16.5 Drag of surface-mounted cuboids (after ESDU, reference 208)

the drag coefficient remained relatively constant between $D/B = 0$ (flat plate) and $D/B = 1$ (square section). There were some reports of much higher values at fineness ratios about $D/B = 2/3$ but, as values reported by different laboratories were not consistent, these were taken as anomalous. However, in 1968 Nakaguchi, Hashimoto and Muto[207] reported that, for two-dimensional ($L/B = \infty$) rectangular cylinders in smooth flow, the drag increases rapidly from $D/B = 0$ to a maximum at $D/B = 0.6$, before falling as expected at higher values of D/B. Some of ESDU's current design data for the drag of surface-mounted cuboids[208] is reproduced as Figure 16.5. This shows that the drag maximum is largest when the cuboid is tall and slender, where vortex shedding is expected to be strong. The apparent inconsistency in the early data was due partly to the use of a range of slenderness ratio, but also to the disturbing effect of turbulence. Scales of turbulence comparable to the breadth B are able to 'wrap around' the building causing earlier shedding of weaker vorticies[209]. It is therefore unlikely that the peak drag is attainable on buildings owing to the high levels of turbulence of this scale in the atmospheric boundary layer. A detailed discussion of vortex shedding and the resulting periodic loading is given in Part 3.

16.1.4 Shielding and shelter

It is helpful to distinguish between the twin effects of **shielding** and **shelter** in order to divide the loading data in a structured manner. It is recognised that there is a range of structure for which the two effects overlap and here the division must be arbitrary.

   Shielding is defined here as where part of the structure is protected or 'shielded' by another part **of the same structure**, so reducing the wind loading. This occurs principally with lattice structures: with multiple lattice frames, loads on downwind frames are reduced as momentum is lost by the wind passing through upwind frames. Ancillaries inside lattice towers are similarly shielded and parts of the tower may be shielded by antennae. Shielding can be conveniently accounted for in the design of the structure and so it is included with the discussion and data for the relevant structural type.

Shelter is defined here as where one structure is protected by or 'sheltered' by **another, independent structure.** This may occur for any bluff structure which is part of a group. Shelter effects are difficult to include in design assessments. This is in part because the designer may have no control of the future development of the neighbouring structures and, if the benefit of shelter is exploited, later demolition may leave the structure vulnerable. It is also because the number of possible combinations of sheltering and sheltered structure is so huge that comprehensive rules are impossible to define. However, the problem is approachable when the sheltering and sheltered structures are similar and form part of the same development: e.g. multiple boundary walls, pairs of similar high-rise blocks and housing spaced in a regular array. The problem of shelter is addressed separately in Chapter 19.

16.2 Line-like structures

16.2.1 Definitions

16.2.1.1 Line-like

Line-like structures were defined in §8.2.2 as structures that have their structural and aerodynamic properties concentrated along a line, i.e. very slender structures, which are well represented by the strip model. Many line-like structures, particularly cantilevers such as stacks or gravity-stiffened structures such as suspension bridges, will be dynamic and should be assessed by the appropriate methods in Part 3 of this Guide. In all except the most exceptional cases, the smaller dimension of the line-like structure will be much smaller than the integral length scales of the incident turbulence (§9.4.4), and the loading of the structure can be assessed by the quasi-steady methods of §12.4.1 in terms of the local mean loading coefficient for the section, $\bar{c}$. In particular, the peak loading caused by gusts is well represented by Eqn 12.40. As strip theory only works well for very slender structures, line-like structures will be defined here as for $L/B \geqslant 8$, that is, for the range of slenderness ratio where the drag coefficient in Figure 16.2 is rising towards the maximum for two-dimensional flow.

16.2.1.2 Local coordinates

As line-like structures may be aligned at any angle to the ground plane, it is more convenient to define a local system of coordinates relative to the structure axis, rather than use the convention of azimuth and elevation angles. It is conventional

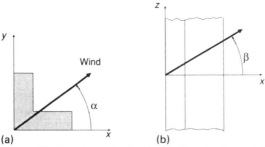

Figure 16.6 Local coordinates for line-like structures: (a) pitch; (b) yaw

to define the x–y plane as the cross-section normal to the long axis of the structure, which becomes the z-axis. Figure 16.6 defines the angles α 'pitch' and β 'yaw' relative to the principal section axes. For vertical cantilever structures pitch, α, corresponds to the conventional wind direction (azimuth angle), θ, and yaw, β, is generally zero.

16.2.1.3 Loading coefficients

The relevant aerodynamic parameters for design are usually the local mean force coefficients, $\bar{c}_F$ (§12.3.6), for the cross-section of the structure:

(a) the force coefficients in **wind axes,**
   $\bar{c}_D$, the local mean drag coefficient, and
   $\bar{c}_L$, the local mean lift coefficient; or
(b) the force coefficients in **body axes,**
   $\bar{c}_{F_x}$, the local mean x-axis force coefficient, and
   $\bar{c}_{F_y}$, the local mean y-axis force coefficient.

In general, the latter are the more convenient to the designer since these are related to the principal axes of the structure. The formal definitions are given by Eqns 12.13 and 12.14 (with the global coefficient C replaced by the local coefficient c), **using the projected area along the relevant axis, A_x and A_y, as the reference areas.** This latter point is most important. It is the policy of the Guide to make the reference area the same as the loaded area, so that the force coefficient is directly equivalent to the global pressure coefficients acting across opposite faces as given by Eqn 12.16. This has the additional advantage of collapsing many data together, reducing the amount that must be presented. (**Note:** the user should be particularly cautious when using external reference sources, many of which (including the ESDU data items) use a fixed reference area which is not always the loaded area.)

Global forces and moments are obtained from the local sectional force coefficients by integration along the long z-axis. Thus the mean global force in the x-axis is obtained from:

$$\overline{F}_x = \int_{z_1}^{z_2} \bar{q}\{z\}\, \bar{c}_{F_x}\{\alpha,\beta,L/B,D/B,Re\}\, B\{z\}\, dz \tag{16.1}$$

As the quasi-steady model is expected to hold well, the corresponding peak force is given by:

$$\hat{F}_x = \int_{z_1}^{z_2} \hat{q}\{z,t\}\, \bar{c}_{F_x}\{\alpha,\beta,L/B,D/B,Re\}\, B\{z\}\, dz \tag{16.2}$$

i.e. with the peak dynamic pressure of a duration, t, given by the TVL-formula of Eqn 12.37 in place of the hourly-mean value. In both Eqns 16.1 and 16.2 the dynamic pressure, q, and the breadth, B (replaced by depth, D, in corresponding equations for y-axis force, F_y) are included within the integral. This is because both may change with position: q with height for all but horizontal structures; and B (or D) if the section dimensions change. The local sectional force coefficient is represented in the equations as a function of orientation, α and β, proportions, L/B (or H/B) and D/B, and Reynolds number, Re. This dependence is discussed below for each cross-section type.

Occasionally, despite the small cross-section size of most line-like structures, it may be convenient to give the distribution of surface pressure in terms of the mean pressure coefficient, $\bar{c}_p$, of Eqn 12.5 (§12.3.4).

16.2.2 Curved sections

16.2.2.1 Circular cylinders

16.2.2.1.1 **Long smooth cylinders.** Being a simple axisymmetric shape, the flow characteristics of very long (two-dimensional) cylinders normal to the flow have been studied in detail by many researchers. Notable early work is that of Fage and Warsap [210] in 1929, which included study of the way surface roughness affects the drag, and the later work at NACA [211]. Unless some asymmetry is introduced into the flow by surface roughness or protrusions, the axial symmetry of the circular cylinder ensures that the net pressure force gives a mean drag, but no mean lift. Hence the mean forces in the principal body axes can be defined in terms of the drag coefficient for flow normal to the cylinder axis, $\bar{c}_{D_o}$ and the pitch angle, α:

$$\bar{c}_{F_x} = \bar{c}_{D_o} \cos\alpha \tag{16.3}$$

$$\bar{c}_{F_y} = \bar{c}_{D_o} \sin\alpha \tag{16.4}$$

The sub-subscript 'o' to $\bar{c}_{D_o}$ is a common convention that indicates the value is for normal flow ($\alpha = \beta = 0°$) and two-dimensional flow ($L/B \to \infty$). There can be no confusion of reference areas in this case, since $B = D$ the diameter of the cylinder.

The instantaneous forces in both x- and y-axes caused by incident large-scale turbulence are obtained from the general quasi-steady model equation, Eqn 12.24, in terms of the instantaneous dynamic pressure and the corresponding mean force coefficients for the instantaneous pitch and yaw angles, α and β. The resulting peak load is expected to occur in the mean wind direction and, since the quasi-steady model holds, the pseudo-steady force coefficient will be identical to the mean. Thus:

$$\hat{F}_x = \tfrac{1}{2}\,\rho\,\hat{V}^2\,\bar{c}_{F_x} = \tfrac{1}{2}\,\rho\,\hat{V}^2\,\bar{c}_{D_o}\cos\alpha \tag{16.5}$$

from Eqn 15.26, and similarly for $\hat{F}_y$. However the periodic forces caused by vortex shedding are not included by the quasi-steady model.

Figure 16.7 shows the data available to Hoerner [206] in 1958 when he collated the results from experiments over a wide range of Reynolds number, Re_D, based

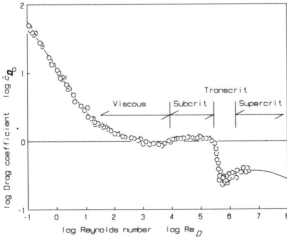

Figure 16.7 Drag of long, smooth, circular cylinders normal to flow (after reference 206)

on the cylinder diameter, on which the flow is strongly dependent. This dependence was described in §2.2.10.2 of Part 1. The drag coefficient is high in the viscous range at very low Re_D, but falls to about $\bar{c}_{D_o} = 1.2$ by $Re_D = 10^3$, but the wind speeds corresponding to this viscous range are far too low for significance in the design of structures. In the subcritical range, $10^4 < Re_D < 4 \times 10^5$, the drag coefficient remains reasonably steady at $\bar{c}_{D_o} = 1.2$, while in the supercritical range, $Re_D > 5 \times 10^6$, the drag coefficient for a smooth cylinder is about $\bar{c}_{D_o} = 0.5$. The drag coefficient varies through the transcritical range, $4 \times 10^5 < Re_D < 5 \times 10^6$, dropping below $\bar{c}_{D_o} = 0.3$ in the middle. This very low drag cannot be exploited in design because it occurs only for a narrow range of Re_D which must be approached via the higher values either side of the range. Only the subcritical and supercritical values are required, the latter being used as an upper bound through the transcritical range. The critical Reynolds number, $Re_{crit} = 4 \times 10^5$, is usually replaced in design guidance by the dimensional product $DV = 6\,m^2/s$, since the density and viscosity of air are taken as constant values (§2.2.6).

16.2.2.1.2 Long rough cylinders.

The roughness of the cylinder surface has little effect on the drag in the subcritical range, but has a dramatic effect on the critical Re_D and the drag at higher values. Fage and Warsap's results[210] for the transcritical range are reproduced in Figure 16.8, replotted to the modern convention. The surface roughness is expressed in terms of the 'equivalent sand

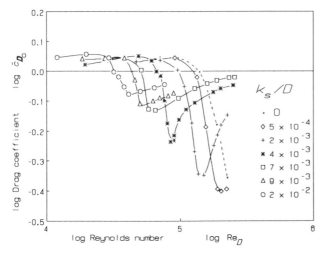

Figure 16.8 Drag of long rough circular cylinder normal to flow (after reference 210)

grain' roughness (§2.2.7.2), k_s. As the surface roughness increases, transition occurs at a lower critical Reynolds number, the drag minimum increases and the drag recovers to a higher supercritical value.

Fage and Warsap experimented by roughening different parts of the cylinder and showed that only the region between 35° and 120° from the front stagnation point, the region between the subcritical and supercritical separation points SP_1 and SP_2 in Figure 2.22, was important. Roughening the front or the back of the cylinder, or both front and back, while leaving this region bare, has no significant effect on the drag. Roughness increases the turbulence in the boundary layer around the cylinder, promoting earlier transition (§2.2.10.2) provided the laminar boundary layer is sufficiently close to natural transition. A similar effect is obtained by small-scale turbulence in the incident wind or by placing a trip wire [210] or other axial protrusion in the critical region (§16.2.2.1.5).

Roughening on only one side of the cylinder can lead to supercritical flow on one side and subcritical flow on the other, producing a net lift force. This is the effect, described in §2.2.10.2, that occurs on stranded cables when the wind is yawed and can lead to galloping oscillations (§8.6.4.2). Design values of local mean drag coefficient, $\bar{c}_{D_o}$, for a range of roughness, together with advice on typical values for k_s, are given in Chapter 20.

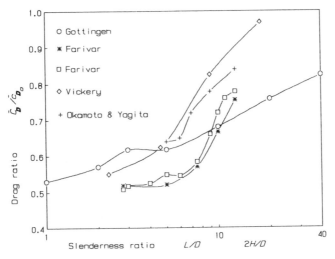

Figure 16.9 Variation of cylinder drag with slenderness at subcritical Re (from reference 212)

16.2.2.1.3 Effect of finite length.

Figure 16.9 shows the effect of finite length on the overall global drag coefficient, $\overline{C}_D$, of a circular cylinder normal to the flow at subcritical Reynolds numbers, from the various sources collated by Basu [212]. The drag coefficient is expressed as the ratio of the two-dimensional subcritical value, $\bar{c}_{D_o} = 1.2$. Below a slenderness ratio of about 8 the drag is low and close to the two-dimensional supercritical value, but at higher slenderness ratios the drag rises towards the two-dimensional subcritical value. The data on cylinders at supercritical Reynolds numbers, from studies in the NPL compressed-air tunnel [213] and measurements at full scale, indicate that the overall global drag does not vary much from the value of $\bar{c}_{D_o} = 0.5$ with varying slenderness, except when the cylinder is very short.

As the critical Reynolds number corresponds to $DV = 6$ m^2/s, or less for rough cylinders, most typical cylindrical structures will be supercritical at the design wind speed. Common exceptions are wires and cables used as guys or for electricity transmission. Figure 16.10 [213] shows how the local drag of a circular stack of slenderness $H/D = 12$ varies with position when it is just supercritical ($Re_D = 2 \times 10^6$). In the middle, $2 < z/D < 10$, the flow is expected to be two-dimensional and the drag is indeed fairly constant. The flow in the top two

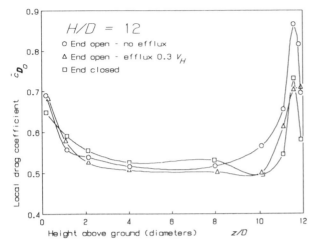

Figure 16.10 Effect of end conditions on local drag coefficient of a supercritical cylinder (from reference 213)

diameters is expected to be three-dimensional, as shown in Figure 8.2 (§8.2.3), and here the drag is locally high. The same effect is found with subcritical cylinders [212, 214, 215, 216], but the increase is not so large. This effect is sensitive to the form of the tip: least with the end closed or open but with an efflux; greatest with the tip open but with no efflux. The critical design condition for a stack is therefore when it is not in use. With a free cylinder, the same condition would occur at the other free end. In this case, however, the 'reflection' in the ground plane at the bottom is not perfect owing to the three-dimensional effects induced by the atmospheric boundary layer (Figure 8.5, §8.3.2.1.1). The local drag rises in the bottom two diameters to a value similar to the closed tip.

16.2.2.1.4 Effect of changes in diameter.   Nearly all the practical design data for circular cylinders are for a constant diameter. When changes of diameter are small, as in Figure 16.11, with the straight taper of (a) or the small steps of (b), then the local drag coefficient at any point can be taken as equivalent to a constant diameter cylinder of the slenderness given by the average diameter. The case when the steps are very large, as in Figure 16.11(c), might represent a communications tower with a platform and cylindrical antenna. By taking each section in turn it is possible to make deductions about the expected behaviour.

1 The bottom cylinder has the ground plane at its base and a much larger cylinder equivalent to a ground plane at its top, so should be treated as two-dimensional, with the slenderness ratio effectively infinite, $L/D = \infty$.
2 The middle cylinder has much smaller cylinders above and below, so has two free ends. Slenderness ratio is therefore $L/D = L_2/D_2$. However, as shown here, it is less than 8, the cylinder is too short to be line-like.
3 The top cylinder has the much larger cylinder at its base, acting as a ground plane, but a free end at the top. The effective slenderness ratio is therefore $L/D = 2H_3/D_3$.

The biggest problem the designer faces with stepped-diameter cylinders is deciding whether a step change in diameter is big or small. Typically, proportions

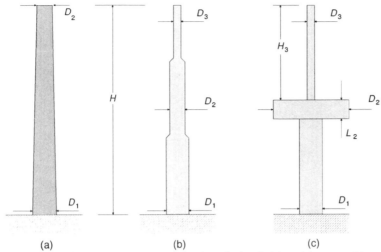

Figure 16.11 Changes in diameter of circular cylinders in (a) straight taper; (b) small steps and (c) large steps

of 1:1.25 (changes of 0.25D) or less can always be taken as being small steps, while proportions of 1:4 can always be taken as being large. In the 'grey area' between these proportions, the designer will always be conservative if he considers both options for each diameter and applies the most onerous result.

The designer must also take care to keep track of the critical Reynolds number, Re_{crit} (or the dimensional product DV), because some parts of the cylinder may be subcritical while others are supercritical, depending on the local diameter and wind speed.

16.2.2.1.5 Effect of axial protrusions.

In attempting to dissect the cumulative effect of surface roughness, Fage and Warsap[210] also studied the effect of small-diameter wires aligned axially along the cylinder at various positions around the cylinder circumference. They found that placing a wire at 65° either side of the front stagnation point and increasing the diameter of the wires produced similar effects, as shown for general surface roughness in Figure 16.8, i.e. promoting transition at lower Re. These wires were all very small compared with the main cylinder diameter, less than 0.3%, so that they affected the transition of the boundary layer on the cylinder without directly affecting the external flow. Large protrusions will affect the flow directly, often causing the flow to separate at the protrusion. In effect the shape of the structure is no longer a pure cylinder. With multiple large protrusions, such as the strakes fitted to stacks to suppress vortex shedding, flow will separate from the strakes at either side to give a wide wake. The supercritical flow regime is suppressed and the drag coefficient remains at about $C_D = 1.2$.

A single axial protrusion on one side of the cylinder affects the flow on that side only, resulting in asymmetric flow and a cross-wind lift force. With a small protrusion, the effect is limited to the range of Reynolds number just below critical, when early transition occurs on one side if the protrusion is in the critical region about 65° from the front stagnation. With a large protrusion, the effect will occur at all Reynolds numbers. Either case may occur in practice for a variety of reasons.

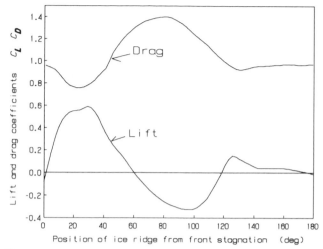

Figure 16.12 Lift and drag of circular cylinder due to ice ridge position (from reference 217)

Large chimney stacks are often fitted with permanent access ladders, the relative location of which will vary with wind direction. Smaller stacks and horizontal cylinders may have prominent weld lines. Stranded power lines are often smoothed by rolling to remove the differential effects of roughness in yawed winds described above and in §2.2.10.2 but, in spite of this, ridges of ice may form in conditions of freezing rain or sleet. Figure 16.12 shows measurements [217] of lift and drag coefficient for a smooth circular cylinder with a simulated ice ridge.

16.2.2.1.6 Effect of ground plane. Flow around long horizontal cylinders, such as pipelines, is influenced by the proximity to the ground. The effect on the pressure distribution around the cylinder and along the ground is shown in Figure 16.13.

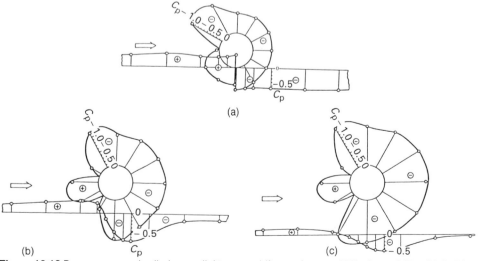

Figure 16.13 Pressure around cylinder parallel to ground (from reference 218) where $G/D = $ (a) 0, (b) 0.4 and (c) 1.0

These data [218] were obtained at Re $= 4.5 \times 10^4$ in the subcritical range. The width of gap between the cylinder and the ground is denoted by G.

In (a) the cylinder is touching the ground, $G/D = 0$, so that all the flow must rise over it. The normal flow separation can still occur on the top surface but the pressure distribution is rotated forward into the rising flow. With no flow underneath the cylinder, the positive pressure lobe at the front and the wake at the back both extend down to the ground. On the ground the pressure is positive in front and negative behind the cylinder, where it remains constant for a long distance downwind, indicating that the wake remains attached to the ground. With only one separating shear layer, regular vortex shedding is suppressed and the drag coefficient falls to $C_D = 0.8$. Owing to the asymmetry of the flow there is a corresponding mean lift coefficient of $C_L = 0.6$. ESDU [219] suggest that the drag falls to $C_D = 0.6$ and the lift rises to $C_L = 0.84$ at supercritical Re.

Moving the cylinder away from the ground allows flow between the bottom of the cylinder and the ground. The position of maximum flow occurs at about $G/D = 0.4$, as shown in (b). Less air rises to go over the cylinder and the front positive lobe and the pressure distribution over the top rotates back slightly. The position of minimum pressure (maximum suction) underneath the cylinder coincides with the narrowest point, but is about the same value as in the upper lobe. On the ground the minimum pressure occurs slightly downwind but now recovers downwind of the cylinder, indicating that the wake is detached from the ground (although widening of the wake by entrainment should make the wake attach to the ground at some distance far downwind). The drag coefficient at $G/D = 0.4$ is about 20% higher than the isolated value, $C_D = 1.45$ for subcritical flow, while the lift coefficient is now quite small, $C_L = 0.15$.

At $G/D = 1$ and greater, the flow is almost restored to the isolated cylinder case, with normal regular vortex shedding, $C_D = 1.2$ and $C_L \simeq 0$ in subcritical flow and the corresponding drag coefficient in supercritical flow.

16.2.2.1.7 Effect of porosity.

Introducing some porosity is an effective way of suppressing vortex shedding, as will be demonstrated in Part 3. Porous cylinders have been suggested as dampers to suppress power-line oscillations [220] and as shrouds around stacks [221, 222, 223, 224]. Whereas introducing porosity to sharp-edged structures generally decreases the drag, introducing porosity in curved structures initially **increases** the drag before, with enough porosity, the structure acts as a lattice and the drag is proportional to the solidity ratio (§16.3). The lower drag of curved structures depends on the flow remaining attached to the surface, longer in supercritical flow than in subcritical flow. Even a small amount of air entering and leaving through a porous cylinder surface upsets this process and the supercritical flow pattern cannot be obtained.

Measurements [220] on a 60% porous cylinder (solidity 0.40) over a range of slenderness $2.7 < L/D < 7.9$ in subcritical flow showed that the drag coefficient based on projected area was consistently about 20% greater than for a solid cylinder of the same slenderness. Extrapolation to $L/D = \infty$ suggests $\bar{c}_{D_o} = 1.44$.

16.2.2.2 Elliptical cylinders

The circular cylinder is just the special case of an elliptical cylinder with $B = D$. The effect of Reynolds number on the drag of a circular cylinder and 2:1 elliptical cylinder at zero yaw is shown in Figure 16.14. Each set of data is a composite from a

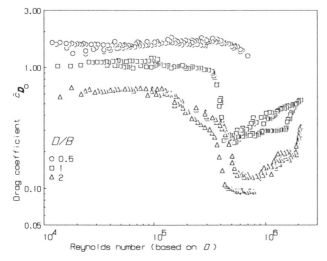

Figure 16.14 Drag of long smooth elliptical cylinders normal to the flow (after reference 211)

range of cylinder sizes of the same proportions, hence the appearance of multiple lines. The local drag coefficient is based on the projected area, i.e. on crosswind breadth B in all cases, corresponding with the standard definition of local x-force coefficient, $\bar{c}_{F_x}$, from Eqn 12.13.

As would be expected, the drag of the elliptical cylinder is greater than the circular cylinder with the major section axis normal to the flow, $D/B = 0.5$, and less than the circular cylinder with the minor section axis normal to the flow, $D/B = 2$. Although this seems obvious at first sight, it should be noted that the variation of projected area is already accounted for in the drag coefficient, so the effect is purely due to the fineness ratio, or the 'streamlining'. Transition also varies with fineness ratio, occurring later for low fineness and earlier, but less abruptly, for high fineness.

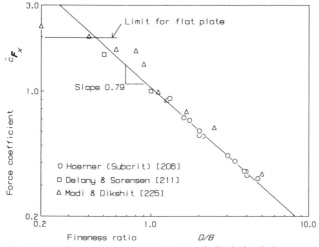

Figure 16.15 Coefficient for x-axis force of elliptical cylinders at zero pitch and yaw angles

Figure 16.15 shows data from various sources [206, 211, 225] for $\overline{c}_{F_x}$ at $\alpha = 0°$, $\beta = 0°$, plotted against fineness ratio D/B, both on logarithmic axes, indicating that $\overline{c}_{F_x}$ tends to stop increasing for $D/B < 0.4$. In the limit as $D/B \to 0$, the elliptical cylinder becomes a flat plate normal to the flow, for which a maximum of $\overline{c}_{F_x} \simeq 2.0$ is expected, as will be seen in §16.2.3.1 later. In the other limit as $D/B \to \infty$, the elliptic cylinder becomes a flat plate parallel to the flow, for which $\overline{c}_{F_x} = 0$ but friction acting on both sides of the plate produces a value for the shear stress coefficient, c_τ (§12.3.5). The data of Figure 16.15 are a reasonable fit to a straight line of slope -0.79 for $D/B > 0.4$, so the effect of fineness ratio can be modelled by the equation:

$$\overline{c}_{F_x}\{D/B\} = \overline{c}_{D_o}(D/B)^{-0.79} \leqslant 2.0 \qquad (16.6)$$

where $\overline{c}_{D_o}$ is the drag coefficient for a circular cylinder and the maximum value of $\overline{c}_{F_x}$ is limited to 2.0 at small D/B.

Body axes are preferred in the Guide, because this gives the relevant structural loads directly. However, some other sources give data in wind axes, for example ESDU, and a practical example of this form of data is useful. Often, particularly with shapes that are more nearly streamlined, the flow characteristics are better described in wind axes. Fine elliptical cylinders behave like inefficient wings and are just such a case. Accordingly, it is convenient here to describe the variation of load with pitch angle in terms of lift and drag coefficients as defined in Eqns 12.11 and 12.12. The variation of drag and lift with pitch angle, α, for flow normal to the cylinder axis, $\beta = 0°$, is shown in Figures 16.16 and 16.17 for fineness ratio and pitch angle over the required range [225, 226, 227]. Here both the coefficients are based on the fixed area normal to the major axis, A_x. This means that the comparisons between different fineness ratios reflect the difference in actual loads for elliptical cylinders with the same size major axis.

In Figure 16.16, the drag varies smoothly from the minimum at $\alpha = 0°$ (minor axis normal to the wind) to the maximum at $\alpha = 90°$ (major axis normal to the wind). The variation is reasonably well fitted by the empirical equation:

$$\overline{c}_{D_o}\{\alpha\} = \overline{c}_{D_o}\{\alpha = 0°\}\cos^2\alpha + \overline{c}_{D_o}\{\alpha = 90°\}\sin^2\alpha \qquad (16.7)$$

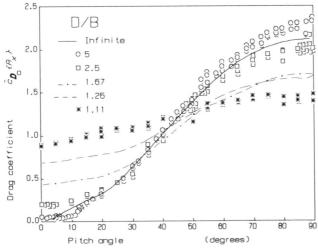

Figure 16.16 Variation of drag with pitch angle of long elliptical cylinders

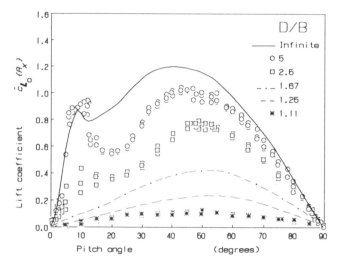

Figure 16.17 Variation of lift with pitch angle of long elliptical cylinders

On the other hand, in Figure 16.17, the variation of lift with pitch angle is more complex. For the bluffer sections with fineness $D/B < 2$ the variation is adequately fitted by the empirical equation:

$$\bar{c}_{L_o}\{\alpha\} = \bar{c}_{L_o}\{\alpha = 45°\}\sin2\alpha = 2\,\bar{c}_{L_o}\{\alpha = 45°\}\sin\alpha\,\cos\alpha \qquad (16.8)$$

except that the actual variation is not truly symmetrical about $\alpha = 45°$ and the maximum value occurs near $\alpha = 50°$. The finer sections with fineness $D/B > 2$ behave like a crude wing at small pitch angles, $\alpha < 15°$. At $D/B = 5$ the effect is quite marked: the lift force increases rapidly with pitch angle up to $\alpha = 10°$, the 'stall' angle at which flow separates from the upper surface, then drops back to follow the variation of Eqn 16.8. This wing-like lift force acts at about $0.25B$ in front of the centre of the cylinder, giving a pitching moment [225, 226] which tends to increase the pitch angle, further increasing the lift. This unstable behaviour makes elliptic cylinders susceptible to galloping and stall flutter, described in §8.6.4.2.

In effect the fine elliptical cylinders show two flow regimes: 'streamlined' at small pitch and 'bluff' at large pitch angles (§2.2.10). An expression for $\bar{c}_{L_o}\{\alpha = 45°\}$ is required to apply Eqn 16.8, which starts at zero for $D/B = 1$ and converges to 1.2 as $D/B \to \infty$. An empirical fit gives a modified equation, 16.8, of the form:

$$\bar{c}_{L_o}\{\alpha\} = 10^{[-1.70 + 1.78\,e]}\sin2\alpha \qquad (16.9)$$

where e is the eccentricity of the ellipse given by:

$$e = [1 - (B/D)^2]^{1/2} \qquad (16.10)$$

The change to 'eccentricity', e, from fineness ratio, D/B, provides the required convergence as $D/B \to \infty$, since $e \to 1$. Being empirical, Eqn 16.9 is not exact and gives a value of 0.02 for $D/B = 1$, instead of zero. It can be used at all values of pitch angle, α, for bluff elliptical cylinders, $D/B < 2$, but should only be used in the range $\alpha > 20°$ for finer sections.

Body axis force coefficients, $\overline{c}_{F_x}$ and $\overline{c}_{F_y}$, for elliptic cylinders should be obtained from the lift and drag coefficients by the normal trigonometrical relationships, i.e. using the complementary equations to Eqns 12.17 and 12.18.

16.2.2.3 Other curved sections

There are far too many possible curved section shapes to give anything other than general advice on design loads. Fortunately, it is found that nearly every shape gives wind loads that are transitional between an elliptical cylinder (§16.2.2.2) and a rectangular section (§16.2.3.2) of the same fineness ratio.

Data for stranded cables indicate that they should be treated as rough circular cylinders. Rules to determine the equivalent roughness, k_s, are given in Chapter 20. Note that the stranding will produce a difference in effective roughness on each side of the cable when the wind direction is yawed.

The transition from circular cylinder to sharp-edged square section cylinder was studied in detail by Delany and Sorensen [211], in terms of the corner radius in the range $0 < r < B$. The effect of Reynolds number on drag when the flow is normal to a face (x-axis force) is shown in Figure 16.18. At a corner radius one-third the

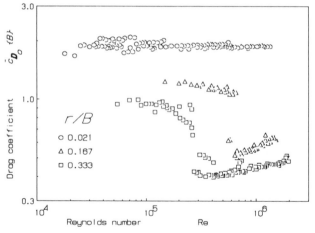

Figure 16.18 Drag of square cylinders with rounded corners and face normal to flow (after reference 211)

breadth, $r/B = 0.333$, or larger, the effect is very similar to the circular cylinder giving a transition from subcritical to supercritical flow in the range $10^4 < Re < 10^5$, with drag coefficients in either case close to the circular cylinder values. As the corners become sharper, transition becomes more abrupt and occurs at higher values of Re, and the drag coefficient for both flow regimes increases. Eventually, the transition no longer occurs, the flow is independent of Reynolds number and the drag coefficient converges to a constant maximum value of $\overline{c}_{D_\theta} = \overline{c}_{F_x}\{\alpha = 0°\} = 2.0$. In this final case, the flow separation points are fixed at the sharp corners of the front face and the front stagnation point is near the centre of the face, as in Figure 2.26(b).

Figure 16.19 shows the corresponding effects when the flow is parallel to the diagonal of the square. In this case the drag coefficient is still based on the side of

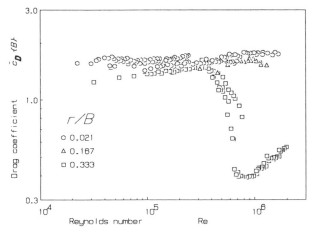

Figure 16.19 Drag of square cylinders with rounded corners, face 45° to flow and zero yaw (after reference 211)

the square, A_x, and **not** the projected area, hence the data of Figures 16.18 and 16.19 can be directly compared on abolute terms. The general variation is similar to before, except that they occur at lower r/B and transition is advanced to higher Re.

16.2.3 Sharp-edged sections

16.2.3.1 Flat plates

Here the interest is only in slender plates that meet the earlier definition of line-like, $L/B \geqslant 8$ (§16.2.1.1). Other plate-like structures, such as free-standing walls, hoardings, fences and signs are discussed later in §16.4.

The loading of slender plates expressed in wind axes [227], i.e. lift and drag, has already been described in the section on elliptical cylinders, §16.2.2.2, since the flat

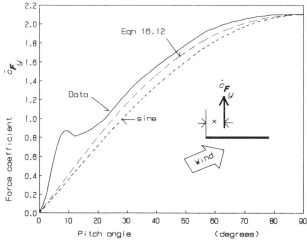

Figure 16.20 Normal force coefficient for flat plate of infinite slenderness at zero yaw (after reference 227)

plate is the limit as $D/B \to \infty$ and $e \to 1$. These data have been converted to body axes in Figure 16.20. To keep the pitch angle, α, compatible with the previous data, the x-axis has been defined as parallel to the surface of the plate.

The normal force caused by pressure acting on the plate can only occur in the y-axis to give $\bar{c}_{F_y}$. Accordingly, we would expect the resultant of the empirical lift and drag equations, Eqns 16.7 and 16.8, to act normal to the plate. This gives an alternative, but more approximate, expression to Eqn 16.8 for the lift coefficient of a flat plate:

$$\bar{c}_{L_o}\{\alpha\} = \bar{c}_{D_o}\{\alpha = 0°\} \sin^2\alpha \tan\alpha \qquad (16.11)$$

Now there are two alternative empirical equations for the normal force coefficient $\bar{c}_{F_y}$:

$$\bar{c}_{F_y}\{\alpha\} = [(2.1 \sin^2\alpha)^2 + (1.2 \sin2\alpha)^2]^{1/2} \qquad (16.12)$$

from Eqns 16.7 and 16.8, or from Eqns 16.7 and 16.11 the simpler equation:

$$\bar{c}_{F_y}\{\alpha\} = \bar{c}_{F_y}\{\alpha = 90°\} \sin\alpha = 2.1 \sin\alpha \qquad (16.13)$$

It is clear from the comparison in Figure 16.20 that Eqn 16.13, indicated by 'sine', underestimates more than Eqn 16.12. The 'wing effect' at small pitch angles (§16.2.2.2) is not represented at all by either equation.

The position of action of the normal force, measured as x from the leading edge, varies with pitch angle, α. At small pitch angles, $\alpha < 10°$, where the 'wing effect' is dominant, the normal force acts at $x = 0.25D$. At higher pitch angles when the flow is fully 'stalled', the normal force acts near the middle of the plate [228], from $x = 0.4D$ at $\alpha = 20°$ to $x = 0.5D$ at $\alpha = 90°$.

In addition to the normal pressure force, friction acting on both sides of the plate produces a small value for the x-axis force equal to the shear stress coefficient, $\bar{c}_{F_x} = \bar{c}_\tau$ (§12.3.5), but this problem is addressed later in §16.4.4.8.

16.2.3.2 Rectangular sections

The drag of rectangular prisms was given in Figure 16.5 and discussed in §16.1.3 'Fineness ratio'. The maximum force coefficients occur when the flow is flow normal to either face, i.e. $\bar{c}_{F_x}\{\alpha = 0°\}$ and $\bar{c}_{F_y}\{\alpha = 90°\}$. For rectangular sections of infinite slenderness, these local sectional coefficients are equal to the global drag coefficient, $\bar{C}_D$, for $D/B = \infty$, given by the upper curve in Figure 16.5.

The peak at the critical fineness ratio of $D/B = 0.6$ is due to vortex shedding as described in §16.1.3, earlier. Even higher values of drag coefficient can be obtained in very smooth and uniform flow, when the vortex shedding is stronger. However, vortex shedding is also easily disrupted by various means, including three-dimensionality (low slenderness) and turbulence (see §2.2.10.4 and Part 3), resulting in reduced drag. Hence the effect of slenderness ratio, H/B, is greater on the peak drag at $D/B = 0.6$ than at the less critical values. Figure 16.21 shows the effect of turbulence intensity, S_u, on the drag [229] for fineness ratios between the critical value and unity. The high values of Figure 16.5 are not sustained as the turbulence intensity rises to values typical of the atmosphere (see Table 9.10). The force coefficients are less at other pitch angles. As $B/D \to 0$ and $B/D \to \infty$ the rectangular section converges to a flat plate and the normal force is given by Figure 16.20. The least plate-like rectangular shape is the square, $D/B = 1$. Figure 16.22 shows the variation of $\bar{c}_{F_x}\{\alpha\}$ of a square-section cylinder [230] with pitch angle, α, for all possible angles. (Data are also shown for an equal-angle structural section,

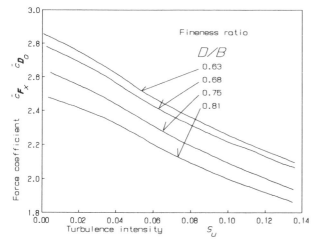

Figure 16.21 Effect of turbulence intensity on maximum drag of rectangular cylinders (from reference 229)

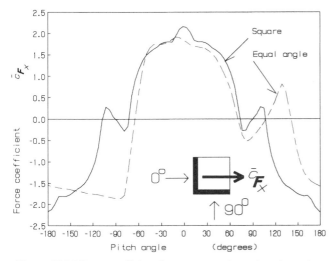

Figure 16.22 Force coefficient for square- and equal-angle sections of infinite slenderness

see below.) The variation is **very** approximately sinusoidal, i.e. with four antisymmetric lobes, each 90° wide, corresponding to the two degrees of symmetry of the square. Owing to this antisymmetry, the other body-axis force is given by:

$$\bar{c}_{F_y}\{\alpha\} = \bar{c}_{F_x}\{90° - \alpha\} \qquad (16.14)$$

This antisymmetry and the roughly sinusoidal form implies that the resultant drag is approximately constant at $\bar{c}_D \simeq 2$ and the lift coefficient is small for all pitch angles.

Compare the lobe of $\bar{c}_{F_x}\{0° < \alpha < 90°\}$ for the flat plate in Figure 16.20 with the lobe of $\bar{c}_{F_x}\{-90° < \alpha < 0°\}$ for the square. The general shape is similar: clearly from the doubly symmetric form of the square, each body-axis force is expected to be a maximum when the flow is normal to the relevant face and zero when the flow is parallel. However, as the square section is bluff, the 'lift effect' of the flat plate does not occur and instead there is a local reversal of the force. Although small, the effect is quite significant, since the reversal of slope makes dynamic structures susceptible to galloping oscillations (§10.6.2.1.3).

Accounting for the combined effects of fineness ratio, turbulence, pitch and yaw (§16.2.4) angles is usually regarded as too complex for codes of practice, so that the maximum value of $\bar{c}_{F_x} = 2.0$ is often adopted as a general value for all 'flat-faced', sharp-edged sections.

16.2.3.3 Polygonal sections

Surprisingly, there is a dearth of reliable data on most of the polygonal shapes, except for the infinitely-slender isosceles triangular wedge, and this has only been studied for the 'face on' and 'corner on' cases [211,231]. Defining the coordinate axes with the origin at the apex and the x-axis passing through the centre of the base (area A_x) gives the following force coefficients. For the 'corner on' case, $\alpha = 0°$, $\bar{c}_D = \bar{c}_{F_x}$ and varies with the included angle, θ, of the apex as given in Table 16.1. For the 'face on' case, $\alpha = 180°$, $\bar{c}_D = -\bar{c}_{F_x} = 2.0$ for all included angles, θ.

Table 16.1   Drag of isosceles wedge of infinite slenderness with apex facing the flow

Included angle of apex	θ	15°	20°	30°	45°	60°
Drag coefficient	$\bar{c}_D$	1.0	1.1	1.3	1.6	1.7

There are no reliable data for 5- to 7-sided polygons. The drag of an octagonal cylinder may be taken as $\bar{c}_D = 1.4$. Polygonal cylinders with 12 or more sides may be treated as the equivalent elliptical or circular cylinder, **including** the effects of Reynolds number.

16.2.3.4 Structural sections

In contrast to the dearth of data for polygonal cylinders, the common structural steelwork sections have been extensively studied, starting with the work of Prandtl and Betz [232] and others [43] at Göttingen in the 1920s and continuing up to the present day [233,234,235,236,237,238,239]. The available data have been collated by ESDU into data item 82007 [240], which provides a much more comprehensive range of data than can be included here.

Figure 16.22 shows Modi and Slater's data [233] of x-force coefficient, $\bar{c}_{F_x}$, for an equal-angle section of infinite slenderness compared with the earlier square section cylinder data for all pitch angles and zero yaw. In both cases, the corresponding y-force coefficient, $\bar{c}_{F_y}$, is given through antisymmetry by Eqn 16.14. The general shape of both curves is remarkably similar, each with a maximum value of $\bar{c}_{F_x} \simeq 2$. While there is exact symmetry about $\alpha = 0°$ for the square, reflecting the symmetry of the section, this does not occur for the angle section. Local force reversal occurs only once in the range of pitch, instead of the two symmetrical cases of the square. Here the flat face of the angle is upwind and the flow around the front is similar to that of the square. The effect is increased and is also displaced from around $\alpha = 90°$ to around $\alpha = 110°$.

Similarly, the corresponding effect on $\bar{c}_{F_y}$ occurs only around $\alpha = -20°$, and this is best seen in Figure 16.23 where both $\bar{c}_{F_x}$ and $\bar{c}_{F_y}$ are plotted together for equal angle sections and for unequal 2:1 angle sections [232, 234, 235, 236, 238]. The similarity between these two pairs of curves for angle sections of very different proportions demonstrates that the definition of the force coefficients using the relevant loaded areas, A_x and A_y, in Eqns 12.13 and 12.14 really does collapse the data in practical cases.

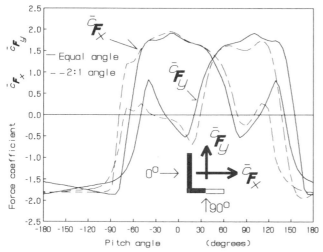

Figure 16.23 Force coefficients for equal- and unequal-angle sections of infinite slenderness

At the opposing pitch angles, a limb of the angle section points into wind making the flow quite unlike that around the square. This is reflected in the difference between the loads in Figure 16.22, particularly in the region $-180° < \alpha < -90°$. Instead of the approximately sinusiodal variation, $\bar{c}_{F_x}$ changes rapidly to a peak (minimum) of $\bar{c}_{F_x} = -2.0$ around $\alpha = -90°$ which reduces only slowly to $\bar{c}_{F_x} = -1.6$ at $\alpha = -180°$, where the peak occurs for the square. In this range, a large bubble of stagnant air near the total pressure (§2.2.2) is trapped in the re-entrant angle. Figure 16.23 shows that $\bar{c}_{F_x}$ and $\bar{c}_{F_y}$ are simultaneously near the peak value over most of this range. At $\alpha = -135°$, when the flow is directly into the angle, $\bar{c}_{F_x} = \bar{c}_{F_y} = -1.6$, giving a resultant drag of $\bar{c}_D = 2.3$, i.e. higher than the 'face on' case.

Body axis force coefficients $\bar{c}_{F_x}$ and $\bar{c}_{F_y}$ are presented in Figures 16.24 and 16.25, respectively, for a number of other common structural sections. These include a 2:1 channel section [238, 239], 2:1 I-beam and 1:1 H-column [232, 237, 238], 1:1 T-section [241], and a 1:1 X-section [238, 241] formed from two structural angles (with a small gap between). Given the alignment of the sections shown in the key to each figure, the definition of the force axes is the same in all cases to the key in Figure 16.23. Taking $\bar{c}_{F_x}$ in Figure 16.24 first, all the sections are symmetric about the x-axis so that the variation of $\bar{c}_{F_x}$ is expected to be symmetric about $\alpha = 0°$ in each case and about $\bar{c}_{F_x} = 0$ for all except the T-section. The channel, I-beam and H-column are all very similar, as would be expected from their similarity of form, and display the same general characteristics as the square section. The T-section shows the effect of the re-entrant angle on both sides of $\alpha = 180°$, instead of just one side for the structural angle. The characteristics of the X-section are more

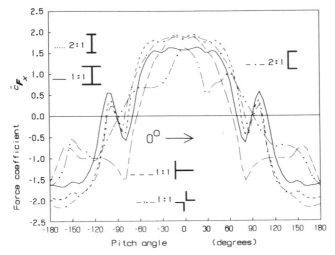

Figure 16.24 x-force coefficients for various structural sections of infinite slenderness

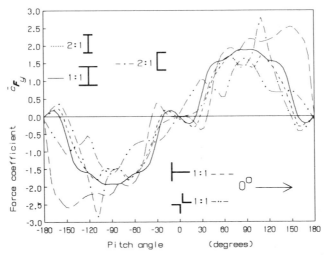

Figure 16.25 y-force coefficients for various structural sections of infinite slenderness

complex, with four re-entrant angles the variation is expected to be doubly symmetric, however this is only approximately true owing to the small gap between the two angles forming the X which allows flow to leak from the stagnant front bubble through to the wake (see §16.2.3.6, below).

Figure 16.25 shows the corresponding variation of $\bar{c}_{F_y}$, which is expected to be antisymmetric about $\alpha = 0°$ and symmetric about $\bar{c}_{F_y} = 0$ for all cases. Now the channel differs from the I-beam and H-column in that the displacement of the web makes the re-entrant region bigger and on only one side, while additionally both flanges of the channel are exposed to the wind, giving a peak of $\bar{c}_{F_y} = \pm 2.9$. The T-section differs significantly from the other sections when the 'leg' of the T points upwind: at $\alpha = \pm 180° \bar{c}_{F_y} = 0$ as expected by symmetry, but for pitch angles either side the 'leg' generates lift like the flat plate which, together with the effect of the re-entrant angle, produces a peak of $\bar{c}_{F_y} = \pm 2.6$.

Accounting for all the variety of possible section shapes and yaw angles is regarded as too complex for codes of practice. The most that codes usually include is a table of force coefficients for wind aligned to the principal axes, i.e. $\alpha = 0°$, $\pm 90°$, $\pm 180°$, plus additional values for intermediate 'skewed' directions when these give larger resultant loads, e.g. at $\alpha = -135°$ for the structural angle section.

16.2.3.5 Effect of length

Figure 16.26 shows the effect of finite length on the overall global drag coefficient, $\bar{C}_D$, of various sharp-edged structural sections of finite length at zero yaw angle, from the collation by ESDU [208, 240] of the available data. The drag coefficient is expressed as the ratio of the corresponding two-dimensional value, $\bar{c}_{D_0}$. The form is similar to the experimental data for circular cylinders in Figure 16.9, but appears more orderly. This is because the data in Figure 16.26 have been averaged from several data sets, whereas the data in Figure 16.9 are raw experimental data.

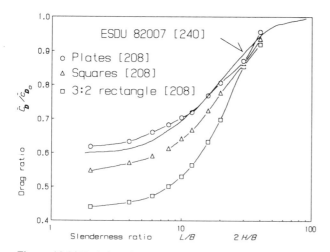

Figure 16.26 Variation of drag of sharp-edged sections with slenderness ratio

The curve marked 'ESDU 82007' is recommended by ESDU for general use. It fits the data for flat plates quite well and is a reasonable upper bound for square-section cylinders. However, the curve for the critical high-drag rectangular section is well below the recommended curve because the drag of the critical section is more sensitive, as indicated earlier (§16.2.3.2), but when the typical code value is used, the recommended curve fits better over the range of practical slenderness ratios.

16.2.3.6 Effect of porosity

Porosity generally always decreases the loads on sharp-edged sections, unless the fineness ratio is large enough to give reattachment and a narrow wake. In this case, porosity in the sides may lead to separation, a wider wake and increased drag. Air injected into the wake increases the momentum in the wake and suppresses vortex shedding. This injection is called 'base bleed' and, for research purposes, is usually forced mechanically so it can be controlled[242]. Passive injection can be introduced in practical structures by a duct connecting front and rear faces[243].

Porosity is usually considered only as a means of reducing vortex shedding[223], where as little as 10% porosity is very effective (as described in Part 3), and the accompanying reduction in drag is seldom stressed. Figure 16.27 shows the

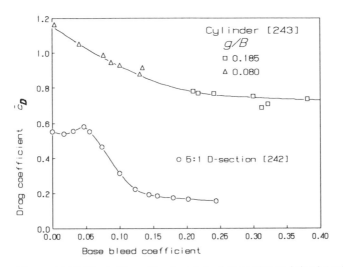

Figure 16.27 Effect of base bleed on drag on D-section and circular cylinder

reduction in drag of a fine $(D/B = 5)$ D-section[242] and a circular cylinder[243] as base bleed is increased. The base bleed coefficient is the ratio of the injected flow to the incident flow on the projected area of the body, A_x. Injection was passive in the case of the circular cylinder by means of a slot of width, g, cut across the diameter. The base bleed was adjusted by rotating the cylinder, with the greatest bleed occuring when the slot was aligned parallel with the flow, between the front stagnation point (§2.2.10.2) and the centre of the wake.

Base bleed is expected to be effective for all bluff shapes in the range of fineness $1 < D/B < 5$ covered by the examples. Neither of the examples in Figure 16.27 is a sharp-edged body and both would be difficult to apply in practice. The slot in the circular cylinder would work only if the cylinder were horizontal since, for vertical cylinders, it would be properly aligned only in one wind direction, while overall porosity would increase the drag, as indicated in §16.2.2.1.7. Holes are effective in the webs of channel or I-beams, since they would reduce the load in the weak bending axis. A value of $g/B = 0.1$, i.e. 10% porosity is about the optimum. Vortex induced oscillations are generally a fatigue or serviceability problem so that a temporary failure of suppression, perhaps by the porosity being blocked by ice, may not be serious. However, if account is taken of the reduction in drag for the ultimate load condition, loss of this effect by blockage could be catastrophic, so that reliance on base bleed to control the drag is not recommended.

16.2.4 Effect of yaw angle

16.2.4.1 Prediction from potential flow theory

Flows around streamlined bodies, when there is no flow separation and wake, can be reasonably well represented by potential flow theory. This allows the incident wind speed vector to be resolved into its components along the three orthogonal body axes, u, v and w :

$$u = V \cos\alpha \cos\beta \tag{16.15}$$
$$v = V \sin\alpha \cos\beta \tag{16.16}$$
$$w = V \sin\beta \tag{16.17}$$

The effect of each component can be assessed separately, and then their combined effect found by superposition, as in §2.2.8.4 and §2.2.8.5. The situation is analogous to the superposition of deflections due to a number of loads acting on a structure, but this is valid only as long as the structure remains elastic. In potential flow, superposition is invalid whenever the flow separates from the body. Thus it is **always invalid in bluff body flows.** Nevertheless, potential flow theory is sometimes used to justify a 'cos² law' for the effect of yaw, so it is worth pursuing this point further.

Assuming that the local force coefficients for flow parallel to each of the three body axes are known, superposition yields equations for the three body-axis local force coefficients for any arbitrary wind direction:

$$\bar{c}_{F_x}\{\alpha, \beta\} = \cos^2\alpha \cos^2\beta\, \bar{c}_{F_x}\{\alpha = 0°, \beta = 0°\} \tag{16.18}$$
$$\bar{c}_{F_y}\{\alpha, \beta\} = \sin^2\alpha \cos^2\beta\, \bar{c}_{F_y}\{\alpha = 90°, \beta = 0°\} \tag{16.19}$$
$$\bar{c}_{F_z}\{\alpha, \beta\} = \sin^2\beta\, \bar{c}_{F_z}\{\alpha = 0°, \beta = 90°\} \tag{16.20}$$

Data exist to test these predictions for the special cases of any pitch at zero yaw and any yaw at zero pitch. For example, at zero pitch angle, $\alpha = 0°$, Eqn 16.18 for the x-force at any given yaw angle, β, simplifies to:

$$\bar{c}_{F_x}\{\alpha = 0°, \beta\} = \bar{c}_{D_o} \cos^2\alpha \tag{16.21}$$

which is the 'cos² law' referred to above. The same argument leads to a 'sin² law' for the effect of pitch angle, α, on $\bar{c}_{F_y}$ from Eqn 16.19, but it is was seen in Figure 16.20 that the variation of $\bar{c}_{F_y}$ of a flat plate with pitch angle is close to a sine.

There is no match with the potential flow predictions for pitch, so why is Eqn 16.21 sometimes promoted for yaw? The answer is that the yaw response of circular cylinders appear to follow Eqn 16.21 quite well. Hoerner[206] describes its use for airships at small angles of incidence, i.e. yaw angle $\beta \rightarrow 90°$. Ramberg[244] concentrated on the differences between these predictions and reality, principally considering the vortex-shedding characteristics, but also considering Reynolds number effects and the base pressure (§16.1.2) which determines the drag of the cylinder. Ramberg's data for the effect of yaw on the drag of circular cylinders are plotted in Figure 16.28, where the potential flow model is represented by the curve marked 'cos²', i.e. by Eqn 16.21. This forms a lower bound to the data, but the fit is quite good, contrary to expectation for bluff bodies.

16.2.4.2 Wake momentum loss model

As force is equal to rate of change of momentum, the drag force on a bluff body can be deduced from the change of momentum in the wake. Consider an infinitely long

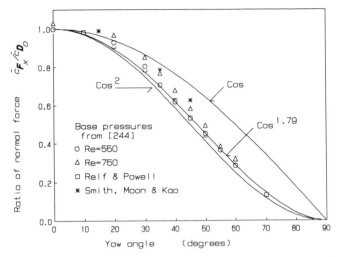

Figure 16.28 Effect of yaw angle on the normal force on a circular cylinder

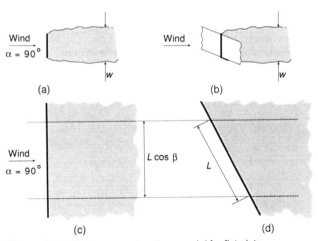

Figure 16.29 Wake momentum loss model for flat plate

flat plate aligned normal to the flow in elevation, as in Figure 16.29(a), for which $\bar{c}_{F_y}\{\alpha = 90°, \beta = 0°\} = \bar{c}_{D_0}$, using the axis convention used for flat plates in §16.2.3.1 and Figure 16.20. This produces a wake of nominal width, w, between the two shear layers which spring from the sharp edges of the plate. Assume that yawing the plate does not change the characteristics of these shear layers and the wake, as in Figure 16.29(b), the width w is unchanged and the momentum loss in the wake is the same as in (a). Now consider the views in plan, Figures 16.29(c) and (d). The momentum lost per unit width across the flow will be the same, so that the force on the length, L, of the yawed plate is the same as that on the shorter length, $L \cos\beta$, of the normal plate. This predicts the force on the plate to be:

$$\bar{c}_{F_y}\{\alpha = 90°, \beta\} = \bar{c}_{D_0} \cos\beta \tag{16.22}$$

which is a simple cosine law.

Figure 16.30 Effect on yaw on effective fineness ration of cylinder: (a) no yaw; (b) yawed

It will be shown in §16.4 'Plate-like structures' that the normal force on walls, hoardings and other plate-like structures follow the prediction of Eqn 16.22 closely. However, this is not true of the circular cylinder, as shown in Figure 16.28, where the data do not fit the line marked 'Cos'. Consider the assumption that the wake is unchanged by the yaw angle. This is reasonable for the flat plate of Figure 16.29 because the yawed flat plate remains flat in cross-section. As the cylinder is yawed, the cross-section in the flow direction is circular at zero yaw, Figure 16.30(a), but becomes more elliptical with increasing yaw, Figure 16.30(b). The effective fineness ratio, B/D, for a yawed circular cylinder becomes:

$$D/B = 1 / \cos\beta \tag{16.23}$$

The drag coefficient of an ellipse, $\bar{c}_{D_o}\{D/B\}$, reduces with fineness ratio, as given by Eqn 16.6. Substituting Eqn 16.6 into the wake momentum model equation, Eqn 16.22, gives:

$$\bar{c}_{F_y}\{\alpha = 90°, \beta\} = \bar{c}_{D_o} \cos^{1.79}\beta \tag{16.24}$$

for the yawed circular cylinder. This is plotted in Figure 16.28 as the curve marked 'cos$^{1.79}$' and is a better fit to the data than the nearby 'cos^2' curve from the potential flow prediction. It is a pure coincidence that the potential flow prediction gives a reasonable fit in this particular case.

For the particular case of the circular cylinder, the axisymmetry allows Eqn 16.24 to be combined with Eqns 16.3 and 16.4 (§16.2.2.1.1) to give the local body axis force coefficients for any orientation:

$$\bar{c}_{F_x} = \bar{c}_{D_o} \cos\alpha \cos^{1.79}\beta \tag{16.25}$$

$$\bar{c}_{F_y} = \bar{c}_{D_o} \sin\alpha \cos^{1.79}\beta \tag{16.26}$$

It is anticipated that the same effect should occur with elliptical cylinders and Eqns 16.24, 16.25 and 16.26 should apply, provided the fineness ratio for normal flow is not less than about $D/B = 0.4$, which is the limit for Eqn 16.6 (see Figure 16.15, §16.2.2.2), but no data have been found to confirm this view. Similarly, it is expected that the general model equation, Eqn 16.22, will apply to sharp-edged sections, and the flat-plate data in §16.4 'Plate-like structures', confirms this particular case gives a simple cosine variation. It will be seen in §16.3 'Lattice structures' that adoption of the simple cosine variation assists the implementation of design load assessments for lattice structures.

16.2.4.3 Finite length structures

When long, but finite length structures are yawed so that the free end points upwind, some peculiar phenomena occur in the region of the free end which lead to asymmetrical flow conditions and large cross-wind forces near the free end. This has been studied on many occasions because the effect is detrimental to the control of guided missiles and the stability of high-speed aircraft and trains[245].

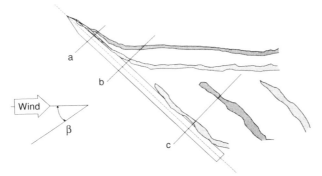

Figure 16.31 Flow around yawed circular cylinder of finite length (schematic)

The flow conditions near the free end of a yawed circular cylinder with a tapered free end are represented in Figure 16.31. The shaded regions represent vortices with the two opposite signs of the vorticity denoted by the two densities of shading. It was suggested by Allen and Perkins[246] in 1951 that the development of the flow from the free end was analogous to the development of vortex shedding from a cylinder started suddenly from rest, with the distance from the free end being linearly related to the time elapsed from starting the flow. As the suddenly started cylinder passes through all the values of Reynolds number from zero to the final steady value, the distance from the end is analogous to increasing Re as described in §2.2.10.2 and §2.2.10.4.

Near the tip, at 'a' in Figure 16.31, there is a symmetrical pair of vorticies of opposite sign as in Figure 2.20 for low Re. At 'b', further from the end, one of the two vortices is bigger than the other and the flow is asymmetrical, as in Figure 2.27. In smooth flow and with a nearly perfectly axisymmetric body, the sign of this asymmetry is random but usually remains stable once established. Disturbances in the incident wind may cause the flow to oscillate randomly between the two signs[247], giving a switching flow (§8.4.2.4). However, any small asymmetry in the body near the tip may cause the asymmetry of the flow to lock to a particular sign. From the tip up to 'b', these vortices remain fixed or 'bound' to the cylinder. Further from the end these vortices stream downwind and by 'c', regular shedding of vortices of alternate sign is established, as in Figures 2.28 and 2.29.

The corresponding distributions of mean local forces along the cylinder are shown in Figure 16.32 for the inwind x-axis force, $\bar{c}_{F_x}$, and Figure 16.33 for the cross-wind y-axis force, $\bar{c}_{F_y}$, at zero pitch, from the data of Mair and Stewart[245]. The mean inwind x-force is initially high, corresponding to the symmetric pair of vortices, and rises to a second peak corresponding to the maximum asymmetry. Similarly, the mean cross-wind y-force is initially zero due to the initial symmetry, rises to a maximum at the position of maximum asymmetry. After the larger first vortex is shed, the second vortex of opposite sign dominates and the y-force changes sign to form a second peak. (This occurs off the region of measurements in Figure 16.33, but the change of sign does appear in the data for $\beta = 65°$.) For the example data of Figure 16.33 the sign of $\bar{c}_{F_y}$ was forced negative by adding a small excrescence to the tip[247], so the dotted curve has been added to show what the data for $\beta = 65°$ would have been if the opposite sign had been forced. By position 'c', $\bar{c}_{F_x}$ levels off at a constant value and $\bar{c}_{F_y}$ falls to zero: but now there are fluctuating components, c'_{F_x} and c'_{F_y} from the regular vortex shedding.

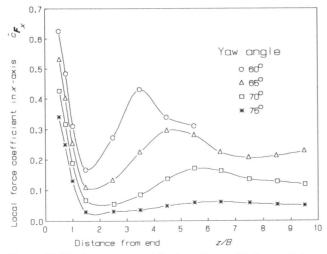

Figure 16.32 x-force on yawed circular cylinder of finite length (from reference 245)

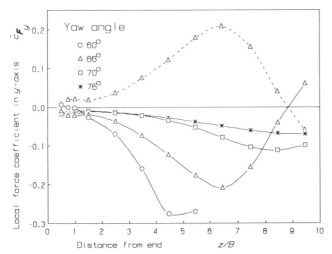

Figure 16.33 y-force on yawed circular cylinder of finite length (from reference 245)

The development of the flow is sensitive to the tip conditions, with the effects being stronger and more stable when the tip is tapered than when it is blunt. The data in Figures 16.32 and 16.33 are from a circular cylinder with a rounded end [245]. Although these are interesting flow phenomena, they will be rare in building applications and no design data are given in Chapter 20. If this problem is thought to be a significant possibility, the designer should seek expert advice.

16.3 Lattice structures

16.3.1 Introduction

Solidity ratio, s, was defined in Eqn 13.17 of §13.5.4.2 as the ratio of the total projected area or 'shadow area' of the individual members to the projected area of the outside shape or 'envelope' of the structure. Solidity ratio must lie in the range $0 \leqslant s \leqslant 1$, with $s = 1$ representing a solid body. The effects of solidity ratio divide approximately into four overlapping ranges:

1    $0 \leqslant s \leqslant 0.2$ – lattice of bluff members with independent wakes.
2    $0.1 \leqslant s \leqslant 0.6$ – lattice of bluff members with interacting wakes.
3    $0.4 \leqslant s \leqslant 0.9$ – porous body with interacting jets.
4    $0.8 \leqslant s \leqslant 1.0$ – porous body with base bleed.

Each range corresponds to different aerodynamic characteristics and is assessed in a different manner, but the change with solidity is so gradual that either of the adjacent methods can be used in the wide overlap range.

This gradual change with solidity is reflected by a change in the physical form of the lattice structures, as illustrated by Figure 16.34. Here five beams with the same envelope dimensions: overall length, L, and breadth, B, are represented. When the solidity ratio is small, the individual lattice elements must be slender as in (a). As the solidity increases, the lattice elements become less slender, through (b) to (c), although the form is still clearly a lattice of elements. However, at solidity ratios above $s = 0.5$, the form becomes much more like a plate perforated by holes, as in (d). Finally, the solidity becomes unity and the structure is solid, as in (e).

In the first range, $s \leqslant 0.2$, the flow divergence is small and both the lattice-plate (§8.3.1) and quasi-steady (§12.4.1) models are expected to apply well. The

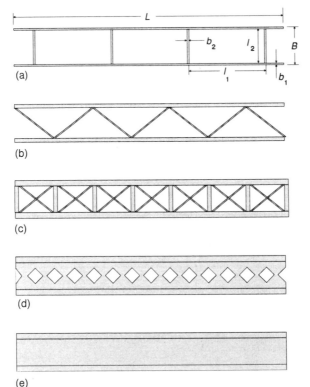

(a)

(b)

(c)

(d)

(e)

Figure 16.34 Beams of various solidities: (a) Vierendeel girder, $s = 0.122$; (b) Warren truss; $s = 0.244$; (c) tension cross-brace, $s = 0.460$; (d) castellated beam, $s = 0.84$; (e) solid beam, $s = 1.00$

elements are sufficiently separated to be independent of each other and the loads on the whole structure are obtained by summation of the element loads. Thus:

$$F_x = \Sigma \, (q\{z\} \, l \, b \, \bar{c}_{F_x}\{l/b\}) \tag{16.27}$$

gives the overall x-axis load directly. The local force coefficient, $\bar{c}_{F_x}\{l/b\}$, depends on the slenderness ratio of the particular element, set by the length, l, and breadth, b, as indicated in Figure 16.34(a). The proportions of the envelope, set by L and B, are unimportant (except in determining the 'wind shadow' cast downwind and hence the degree of shielding offered to downwind elements – see §16.3.5 and §16.3.6 later). The local dynamic pressure, q, remains inside the summation to allow for variation of incident wind speed over the lattice, principally with height, z, as indicated. Alternatively, an overall force coefficient can be defined by:

$$C_{F_x} = \Sigma \, (\phi_q\{z\} \, l \, b \, \bar{c}_{F_x}\{l/b\}) \, / \, A_x \tag{16.28}$$

where $\phi_q\{z\}$ is an influence function[*] defining the weighting of the dynamic pressure over the lattice in the manner of Eqn 12.4 (§12.3.2). Typically, this influence function will be given by:

$$\phi_q\{z\} = (z/H)^{2\alpha} \tag{16.29}$$

when the reference dynamic pressure is taken at the top of the lattice, $z = H$, and where α is the exponent of the power-law model for the wind speed profile (§7.2.1.3.1, §8.6.2.1). With most three-dimensional lattices in this range, shielding of downwind elements will not occur unless they happen to fall in one of the individual wakes and, even then, a small change of wind direction will remove the effect. However, shielding does becomes significant when the lattice is very deep and composed of many elements.

In the second range, $0.1 \leqslant s \leqslant 0.6$, the drag of each element is still small and depends on its proportions. The lattice-plate (§8.3.1) and quasi-steady (§12.4.1) models are still expected to apply well, but less well towards the upper limit. However, the total drag gives sufficient divergence to drive a significant proportion of the flow around, instead of through, the lattice. This gives a region of 'wind shadow', defined approximately by the projection of the envelope area in the wind direction, in which the momentum of the wind is reduced, providing significant shielding for downwind elements of three-dimensional lattices. The greater number of elements and the variety of their proportions makes the summation of Eqns 16.27 and 16.28 increasingly complex, so that an overall approach based on the solidity ratio is usually used. The force coefficients remain based on the solid area, which will be $A_x = s_x LB$ for the x direction. This is the range for which lattice plate theories and empirical design methods are most appropriate, and to which the majority of this section is devoted.

In the third range, $0.4 \leqslant s \leqslant 0.9$, dependence on proportions is transitional between the elements and the overall envelope. At the more porous end of the range, the elements dominate and the lattice-plate (§8.3.1) and quasi-steady

---

[*] Influence functions were introduced in §8.6.2.1 of Part 1. They are employed whenever factors are required to adjust an action for the influence of a parameter. The general function, denoted by Φ, is the ratio of the specific value to the reference value: $\Phi_q\{z\} = q\{z\}/q\{z_{ref}\}$, so may take any value. The corresponding influence coefficient, denoted by ϕ, is the ratio of the specific value to the *maximum* (positive or negative) value and must therefore always lie in the range $-1 \leqslant \phi \leqslant 1$. Eqns 16.28 and 16.29 use the coefficient, ϕ, since $z = H$ is the top of the lattice and here the dynamic pressure is greatest.

(§12.4.1) models are still reasonably applicable. As the solidity increases through the range, the elemental wakes merge to form a single wake, so that the overall envelope dimensions dominate at the more solid end of the range. Although the empirical design methods for the second range can be extended to cover this range adequately, the design methods for solid structures of the same overall dimensions often give better representation, particularly at the more solid end of the range. Here the quasi-steady model does not hold well and must be replaced by one of the bluff-body models: the peak-factor method (§12.4.2), the quantile-level method (§12.4.3), or the extreme-value method (§12.4.4). (In this Guide the pseudo-steady format of the fully probabilistic extreme-value method described in Chapter 15 will be used.) Sometimes these methods are implemented in terms of force coefficients based on the envelope area and are therefore incompatible with the definition used in this Guide (§12.3.6). The only advantage of such 'envelope coefficients' is that they display the effects of flow parameters in absolute terms, which can be very valuable in discussion. To avoid any confusion in the definition of coefficients, the Guide will maintain the convention that force coefficients C_F always refer to the area of solid members, maintaining the equivalence between force and pressure coefficients defined in Eqn 12.16. Thus 'envelope coefficients', which are always smaller in value than the Guide definition by the factor of the solidity ratio, s, are always equal in value to the product $s\,C_F$ and will be represented as such in the Guide.

In the fourth range, $0.8 \leqslant s \leqslant 1.0$, the loading is essentially that for the solid body, modified by the effects of the 'base bleed' (§16.2.3.6) through the small porosity. These effects are typically suppression of vortex shedding accompanied by a reduction in drag, but may produce an increase in drag for supercritical curved sections and for bluff bodies of high fineness where the base bleed induces reattached flow to separate. As solidity ratio tends to unity, coefficients based on solid or envelope areas converge to the same value and the distinction between them is lost. These effects will be discussed in the sections appropriate to the solid form, usually under the heading 'Effect of porosity', as in §16.2.2.1.7 and §16.2.3.6.

In the following sections discussion of the loading of lattice structures is developed in terms of the form of the structure, starting with simple plane frames and culminating in three-dimensional lattice arrays. While all current codes of practice and other design guidance cover the simpler lattice structures and specific common lattice forms, such as towers[189], the problem of large three-dimensional lattices has not been previously covered by adequate guidance. It is quite common for current codes, such as the UK code CP3 Chapter V Pt2[4] to overestimate the loading of large unclad frames by a factor of ten and predict loads several times greater than the fully-clad building. In 1972, Moll and Thiele[248] were obliged to make wind-tunnel model tests for the design assessment of an extensive unclad storage rack, the only possible solution at that time. Development of a shielding theory, backed up by recent model test results, has enabled design guidance to be formulated in this Guide.

16.3.2 A steady theory for lattice plates

16.3.2.1 Two-step approach

As indicated above, a lattice plate can be regarded as an array of interacting wakes at low solidities and an array of jets at high solidities. In free air, the flow approaching the lattice has the choice of flowing through or around the lattice, and in practice a gradual change occurs from 'through' to 'around' with increasing solidity. Akin to the principle of minimum strain energy in structures, the balance between 'through' and 'around' is set by the minimum rate of strain energy in the flow.

The problem can be studied in two steps:

1 Constrain all the flow to pass through the lattice and obtain a model for the lattice loading in terms of solidity ratio, s.
2 Remove the constraints and assess the reduction in loading as some of the flow passes 'around' instead of 'through'.

16.3.2.2 Resistance coefficient

Many experiments are concerned with the first step, since this corresponds to flow in a duct such as a wind tunnel, and gauzes are frequently used to smooth the flow, or grids are used to generate turbulence in wind tunnels. The pressure drop through such a grid or gauze is conventionally described by a 'resistance coefficient', K, defined by:

$$K = \Delta \bar{p} / \bar{q} \qquad (16.30)$$

where $\Delta \bar{p}$ is the mean pressure drop across the grid and $\bar{q}$ is the mean dynamic pressure in the approaching flow. In terms of the load on the plate, K is an 'envelope coefficient' equivalent to the product $s_x\, C_{F_x}$ for an infinitely large plate.

With all the flow constrained to pass through the grid, the wind speed in the holes of the grid, V_{hole}, is increased by the continuity equation, Eqn 2.4 to:

$$V_{hole} = V / (1 - s_x) \qquad (16.31)$$

as the air is forced to squeeze through the reduced open area. Assuming that the sectional drag coefficient of the members is unchanged in this faster flow, the resistance coefficient, K, is given by:

$$K = \bar{c}_D\, s_x / (1 - s_x)^2 \qquad (16.32)$$

Measurements to test Eqn 16.32 were first made by Taylor and Davies[249] in 1944, and the comparison in Figure 16.35 shows that their data for various forms of

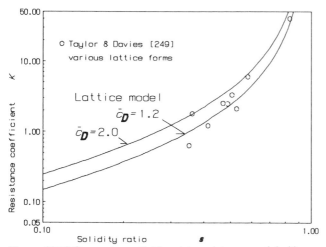

Figure 16.35 Comparison of lattice plate resistance model with measurements

lattice all lie around the model for the expected range of sectional drag coefficient. Later, more reliable data all conform to this model.

16.3.2.3 Drag coefficient C_D

In 1944 Taylor[250] proposed a potential flow model for the flow everywhere outside the actual wakes of the elements. This predicts the envelope drag coefficient when the flow constraints are removed, $s\ C_D$, in terms of the constrained resistance, K, to be:

$$s\ C_D = K / (1 + K/4)^2 \tag{16.33}$$

Taylor also showed this result is obtained by considering the momentum loss in the wakes behind the lattice. Actually, Glauert, Hirst and Hartshorn[251] had produced exactly the same result using a third independent theory 12 years earlier. Taylor's two methods are the more useful because they provide a shielding theory described later (§16.3.5). Figure 16.36 shows Taylor's model equation compared

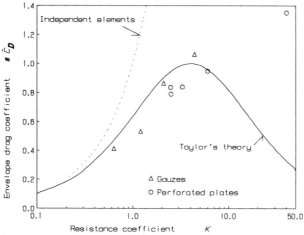

Figure 16.36 Comparison of Taylor's lattice plate drag theory with measurements

with the result from Eqn 16.28 for independent elements of constant sectional drag, showing that they match at low solidity but diverge at $s > 0.2$, as expected. Also shown on the diagram is Taylor's own data when testing the theory by experiment with Davies[249]. The data match at moderate solidity, but the measured envelope drag continues to rise at high solidity while the theoretical values fall. None of the three theories leading to Eqn 16.33 accounts for the flow reversal in the wake at high solidity. A limitation of the theory is that C_D from Eqn 16.33 reaches a maximum of $C_D = 1$ at $K = 4$, corresponding to $s = 0.6$. In fact C_D must tend to the value for a solid plate of the same slenderness, so can reach $C_D = 2$ for infinitely slender plates. Thus Taylor's theory is only useful in the first two ranges of solidity, $s \leqslant 0.6$.

The result of the balance between 'through' and 'around' can be seen by combining Eqns 16.32 and 16.33 to give the normal drag coefficient, C_D, in terms of the solidity. Figure 16.37 shows the predicted values for various constant values of sectional drag coefficient, $\bar{c}_D$. In practice, $\bar{c}_D$ cannot be constant since it is a

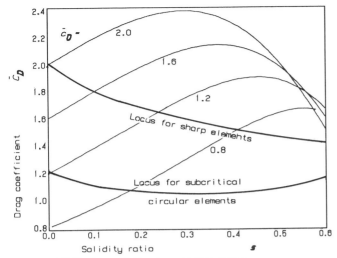

Figure 16.37 Taylor model for drag of lattice plates

function of the slenderness of the elements and, in turn, the slenderness is dependent on the solidity. The drag coefficient of practical lattices is expected to follow a locus across the theoretical lines, starting from the sectional coefficient for infinite slenderness ($\overline{c}_D = 2.0$ for flat sections, $\overline{c}_D = 1.2$ for subcritical circular sections) as indicated. Because of this effect, Taylor's theory cannot be of direct use in setting design values for force coefficients and it is necessary to resort to empiricism.

16.3.3 Single plane frames

16.3.3.1 Flow normal to frame

16.3.3.1.1 **Sharp-edged members.** The first comprehensive experimental reviews of the loads on plane lattice frames were started by Flachsbart [41, 42, 43] in 1934. Figure 16.38 shows his data for the normal force coefficient of single plane lattice frames of various types, together with some contemporary data by Georgiou [252] which will be discussed later. Flachsbart's work has remained the standard datum for lattices for many years, although the pool of data has been considerably increased by later studies. Many codes of practice have used the data directly: the UK building code, CP3 ChV pt2 [4] and others in steps; and the UK code for lattice towers, BS 8100 [189] as a continuous curve. Both are conservative upper bounds to the data, as shown on Figure 16.38.

Flachsbart's data form two groups: one for a very slender envelope, becoming a slender beam as $s \rightarrow 1$; the other for a near-square envelope, becoming a solid square plate as $s \rightarrow 1$. It is convenient to define the drag coefficient for normal flow at the high solidity limit, $s = 1$, as $\overline{C}_{D_1}\{L/B\}$, i.e. subscript '1' denoting 'solid'. This converges towards 2.0 for the slender envelope and towards 1.2 for the square envelope, as discussed earlier in §16.1.2.

At the low solidity limit, $s = 0$, both groups of data tend towards the sectional drag coefficient for infinite slenderness, $\overline{c}_{D_0} = 2.0$, as expected. Empirical curves are shown in Figure 16.38 which have been fitted through both these groups

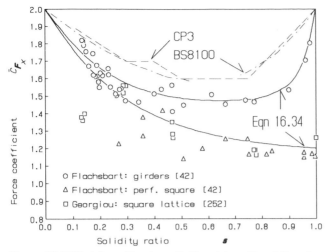

Figure 16.38 Variation of the drag of lattice plates with solidity ratio

without conservatism and with the expected values at the limits. All lattices of intermediate envelope proportions are found to have drag coefficients between these two curves. The group of data points at Georgiou's [252] lowest solidity does not follow the trend established by the other data and the author recognises this as being anomolous. This a common problem when assessing experimental data for use in design and it is necessary to achieve a consensus from independent sources so that such anomalous 'outliers' can be recognised.

For plane lattices of flat-sided members, the normal force coefficient is well fitted by the empirical equation:

$$\overline{C}_{F_x}\{\text{flat}\} = 1.17 + 0.83e^{-3.3s} + 0.4\,s\,[\overline{C}_{D_1} - 1.2]$$
$$+ 0.6\,s\,[\overline{C}_{D_1} - 1.2]e^{20(s-1)} \tag{16.34}$$

where $\overline{C}_{D_1}\{L/B\}$ is the drag coefficient of the limiting solid form. This equation describes both the empirical curves shown in Figure 16.38 and all the intermediate values appropriate to the value of $\overline{C}_{D_1}\{L/B\}$, which form a 'fan' of curves between the two limits shown. The first two terms of Eqn 16.34 describe the transition from porous lattice to solid square plate, the lower empirical curve in Figure 16.38, i.e. independently of the envelope proportions. The last two terms add the effect of slenderness of the envelope, L/B, as $s \to \infty$: the third term gives the general effect of form described in §16.3.1, and the final term gives the specific effect of base bleed (§16.2.3.6) on the vortex-shedding induced drag. (The final term is insignificant for $s < 0.8$.)

16.3.3.1.2 Circular-section members.
A similar empirical form is obtained for plane lattices of circular members, except that the local sectional drag coefficient of the members, $\overline{c}_{D_0}$, also varies with Reynolds number and surface roughness (§16.2.2.1). The force coefficient is well fitted by the empirical equation:

$$\overline{C}_{F}\{\text{circ}\} = \overline{C}_{F_x}\{\text{flat}\}\,(\overline{c}_{D_0}/2 + [1 - \overline{c}_{D_0}/2]\,s^{1.5}) \tag{16.35}$$

which determines the value for circular members from the flat-member value of Eqn 16.34. A similar approach is used by BS 8100 [189]. For subcritical members,

$\bar{c}_{D_o} = 1.2$, a set of curves which form a 'fan' similar to Figure 16.38 is obtained. For supercritical members, $\bar{c}_{D_o}$ depends on the surface roughness, k_s (§16.2.2.1.2), so Equation 16.35 gives a set of curves which form a 'fan' at both ends. A selection of typical curves from Eqn 16.35, together with the BS 8100 curves for circular sections, is shown in Figure 16.39. A fuller set of design curves is given in Chapter 20. Note that BS 8100 overestimates for smooth cylinders to a smaller degree than it does for the sharp-edged elements in Figure 16.38, but underestimates for rough supercritical elements (an unlikely combination that BS 8100 does not consider).

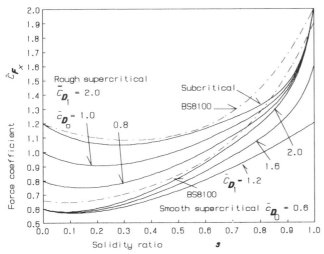

Figure 16.39 Variation of the drag of lattice plates of circular elements with solidity ratio

16.3.3.1.3 Members of mixed form.

Many practical lattice frames will be comprised of a mixture of flat and circular elements. The UK codes CP3 ChV Pt2 [4] and BS 8100 [189] make the pragmatic assumption that the elements of each form contribute to the total force in the proportion of their respective areas, giving:

$$\overline{C}_{F_x}\{\text{mixed}\} = \Sigma(\overline{C}_{F_x}\{\text{form}\}\ \phi_A\{\text{form}\}) \tag{16.36}$$

where:

$$\phi_A\{\text{form}\} = \Sigma(l\,b)\{\text{form}\}\ /\ A_x = s\{\text{form}\}\ /\ s \tag{16.37}$$

is the proportion of the area of the elements of that form to the total area of elements. In this context, $\phi_A\{\text{form}\}$ is a weighting function in the same manner as $\phi_q\{z\}$ in Eqn 16.28. Indeed, because Eqns 16.34 to 16.36 are derived from tests in uniform flow, $\phi_q\{z\}$ may also need to be included if there is significant variation of wind speed across the lattice. Most codes simplify the problem to only the upper bound envelopes for flat, subcritical and supercritical smooth elements, with $\overline{C}_{D_1}\{L/B = \to \infty\} = 2.0$ at the high solidity limit, in which case Eqn 16.36 simplifies to the form:

$$\begin{aligned}
\overline{C}_{F_x}\{\text{mixed}\} = {} & \overline{C}_{F_x}\{\text{flat}\}\ A_x\{\text{flat}\}\ /\ A_x \\
& + \overline{C}_{F_x}\{\text{subcrit}\}\ A_x\{\text{subcrit}\}\ /\ A_x \\
& + \overline{C}_{F_x}\{\text{supercrit}\}\ A_x\{\text{supercrit}\}\ /\ A_x
\end{aligned} \tag{16.38}$$

found in CP3 ChV Pt2 [4] and BS 8100 [189].

16.3.3.1.4 Angle-section members.

By simplifying the summation of forces on the individual elements to a single coefficient based on solidity, most individual flow effects are averaged out over the lattice and become unimportant. However, there is a notable exception to this general rule. Figure 16.23 shows that there is a considerable cross-wind force on structural angle sections when one angle limb faces forwards into the flow. As long as structural angle booms are aligned in opposing directions and structural angle bracings are aligned in alternating directions, the cross-wind forces will cancel out. But if structural angle booms and bracing are all aligned in the same direction there will be a significant net cross-wind force which must be accounted for.

16.3.3.2 Effect of wind direction

So far the discussion has been only for plane lattice frames normal to the flow, for which all members are also normal to the flow. Consider the effect of wind direction, θ, from normal on a lattice frame in the vertical plane. Vertical members of the lattice will see θ as equivalent to the pitch angle, α, in Figure 16.6, while horizontal members will see θ as equivalent to the yaw angle, β. Skewed members will see θ as a combination of α and β. To account for pitch and yaw for an extensive lattice with a mixture of horizontal, vertical and skewed members by summation of the individual member loads through Eqn 16.28 is an onerous task to perform by hand. Fortunately, the assumption of the simple cosine variation with either angle (§16.2.4.2) simplifies the problem considerably.

Figure 16.40 shows BRE data for a plane square-meshed lattice plate of square-section members. The square mesh results in half the members being vertical and half being horizontal, and the square-section members give a total area in the y-direction which is half the normal area, $A_y = A_x / 2$. The expectation is that the x-axis force, $F_x\{\theta\}$, will decrease from the normal value following a cosine variation while the y-axis force, $F_y\{\theta\}$, will rise from zero to half the normal value F_x, following a sine variation. The data show that the x-axis force follows this

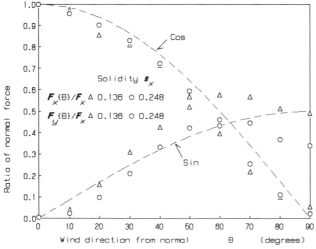

Figure 16.40 Effect of wind direction on forces on plane square-meshed lattice plate of square section members

expectation very well and the y-axis force follows adequately. As the wind direction approaches $\theta = 90°$ parallel to the plane of the frame, the y-axis force decreases as downwind members tend to become shielded. The ratio of the areas A_x and A_y and the consequential loads depend on the cross-section of the members as well as their orientation. With flat-plate members, $A_y = 0$, giving no y-axis force except for a small contribution from surface friction.

16.3.4 Lattice towers and booms

16.3.4.1 Dominance of form

Lattice towers and booms are usually confined to specialised structures, principally radio and microwave transmission towers and masts, electricity pylons, lighting masts, and cranes. The common factor for these specialised structures is that they are constructed from long 3-or 4-boom lattice trusses. The design assessment of lattice towers and masts is well served by several sources of modern design guidance. The principal external source for towers and masts is the UK code of practice BS 8100[189], first issued in 1986 and described earlier in §14.1.3.3. The loading coefficient part of this code is also relevant to other lattice structures comprised of 3- or 4-boom trusses. Lattice cranes are also covered by the UK code BS 2573[253], but the guidance here is much simplified. These codes supersede earlier UK guidance such as CP3 ChV Pt2[4] and reference [254] largely based on smooth uniform flow tests[41, 255]. Other codes, e.g. ANSI[177] are simpler but adequate[256]. In the ESDU series, Data Item 81028[257] specifically addresses the loading of this form of lattice.

The consistent form means that the range of behaviour is small and a simple empirical approach can work well. The majority of available data comes from model tests in both smooth and turbulent flow. The expectation from quasi-steady theory is that both should yield similar results, but it is known that the local sectional force coefficients are affected by small-scale turbulence, particularly at the very low Reynolds numbers that are typical when lattice elements are modelled at small scale factors. For this reason, BRE commissioned a series of tests at the

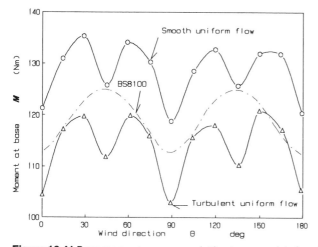

Figure 16.41 Base moment on a square lattice tower model of angle-section members

National Maritime Institute (NMI) for the drafting of BS 8100 using very large models [191, 192], including those shown in Figures 13.33 and 13.34. The results for the tower shown in Figure 13.34, without the microwave dishes, are compared in Figure 16.41 with the prediction of BS 8100. There is about 10% difference between the smooth and turbulent flow results, with the BS 8100 prediction lying between. Other similar comparisons were made in a comprehensive calibration exercise [258] before the code was adopted. The guidance in this section and the corresponding design data in Chapter 20 are largely based on this work, and so are effectively identical to that given in BS 8100. However, an alternative approach, more recently developed at BRE for the design assessment of crane structures, is also discussed.

16.3.4.2 Drag coefficients

A consequence of the form of lattice trusses is that the resultant force is always closely aligned to the wind direction. It is therefore convenient to work in terms of the overall drag coefficient, $\overline{C}_D$, instead of the force coefficients for the three axes, since these are obtained from the drag by simple trigonometry. This applies to both the approaches described below: the 'Reference Face' approach used by BS 8100 [189, 190] and the method proposed by Eden, Butler and Patient [259].

16.3.4.3 The 'Reference Face' approach

16.3.4.3.1 Principle of the approach.
This method was developed by the Flint and Neill Partnership using the results of the BRE/NMI tests specifically for BS 8100 [189, 190]. The empirical method is based on the solidity ratio, s_{face}, and the 'shadow area', A_{face}, of the members of **one reference face** of the tower or truss, denoted by the subscript 'face'. This limits the applicability to symmetrical trusses: equilateral 3-boom trusses or square 4-boom trusses, since the method assumes the other faces to be identical. The corresponding drag coefficient, $\overline{C}_{D_{\text{face}}} = D / (q_{\text{ref}} A_{\text{face}})$, is normalised by the reference face area, A_{face}, **not** the actual loaded area of the section, so violating the standard convention of the Guide (§16.2.1.3). This non-standard definition is retained for compatibility with BS 8100 and because it involves less work for the user who needs only to determine the reference area once. As a reminder of this departure from standard, the subscript 'face' will always be used.

On this basis, the drag coefficients for the complete section must be greater than that for the single plane frame and, with more members, a 4-boom truss must give a higher force coefficient than a 3-boom truss for the same face solidity. The empirical determination of the value of these coefficients automatically accounts for the degree of shielding of the elements of the downwind faces, although this shielding is not always necessarily included in the design guidance. For example, the data shown in Figure 16.41 were obtained at 15° intervals of wind direction and, being a symmetrical square truss, downwind booms and bracing members move into the wakes of the corresponding upwind members every 45°. The method rejects this direct shielding because it occurs only over small range of direction in each case, which is not apparent in the 15° intervals shown here.

16.3.4.3.2 Flow normal to reference face.
The method works in a similar manner to the single plane frames, except that the truss is assumed always to be slender overall, $L/B = \infty$, and circular-section members are assumed always to be smooth.

This simplifies the drag coefficients at the two limits of solidity to the standard values:

$$\overline{c}_{D_0}\{\text{flat}\} = 2.0,$$
$$\overline{c}_{D_0}\{\text{subcrit}\} = 1.2 \text{ and}$$
$$\overline{c}_{D_0}\{\text{supercrit}\} = 0.6$$

at the low solidity limit, and to

$$\overline{C}_{D_1}\{L/B = \infty\} = 2.0$$

for all members at the high solidity limit. Except for the omission of rough supercritical members (an unlikely combination), this results in an upper bound envelope.

The drag coefficient, $\overline{C}_{D_{\text{face}}}\{\theta = 0°\}$, which occurs when the wind direction $\theta = 0°$, along the x-axis is used as the datum value. This is well fitted by the empirical equations:

$$\overline{C}_{D_{\text{face}}}\{\text{flat}, \theta = 0°\} = 1.76\, C_1\left[1 - C_2\, s_{\text{face}} + s_{\text{face}}^2\right] \tag{16.39}$$

$$\overline{C}_{D_{\text{face}}}\{\text{subcrit}, \theta = 0°\} = C_1\,(1 - C_2\, s_{\text{face}}) + (C_1 + 0.875)\, s_{\text{face}}^2 \tag{16.40}$$

$$\overline{C}_{D_{\text{face}}}\{\text{supercrit}, \theta = 0°\} = 1.9 - \left[(1 - s_{\text{face}})\,(2.8 - 1.14\, C_1 + s_{\text{face}})\right]^{1/2} \tag{16.41}$$

where the coefficients C_1 and C_2 are given in Table 16.2. The two curves for square and triangular trusses with flat members given by Eqn 16.39 are shown with the experimental data in Figure 16.42. When there is a mixture of flat and circular members Eqn 16.38 again applies.

Table 16.2   Coefficients C_1 and C_2 in Eqns 16.39–16.41

Coefficient	Square towers	Triangular towers
C_1	2.25	1.9
C_2	1.5	1.4

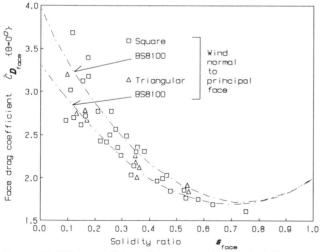

Figure 16.42 Drag coefficients for square and triangular lattice trusses of flat elements

16.3.4.3.3 Effect of wind direction.
The drag coefficient in any wind direction, θ, is given by factoring the drag coefficient at $\theta = 0°$, thus:

$$\overline{C}_{D_{face}}\{\theta\} = \overline{C}_{Dface}\{\theta = 0°\} \, \Phi_\theta \tag{16.42}$$

where the factor Φ_θ is a weighting function for direction. The empirical formulae for Φ_θ for square and triangular towers, developed for BS 8100 are:

$$\Phi_\theta\{\text{square}\} = 1.0 + (0.55 + 0.25\,\phi_{circ})\,\phi_s \sin^2 2\theta \tag{16.43}$$

$$\Phi_\theta\{\text{triangular}\} = \phi_{circ} + (1 - \phi_{circ})(1 - 0.1\sin^2 1.5\theta) \tag{16.44}$$

where $\phi_{circ} = s_{face}\{circ\}/s_{face}$ is the proportion of circular-section members: $\phi_{circ} = 0$ being all flat, $\phi_{circ} = 1$ being all circular, and ϕ_s is an influence coefficient for solidity given in Table 16.3.

Table 16.3   Coefficient ϕ_s in Eqn 16.43

Solidity of face	Coefficient ϕ_s
$s_{face} < 0.2$	0.2
$0.2 \leq s_{face} < 0.5$	s_{face}
$0.5 \leq s_{face} < 0.8$	$1 - s_{face}$
$s_{face} \leq 0.8$	0.2

In most design assessments only the critical cases need be considered. Typically these are: For square towers and masts:

1 Face on: $\theta = 0°$, for maximum shear; at which $\Phi_\theta\{\text{square}\} = 1.0$, (corresponding to the reference value).
2 Corner on: $\theta = 45°$, for maximum compression in downwind boom and maximum tension in guys; at which $\Phi_\theta\{\text{square}\} = 1.4$ is the maximum possible value, (corresponding to circular-section members and solidity $s_{face} = 0.5$).

For triangular towers and masts:

1 Face on: $\theta = 0°$, for maximum shear and maximum compression in downwind boom.
2 Corner on: $\theta = 60°$, for maximum tension in guys; at which $\Phi_\theta\{\text{triangular}\} = 1.0$ is the maximum possible value for both cases, (corresponding to circular-section members at any solidity).

These values give upper bounds that contain the worst action effects at intermediate wind directions [258].

As BS 8100 deals specifically with vertical towers and masts which are line-like, the wind direction, θ is the angle in the plane normal to the axis of the truss, corresponding to the pitch angle, α, in the local coordinates defined in Figure 16.6 (§16.2.1.2). The implication is that the method is also applicable to long trusses in general at flow angles normal to the axis of the truss.

16.3.4.3.4 Effect of ancillaries.
The wind loads on any ancillaries contribute to the total loading of a tower. The 'reference face' method can cope with ancillaries limited to an additional solidity, $s_{ancillaries}$, when $s_{ancillaries} < s_{face}$, by adding the ancillary loading to the bare tower loading. Linear ancillaries: ladders, waveguides,

etc., should be treated as elements of a lattice frame. Discrete ancillaries: microwave dishes, cylindrical antennae, etc., should be treated as individual components. Symmetrical towers and trusses with many ancillaries, e.g. pipe-bridges, should be treated as multiple lattice frames as in §16.3.6, provided $s_{\text{ancillaries}} < 0.6$, otherwise they should be treated as porous bluff bodies of the corresponding envelope dimensions.

Shielding of the tower structure becomes very significant with large ancillaries, but the ancillary loads are very large and dominate. The Guide does not give design data to perform this assessment and the designer is directed to BS 8100[189,190] for the implementation of the method. Alternatively, the designer can use ESDU 81028[257] which also gives a procedure for estimating the torque induced by large external ancillaries.

16.3.4.3.5 Asymmetrical trusses.
The method for asymmetrical trusses used in BS 8100 is identical in principle to the method used in §16.3.5 'Pairs of frames'.

16.3.4.4 The method of Eden, Butler and Patient

16.3.4.4.1 Principle of the approach.
This method was developed specifically for cranes using specialist data from model tests performed at NPL/NMI for BRE between 1953 and 1977[259,260,261], so is restricted to lattice trusses without large ancillaries. (The current UK code of practice for cranes BS 2573[253] still uses a very simple approach based on the UK code of practice for buildings CP3 ChV pt2[4].) This empirical method is based on the sum of the projected areas in the wind direction of **all** members of the lattice, even if they are 'hidden' in the 'shadow' of other upwind members, denoted by ΣA, and this corresponds to the Guide definition of loaded area. The solidity corresponding to this area is the sum of the solidities of the faces, hence $\Sigma s = \Sigma A / A_{\text{proj}}$, so can take values greater than unity. The corresponding drag coefficient complies with the standard convention of the Guide, but will be denoted by the subscript 'total' in this section to distinguish it from the 'face' coefficient of the previous reference face method.

Thus we have $\overline{C}_{D_{\text{total}}} = D / (q_{\text{ref}} \Sigma A)$. On this basis the value of $\overline{C}_{D_{\text{total}}}$ for a truss must be **less** than that for the single plane frame of the same solidity because all the members are included in the reference area ΣA_{proj}, and those in the downwind faces are shielded to some degree.

16.3.4.4.2 Flow normal to axis of truss.
Eden, Butler and Patient[259] proposed the following empirical equations from the model crane truss data when the wind is at any pitch angle, α, in the plane normal to the axis of the truss ($\beta = 0°$):

$$\overline{C}_{D_{\text{total}}}\{\text{flat}\} = 1.21\,[1 - (1 - 0.16/\Sigma s)^{4.4}] \tag{16.45}$$

$$\overline{C}_{D_{\text{total}}}\{\text{subcrit}\} = 0.84\,[1 - (1 - 0.21/\Sigma s)^{6.0}] \tag{16.46}$$

for trusses of flat and subcritical circular members, respectively. The data supporting Eqn 16.45 for flat members is shown in Figure 16.43, where the reciprocal of the solidity is used because of the form of the equation. These data comprise the results from the NPL/NMI crane truss models and from BRE data for multiple lattice arrays (see §16.3.5 and §16.3.6). The data supporting Eqn 16.46 for subcritical circular members came from the NPL/NMI tests alone. None of these tests were performed at supercritical Reynolds numbers, so there is no

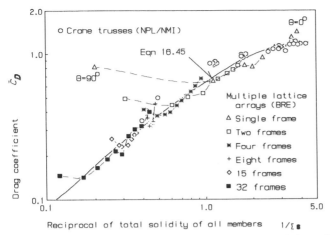

Figure 16.43 Comparison of Eden, Butler and Patient's model equation with force measurements on lattice towers and arrays of frames

corresponding equation, but it would be expected from the behaviour of single plane frames that $\overline{C}_{D_{\text{total}}}\{\text{supercrit}\} \rightarrow \frac{1}{2}\overline{C}_{D_{\text{total}}}\{\text{subcrit}\}$ as $\Sigma s \rightarrow 0$ and $\overline{C}_{D_{\text{total}}}\{\text{supercrit}\} \rightarrow \overline{C}_{D_{\text{total}}}\{\text{subcrit}\}$ as $\Sigma s \rightarrow 1$. Similarly, for trusses of mixed member forms, Eqn 16.38 is again expected to apply.

Equations 16.45 and 16.46 predict the total drag coefficient from Σs alone. This works only because of the dominance of form (§16.3.4.1). In the multiple array data in Figure 16.43, each set of data forms an individual 'J' shaped curve that cuts across the curve of Eqn 16.45. By lumping all the member areas together, the parameter Σs has no knowledge of how deeply shielded the members are. Eqn 16.45 describes a locus through the data which connects the 'truss-like' forms together, so the applicability of the method is restricted to this form of lattice [261].

16.3.4.4.3 Effect of wind direction.

The method was designed to be used for cranes that can 'slew' (rotate around vertical axis) and, for jib cranes, can also 'luff' (raise and lower the jib), hence must also cope with all possible combinations of pitch and yaw angles. Luff and slew angles represent the position of the truss in wind axes, whereas the notation of pitch and yaw angle used in the Guide represents the angle of the incident wind in body axes. Solely for this application luff angle, A, and slew angle, B, are defined as zero when the axis of the truss is aligned in the wind direction. Thus when the truss is horizontal and normal to the wind A = 0° and B = 90°, and when vertical A = 90°.

In a procedure similar to BS 8100 for wind direction, Eden, Butler and Patient proposed that the drag at any combination of slew and luff can be predicted from the values with flow along each of the three orthogonal axes using the empirical equation:

$$D\{A,B\} = (D\{A = 90°, B = 0°\} \sin^n A + D\{A = 0°, B = 0°\} \cos^n A) \cos^m B$$
$$+ D\{A = 90°, B = 90°\} \sin^m A \tag{16.47}$$

In their first paper[259], the coefficients n and m were set to be the same, $n = m = 1.5$. The predictions for a model crane jib are compared with measurements in Figure 16.44 and show good agreement between the measure-

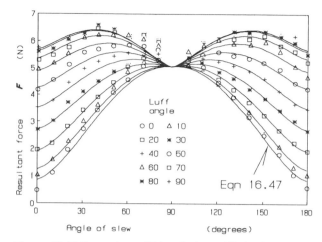

Figure 16.44 Comparison of Eden, Butler and Patient's model equation with force measurements on models of crane jibs

ments and the family of curves produced by Eqn 16.47 for the range of skew and luff angles. The top curve in Figure 16.44 represents a vertical crane jib rotated about its axis, so is directly equivalent to the vertical lattice tower. The bottom curve represents a horizontal jib slewed from pointing into wind, through normal to pointing downwind, so is directly equivalent to a horizontal lattice gantry or pipe-bridge.

In their second paper[260], they reported that a better fit is obtained with $n = 1.8$ and $m = 1.4$. A major problem with Eqn 16.47 is that each different crane jib produces different best fit for the coefficients n and m. This reduces the usefulness of the approach in design, since values for the coefficients should be determined for each individual case. Also, Eqn 16.47 is framed in terms of drag, D, for compatibility with the BS 8100 approach. While this is appropriate for flow normal to the axis of the truss ($A = 90°$), Eden, Butler and Patient note that the direction of the resultant force is rotated significantly out of the wind direction when the wind is skewed. Strictly, $D\{A,B\}$ in Eqn 16.47 should be the resultant force, but the direction of the resultant is not given. Discussion of the method[261] includes the view that the additional complexity of determining the projected member areas in every wind direction outweighs the benefits.

16.3.5 Pairs of frames

16.3.5.1 Momentum loss far behind frame

Consider a fine-mesh plane lattice frame of many members, so that the spacing of the members in the mesh, b, is very much smaller than the overall breadth of the frame, B. The wind force on the frame reduces the momentum in the 'wind shadow' behind the frame by an equal amount, because force is equal to the rate of change of momentum, giving:

$$F_x = \rho A_x V \, \Delta V \tag{16.48}$$

where $\rho A_x V$ is the rate of mass flow through the lattice and ΔV is the change in velocity. Because $b < < B$, this momentum deficit can be assumed to mix and be distributed uniformly over the wind shadow defined by the projection of the frame envelope in the wind direction (A_{proj} in Figure 12.8). This leads directly to an expression for the change of velocity in the 'wind shadow' behind the lattice:

$$\Delta V / V = \tfrac{1}{2}\, s\, C_{F_x} \qquad\qquad (16.49)$$

A second lattice frame placed in the wind shadow of the first lattice frame experiences a reduced dynamic pressure, so that the drag of the second frame is reduced by the 'shielding factor', η, given by:

$$\eta_\infty = (1 - s\, C_{F_x} / 2)^2 \qquad\qquad (16.50)$$

The 'infinity' symbol subscripting η_∞ indicates the value 'far' downwind. In practice this means sufficiently far downwind for the momentum loss from individual members to be mixed into the flow, but not so far that the entrainment of faster flow from the wind has restored significant momentum to the wake.

Equation 16.50 is exact within the assumptions of uniform mixing over the projected envelope area of the frame. Compare Eqn 16.50 with the empirical expression derived by Flachsbart[42] from his measurements on lattice girders: $\eta_\infty \simeq (1 - s)^2$. Noting that Flaschbart used flat-sided members for his girders, for which $C_{F_x} = 2$, the two equations are identical.

16.3.5.2 Momentum loss close to frame

The loss of momentum and the consquent reduction in wind speed is accompanied by divergence of the flow in accordance with the continuity equation, Eqn 2.4 (§2.2.2). This happens gradually, both upwind and downwind of the frame, as shown in Figure 16.45(a). The gradual decrease in wind speed is reflected as a gradual rise in static pressure through the Bernoulli equation, Eqn 2.6. The force on the lattice frame is balanced by the pressure drop across the lattice, shown in Figure 16.45(b).

Taylor's theory[250], described in §16.3.2.3, also gives expressions for the changes in velocity of the incident wind as it passes through the frame. The in-wind component, u, and the cross-wind component, v, are given by:

$$- u = K (\theta_1 - \theta_2) / [4\pi (1 + K/4)^2] \qquad\qquad (16.51)$$
$$v = K \log(\sin\theta_2/\sin\theta_1) / [4\pi (1 + K/4)^2] \qquad\qquad (16.52)$$

where the angles θ_1 and θ_2 (in radians) are defined in Figure 16.45(c), and K is the resistance coefficient of the frame. These velocity fields remain geometrically similar for all lattice plates.

Values for the local shielding factor at any position close to the frame, $\eta\{x,y\}$ can be calculated from Eqns 16.51 and Eqn 16.52, to give a weighting function for dynamic pressure analogous to that for the wind speed profile in Eqn 16.29. Practical structures usually consist of a number of similar parallel frames, so that the effect can be integrated over the whole frame to give a shielding factor, $\eta\{x\}$, dependent on the separation, x, between frames. The effects of the cross-wind component, v, integrate to zero over the frame, so can be neglected. The expression for u in Eqn 16.51, acting through the resistance coefficient is inconvenient, since empirical expressions for the force coefficient $\overline{C}_{F_x}$ have been adopted in Eqns 16.34 to 16.38. Eqn 16.51 can be restated in terms of Eqn 16.49 for ΔV, the velocity deficit far downwind as $x \to \infty$, giving:

$$- u/V = sC_{F_x} (\theta_1 - \theta_2) / 4\pi \qquad\qquad (16.53)$$

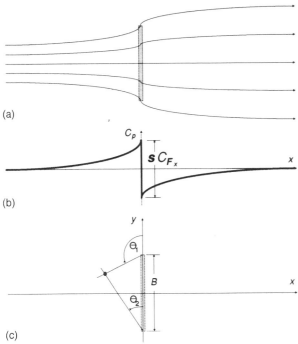

(a)

(b)

(c)

Figure 16.45 Flow near lattice plate: (a) divergence of flow; (b) pressure on centreline; (c) definition of position in flow

The shielding factor, $\eta\{x\}$, for a second frame of the same breadth at x, upwind or downwind of the first frame is determined by integrating the local dynamic pressure over the breadth of the frame:

$$\eta\{x\} = \int_0^1 (1 - u/V)^2 \, \mathrm{d}(y/B) \qquad (16.54)$$

Figure 16.46 shows the results of the integration for a range of solidity. The shielding factor, $\eta\{x\}$, reduces from unity far upwind to the minimum value, η_∞, far downwind given by Eqn 16.50. The majority of this change occurs within a region several breadths either side of the frame. Separation of lattice frames in practice can be very small, typically $B/10$ for unclad building frames, so this effect is significant. Despite the antisymmetry of the velocity deficit, u, either side of the frame, the shielding factor is skewed to give more shielding downwind, owing to the square term in Eqn 16.54.

This family of curves collapse well onto a single curve when normalised to:

$$(1 - \eta^{1/2}\{x\}) / (1 - \eta_\infty^{1/2}) = \tilde{u}\,\{x\} / \tilde{u}_\infty = \phi_u\{x\} \qquad (16.55)$$

Here $\tilde{u}\{x\}$ represents the equivalent uniform velocity deficit over the breadth of the frame and $\tilde{u}_\infty = \Delta V$ is its value far downwind, so that the ratio represents the proportion of the final velocity deficit effective at that location. This is an influence coefficient, $\phi_u\{x\}$, for the velocity deficit. It is again antisymmetric either side of the frame and is well fitted by the empirical equation:

$$|\phi_u\{x\} - 0.5| = 10^{[-0.3983 + 0.5508 \log(\arctan|x/B|)]} \qquad (16.56)$$

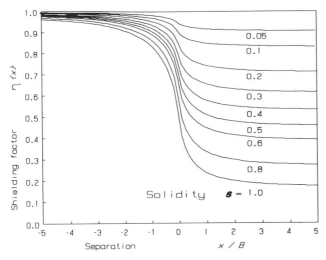

Figure 16.46 Shielding factors from Taylor's theory

Combining Eqns 16.50, 16.55 and 16.56 gives the following empirical expressions for the shielding factor, $\eta\{x\}$:

$$\eta\{x\} = [1 - \tfrac{1}{2}s\, C_{F_x} \phi_u\{x\}]^2 \tag{16.57}$$

$$\eta\{x < 0\} = [1 - \tfrac{1}{2}s\, C_{F_x}(0.5 - 10^{[-0.3983 + 0.5508\, \log(\arctan[-x/B])]})]^2 \tag{16.57a}$$

$$\eta\{x > 0\} = [1 - \tfrac{1}{2}s\, C_{F_x}(0.5 + 10^{[-0.3983 + 0.5508\, \log(\arctan[x/B])]})]^2 \tag{16.57b}$$

16.3.5.3 Pairs of frames normal to the wind

The prediction of $\eta\{x\}$ from Eqn 16.57 is obtained from a single frame in isolation. When a second frame is placed in the wind shadow of the first frame, it is shielded. However, the presence of this second frame produces its own momentum deficit so that it also shields the first frame. In this situation, it is easy to be confused between the two frames. The **position** of the two frames will be distinguished by calling one the 'upwind frame' and the other the 'downwind frame'. When considering the **load** on a particular frame it will be called the 'loaded frame', and the other frame becomes the 'shielding frame'. The shielding factor for each (loaded) frame is set by the envelope drag coefficient, $s\, C_{F_x}$, of the **other** (shielding) frame and their mutual separation, x/B. When the loaded frame is the upwind frame $x < 0$, and when the loaded frame is the downwind frame, $x > 0$.

Assume first that there is no interaction between the frames, i.e. each shields the other, but the degree of shielding is unaffected by the presence of the shielded frame. According to this model, the shielding factor, $\eta\{x\}$, is independent of the solidity of the loaded frame and is given by Eqn 16.57. This hypothesis is tested in Figure 16.47 which shows the predictions of Eqn 16.57, denoted by the solid line, compared with BRE measurements for three square-mesh lattice frames of differing solidities, denoted by the symbols given in the key. Dimensions of the frames used are given in Table 16.4. In addition to the three lattice frames, a solid plate was used only as a shielding frame.

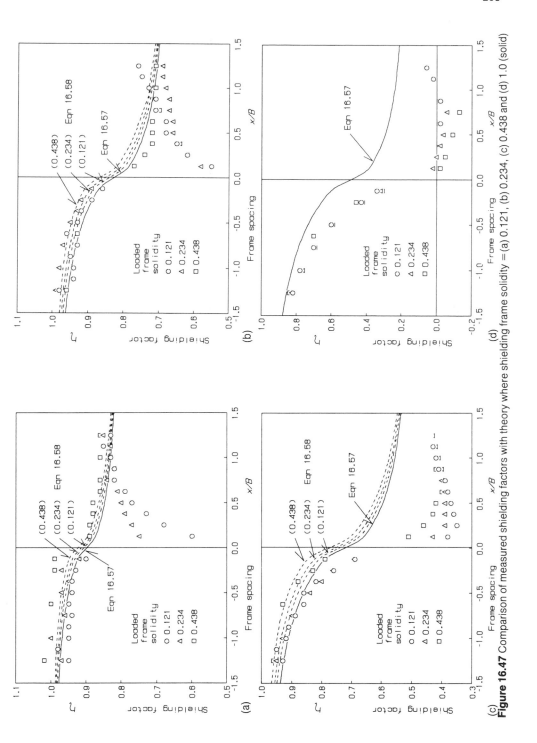

Figure 16.47 Comparison of measured shielding factors with theory where shielding frame solidity = (a) 0.121, (b) 0.234, (c) 0.438 and (d) 1.0 (solid)

Table 16.4   Dimensions of lattice frames used in BRE measurements

Solidity ratio s	Frame breadth B (mm)	Mesh width l (mm)	Bar width b (mm)
0.121	203	50.8	3.18
0.234	203	25.4	3.18
0.438	203	12.7	3.18
1.000	203	0	—

The shielding factor for the one frame is set by the drag of the other frame, but this drag is reduced by the mutual shielding so that the degree of shielding is also reduced. This interaction means that the shielding factor for each frame depends on its own solidity as well as the solidity of the shielding frame. Eqn 16.57 becomes the pair of simultaneous equations:

$$\eta_1\{-x\} = [1 - \tfrac{1}{2}s_2\, C_{F_{x,2}}\, \phi_u\{-x\}]^2 \qquad (16.58a)$$

$$\eta_2\{x\} = [1 - \tfrac{1}{2}s_1\, C_{F_{x,1}}\, \phi_u\{x\}]^2 \qquad (16.58b)$$

where the subscripts '1' and '2' refer to the two frames as indicated in Figure 16.48(a). These can only be solved by iteration and the results for each of the test frames are shown in Figure 16.47 as the dotted lines.

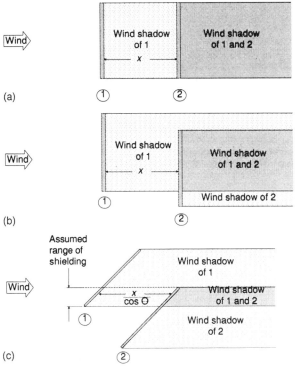

Figure 16.48 Model for pair of lattice frames: (a) wind normal to aligned frames; (b) wind normal to offset frames; (c) wind skewed to frames

The comparisons show that Eqn 16.57, assuming no interaction, is a good first-order model for loaded frame solidity $s < 0.4$, but becomes progressively more conservative as the shielding frame solidity is increased. This is particularly significant when the loaded frame is downwind, where two effects can be seen. Far downwind, $x/B > 1$, where the results for the loaded frames come together, the conservatism is due to the formation of a bluff-body wake behind the upwind shielding frame as it becomes more solid. In the case of the solid shielding frame, Figure 16.47(d), the model still assumes half the velocity deficit occurs upwind and half downwind of the frame, whereas in reality no flow can pass through the frame at all. Closer to the shielding frame, $x/B < 1$, in addition to this first effect, the elements of the downwind loaded frame tend to lie in the wakes of the members of the upwind frame and the assumption of uniform mixing does not hold. This second effect is greatest for the $s = 0.121$ **shielding** frame in Figure 16.47(a). For the $s = 0.121$ **loaded** frame, both frames are identical so all members lie in wakes and this extra wake shielding is greatest. For the $s = 0.234$ **loaded** frame, half the members lie in wakes and so the wake effect is about half the previous case. For the $s = 0.438$ **loaded** frame, only one in eight members lie in wakes and the effect is not significant. This wake effect is quantified later in §16.3.6.2 'Multiple plane frames'.

The dependence on loaded frame solidity through the interative solution of Eqn 16.58 is seen to be a smaller, second-order effect. It is doubtful whether the complexity of Eqn 16.58 is justified in design. Equation 16.57 gives a reasonable model for the shielding both upwind and downwind of lattice frames in the first two ranges of solidity, $s < 0.6$, but is conservative at the solid end of this range. This model is more useful than previous guidance because it partitions the load properly between the upwind and downwind frames and allows them to be of differing solidities.

16.3.5.4 Simplified approach

Previous guidance, as in the UK code of practice [4], uses the simplifying assumption that the upwind frame is unshielded and the downwind frame is shielded to the maximum value, η_∞, given by Eqn 16.50. When the overall loading of both frames is required, perhaps for foundation load or stability calculations, it is quite safe to use this simplifying assumption, since:

$$1 + \eta_{\infty,2} > \eta_1\{-x\} + \eta_2\{+x\} \tag{16.59}$$

Equation 16.57 also gives a safe result for lattice loaded frames in the range $s < 0.6$ when upwind of a solid shielding frame so, for the first time, gives some allowance for the shielding of lattices upwind of a solid wall (e.g. sun or debris screens). Also, since each frame of the pair must be upwind and unshielded for some wind directions, the assumption is usually also safe for the individual frames.

Accordingly, this simplified approach is recommended for use in design. The loading of more solid plates is dealt with in §16.4 'Plate-like structures'. Design values of shielding factor for various spacing and solidity are given in Chapter 20.

16.3.5.5 Effect of wind direction and frame offset

The foregoing assessment applies to a pair of frames that are aligned normal to the wind direction, with the downwind frame entirely within the wind shadow of the upwind frame, as shown in Figure 16.48(a). Figures 16.48(b) and (c) show two other possible alignments.

In (b) the frames are still aligned normal to the wind, but the downwind frame is offset so that it is only partly shielded. In this case, the shielding factor should be applied only to the part within the wind shadow, and the part outside the wind shadow should be assumed to be unshielded. In reality, the transition at the edge of the wind shadow will not be sudden, but the decrease in shielding just inside the wind shadow is balanced by the increase just outside the wind shadow, and the net effect is the same. Downwind of both frames, the wind shadow has regions shielded by either or both of the frames, and this has repercussions on any subsequent frames (see §16.3.6 below).

In (c) the frames are aligned as in (a), but the wind direction is skewed from normal. Now the effect of misalignement as in (b) is combined with the effect of wind direction on each frame as in §16.3.3.2. Assuming that these two effects do not interact, the proportion of the downwind frame that is shielded changes with wind angle and the effective spacing also increases in proportion to $1/\cos\theta$, while the force coefficients $C_{F_x}\{\theta\}$ and $C_{F_y}\{\theta\}$ will vary in proportion to $\cos\theta$ and $\sin\theta$ respectively. The model equation for the normal force on the downwind frame is therefore:

$$C_{F_x}\{\theta, x/B, \eta\} = C_{F_x}\{\theta = 0\} \cos\theta \, (\tan\theta \, x/B + \eta\{x/(B \cos\theta)\} \, [1 - \tan\theta \, x/B]) \quad (16.60)$$

where $\tan\theta \, x/B$ is the unshielded proportion of the frame and $x/(B \cos\theta)$ is the effective separation. Eqn 16.60 is valid while part of the downwind frame lies in the wind shadow, i.e. for the range $\tan\theta \, x/B < 1$.

The prediction of Eqn 16.60 is compared with BRE measurements using the frames of Table 16.4 at a spacing of $x/B = 1$ in Figure 16.49, where the model and the measurements correspond well. As the solidity of the upwind shielding frame increases, so the load on the downwind frame decreases at small angles. At this spacing the downwind frame remains in the wind shadow until $\theta = 45°$, so the curves coincide with the unshielded curve at larger angles. The maximum normal load occurs at an wind angle that increases from $\theta = 0°$ at low shielding to $45°$ at high shielding. An important aspect of shielding that should be remembered is that

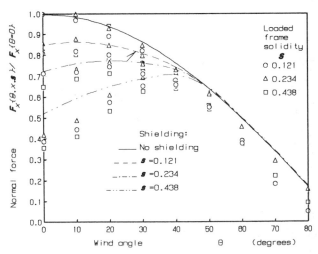

Figure 16.49 Effect of wind angle on shielding of downwind frame of pair of lattice frames

the wind direction for the maximum load case when shielded is usually different from the maximum unshielded load case.

In (c) the wind angle of both frames is the same. Other possibilities include pairs of frames that are not parallel as well as being offset and skewed from normal, although these are likely to be rare in practice. It is expected that this model will still apply, with the shielding factor and the extent of the shielded region obtained from the upwind frame solidity and its orientation, and the force coefficient obtained from the downwind frame solidity and orientation. The only difference is that the effective separation between the frames will vary across the shielded region.

The 'wind shadow' model of Figure 16.48 is valid for the first two ranges of solidity (§16.3.1), $s < 0.6$. At higher solidities, each frame begins to act like a flat plate with porosity. As the wind direction approaches parallel to the frames, $\theta \to 90°$ ($\alpha \to 0°$ in §16.2.3.1), the 'wing effect' will produce a lift force on both frames which will deflect the wake and the 'wind shadow' will not occur. In this state a pair of nearly solid frames will be acting like the wings of a biplane aircraft, both generating a similar amount of lift. Accordingly, this shielding model should not be used for $s > 0.6$ and the designer should refer instead to §18.6.3 and §19.2.2.

16.3.6 Multiple lattice frames

16.3.6.1 Introduction

Unclad building frames and scaffolding arrays supporting falsework can consist of so many frames that the entire array appears almost solid, as in Figure 16.50,

Figure 16.50 Scaffolding supporting falsework for deck of motorway bridge

although each individual frame is of very low solidity. In this case, shielding becomes very important because the total area of all the individual elements far exceeds the envelope area and, if shielding were ignored, the predicted loads would greatly exceed the loads for a fully clad, solid structure of the same dimensions. With a large number of frames, the whole assembly becomes effectively bluff at lower individual frame solidities, owing to the accumulation of drag from each frame. This further restricts the range of solidity for which the lattice-plate and quasi-steady models apply, so that for many frames a maximum limit of $s < 0.3$ is more appropriate.

Two approaches to the problem are current in design practice, both assuming that all the frames have the same solidity. In the absence of a suitable theory for multiple frames, it can safely be assumed that all downwind frames are shielded at least as well as the second frame of a pair. The UK building code, CP3 ChV Pt2 [4], takes the shielding factor for the downwind frame of a pair and applies this to all subsequent frames, thus: $\eta_1 = 1$, $\eta_2 = \eta_\infty$ and $\eta_n = \eta_\infty$. While this is satisfactory for a small number of frames, the predicted total load builds up rapidly to exceed the fully-clad loads. As the shielding factor represents the reduced dynamic pressure behind the shielding frame, at first sight it seems appropriate that every subsequent frame will reduce the dynamic pressure by the same factor, thus: $\eta_1 = 1$, $\eta_2 = \eta_\infty$ and $\eta_n = \eta_\infty^{n-1}$. This model, in which the shielding factor reduces geometrically with the number of frames, is used by the UK code of practice for cranes BS 2573 Pt1 [253], but is limited there to ten frames. This limit is necessary because the geometric model reduces the shielding factor too quickly, so is not conservative for large numbers of frames. This geometric model is not really restricted to identical frames, since $\eta_n = \eta_\infty^{n-1}$ is just a special case of $\eta_n = 1 \cdot \eta_2 \cdot \eta_3 \cdot \eta_4 \ldots = \Pi_{i=1}^{n-1} \eta_i$, where each individual frame shielding factor, η_i, may be different.

16.3.6.2 Multiple plane frames

16.3.6.2.1 Frames normal to wind.
The momentum loss model for a pair of fine-mesh lattice frames in §16.3.5.1 extends to multiple frames by accumulating the momentum loss behind each successive frame, as indicated in Figure 16.51(a) by the increasing density of shading in the 'wind shadow'. Equation 16.48 still holds, in which F_x is now the accumulated normal force for all upwind frames. Equation 16.50 then becomes:

$$\eta_{\infty,n} = \left(1 - \sum_{i=1}^{n-1} [s_i\, \eta_i\, C_{F_{x,i}}]/2\right)^2 \qquad (16.61)$$

for the shielding factor of nth frame. Eqn 16.61 contains the shielding factors for all the previous frames. Assuming no interaction of the downwind frames on the upwind frames, Eqn 16.61 can be solved by starting with the upwind frame and working downwind. However, if the upwind divergence of §16.3.5.2 were included, Eqn 16.61 would be very difficult to solve.

Accordingly, only the simplified approach of §16.3.5.4 is used here and extended to multiple frames. Upwind frames are assumed to be unaffected by downwind frames, while downwind frames are assumed to experience the far downwind shielding whatever the actual frame separation. Multiple frames are often placed close together, so this action needs justification by example. Indeed, building frame separation, x, is typically similar to the mesh spacing, l. BRE measurements on individual frames spaced at $x/B = 0.125$, in Figure 16.52, show that the upwind effect is indeed small, so that the simplified approach can be expected to give good

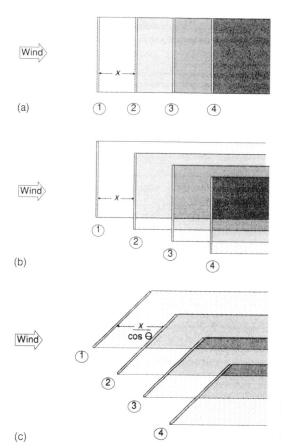

Figure 16.51 Model for multiple lattice frames: (a) wind normal to aligned frames; (b) wind normal to offset frames; (c) wind skewed to frames

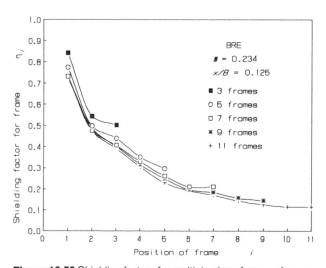

Figure 16.52 Shielding factors for multiple plane frames of square mesh and solidity **s** = 0.234

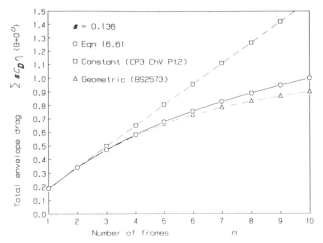

Figure 16.53 Comparison of models for total drag on multiple lattice frames normal to the wind

results. Note that the shielding factor continues to decrease towards zero with increasing number of frames, as predicted by Eqn 16.61.

The accumulated normal force predicted from Eqn 16.61 for up to ten identical frames, each with a solidity ratio of $s = 0.136$, plotted in terms of the sum of the envelope normal force coefficients of each frame, $\Sigma[s_i\,\eta_i\,C_{Fx,i}]$, are compared with the constant shielding factor of CP3 ChV Pt2 [4] and the simple geometric model of BS 2573 Pt1 [253] in Figure 16.53. (For flow normal to the frames, $\theta = 0°$, the normal force is also the drag, as indicated on the y-axis.) This particular solidity was selected because it was the lowest value used in the extensive experimental study by Georgiou [252, 262]. The constant shielding factor model used by CP3 ChV Pt2 [4] quite unreasonably continues to accumulate drag, so that after eight frames the predicted load exceeds the value of 1.2 expected for a solid frame. On the other hand, the simple geometric model used by BS 2573 [253] follows the prediction of Eqn 16.61 reasonably well, but tends to underestimate for large numbers of frames.

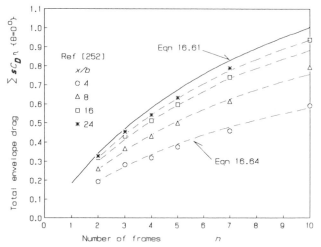

Figure 16.54 Comparison of total drag on multiple frames normal to flow for solidity $s = 0.136$

Georgiou[252,262] measured the **total** normal force for a range of wind directions on up to ten parallel frames, aligned like a 'toast rack' as in Figure 16.51(a), for solidities of $s = 0.136, 0.286, 0.464, 0.773$ and 1.0, and for spacing between frames in the range $0.06 < x/B < 1.2$. In measuring the total force, it is not possible to infer the individual frame loads, except through the simplified approach. Only the lowest two solidities are in the range that the lattice plate and quasi-steady models will hold for all ten frames. The remainder act more like fences or walls than like lattice frames (§16.4.2). Although there are other similar studies[41,42,263,264], none match Georgiou's range and quantity of data. Georgiou's data are compared with the prediction of Eqn 16.61 in Figure 16.54 for a wide range of frame separations. The experimental data all lie below the model prediction, the data for the largest separation being quite close, but the values reducing as the separation is decreased. For this normal wind angle all frame members, except those of the upwind frame, lie directly in the wakes of the members of the previous frame.

At the small separations, the wake effect is very significant, particularly as every frame except the upwind frame is affected. Dimensional analysis (§2.2.5) leads to the expectation that bluff-body wakes develop in proportion to $(x/b)^{1/2}$, where b is the breadth of the element (Figure 16.34), so that the additional shielding by direct wake action should be proportional to $(x/b)^{-1/2}$. Defining the ratio of the measured to predicted shielding factors as the weighting function for wake shielding, $\phi_n\{x/b\}$, gives:

$$\eta_n\{x/b\} = \phi_n\{x/b\}\,\eta_{\infty,n} \tag{16.62}$$

for the nth frame. As all the frames in Georgiou's study were identical, this is given by the ratio of the measured to predicted total loads as $n \to \infty$. This ratio is plotted in Figure 16.55 against $(x/b)^{-1/2}$, and confirms that the data for large n collapse. This curve is given by the empirical equation:

$$\phi_n\{x/b\} = 1.18 - 1.2\,(x/\,[b\,\cos\theta])^{-1/2} \quad \text{for } x/\,[b\,\cos\theta] < 25$$
$$= 1 \qquad\qquad\qquad\qquad\qquad \text{for } x/\,[b\,\cos\theta] \geqslant 25 \tag{16.63}$$

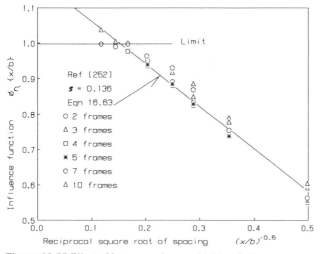

Figure 16.55 Effect of frame spacing on shielding factor

where the $\cos\theta$ term allows for wind directions other than normal (Figure 16.51(c)), but is unity in this instance. When the frames are not identical, not all elements of a frame may lie in wakes. This will also occur for identical frames with changes in wind direction, where the horizontal members remain in wakes while the vertical members do not. Thus the weighting function for wake shielding given by Eqn 16.63 should be applied only to the proportion of the elements in wakes.

Allowing for the proportion of elements shielded in wakes, s_{wakes}/s, gives a composite theoretical/semi-empirical model for the shielding of multiple lattice frames normal to the wind:

$$\eta_n\{x/b\} = (1-[1-\phi_n\{x/b\}]\,s_{\text{wakes}}/s)\,\eta_{\infty,n} \tag{16.64}$$

where the parameters in Eqn 16.64 are given by Eqns 16.61 to 16.63. The predictions of Eqn 16.64 are included in Figure 16.54 for each spacing as the family of dashed curves. Note that $s_{\text{wakes}}/s = 1$ in this case as all elements of the second and downwind frames lie in wakes. Here the variation with number of frames is purely theoretical, while the experimental data set just the intercept of the $(x/b)^{-1/2}$ model for the wake effect, so that the observed excellent fit denotes good correspondence with expected theory.

16.3.6.2.2 Effect of wind direction and frame offset.

As with pairs of frames (§16.3.5.5), the foregoing assessment in which the downwind frames are entirely within the accumulating wind shadow can be extended to allow for misalignment as in Figure 16.51(b), wind direction as in Figure 16.51(c), or a combination of both.

Consider the four parallel frames of Figure 16.51(c) at wind angle θ. The first frame is unshielded and is treated as an isolated frame. The second frame is unshielded over a width of $x\tan\theta$ and shielded by the first frame over the remainder, so that Eqn 16.60 for the downwind frame of a pair again applies, but with η_2 from Eqn 16.64 as the shielding factor for the shielded part. The third frame is also unshielded for a width of $x\tan\theta$, is shielded by the first frame over the same width and is shielded by the first two frames for the remainder, thus:

$$C_{F_{x,\,n\,=\,3}}\{\theta\} = C_{F_x}\{\theta = 0\}\cos\theta\,([1+\eta_2\{x/(B\cos\theta)\}]\tan\theta\,x/B$$
$$+\,\eta_3\{x/(B\cos\theta)\}\,[1-\tan\theta\,2x/B]) \tag{16.65}$$

The fourth frame is outside the wind shadow of the first frame at the wind angle shown, so that this and subsequent frames are loaded identically to the third frame. In the general case, however, the nth frame can have up to n differently shielded regions.

Figure 16.56 shows the results of this model compared with Georgiou's data [252] for solidity $s = 0.136$ with up to ten frames at a spacing of $x/B = 0.25$. This particular set of data was chosen for comparison because the spacing was equal to the mesh width, l, so is typical of many lattice structures, but his other data give similar results. Here the total normal force for all frames is given as the ratio of the normal force on a single frame at $\theta = 0°$. The wake effect was calculated on the assumption that all elements lie in wakes at $\theta = 0°$, but that only the horizontal members lie in wakes for all other directions. These frames are quite coarse, with only four mesh widths across the frame, so the load value should change suddenly as each vertical member moves in or out of a shielded region and in or out of wakes. Thus some variation is to be expected between the data and the model, with its fine-mesh assumption. Considering the errors that could accumulate with a large number of members, the match is very good. At this spacing, the wake effect is

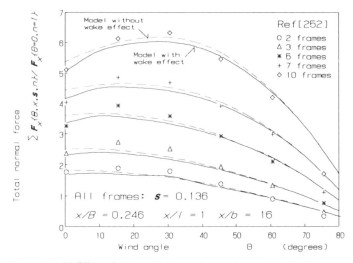

Figure 16.56 Effect of wind angle on total normal force on multiple plane lattice frames

small and neglecting the wake weighting function of Eqn 16.63 simplifies the problem and gives an additional small degree of conservatism.

16.3.6.3 Three-dimensional arrays

The scaffolding array of Figure 16.50 is a three-dimensional array of frames in both x and y directions on a rectangular-plan grid. This is a very common form of lattice in building applications, principally for falsework, scaffolding and unclad building frames. It is also common in industrial applications, as in the storage rack investigated by Moll and Thiele [248], except that this had the extra complication of a sheeted roof and sheeted walls at the two short faces (§16.3.6.4).

The principle of 'wind shadow' also extends to these three-dimensional arrays of low solidity, $s < 0.3$, as indicated in Figure 16.57. Here the 'x-frames' normal to the x-axis are numbered as in Figure 16.51, while the additional 'y-frames' are labelled alphabetically. When the wind is normal to either face as in (a), the frames in one set are parallel to the wind and cast no shadow, while those in the other set are normal to the wind and are loaded equivalent to the normal case for multiple plane

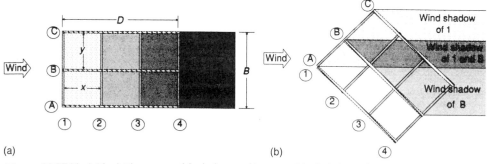

Figure 16.57 Model for lattice arrays: (a) wind normal to array; (b) wind skewed to array

frames given by Eqns 16.61–16.65. When the wind is skewed as in (b), both sets of frames cast wind shadows: the shadow from upwind frames of the same parallel set contribute shielding in exactly the same manner as before (§16.3.6.2.2) at a constant spacing, as for 'Wind shadow of 1' in (b); but each frame is cut through by every frame of the other set, only the upwind part of these contributes shielding and the spacing varies with position as for 'Wind shadow of B'.

The previous approach extends to these three-dimensional arrays, but is complex to apply without recourse to computer-aided design methods. Accordingly, it is fortunate that an empirical approach works well for the total load in the commonest situation where the envelope is cuboidal, and the solidity and spacing of the frames are similar in the x and y directions, e.g. typical unclad building frames. In this case the total load in each horizontal building axis for any direction is given in terms of the loads when the wind is aligned along the respective axis, thus:

$$\Sigma\, C_{F_x}\{\theta\} = \Sigma\, C_{F_x}\{\theta = 0°\}\,(\cos\theta + 0.12\,[D/B]^{0.8}\,\sin2\theta) \tag{16.66}$$

$$\Sigma\, C_{F_y}\{\theta\} = \Sigma\, C_{F_y}\{\theta = 90°\}\,(\sin\theta + 0.12\,[B/D]^{0.8}\,\sin2\theta) \tag{16.67}$$

where B and D are the envelope breadth and depth for wind at $\theta = 0°$ as indicated in Figure 16.57(b). Thus D/B in Eqn 16.66 is the fineness ratio (§16.1.3) for $\theta = 0°$, and B/D in Eqn 16.67 is the fineness ratio for $\theta = 90°$.

Experimental results for a set of relatively solid ($s = 0.25$) uniform lattice arrays for the range $1 < D/B < 8$ are compared with the empirical model in Figure 16.58. The model is unbiased, so that it can overestimate or underestimate slightly. The model satisfactorily reproduces the maximum load which occurs in skewed winds when the array is long.

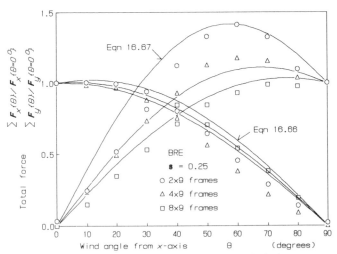

Figure 16.58 Empirical model for uniform rectangular lattice arrays

16.3.6.4 Partly clad lattice arrays

The progressive increase in shielding through many frames with the corresponding decrease in shielding factor, η, as in Figure 16.52, occurs because the flow is able to diverge, leaving the envelope of the array of frames through the sides and top. Cladding the roof or sides, or a combination of both, restricts this divergence.

If just the roof is clad, the flow is able to diverge through the sides and will remain approximately in the wind direction. If one side is clad, the flow must pass around this wall in a similar manner to an isolated wall. This will generate large forces normal to the solid wall (§16.4.2) which will be the dominant loading for all wind directions except parallel to the wall, when the loads on the wall are only small friction loads and the lattice loads are not significantly affected by its presence.

Consider the case of two opposite walls and roof fully clad with the incident wind aligned along the open axis of the array. Flow entering one end must pass through all frames since it cannot escape. Instead of the gradual increase of shielding for the unclad array, the flow must decide on the balance between 'through' or 'around' the partly clad array before entering the array, after which each frame must be similarly loaded. Measurements at BRE show that the new balance results in a total load virtually the same as the unclad case and confirm that this total load is shared equally between the frames. When the wind is skewed to the open axis of the array, the flow through the array is steered by the side walls and remains parallel to the axis. There will be large forces on the side walls and this is one of the partly clad building cases considered later (§18.6.3). Removing the roof while retaining the walls releases some of the constraint on the divergence, but retains the steering effect.

The practical applications for which these cases are relevant are the permanent cases of industrial racking or plant provided with some shelter from the weather, and the temporary cases of bridge deck falsework and building frames during the cladding process. In these cases the loading on the clad parts should be determined as described for canopy roofs (§16.4.4) or open-sided buildings (§18.6), as appropriate, and these forces will be transferred to the lattice through the cladding fixings. The total **direct** loading of the lattice array by wind acting on the lattice elements will not exceed the unclad case, but steering of the flow by walls makes only the parallel component significant and containment of the flow by walls and roof shares this total load equally among the frames.

16.4 Plate-like structures

16.4.1 Introduction

Plate-like structures are bluff in most wind directions, producing normal pressure forces over the surface of the plate. This case was used to demonstrate the effects of slenderness ratio in §16.1.2. Here the quasi-steady model does not hold well and must be replaced by one of the bluff-body models: the peak-factor method (§12.4.2), the quantile-level method (§12.4.3), or the extreme-value method (§12.4.4), which allow for the action of structure-generated turbulence. (In this Guide the pseudo-steady format of the fully probabilistic extreme-value method described in Chapter 15 will be used.) The exception is when the wind is parallel to the plate, the plate has a high fineness ratio (§16.1.3) and flow is streamlined, passing either side of the plate where friction produces a shear stress (§12.3.5) on both faces of the plate, which is covered in §16.4.4.8 later.

Practical structures of this form are chiefly:

(a) boundary walls, hoardings, fences and other vertical plates with one edge on the ground;

(b) signboards and other vertical plates mounted clear of the ground; and
(c) canopy roofs and other nearly-horizontal plates mounted clear of the ground.

As there were very little design data and those for canopy roofs were inconsistent, BRE commissioned a number of wind tunnel studies for cases (a) and (c) at the Oxford University Department of Engineering Science between 1981 and 1986. In every case the peak loadings were assessed in terms of the pseudo-steady coefficients for durations of $t = 1$, 4 and 16 s adopted for this Guide (§15.3), as well as the previous conventional mean values. This work, augmented by additional BRE studies, is the basis of the following sections.

16.4.2 Boundary walls, hoardings and fences

16.4.2.1 Long solid walls normal to the flow

Overall force and moment coefficients for a wall are obtained by integrating the difference in pressure between the two faces over the surface of the wall, thus:

$$C_{F_x} = \iint [c_{p_{front}} - c_{p_{rear}}] \, \mathrm{d}(z/H) \, \mathrm{d}(y/B) \tag{16.68}$$

for the x-axis force coefficient, giving the base shear, and:

$$C_{M_y} = \iint [c_{p_{front}} - c_{p_{rear}}] \, z/H \, \mathrm{d}(z/H) \, \mathrm{d}(y/B) \tag{16.69}$$

for the corresponding y-axis moment, giving the base bending moment. These two equations differ only by the z/H term in Eqn 16.69, the height above the base. Thus the base bending moment for walls is more conveniently given in terms of the normal force, C_{F_x}, and the height of its point of application, z_{F_x}, often called the 'centre of pressure'. For walls of finite length, the local force coefficient can be given for any location, y, along the wall:

$$c_{F_x}\{y\} = \int [c_{p_{front}}\{y\} - c_{p_{rear}}\{y\}] \, \mathrm{d}(z/H) \tag{16.70}$$

by integrating a narrow vertical strip. For very long walls, two-dimensional flow conditions make the local force coefficient constant along the wall and equal to the overall force coefficient.

For walls normal to the flow, where the slenderness ratio is unity or less, the drag coefficient was stated earlier (in Figure 16.1, §16.1.2) to have a value of $C_D = 1.2$. This is in the middle of the range obtained from measurements on two-dimensional walls immersed in turbulent boundary layers [206, 265, 266] and corresponds a mean local value, $\bar{c}_D = \bar{c}_{F_x}\{\theta = 0°\}$, based on the mean wind speed at the top of the wall, $\bar{V}\{z = H\}$. This drag might reasonably be expected to depend on the particular wind speed profile generated by the rough ground surface, and so be dependent on Jensen number, $Je = H / z_0$. Indeed the detailed study by Ranga Raju, Loeser and Plate [266] shows an excellent collapse of all their data to a single curve with Je when the drag is normalised by the friction velocity, u_* (§2.2.7.1), which seems to reinforce this view. However, the velocity profile is itself dependent on u_* through the log-law model, Eqn 7.7 (§7.2.1.3.2). When their data are replotted in terms of the drag coefficient based on the wind speed at $z = H$ to remove this dependence, as shown in Figure 16.59, the remaining variation with Je is seen to be small over a very wide range. The typical range of Je for boundary walls is $10 < Je < 100$, for which $\bar{c}_D = 1.2$ is seen to be a good working value.

The height of application of the normal force depends on the distribution of pressure up the wall. The Oxford–BRE measurements of $\bar{c}_p$ for peak load

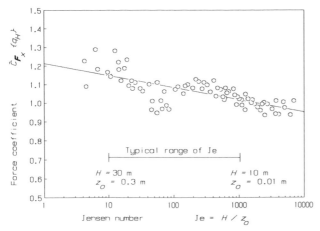

Figure 16.59 Effect of ground roughness on the normal force on an infinitely long solid wall (from reference 266)

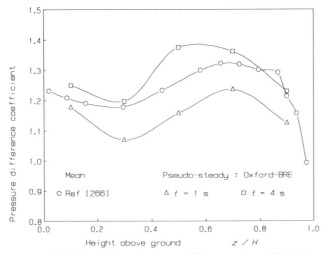

Figure 16.60 Distribution of pressure difference down infinitely long wall with wind normal

durations of $t = 1\,\text{s}$ and $16\,\text{s}$ on an infinite wall, Figure 16.60, bracket the mean data from Ranga-Raju *et al.* [266]. The variation of pressure difference is quite small, except very close to the top of the wall, so the centre of pressure is expected to be near mid-height. In fact, the height of the centre of pressure was found to vary only in the narrow range $0.49 < z_F /H < 0.53$. In view of this small variation, only one load parameter: the base shear, $\bar{c}_{F_x}$, **or** the base moment, $\bar{c}_{M_y}$, need be given with a constant value for z_{F_x}. As the base moment is usually the more important design parameter, this was measured in both the Oxford and the BRE studies; by pneumatic averaging (§13.3.3.1) at Oxford, and directly by a balance (§13.3.3.2) at BRE. In both cases, to comply with the convention of this Guide (§12.3.8), this base moment has been converted back to base shear assuming $z_{F_x} = 0.50\,H$, i.e. weighted by the linear influence coefficient $\phi\{z\} = z/H$.

16.4.2.2 Effect of length and wind direction

On this basis, Figure 16.61 shows the variation of $\tilde{c}_{F_x}$, the normal force coefficient, with wind angle and position from the end of a semi-infinite wall (a wall with one end). For an infinitely long wall, the variation with wind direction is close to cosine as already established for a line-like plate[*] in §16.2.4. A major difference is that $\tilde{c}_{F_x} > 0$ at $\theta = 90°$, where the mean is zero. This occurs because of the action of turbulence on the wall, which causes fluctuations of load about zero when the flow is parallel to the wall (see also §15.3.3, Figure 15.12).

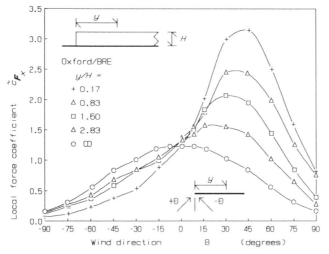

Figure 16.61 Effect of wind direction on the normal force on a semi-infinite solid wall

This cosine variation changes radically towards the free end of a wall. With the end tending to point downwind (θ negative) the loading is reduced as wind slips around the end instead of rising over the wall. With the end tending to point upwind (θ positive) a high suction forms on the rear face which is greatest near the end and remains significant for several wall heights. The effect is greatest at $\theta = 45°$, when local force increases rapidly towards the free end, reaching almost three times the value at $\theta = 0°$. This loading is reflected in a common mode of failure for brick boundary walls, where the end region fractures and falls, leaving the remainder of the wall standing with typically a 45° taper to the ground.

16.4.2.3 Effect of corners

In order to enclose roughly rectangular areas, boundary walls commonly have nearly right-angle 'L'-shaped corners and 'T'- or 'X'-shaped junctions. With 'L'-shaped corners there are two principal ranges of direction, wind onto the external corner, '1' in Figure 16.62(a), and wind into the internal corner, '2'. In case '1', the load increases approaching the corner less quickly than for the free

---

[*] Note that the axis conventions have changed by 90°. The line-like plate was defined with the x-axis parallel to the plate because it was the limiting form with increasing fineness of an ellipse. The wall is defined with the x-axis normal to the wall to comply with the convention of body axes with the x-axis normal to the principal building face, as established in §12.3.6, Figure 12.7.

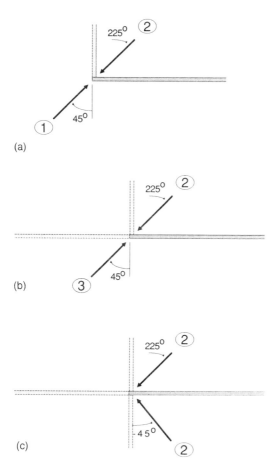

Figure 16.62 Effect of (a) L-shaped corner, (b) T-shaped junction and (c) X-shaped junction on boundary walls

end, and the peak value is reduced to about $\tilde{c}_{F_x} = -2.1$. In case '2', the other limb of the wall tends to contain the wind, increasing the loading in the corner. With 'T'-shaped junctions, there are also two ranges as shown in Figure 16.62(b). Wind directions into the corner act similarly to case '2' for 'L'-shaped corners. In wind directions onto the plane side of the wall, case '3', the junction is not 'recognised' by the flow and the wall acts like an infinitely long wall. With 'X'-shaped junctions, every wind direction has a component into an internal corner and behaves like case '2' of the 'L'-shaped corner.

Unfortunately, there are no experimental data currently available for corner angles other than right-angles. Corner angles between 90° and 180° may be expected to be transitional between the cases given above and the infinite plane wall.

16.4.2.4 Effect of porosity

A porous plane fence is essentially a long low lattice plate, thus fence loads should be given by Eqn 16.34 (Figure 16.38) when composed of flat elements and by

Eqn 16.35 (Figure 16.39) when composed of circular elements. However, as the slenderness ratio H/B is always very small, the last two terms of Eqn 16.34 are insignificant and it reduces to:

$$\overline{C}_{F_x}\{\text{flat}\} = 1.17 + 0.83e^{-3.3s} \tag{16.71}$$

which is now only a single curve, the lower curve of Figure 16.38. Similarly, as only subcritical elements are likely, Eqn 16.35 also reduces to a single curve with $\overline{C}_{F_x}\{\text{circ}\} = 1.2$ at both the low and the high solidity limits. As the variation with solidity is small between these limits, a simplification to:

$$\overline{C}_{F_x}\{\text{circ}\} = 1.2 \tag{16.72}$$

for all solidities gives a conservative value. The variation with wind angle follows the cosine model established for plane lattice frames (§16.3.3.2).

Typically Eqn 16.72 applies to wire-mesh and chain-link fencing, while Eqn 16.71 applies to expanded metal and slatted fences. For this last case, experiments were again made at Oxford and BRE to determine whether the orientation of the fence slats was significant. Base bending moment balances were used in both sets of tests and $\tilde{C}_{F_x}$ was inferred from the measured moment assuming $z_{F_x} = H/2$ as before (§16.4.2.1). Data from the Oxford tests using a small number of slats, plotted in Figure 16.63, appear to show that vertical slats follow Eqn 16.72 and the cosine

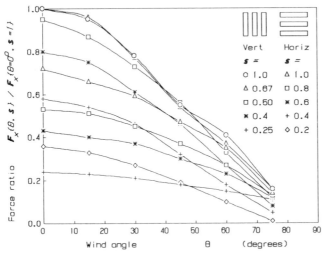

Figure 16.63 Effect on porosity and wind direction on normal force on slatted fences – Oxford data

model well, while horizontal slats apparently gave larger values. The tests at BRE using large numbers of slats found no significant difference between horizontal and vertical slats. The reason for the discrepancy for the same porosity was found to be due to the coarseness of the fences as shown in Figure 16.64. Where Oxford used typically three slats as in (a), BRE used 10, as in (b). The centre of pressure acts in the middle of the envelope of the fence, as indicated by 'CP' in Figure 16.64. In the Oxford tests the lowest slat was some distance above the ground, raising 'CP' above

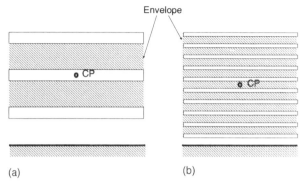

Figure 16.64 Effect of number of slats on base moment for horizontally-slatted fences: (a) Oxford tests; (b) BRE-tests

$H/2$, increasing the moment arm and so exaggerating the apparent value of $\tilde{C}_{F_x}$. Eqn 16.71 works well for all forms of slatted fencing provided the centre of pressure is taken at the centre of the physical envelope.

The BRE tests also showed that the high loads at the end of solid walls in skew winds, described in §16.4.2.2 and Figure 16.61, are completely suppressed when the solidity of the fence is $s = 0.7$, or less. No complimentary data exist for the range $0.7 < s < 1.0$, but a sharp increase to the solid values as $s \to 1$ in a manner similar to the upper curve of Figure 16.38 is expected.

16.4.3 Signboards

There have been a number of studies of the drag of plates of various shapes held clear of the ground and normal to the flow. Most of these are fundamental studies in uniform smooth or turbulent flow. Bearman's study [56] for square plates is typical and gives $\overline{C}_D \simeq 1.2$ for turbulent flow. These available data have been collated by ESDU in Data Item 70015 [228].

The problem of rectangular signboards was briefly studied as part of the Oxford and BRE tests, for the range of proportions $0.25 < H/B < 2$. These tests gave pseudo-steady values for the peak loads consistently greater than the mean values for earlier studies. The variation of load with wind direction was found to diverge from the cosine model, remaining close to the maximum value for 45° each side of normal $(-45° < \theta < 45°)$ before falling almost linearly to near zero at $\theta = 90°$. The clearance from the ground in the tests was equal to the height of the board when $H/B < 1$ and equal to the width of the board when $H/B > 1$, so that the data are typical of the range of practical signboards.

The height of centre of pressure remains on the centreline of the signboard for all wind angles. The horizontal position changes with wind direction, being in the centre for wind normal to the board, but moving to $y = B/4$ from the centre for $\theta > 45°$. This displacement can be a problem with signboards mounted on a single central post, leading to the possibility of divergence or stall flutter instabilities (§8.6.4.2, §10.6.2.2).

Design data for various proportions and wind directions are given in Table 20.4. However, taking a fixed design value of $\tilde{C}_{F_x} = 1.8$ acting either at the centre of the board or at $y = B/4$ from the centre, whichever is the more onerous, will give a safe design.

16.4.4 Canopy roofs

16.4.4.1 Introduction

This section applies to free-standing canopy roofs, such as dutch barns, petrol station canopies and similar shelters that do not have permanent walls. An empty free-standing canopy has no side walls to restrict the flow and the wind is free to pass above and below the canopy. In this situation the principal loads on the canopy are normal forces obtained by integrating the difference in normal pressure between top and bottom faces, and tangential forces from shear stresses by the action of friction on both faces and from normal pressure forces on any vertical facias. However, the volume under such a canopy can become partially or totally blocked with stacked contents, which restricts flow under the canopy and changes the loading. A fully blocked canopy behaves like an open-sided building, such as a grandstand, and this is described in §18.6. However, partially blocked canopies, which are transitional between empty and fully blocked, are discussed in this section.

Prior to 1982, no systematic study of canopies had been performed, either in full or in model-scale, except for some model studies in smooth uniform flow. Nevertheless, design loading coefficients for canopies were given in most national codes of practice, as indicated in Table 16.5. Fifteen different conflicting sets of values were used by 23 countries. Concerned with the disparity, BRE commissioned a comprehensive experimental study of canopy roofs at model-scale in the Oxford University boundary-layer wind tunnel and the data from this study forms the main basis of this section. A much simplified sub-set of these data was put into immediate use by BRE Digest 284[267] and by an amendment to the UK code [4]. There is some difficulty in measuring canopy loads at model scale. Overall loads may be determined by mounting the entire canopy on an under-floor balance (§13.3.3.2). However determining the loading distribution requires pressure tappings to be inserted between the top and bottom surfaces of the canopy. A model used in the Oxford study was shown in Figure 13.36. Pneumatic averagers (§13.3.3.1) were used to integrate the pressure distribution over relevant areas of the canopy, reducing the number of pressure tubes leaving the model through its legs. Although the measurements were made in terms of pressure coefficient, the net difference between the top and bottom surface pressures are required for design, and these are equivalent to force coefficients as described by Eqn 12.16 (§12.3.6).

By this means, the data were acquired in a format directly suitable for this Guide. Accordingly, the data figures have all been included with the rest of the design data in Chapter 20, making it necessary for the reader to search forward to the relevant figures when following the discussion of this section. The alternative, to make the reader search back to this section for the design data when using Chapter 20, was considered less acceptable. On the other hand, figures showing the sparse data for curved canopies are included in this section because they were not considered sufficiently comprehensive to include with the Chapter 20 data.

Canopy roofs have also been a subject for study at full scale by the National Institute for Agricultural Engineering (now AFRC Engineering Silsoe), with measurements reported in 1985[268] and 1986[269]. These full-scale data are valuable as benchmarks to assess the validity of the model data.

Table 16.5   Code of practice data for canopy roofs current in 1982

Pitch Region*	Duo 10° U	 D	Duo 30° U	 D	Duo −10° U	 D	Duo −30° U	 D	Mono 10° U	 D	Mono 30° U	 D
USSR German DR Czechoslovakia	−0.3 1.3	−1.3 0.3	−0.2 1.4	−1.4 0.2	−1.3 0.3	−0.3 1.3	−1.4 0.2	−0.2 1.4	1.3 0.4	0.4 1.3	1.4 0.4	0.6 1.4
Switzerland New Zealand Canada	−1.3	−0.7	1.6	0.4	1.0	1.1			1.3		1.5	
Australia	−1.3	−0.7	1.6	0.4	1.0	1.1			1.1		1.3	
Norway, Sweden Denmark, UK	1.4 −0.6	−1.0	2.0	0.0	−1.4 −0.6	1.0	−2.0	0.0	1.1		1.3	
Chile Uraguay	−0.7	−0.7	1.6	0.5	0.7	0.7	−1.6	−0.5	0.2		0.5	
France	0.5	−0.2	0.8	−0.4	0.1	−0.3	0.5	−0.5	0.4		1.3	
India	−0.6	−0.6	1.3	0.4	0.6	0.6	−1.3	−0.4	0.2		0.5	
Israel	0.6 −0.1	−0.3 −0.8	0.8 −0.2	−0.4 −1.0	−0.6 0.1	0.3 0.8	−0.8 0.2	0.4 1.0	0.5		1.3	
Italy	0.7	−0.6	0.9	−0.6	−0.7	0.6	−0.9	0.6	0.8			
Japan	1.3	−1.0	2.0	0.0	−1.3	1.0	−2.0	0.0				
Poland	1.4 −0.4	0.4 −1.4	1.6 −0.2	0.2 −1.6	−1.4 0.4	−0.4 1.4	−1.6 0.2	−0.2 1.6	1.4 −0.2	0.2 −1.4	1.8 −0.6	0.6 −1.8
Portugal	−1.0	0.5	−1.0	0.5	−1.0	0.5	−1.0	0.5				
Romania	−0.4	−1.0	1.7	−0.4	0.2	0.7	−1.6	−0.8	0.4 −1.0		−1.6 0.5	
Spain	0.8 0.0	0.0 −0.8	1.2 0.4	0.0 −0.8	−0.8 0.0	0.0 0.8	−1.2 −0.4	0.0 0.8	0.4		1.2	
USA									0.6		1.2	

*Key to Regions: U = upwind slope, D = downwind slope.

16.4.4.2 Monopitch canopies

An empty free-standing monopitch canopy is essentially a flat plate held at some pitch angle, α, to the horizontal by support legs. When the pitch angle is zero, the wind passes above and below the canopy with little resistance: the principal vertical loads are normal pressure fluctuations from atmospheric turbulence, and the principal horizontal loads are from friction. At pitch angles other than zero, the inclination of the plate to the wind produces large forces normal to the plate by the 'wing effect' (§16.2.3.1).

Figure 20.17 gives the key to the definitions used for monopitch canopies. The highest loading occurs near the upwind edges of the canopy, so the loading was determined for the areas shown in (a). The 'eave' is defined as the horizontal edge and the 'gable' as the sloping edge. Areas are divided into the upwind half 'A' and downwind half 'B', with two local regions, 'C' along the upwind eave and 'D' along the upwind gable. Hence positive pitch angles, $+\alpha$, are when the low eave is upwind and negative angles, $-\alpha$, are when the high eave is upwind as in (b). The reference height, z_{ref}, for the reference peak dynamic pressure, $\hat{q}_{ref}$, has been taken at the mean height of the canopy as in (b). The pseudo-steady force coefficients for each area, $\tilde{c}_{F_A}$ to $\tilde{c}_{F_D}$ are assumed to act through the centroid of each area, A_A to A_D, with the positive direction downwards as in (b). The overall forces and moments are obtained by summation, including the local regions (see §20.2.5.3).

Data for monopitch canopies are unique only over the range of wind direction $0° < \theta < 90°$. Symmetry for wind angle and roof pitch gives values for all the other wind directions, hence:

$$\tilde{c}_{F_A}\{\alpha, 180° - \theta\} = \tilde{c}_{F_A}\{-\alpha, \theta\} \tag{16.73a}$$

$$\tilde{c}_{F_A}\{\alpha, 180° + \theta\} = \tilde{c}_{F_A}\{-\alpha, \theta\} \tag{16.73b}$$

$$\tilde{c}_{F_A}\{\alpha, 360° - \theta\} = \tilde{c}_{F_A}\{\alpha, \theta\} \tag{16.73c}$$

The local regions move around the canopy to remain along the upwind edges as shown in (c). Note that the pitch angle is reversed as the upwind eave changes from low eave to high eave and vice-versa. This use of symmetry keeps the amount of presented data to the minimum necessary and is common elsewhere in this Guide (see §20.2.5).

Figures 20.18 and 20.19 show the variation of the main region coefficients, $\tilde{c}_{F_A}$ and $\tilde{c}_{F_B}$, respectively, with pitch angle, α, and wind direction, θ. The presentation is as contours of $\tilde{c}_F$ on the θ–α plane. The value for any pitch and wind direction is found by interpolation between the contours. The relevant maximum **or** minimum pseudo-steady value has been given according to the sign of the mean (see §15.3.2), where this is clearly the design value, to give a single design loading case. However, both values have been given in the 'change-over region' where the mean is near zero where both maximum and minimum design load cases should be considered. An important aspect of this form of presentation is that high-load cases are immediately apparent as contour 'mountains' and low-load cases as 'valleys', which is useful at the early design stage to avoid problems and allow optimisation of the design (§19.5).

It is clear that the loading for the two main halves of the monopitch canopy is always greatest with the wind normal to the eave and least with the wind parallel to the eave. These cases correspond to the maximum and minimum flow incidence angles, α at $\theta = 0°$ and to zero incidence at $\theta = 90°$. The loads increase almost linearly with flow incidence angle, hence the contours of Figures 20.18 and 20.19 approximate to lines of constant arctan(tanα cosθ).

Figures 20.20 and 20.21 show the variation of the local region coefficients, $\tilde{c}_{F_C}$ for the eave region and $\tilde{c}_{F_D}$ for the gable region, repectively, in the same format. The eave region behaves like the two main regions, but with larger values. The combined action of $\tilde{c}_{F_A}$, $\tilde{c}_{F_B}$ and $\tilde{c}_{F_C}$ gives a resultant acting between 0.25W and 0.5W from the upwind edge, which compares to the line-like plate in §16.2.3.1. This is always upwind of the canopy centreline, leading to the possibility of divergence or stall flutter instabilities when supported in the centre (§8.6.4.2, §10.6.2.2). The gable region acts quite differently, because here the incidence of the canopy to the wind decreases to zero as the wind angle becomes normal to the gable edge ($\theta = 90°$), so that the maximum value occurs at some intermediate wind angle. The high loading along this edge is due to a 'delta-wing' vortex (§8.3.2.2.2) which creates high suction underneath the canopy at positive pitches and above the canopy at negative pitches.

Inclusion of gable ends was investigated for a 15° monopitch canopy, whereby triangular plates were added at either gable edge to form a wall from the roof to the level of the low eave. This was found to increase the loads considerably and is therefore not recommended.

16.4.4.3 Duopitch canopies

An empty free-standing duopitch canopy is essentially two flat plates, usually at equal pitch angles, joined along one edge to form a ridge or trough. Figure 20.22 gives the key to the definitions used. The two main faces are subdivided similarly to the monopitch canopy, as shown in (a), except that two gable local regions 'D' and 'E' are now required and an additional region 'F' immediately downwind of the ridge or trough line. Positive pitch angles are when the canopy is 'ridged' like a conventional house roof, and negative angles when it is 'troughed'. Otherwise the definitions are similar to the monopitch canopy.

As duopitch canopies are symmetric about the ridge/trough centreline when both faces have the same pitch, instead of antisymmetric as for monopitch canopies, the loading coefficients differ from Eqn 16.73 with wind direction and pitch angle. Data are now unique over the range of direction $0° \leqslant \theta \leqslant 180°$, and the pitch angle no longer alternates in sign, hence:

$$\tilde{c}_{F_A}\{\alpha, 360° - \theta\} = \tilde{c}_{F_A}\{\alpha, \theta\} \tag{16.74}$$

but, owing to the symmetry:

$$\tilde{c}_{F_B}\{\alpha, 180° - \theta\} = \tilde{c}_{F_A}\{\alpha, \theta\} \tag{16.75}$$

The local regions again move around the canopy to remain along the upwind edges as shown in (c).

Because of the symmetry of Eqn 16.75, the data for the main regions 'A' and 'B' can be plotted on the same contour graph, Figure 20.23. The scale of wind angle θ for slope 'A' is marked along the bottom edge of the graph, while the scale for slope 'B' is marked along the top edge. At first thought, it might be expected that the upwind pitch 'A' might behave in a similar way to the corresponding upwind half of a monopitch canopy. That it does not is because the second pitch alters the flow around the whole canopy, and the high forces at the larger incidences are reduced. For 'ridged' canopies, positive (downward) loads occur on the upwind slope in combination with negative (upward) loads on the downwind slope, while for 'troughed' canopies this situation is reversed. With 'ridged' canopies, the maximum downward load occurs when the wind is normal to the upwind eave, whereas the maximum upward load occurs when the wind is skewed by 30°, and the worst combination occurs here too. With 'troughed' canopies the worst cases coincide at the 30° skew angle. Hence duopitch canopies differ from monopitch canopies in that the worst loading case does not occur when the wind is normal to the eave.

Figures 20.24 to 20.26 show the data for the local regions 'C' to 'F'. As these local regions move around the canopy periphery in 90° sectors to remain upwind, as shown in Figure 20.22(c), data for only $0° \leq \theta \leq 90°$ are required. However, the opportunity has been taken to plot the data for regions 'D' and 'E' together in Figure 20.25 in the same format as Figure 20.23 (but if the convention of Figure 20.22(c) is followed, only the $0° \leq \theta \leq 90°$ range is used for each region). The behaviour of regions 'C', 'D' and 'E' are not similar to the same regions of a monopitch canopy for the reason given for region 'A' above, and the range of values is reduced. The worst loads for the new ridge region 'F' occur at the same skew angle of 30° as for the worst loads for the main downwind region 'B'.

The effect of lengthening the dimension L of duopitch canopies was investigated for the $-15°$ and $+15°$ pitches. This revealed reduced loads for the range of wind angle $60° \leq \theta \leq 90°$, i.e. approaching parallel to the ridge/trough line, on the extended downwind end, but here the main loads are at their least in any case. There were no significant changes to the high load range, $0° \leq \theta \leq 60°$, so no benefit can be gained in design.

Inclusion of gable ends to fill the triangle between either eave and the ridge was also investigated for 15° ridged and $-15°$ troughed canopies. The loading of the gables themselves is discussed below. The gables had some effect on the canopy loads, reducing some critical load cases for the ridged canopy by a value of $|\Delta \tilde{c}_F| = 0.2$, but only reducing non-critical load cases for the troughed canopy. As the changes are small and the behaviour at other pitches is unknown, no benefit can be gained in design.

There remains the case of unequal pitch angles for the two slopes of the roof. No suitable data exist for this case. A practical rule-of-thumb will be to use the equal-pitch data corresponding to the pitch angle of each slope. For example, one face horizontal and the other at $\alpha = +15°$ gives a 'ridged' canopy: when the horizontal face is upwind take $\alpha = 0°$ data for regions 'A', 'C' and 'D' with $\alpha = 15°$ data for regions 'B' and 'E'; when the $\alpha = 15°$ face is upwind take $\alpha = 15°$ data for regions 'A', 'C' and 'D' with $\alpha = 0°$ data for regions 'B' and 'E'. Note that the local ridge region 'F' was not included in these rules because more appropriate rules can be deduced. Referring to Figure 20.26, the worst loads for ridged canopies occur on a locus between $\alpha = 15°$ pitch at $\theta = 0°$ to $\alpha = 30°$ pitch at $\theta = 45°$, corresponding approximately to a constant change of slope in the wind direction. This indicates that it is the included angle at the ridge that principally governs these local loads. Flow along the upwind slope is 'asked' to turn by this angle to flow along the downwind slope and its failure to comply, due to the inertia of the flow, causes these local loads. Accordingly, data for the ridge region 'F' should be taken for the average pitch angle, i.e. for $\alpha = 7.5°$ in the above example.

16.4.4.4 Multi-bay canopies

Widening the W dimension of a duopitch canopy can be achieved by reducing the pitch angle α, but very wide canopies commonly consist of multiple pitches.

Loading of multi-bay canopies was investigated at 15° pitch and compared with the duopitch canopy data. In general, the loads were reduced or remained similar, but never increased. The critical high-load cases were always reduced. Corresponding reduction factors for each bay, $\phi\{\text{Bay}\}$, to act on the duopitch loads are given in Table 16.6 according to the key given in Figure 20.27. These reduction factors are partitioned between the main and local regions, and may be assumed to act for all roof pitches in the absence of contradicting data. The force coefficient for the nth bay, $\bar{c}_F\{\text{Bay}\}$, is obtained by factoring the corresponding duopitch canopy value:

$$\bar{c}_F\{\text{Bay}\} = \phi\{\text{Bay}\}\,\bar{c}_F \qquad (16.76)$$

Bays correspond to each duopitch pair, counting from the upwind end. For the main regions, 'A' and 'B', the multi-bay canopy should be regarded as 'ridged' or 'troughed' depending on the pitch of the first bay. Hence the canopy in the key diagram, Figure 20.27, is 'troughed' in winds from left to right and 'ridged' in winds from right to left.

Table 16.6   Reduction factors for multi-bay canopies

	Main regions		Local regions	
	Maxima	Minima	Maxima	Minima
Position	$\phi\{\text{Bay}\}$	$\phi\{\text{Bay}\}$	$\phi\{\text{Bay}\}$	$\phi\{\text{Bay}\}$
Bay 1	1.00	0.81	0.93	0.79
Bay 2	0.87	0.64	0.63	0.71
Bay 3	0.68	0.63	0.56	0.69
et seq				

The gable local regions correspond to the main regions, as above. The eave region 'C' corresponds to the upwind eave only. The ridge/trough region 'F' is alternately a ridge (α positive) and a trough (α negative) region, as indicated in the key.

16.4.4.5 Curved canopies

There are only a small amounts of data for curved canopies and these are all early data obtained in smooth uniform flow. For example Irminger and Nøkkentved [45] included a barrel-vault canopy with a rise of $r = W/4$ in their studies published in 1936, and Blessman [270] included domed canopies with rises of $W/4$ and $W/8$ in his 1971 studies. The validity of the loading coefficient values is open to question in the light of current techniques, however the general loading characteristics may be still be useful.

For a barrel-vault canopy [45] formed from the arc of a circular cylinder and having a rise/width ratio $r/W = 0.25$, Figure 16.65 shows the distribution of local normal force coefficient, $\bar{c}_F$ averaged along the axis of the barrel, with position around the circumference of the barrel for several wind angles, θ, from normal to the axis. In all cases there is a small lobe of positive (inward) load near the front of the canopy, but the majority of the canopy has a negative (outward) normal load. At $\theta = 0°$, the distribution shows the characteristic peak suction near the crest followed by a drop to a constant value, indicating flow separation similar to that for a circular cylinder (§2.2.10.2 and §17.2.2). In the other skew wind directions, for which the effective rise/width ratio is reduced, there is no flow separation. This

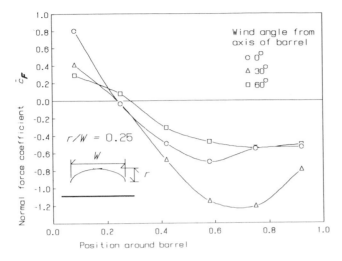

Figure 16.65 Loads on barrel canopy in smooth uniform flow (from reference 45)

indicates that the rise/width ratio of $r/W = 0.25$ is near the critical for separation, with higher rises giving separation and a loading distribution similar to a circular cylinder, and lower rises maintaining attached flow and behaving like a cambered wing. For this example, the maximum normal loads occur with the wind at $\theta = 30°$.

Domed canopies formed from part of a sphere are insensitive to wind direction because they are axisymmetric. Figure 16.66 shows Blessman's data [270] for the top surface (line) and bottom surface (spot values) mean pressure coefficients on the centreline of domed canopies of several rise/width ratios and heights above ground. At the rise/width ratio $r/W = 0.25$ there is no sign of flow separation and the pressure distribution is almost symmetrical, whereas at $r/W = 0.50$ the distribution clearly shows separation and a wake region for the downwind third of the canopy. The critical rise/width ratio for this change is between these two values, higher than for the barrel-vault canopy because for the reduced tendency for flow separation from spheres (see §17.2.1.1). Ground clearance does not appear to be significant.

These sources give no data for the local high load regions. For the barrel-vault, the local regions of the equivalent duopitch canopy can be used, with the exception of the ridge region 'F' which does not exist because there is no abrupt change of slope and the load distribution is continuous over the arc. In the case of the domed canopy, only the upwind eave region exists and rotates around the canopy to remain always upwind.

16.4.4.6 Effect of under-canopy blockage

When a canopy is fully blocked by stacked goods, all the wind must flow around or over the canopy instead of through. The canopy/goods combination becomes bluff instead of plate-like and the external distribution of pressure becomes like a solid building of the same external shape. When the goods are stacked to form a wall up to the eaves on one side of the canopy, typically a dutch barn half full of straw bales, the effect is similar to a grandstand or other open-sided building. Both these cases are covered fully in Chapter 17. However, an overview of the problem is

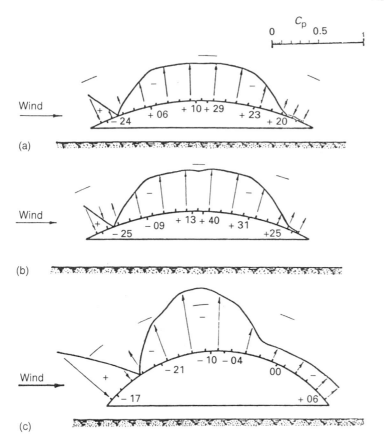

Figure 16.66 Distribution of normal loads on domed canopies (from reference 270) where r/W = (a) 0.25, (b) 0.25 and (c) 0.50

given here so that rules can be derived for the partially blocked case, intermediate between the empty and fully blocked canopies.

The effects of stacked goods was investigated for a range of monopitch, ridged and troughed duopitch canopies at model-scale in the Oxford study. They have also been extensively studied at full scale by NIAE (AFRC Engineering) on two typical duopitch dutch barns with roof pitches of $\alpha = 15°$ and $\alpha = 17°$, the latter with the gable ends sheeted down to the eaves. Data for this barn with wind normal to the eaves are given in Figure 16.67 for a range of stacking regimes using standard size straw bales [268]. In each case, the straw bales were stacked to the eaves, giving 100% blockage ($s = 1.0$). The top surface pressure (chained line), bottom surface pressure (dashed line) and the net difference (solid line) are shown for the section in the centre of the canopy. With the canopy empty (a), the top surface pressure is positive and the bottom surface pressure is negative on the front half of the upwind slope, giving a net downward load; this pattern is reversed for the rest of the canopy, giving a net upwards load. The other seven cases, (b)–(h), show various arrangements of 100% blockage.

Compare this top surface pressure distribution (chained line) of (a) with these blocked cases. Over the first half of the upwind slope the top surface pressure

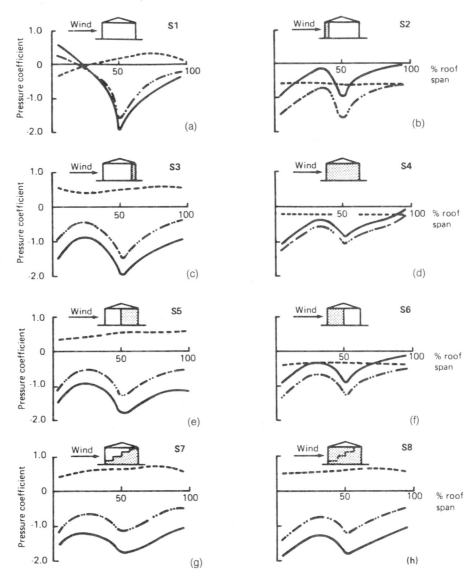

Figure 16.67 Effect of stacked contents on canopy loads (from reference 268)

changes from positive to strongly negative as air is forced over the canopy instead of underneath, but the pressure on the rest of the canopy remains similar to (a). This distribution is very similar for all the stacking arrangements, indicating that it is the degree of blockage that controls the top surface pressures, not the stacking arrangement. This top surface pressure distribution is identical to that on conventional duopitch buildings (§17.3.3.3). Now compare the bottom surface pressures. These depend on the position of the blockage. When the blockage is at the downwind eave, (c), (e), (g) and (h), the dynamic pressure of the wind blowing into the front is recovered as positive pressure, adding to the uplift on the canopy.

When the blockage is at the upwind eave, (b) and (f), the negative wake pressures feed back under the canopy, giving a negative bottom surface pressure, reducing the uplift. These two cases are similar to open sided buildings (§18.6) or the 'dominant opening' case for conventional buildings (§18.5). When the canopy is blocked on all four sides, representing a completely full dutch barn as in (d), the bottom surface pressure is only slightly negative. The canopy is not fully sealed by the contents, so this pressure is established by the balance of flow inwards at the windward eave and outwards at the gables and downwind eave, by the same mechanism by which the internal pressure of a building is established (§18.2). Accordingly, the loading of a fully blocked canopy may be determined in the same manner as an open-sided building, which is by taking the external pressures for the equivalent monopitch or duopitch building in combination with a suitable internal pressure. Table 16.7 gives values of the internal pressure coefficient, c_{p_i}, for various stacking arrangements.

Table 16.7   Internal pressure coefficient for fully blocked canopies

Stacking arrangement	c_{p_i}
Blocked on one side*:	
blocked on upwind side	-0.3
blocked on downward side	$+0.5$
Blocked on three sides:	
open on upwind side	$+0.6$
blocked on upwind side	-0.3
Blocked on all four sides	-0.1

* When the blockage occurs somewhere in the middle of the canopy, instead of at the upwind or downwind eave, take $c_{p_i} = +0.5$ upward of the blockage and $c_{p_i} = -0.3$ downwind of the blockage

    Clearly, if a structure is likely to become fully blocked in use, a dutch barn being typical, then this becomes the most onerous design case. The increase in loading is most critical for nearly-flat canopies, since these are the most lightly loaded when empty and the most heavily loaded when blocked. The contents will also be loaded by the wind (by the the wall loads for the equivalent open-sided building) and this will affect the safety of the structure in use.
    There will be a large number of free-standing canopies that are only ever partially blocked, for example, canopies over petrol stations where the permanent blockage comes from the petrol pumps, kiosks and associated equipment, and temporary blockage is caused by vehicles. For these cases some rules are required in terms of the blockage ratio s. In the Oxford study, the effect of increasing the height of stacked contents to the downwind eave was investigated on an $\alpha = -15°$ monopitch canopy at $\theta = 0°$. The effect of the force coefficient for the main regions, upwind half 'A' and downwind half 'B', is shown in Figure 16.68. The transition from no blockage, $s = 0$, to full blockage, $s = 1$, is not linear. There is little change in the range $0 < s < 0.5$; probably because as the blockage grows upwards from the ground, part of the flow rises over the canopy while the remainder accelerates through the gap and, since both effects tend to lower the pressure, the effects cancel out. Most of the change occurs in the range $0.5 < s < 1.0$ when the majority of the flow is forced over or around the canopy. The uplift on an $\alpha = -15°$ monopitch canopy at $\theta = 0°$ is already high, but the full blockage almost doubles the uplift. In retrospect, it would have been more interesting to see the effect on a flat ($\alpha = 0°$) canopy, where the initial empty loading is small.

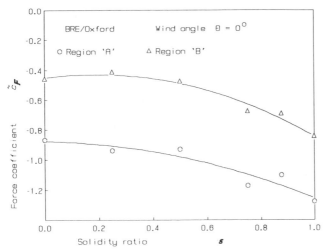

Figure 16.68 Effect of solidity of under canopy blockage at downwind eave of − 15° canopy

Although the knowledge is based only on one canopy pitch, it is reasonable to expect that the observed behaviour is representative of all canopies. Accordingly it is recommended that the loads on partially blocked canopies be interpolated between the two limits **when the blockage grows upwards from the ground** by:

$$\tilde{c}_F\{s\} = \tilde{c}_F\{s = 0\} \qquad \text{for } s \leqslant 0.3$$
$$= [\tilde{c}_F\{s = 0\}\,(1 - s) + \tilde{c}_F\{s = 1\}\,(s - 0.3)]\,/\,0.7 \quad \text{for } s > 0.3 \qquad (16.77)$$

i.e. linearly between $s = 0.3$ and $s = 1$. This means that no corrections for blockage are required for canopies blocked up to 30% from the ground.

16.4.4.7 Loading of facias and gable ends

The action of the wind on vertical facias and gable ends will produce horizontal loads. When the roof pitches are large, the resolved component of the normal canopy loads will dominate the horizontal loads; but when the roof pitches are low, fascia and gable-end loads may give the dominant horizontal load case for stability considerations. The horizontal force coefficient for vertical fascias and gable ends on the upwind edge may be taken as $\tilde{c}_F\{\text{upwind fascia}\} = 1.2$, the value for a long wall, since the canopy forms a ground plane to prevent vortex shedding (§16.1.2). The corresponding fascia or gable end at the downwind edge will be partially shielded by the upwind fascia if the canopy is narrow, or by the effect of friction on the canopy surface if the canopy is wide, so that the force coefficient may be taken as $\tilde{c}_F\{\text{downwind fascia}\} = 0.6$. In the case of 'thick' canopies, i.e. canopies with separate surfaces from the top and bottom of the fascia, the overall horizontal force coefficient can be taken as $C_F = 1.2$ for both x and y directions. In all cases, the cosine model should be used to account for wind direction.

16.4.4.8 Friction-induced loads

The action of the wind blowing parallel to the top and bottom surfaces of the canopy will produce friction loads given by the shear stress coefficient c_τ (§12.3.5).

For low-pitched canopies without significant thickness or fascia, friction forces may produce the dominant horizontal loads, although these will still be small.

Values of mean shear stress coefficient specified by the UK code of practice CP3 Ch5 Pt2 [4] are given in Table 16.8. The derivation of these values is unknown. The use of profiled metal sheets for the decking of canopies and light industrial buildings has become popular in recent years, and it was therefore thought prudent to check that the code values were appropriate. Measurements were made for BRE at the City University on a typical profiled metal sheet, from the pressures acting on the resolved vertical area through Eqn 12.8, as indicated in Figure 12.6 (§12.3.5). The results, shown in Figure 16.69, indicate that the shear stress with the wind normal to the corrugations agrees exactly with the code value in Table 16.8, and that the variation with wind angle follows the cosine model well.

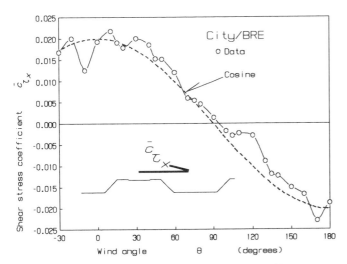

Figure 16.69 Friction forces on typical profiled metal decking

It is also possible to derive better estimates for the case of ribs across the wind direction. Each rib will correspond to a drag coefficient of $c_D = 1.2$, equivalent to a wall, when the wind is normal. If the ribs are closely spaced, then some shelter will reduce this value (§18.2.1). Hence the maximum shear stress will be:

$$\bar{c}_{\tau_{max}} = 1.2 \cos\theta \, a / b \tag{16.78}$$

where a / b is the ratio of vertical to horizontal areas, defined in Figure 12.6.

Table 16.8   Shear stress coefficient for various surfaces

Surface	$\bar{c}_\tau$
Smooth surfaces without corrugations across the wind direction	0.01
Surfaces with corrugations across the wind direction	0.02
Surfaces with ribs across the wind direction	0.04

The shear stress acts on the area of the canopy swept by the wind, which is both top and bottom surfaces when the canopy is empty and the top surface only when the canopy is fully blocked. Fascias produce a separation bubble in their wake which produces reversed flow where the local shear stress acts upwind, as in Figure 12.5(a). To account for this effect, a region $4h$ deep behind the fascia, where h is the fascia height, should be excluded from the area swept by the wind for calculating the friction loads.

17 Bluff building structures

17.1 Introduction

17.1.1 Scope

Chapter 16 dealt with structures that were long and thin (line-like), flat and thin (plate-like) or very porous (lattice). This chapter deals with virtually everything else: structures that are reasonably solid, not long and thin, so occupying a significant volume. Such structures are represented by the bluff-body model of §8.2.3, §8.3.2 and §8.4.2. Nearly all the typical building forms for human occupancy and many industrial structures belong to this group. Although the design data in Chapter 20 contains everything necessary to calculate the wind loading of most common building forms, the additional information given in the more general discussions of this chapter may be directly useful and should also be consulted.

The early wind engineering research at model scale concentrated on building up a library of data on typical building shapes in smooth uniform flow, with the studies of Irminger and Nøkkentved [40,45], of Pris [50] and of Chien et al. [49] (see §13.3.2). Following the work of Jensen [48] (§2.4.3) the model-scaling rules were modified to include Jensen number (Je) scaling (§13.5.1) to match the building modelscale to that of a simulated atmospheric boundary layer in the wind tunnel (§13.5.3). The tendency to collect data for design continued, first in 'velocity profile only' simulations [54] and later in properly scaled boundary layers. Nevertheless uniform flow data still remain in many of the world's codes of practice (§14.1) and some uniform flow tests were still being routinely conducted as late as the early 1980s [271,272]. The data presented in this Guide have been selected as the best currently available. Data from properly scaled boundary layers are always used unless 'velocity profile only' data or, in very rare cases, uniform flow data are the only data available. In the latter case, these data have been assessed for their likely degree of error and included only when their use is expected to be safe.

The variety of shape and form of bluff building structures makes it very difficult to divide the range of data for presentation in a logical fashion. Some specific shapes stand out as clearly different and lead to the expectation that major classes forms could be defined. However, there are always other intermediate forms that fall between such definitions. Nevertheless some division must be attempted and the intermediate structures arbitrarily assigned to the major classes. This chapter and the corresponding sections of Chapter 20 are loosely divided into the two main classes:

1 **Curved structures:** comprising all structures that are predominantly curved, but may have some flat sections.
2 **Flat-faced structures:** comprising all structures that are predominantly flat faced, but may have some curved sections.

These main classes are further sub-divided as seems appropriate, e.g. into 'walls' and 'roofs'.

The variety of shape and form also means that it is impossible to cover the whole range of available data in full detail. Effort has been made to give as much detail as the designer needs. Liberal references have been made to the original sources, but the designer should be cautious about using raw data directly, taking due account of data quality and differences in definitions of coefficients and dimensions from the standard adopted in this Guide. Gaps in the data may be filled from external sources or by specially commissioned tests. External data from uniform flow tests should never be used without first seeking expert advice. The validity of external data from tests in simulated boundary layers can be assessed using the guidance given by Chapter 13. Advice on commissioning tests is given in §13.5.6 and Appendix J.

17.1.2 Pressure-based approach

Typical bluff structures differ from the line-like, lattice and plate-like structures in that they require the design loading data to be in the form of pressure distributions over the loaded surface rather than overall forces and moments. This is to enable the estimation of design loads on small elements such as windows or cladding panels as well as on whole faces or the complete structure. By enclosing an internal volume that may be sub-divided into interconnecting rooms, the problem extends from the distribution of pressure over the external surfaces swept by the wind to include the variation of internal pressure in the rooms, and this aspect is dealt with in Chapter 18.

The peak pressure acting over a surface can be measured directly by pneumatic averaging techniques (§13.3.3.1), but this supposes that the relevant areas can be pre-defined, which is ideal for *ad-hoc* design studies but impractical for general design guidance. Pressures measured sequentially over an array of points suffer from the problem that the peak values do not act simultaneously. A number of approaches have been developed to account for this effect. The covariance technique suggested by Holmes[84,273] requires data to be collected for every combination of pairs of points, which makes the procedure uneconomic except for *ad-hoc* studies. Most other proposed methods, including the integral admittance approach of Greenway[274], can fairly be regarded as too complex for direct use in design except when used by Wind Engineering experts. The method adopted for this Guide is the simple *TVL*-method given by the formula of Eqn 12.37 which relates a characteristic size, l, of the loaded area to the equivalent psuedo-steady load duration, t. This maintains full flexibility in the data which can be given as distributions of pressure, averaged over any relevant area and applied with a peak reference dynamic pressure of the required duration, t. The implementation of this procedure in design is described in §20.2.3 'Influence functions and load duration'.

17.1.3 Influence of wind direction

Figure 17.1 represents a cuboidal building viewed in plan and incorporates some of the major flow structures introduced in Chapter 8 of Part 1 (§8.3.2). The purpose

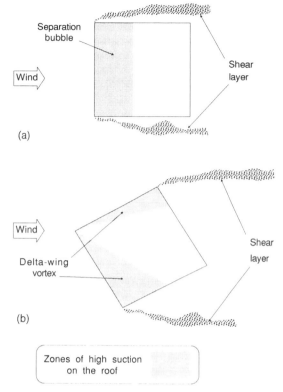

(a)

(b)

Separation bubble

Shear layer

Wind

Delta-wing vortex

Shear layer

Wind

Zones of high suction on the roof

Figure 17.1 Influence of wind direction: (a) normal flow; (b) skewed flow

here is to emphasise the differences in the flow conditions with the wind normal to a face from those when the wind is skewed. Many codes of practice give data only for normal flow on the assumption that this gives the most onerous case. Reinhold *et al.* [275] point out that this is usually true for the overall base shear and overturning moment coefficients, but that the maximum torque coefficient, C_{M_z}, typically occurs near $\theta = 15°$ and that the maximum rate of change of base shear coefficient with wind direction, $dC_{F_x}/d\theta$ (relevant for stability of aeroelastic structures) is greatest at $\theta = 65°$ for a square-section tower. When applied to a real site where the incident wind speed varies with direction due to the wind climate, site exposure and position of neighbouring buildings, the overall forces and moments will depend on the orientation of the structure, and may no longer be greatest when the flow is normal. The maximum local suctions on roofs and walls also tend to occur at skewed wind angles, as will be demonstrated below.

The range of wind angle for which the flow is sensibly the normal case of Figure 17.1(a) is only about $\theta = \pm 12°$. This represents only 27% of the full range of wind direction, so that it is the normal case that should be regarded as 'special' and the skewed flow case as 'typical'.

17.1.4 Influence of slenderness ratio

Figure 17.2 represents the windward or 'front' face of various proportioned cuboidal buildings. The wind can be regarded as 'lazy' because it seeks out the

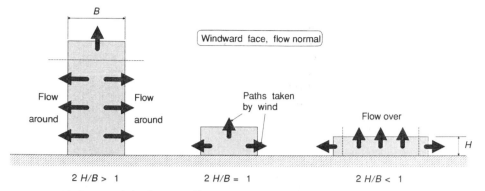

Figure 17.2 Influence of slenderness ratio

easiest path around a building. When the building is tall, $2H/B > 1$, the wind finds it easier to flow around the sides than over the top, except for the zone very near the top. The characteristic size of the building 'seen' by the wind is the cross-wind breadth, B. The flow characteristics and the consequent loading both scale principally to this dimension. This is the case even for the flow over the top, since the size of the top region above the dashed line is set by shortest distance for the flow to travel, which is still the breadth. Conversely, when the building is squat, $2H/B < 1$, the wind finds it easier to flow over the top than around the sides, except for the zones very near the ends. Now the characteristic size to which the flow characteristics and loading principally scale is the height, H. Between these two limits, near $2H/B = 1$, is a regime where both B and H are equally important, but have similar values.

If the characteristic scaling length is denoted by b, then the problem can be conveniently divided into **tall** buildings, with the scaling length $b = B$, and **squat** buildings, with $b = 2H$, depending on the slenderness ratio. This approach has already been used earlier in Guide in that the boundary-wall data of Figure 16.61, where $2H/B < < 1$, are scaled as y/H and is the standard approach in the design data of Chapter 20. However, there may be special exceptions and the data must be seen to comply for the approach to be adopted. For example, in the keys to the canopy roof data, Figure 20.17 and 20.22, the scaling length for the local regions is the smaller of the length, L, or width, W, of the canopy.

17.1.5 Influence of Jensen number

This fundamental scaling parameter, $Je = H/z_o$, established by Jensen [48] §13.5.1, relates the length scale of the structure to the mean velocity profile and the turbulence characteristics of the atmospheric boundary layer near the ground surface. The importance of matching this parameter and its influence on the flow around bluff structures have been discussed earlier in §2.4.3, §8.3.2 and §13.5.1. Later in this chapter, the influence on specific bluff structure forms is discussed individually, but it is convenient here to discuss some general principles.

Jensen's demonstration, described in §2.4.3, established the need for matching Je and this has been confirmed by very many subsequent model studies. The difference between uniform flow ($Je = \infty$) and the range of Jensen number in atmospheric flows, typically $Je = 50 \sim 500$, is very marked, as shown in Figure

2.37. Strictly, data for each shape of structure is required for a range of Je, but usually the variation within the typical range is smaller than the variation from other parameters. In a similar way to Reynolds number, it is sufficient to ensure that Jensen number is in the correct range, then only a single set of data is required for each shape of structure. All the data from BRE studies and most of the external data collated in Chapter 20 have been acquired in the range $20 < Je < 1000$, which covers the range of real buildings.

17.1.6 Reference dynamic pressure

Pressure coefficients are obtained by dividing the pressure by a reference dynamic pressure. The choice of reference height, z_{ref}, for the reference dynamic pressure, q_{ref}, is made to minimise the variation in loading coefficient value. The ideal coefficient is obtained when all data collapse to a single universal curve.

In the case of line-like (§16.2) and lattice (§16.3) structures, the reference dynamic pressure for any location on the structure is the incident dynamic pressure at the local position, i.e. the reference height, $z_{ref} = z$, the local height above ground. This is a feature common to the lattice plate model (§8.2.1) and strip model (§8.2.2) which occurs because the flow divergence remains small. Flow divergence is large for the bluff-body model (§8.3.3), with the consequence that the local dynamic pressure does not give a good collapse. The solid boundary-wall data in Figure 16.59 (§16.4.2.1) collapse to $C_{F_x} \simeq 1.2$ when the reference dynamic pressure is taken at the height of the wall, $z_{ref} = H$. In general, the data for squat structures collapse well with a single fixed value of q_{ref}. The change in slenderness from squat, through tall, to line-like structures represents a gradual transition from local to fixed reference values. Several attempts have been made to define a reference dynamic pressure which gives a perfect collapse. 'In search of a universal pressure coefficient' is how the process was described by Corke and Nagib [276]. Figure 17.3 shows the data of Reinhold *et al.* [275] for the local sectional mean force

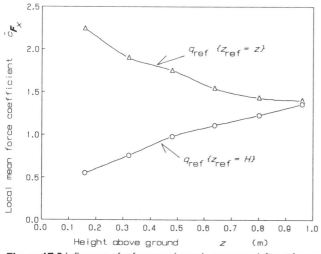

Figure 17.3 Influence of reference dynamic pressure (after reference 275)

coefficient, $\bar{c}_{F_x}$, of a tall square section building normalised in two ways:

(a) by the mean dynamic pressure at the local height, and
(b) by the mean dynamic pressure at the top of the building.

Neither method produces a constant 'universal' coefficient.

Corke and Nagib[276] sought to improve the fit of the first case, using local height z, by adding a fraction, n, of the turbulence intensity to the mean wind speed in calculating the reference dynamic pressure. They concluded that the best collapse was obtained with a value of $n = 1$ for the mean load and $n = 4$ for the peak load, giving a 'universal local coefficient' of $\bar{c}_{F_x} = 0.8$ over the middle region of the building between $z = 0.2H$ and $z = 0.9H$. These values of n compare with the peak factor method (§12.4.2) values of $n = 0$ for the mean and $n \simeq 3.5$ for peaks. This approach exploits the fact that the turbulence intensity increases near the ground, but is a purely arbitrary pragmatic approach because this characteristic has no causal connection with the physical mechanism in Figure 8.5 which causes the loading variation. A better collapse for global force and moment coefficients of cuboidal buildings was derived by Akins et al. [277], based on a fixed reference dynamic pressure calculated from the wind speed averaged over the height of the building. This is described and discussed in §17.3.1 'Cuboidal buildings'.

Local pressures are far more uniform with height in the separated flow regions on side and rear faces of buildings than on the front face, but can vary greatly with horizontal position. The test data for Exchange Square, Hong Kong, shown in Figure 13.47, illustrate this point well. The peak pressures are almost as great near the ground as they are near the top of the building. Pressure coefficients based on a fixed height tend to give nearly vertical contours which enable local regions to be defined as vertical strips.

The search for a truly universal coefficient is quite like the search for the Holy Grail in Arthurian legend – an ideal only to be glimpsed at a distance, but never actually attained. The approach of the Guide remains pragmatic. A fixed reference dynamic pressure is used which gives the optimum collapse of data, as in the special case of Akins et al.'s data [277], but this proves usually to be for a height near the top of the structure.

17.2 Curved structures

17.2.1 Spherical structures

17.2.1.1 Spheres

17.2.1.1.1 Drag in uniform flow.
Spheres have been extensively studied in smooth uniform flow, but unfortunately not in properly scaled atmospheric boundary layers. Like a cylinder (§2.2.10.2, Eqn 2.27), the ideal case of inviscid flow around a sphere can be solved mathematically by potential flow theory, predicting a pressure distribution around any meridian through the front stagnation point given by:

$$c_p = 1 - 2.25 \sin^2\theta \quad \text{(inviscid flow)} \tag{17.1}$$

where θ is the angle measured from the front. Inviscid flow does not allow for any momentum losses, so that this distribution predicts no net drag or lift forces. However, in real viscid flow, momentum loss in the boundary layer on the sphere results in flow separation and a turbulent wake leading to a net drag coefficient, in a similar manner to the circular cylinder (§2.2.10.2, §16.2.2.1). The symmetry of the flow around the sphere produces no net lift.

The variation of drag with Reynolds number follows a very similar form to that of the circular cylinder: with viscous, subcritical, transcritical and supercritical ranges, as in Figure 16.7, except that the drag coefficient is much lower, about 40% of the cylinder values as compared in Table 17.1. The effect of surface roughness on the transition Re and supercritical drag is also very similar to the cylinder (§16.2.2.1.2). Although there are few confirmatory data, it must be expected that the effects of protrusions and porosity bear a similar proportional relationship to the cylinder.

Table 17.1   Mean drag coefficient of sphere and two-dimensional cylinder [206]

C_D	Sphere	Cylinder
Subcritical	0.47	1.2
Transcritical	0.10	0.4
Supercritical	0.19	0.6

17.2.1.1.2 Pressure distribution in uniform flow.   Figure 17.4 gives the experimental data available to Hoerner [206] in 1958, compared with the inviscid prediction of Eqn 17.1. The distribution follows the prediction from the front stagnation point to the separation point with some boundary-layer losses (greatest in the subcritical range). In the subcritical range separation occurs just before $\theta = 90°$, leaving a wide wake and a constant base pressure of about $\bar{c}_p = -0.35$. In the transcritical and supercritical ranges separation occurs much later, leaving a narrow wake and a higher (less negative) base pressure.

Consider the sphere to be a terrestrial globe with the front stagnation point as the North pole. The lines of 'latitude' on this globe are contours of equal pressure, i.e.

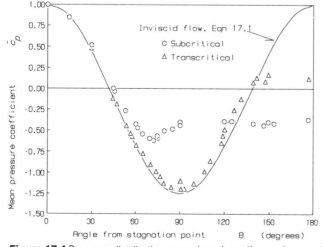

Figure 17.4 Pressure distribution around a sphere (from reference 206)

straight lines when viewed normal to the axis through the poles, in plan or in side elevation. The transcritical case shown in Figure 17.4 represents the drag minimum, where the base pressure recovers to a slightly positive value. As with the cylinder, this minimum drag cannot be exploited in design (§16.2.2.1.1) and only subcritical and supercritical values are required. The maximum suction in supercritical flow is $\bar{c}_p = -1.25$ and occurs along the 'equator', $\theta = 90°$. When all wind directions are considered, all parts of the sphere will experience this suction at some time.

17.2.1.1.3 Effect of ground plane.

Klemin *et al.* measured the drag of a model of the 'Perisphere' of the 1939 New York World's Fair [206,278], again in smooth uniform flow. This was a 200 ft diameter sphere supported just above the ground. This large size gave supercritical flow at all appreciable wind speeds. The measurements were made with no supports between sphere and ground, with four supporting columns and with a solid cylinderical collar about half the diameter of the sphere. The results are given in Table 17.2. The gap, G, between the 'Perisphere' and the ground was only about 1/30th of the diameter, D. The effect of the ground is to restrict flow underneath the sphere, inducing earlier separation and widening the wake towards the ground. The drag increases and the asymmetry of the flow gives a net lift force. Adding support columns increases this effect. The solid collar allows the wake to attach to the ground, increasing the effect still further, as does lowering the sphere until it touches the ground.

Corresponding modern structures on columns include high pressure gas or liquified gas tanks, and those on a collar include radar and astronomical observatory radomes. The multi-faceted forms of radome can be treated as rough spheres. The spherical part of the common European 'sphere on a stick' water towers are sufficiently clear of the ground to be treated as isolated spheres.

Table 17.2   Mean drag coefficient of sphere close to the ground [278]

Gap $G = D/30$	$\bar{C}_D$	$\bar{C}_L$	Ref $\hat{q}$ at centre of sphere
No supports	0.30	0.03	
On 4 columns	0.49	0.29	
On collar (0.5D)	0.58	0.41	
Touching ground	70	?	

17.2.1.2 Domes

17.2.1.2.1 Hemispherical domes.

A hemispherical dome springing directly from the ground is expected to act like half of a sphere. Early measurements [270] in smooth uniform flow confirm this, except that the pressures near the base of the dome are affected by the thin boundary layer that always forms on the ground surface. The boundary-layer has been represented at various depths in later measurements, including thin boundary layers on smooth ground [279] representative of rivet heads, etc., as well as atmospheric boundary-layer simulations [280,281,282], but measuring only mean pressures.

Peak pressure values are only available from *ad-hoc* design studies, such as the nearly hemispherical dome studied at BRE, for which the design pressures (in

Pascals) are shown in Figure 17.5. The contours of constant pressure follow lines of 'latitude' over most of the dome, similar to the sphere, but the pressures near the base of the dome are less, owing to the 'horse-shoe' vortex [279], as in Figure 8.7. The shape of the contours is similar for mean, minimum and maximum values. The working assumption of constant pressure along lines of latitude remains good and this requires only the variation along the centreline to be specified. The values in Figure 17.5 cannot be used directly for design because they are specific to the site environment of the full-scale dome. To obtain the range of verified data required for design, it is necessary to revert to the quasi-steady assumption and use the mean pressures given by the various general studies [270, 279, 280, 281, 282].

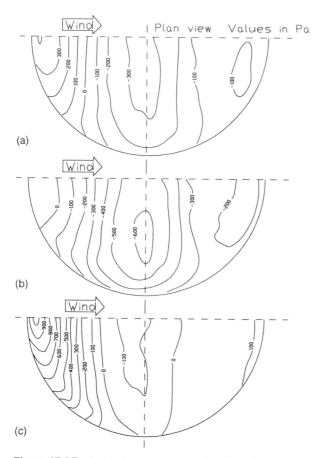

Figure 17.5 Typical design pressures on hemispherical dome: (a) mean; (b) minimum; (c) maximum

Studies in rough- and smooth-wall boundary layers have been made by Toy *et al.* [280], but the boundary layer is only several times deeper than the height of the dome (corresponding to a 1km high dome!). Blessmann's later measurements [281] were made in a variety of boundary layers scaled to the power law. The study by Newman *et al.* [282] was made in a deep atmospheric boundary-layer simulation, but at very low Reynolds numbers. The data, [280] and [281], are presented as

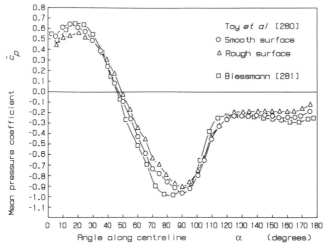

Figure 17.6 Effect of Jensen number on the pressure distribution along the centreline of a hemispherical dome

pressures along the centreline in Figure 17.6. The collapse of data to a single curve is very good. It is clear that the variation with Jensen number and immersion ratio implied by the variety of simulations and depth of the boundary layers is very small.

17.2.1.2.2 Effect of rise ratio.   If the 'rise' of a dome is defined as the height of the dome, H, then the rise ratio is H/D, where D is the diameter of the dome in plan. For a hemispherical dome or taller dome, $H/D \geqslant 0.5$ and D is also the diameter of the parent sphere. Rise ratios greater than 0.5 represent structures between the sphere near the ground (§17.2.1.1.3) and the hemisphere, such as radomes, and the pressures are transitional between these two cases. Rise ratios less than 0.5 represent shallower domes and D is the base diameter, for which data from Blessmann's later studies[281] are given in Figure 17.7. Here the data have been presented in two forms: in (a), the position on the centreline is expressed as the elevation angle, α, from the axis of the parent sphere, as in Figure 17.6; but this may not be as convenient as (b), where the position is expressed as the fraction of the arc of the centreline.

   The trend is for both pressure and suction to reduce with smaller rise ratios and the flow is less prone to separate to form a wake: separation occurring for $H/D = 0.25$, but fully attached flow occurring for $H/D = 0.125$. Low-rise domes are less severely loaded than high-rise domes when they spring directly from the ground. This is not true of domed roofs on cylindrical walls, as demonstrated below. The largest cladding suction occurs along the 'equator' at $s/S = 0.5$, and when all wind directions are considered, all parts of the sphere experience this value.

17.2.2 Cylindrical structures

17.2.2.1 Vertical cylinders

17.2.2.1.1 Scope.   Vertical cylinders include storage tanks, silos, cooling towers and circular-plan buildings. Gentle changes in diameter, such as hyperbolic cooling towers, are included here, but the effect of sudden changes in diameter was

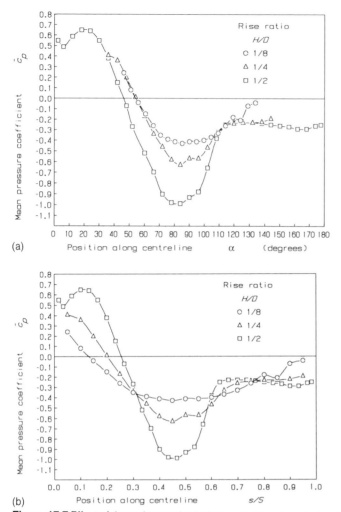

Figure 17.7 Effect of rise ratio on spherical-section dome by: (a) elevation angle to centre of sphere and (b) relative position along arc

discussed earlier in §16.2.2.1.4. The scope includes all cylinders with $H/D < 4$, whether open-topped or fitted with flat, conical or domed roofs. For all practical structures, their diameter, D will be sufficiently large that $DV > 6$ m²/s for any significant wind speed. Hence only super-critical Reynolds number data are considered here. Vortex shedding will be weak from such short cylinders and need not be considered for static effects, however ovalling oscillations may occur if the cylinder walls are very flexible (see Part 3). Taller structures, such as chimney stacks, which are line-like (§16.2.2.1, §20.3.3.1) are outside the scope of this section.

This is one of the few applications where sufficient full-scale data exist to confirm the validity of model studies, of which there are so many that only the most contemporary are considered here. These data are still mostly mean values, but some peak values are available for comparison.

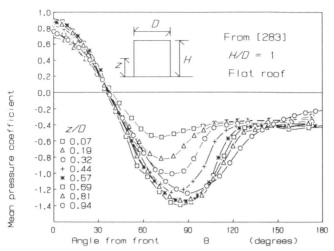

Figure 17.8 Variation of circumferential pressure around cylinder with height above ground

17.2.2.1.2 Walls. Static buckling of the walls in response to the circumferential pressure distribution is usually an important design consideration. For tall cylinders, $0.5 < H/D < 4$, the circumferential pressure distribution does not vary much with height, except in the region close to the ground, $z/D < 0.5$, as shown in Figure 17.8[283]. Adopting a single circumferential distribution based on the constant region is conservative. Figure 17.9 compares the Australian code proposals for several values of slenderness with the parametric model tests[283] on which they are based. Also shown is the mean pressure distribution compiled by Briassoulis and Pecknold[284] from available full-scale data on cooling towers. The trend for the suction lobe on the side to reduce with slenderness ratio as more of the flow goes over the top is confirmed by Maher[285]. The form of the roof makes no appreciable difference in this example, where both a flat and a 25° conical roof are

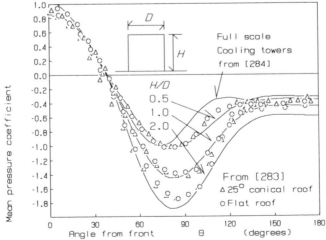

Figure 17.9 Effect of slenderness on the circumferential pressure around a vertical circular cylinder

shown. The sparser data for very low cylinders [285, 286], suggest that the suction lobe does not continue to reduce and the distribution remains close to that for $H/D = 0.5$.

Pseudo-steady pressure coefficients from peak load measurements by BRE on a full-scale silo with $H/D = 1.14$ were compared in Figure 15.16 with estimates from the measured mean values using Eqn 15.25. Further comparison with the corresponding mean pressure coefficients for $H/D = 1$ in Figure 17.9 confirms that the peak positive and negative pressures match well. However, where the mean values in Figure 17.9 are zero and the quasi-steady model predicts no load, the pseudo-steady coefficients indicate a range of $\tilde{c}_p = \pm 0.7$. This is a common feature with most curved structures, including the earlier spherical structures of §17.2.1. Although the values in this region are not the maximum experienced, so are less important for static design, they have the largest range. The random cycling of the pressure through zero may have a significant effect on the fatigue resistance of fastenings (see §19.5.2).

17.2.2.1.3 Flat and monopitch roofs.

Holroyd's sketch of the flow around a flat-topped tank, from flow visualisation studies in a scaled atmospheric boundary layer [286], is shown in Figure 17.10(a). The separation bubble at the front edge is followed by flow reattachment at the rear. This scales to D, i.e. a fixed proportion of the roof, as shown in Figure 17.10(b), for $H/D > 0.5$ and to H, i.e. a reducing proportion of the roof, as in Figure 17.10(c), for $H/D < 0.5$ according to the principles in §17.1.4.

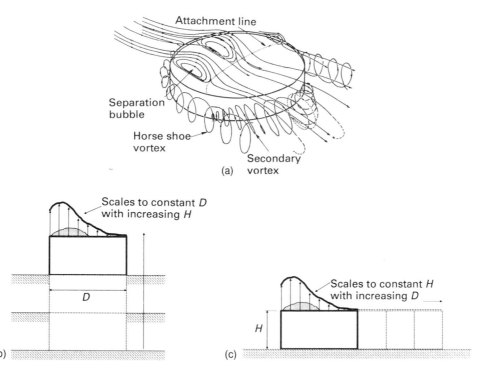

Figure 17.10 Flow over flat cylinder roof: (a) flow around squat flat-roofed cylinder (from reference 286); (b) $H/D > 0.5$; (c) $H/D < 0.5$

Almost exactly the same behaviour occurs with cuboidal buildings when normal to the flow (§17.3.3.2.1). As there are no corners, the wind is always normal to the front edge and 'delta-wing' vortices never form. Thus the special case established in §17.1.3 is the typical case in this instance. The similarity is so close, even in uniform flow, that Yoshida and Hongo[272] suggest that the pressure distribution for the flat cylinder roof can be deduced directly from the distribution on the square cuboid of the same height, as indicated in Figure 17.11. A corollary is that monopitch roofs on vertical cylinders might be expected to be similar to monopitch roofs on cuboidal buildings when normal to the high or low eave ($\theta = 0°$ or $180°$) and this model is used in Chapter 20.

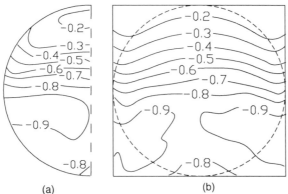

(a)   (b)

Figure 17.11 Model for pressure on flat cylinder roof (from reference 272): (a) cylinder and (b) equivalent cuboid where $H/D = 0.5$

17.2.2.1.4 Domed roofs.

Parametric studies of domed roofs on cylinders have been made in uniform flow only [270,281,285]. The tests are sufficient to deduce flow behaviour, but can also be used in design if care is taken to account for the expected differences in the atmospheric boundary layer.

Hemispherical and taller domed roofs, $R/D > 0.5$ where R is the rise of the dome above the top of the cylinder, behave like the sphere and the dome on the ground. Below $R/D = 0.5$ a sharp edge forms at the junction of the walls and roof. Eventually, the positive pressure lobe at the front is replaced by a separation bubble, with negative pressures, in an arc around the upwind eave. The rise ratio at which this occurs depends on the slenderness of the cylinder, but corresponds to an equivalent pitch at the eave in the range $30° < \alpha < 40°$. For lower domes, the distribution converges towards the flat roof values.

17.2.2.1.5 Conical roofs.

Conical roofs are more common than domed roofs owing to their ease of construction. Studies for silos with $H/D \geqslant 0.5$ have been made for a range of cone pitch, α, between $15°$ and $45°$ in properly scaled atmospheric boundary layers [283,287], supplemented by a number of full-scale studies [287]. The typical distribution of pressure shown in Figure 17.12 for $H/D = 1$ and $\alpha = 25°$ is similar in form to the dome in Figure 17.5, except that the contours are 'pulled' towards the central apex of the cone and the positive pressure lobe is replaced by suctions in the arc-shaped separation bubble around the front

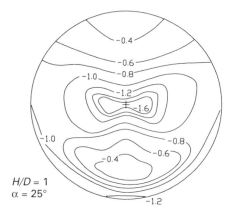

$H/D = 1$
$\alpha = 25°$

Figure 17.12 Pressure on conical roof of cylinder
(from reference 283)

edge. For higher cones, the positive presure lobe at the front is restored around $\alpha = 35° \sim 40°$, and for cones taller than $\alpha = 45°$ the distribution converges towards that around a (tapered) cylinder, e.g. church spires. For lower cones, the distribution converges towards the flat roof values.

17.2.2.1.6 Open roofs. With tall cylinders when there is no roof, the hole in the top acts as a dominant opening which transfers the external pressure at the hole uniformly to the inside of the cylinder, as discussed later in §18.5. With squat cylinders, the situation is far more complex, the flow over the top inducing a circulation inside the cylinder which results in non-uniform pressures on the inner face of the cylinder walls. This is illustrated in Figure 17.13 for $H/D = 0.2$, which may be compared with the flat roof case of Figure 17.10(a) [286]. In the limit as $H/D \to 0$, the walls of the cylinder will act like a circular boundary wall (§16.4.2).

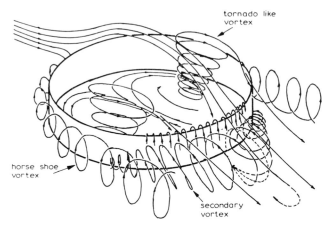

Figure 17.13 Flow around squat open-topped cylinder (from reference 286), to be compared with Figure 17.10(a)

The distribution of pressure difference across the wall does not vary much with height, as in the general case of the external pressure (§17.2.2.1.2). Figure 17.14 presents the available data [283, 286] for the pressure difference across the wall as a

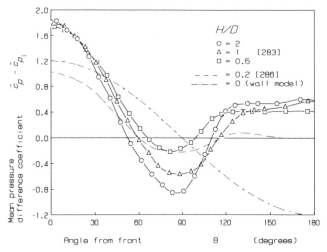

Figure 17.14 Pressure around open-topped vertical cylinder

function of circumferential position. The curves for tall cylinders [283], $H/D \geqslant 0.5$, are almost identical to the corresponding curves of external pressure in Figure 17.9, except that they are offset to positive values by the negative internal presure. The cylinder tends to be compressed along the x-axis, resulting in the risk of a buckling failure as shown in Figure 3.17. The distribution for squat cylinders [286] is transitional between the tall case and the model for a boundary wall ($H/D = 0$) from §16.4.2.

The completely open-topped tank is frequently used to hold water or farm slurry. Volatile liquids are sometimes held in tanks with floating lids that rise and fall with the contents of the tank. Such tanks are 'variable-geometry' structures, as discussed in §19.7. This was the main subject of Holroyd's study [286], for which Figures 17.10(a) and 17.13 represent the upper and lower limits of the roof. The critical design case is always the empty tank, since the loading reduces and the tank becomes stiffened by the ring tension caused by the weight of the contents as the tank is filled.

17.2.2.2 Horizontal cylinders

17.2.2.2.1 Cylindrical tanks. Data exist for short horizontal circular cylinders clear of the ground towards the $L/D \rightarrow 0$ limit of the line-like cylinder data, but only for the case of flow normal to the cylinder axis and flat ends. Unfortunately, no data exist for short cylinders in proximity to the ground in properly scaled atmospheric boundary layers. Structures of this type are mostly pressurised gas or liquid storage tanks, which are inherently strong and heavy, so that wind-loading calculations are less important, and this is probably why no adequate data exist. Nevertheless, the discussion of line-like cylinders and the corresponding design data are expected to be relevant.

17.2.2.2.2 Arched structures. In contrast to the complete horizontal circular cylinder, the hemicylinder and lower cylindrical-section arched structures are a common building form. It is particularly used for covering sports arenas, swimming pools, etc., because it is a structurally efficient way of roofing a rectangular area and because it is also a natural shape for air-supported structures. The form was among the earliest full-scale studies, on the Akron airship hangar [38,288], and has been studied more recently at full scale [289,290] and in the wind tunnel [291,292].

Figure 17.15 shows the mean pressure distribution around the centreline arc of a hemicylinder normal to the flow [292], $\theta = 0°$ for various lengths, L/D. This is similar to the distribution around a hemispherical dome when L/D is small, but the suction peak increases as $L/D \to \infty$. Roughening the surface, Figure 17.16, reduces this effect [291]. This centreline distribution is typical over most of the length for $0° \leqslant \theta \leqslant 45°$, as shown in Figure 17.17(a) and (b) [292]. As the wind turns further towards parallel with the axis, $\theta = 90°$, the pressures around the arc depend

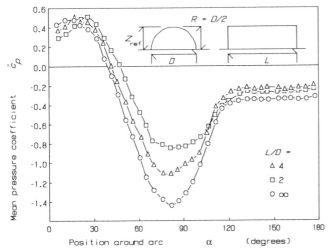

Figure 17.15 Mean pressure distribution around a hemicylinder normal to the flow (from reference 292)

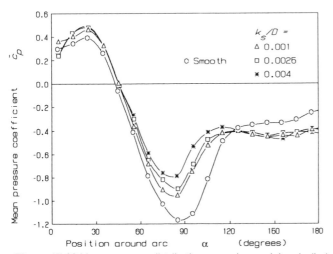

Figure 17.16 Mean pressure distribution around a rough hemicylinder normal to the flow (from reference 291)

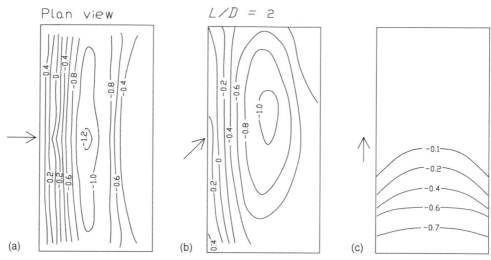

Figure 17.17 Effect of wind angle on pressure distribution on hemicylinder (from reference 292): (a) 0°; (b) 45°; (c) 90°

strongly on the form of the ends. Flat ends produce a separation bubble at the end resulting in a band of high local suctions, followed by reattachment and small suctions over most of the length, as in Figure 17.17(c). Domed ends produce a flow similar to the hemispherical dome, with the suction lobe at the 'equator', followed by attached flow over the whole length (see Figure 17.21 in §17.2.3).

Pressures on the ends themselves obviously depend on their form. Flat ends behave much like the gable ends of buildings (§17.3.2.1), with the face under suction with a high local suction region at the upwind edge for $\theta = 0°$ and the face in positive pressure for $\theta = 90°$. Domed ends act as stated above, with the pressure distribution over the dome end blending smoothly into the centreline arc distribution on the hemicylinder.

There are few data for lower-rise cylinder sections, most being for cylindrical-section curved roofs on cuboidal buildings, which are considered later in §17.3.3.5 Barrel-vault roofs. The expectation is that the pressure distribution will change in an analogous manner to the dome, as in Figure 17.19. Toy and Tahouri [292] investigated variations on the hemicylinder, where the cross-section is either lengthened to produce a flat top, or shortened to produce a sharp ridge. In the first case, a short flat top interrupts the hemicylinder distribution, inserting a 'plateau' where the pressure remains reasonably constant. Figure 17.18(a) shows the distribution for a wind angle of $\theta = 45°$, when a $0.25\,D$ wide flat top is inserted. Comparing this with Figure 17.17(b) for the hemicylinder, shows the suction lobe is slightly reduced in value, but spread out almost uniformly over the 'plateau'. If the flat top is extended, further reduction and extension of the suction lobe occurs, resulting eventually in a flat-roofed building with curved eaves, with attached flow and very small suctions over most of the surface; this is discussed in §17.3.3.2.7 and §19.3. In the second case, flow separates from the sharp ridge, causing a deeper wake and a more extensive positive pressure lobe on the front face. At skewed wind angles a vortex forms behind the ridge in a similar manner to duopitch roofs (§17.3.3.3.2), creating a region of high local suction just behind the ridge. Figure 17.18(b) shows the corresponding pressure distribution for $\theta = 45°$. In general,

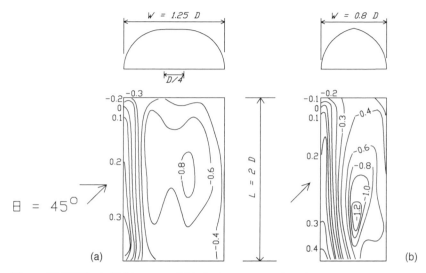

Figure 17.18 Effect of (a) longer and (b) shorter cross-section shape on cylindrical-section arched buildings (from reference 292)

widening the section reduces both the overall drag and the local cladding suctions and narrowing the section has the opposite effect, particularly with a sharp ridge. However, the lift force is greatest for the pure hemicylinder, as given in Table 17.3. Adding a ridge ventilator lantern to an otherwise smooth curved profile has a similar detrimental effect to the sharp ridge [293].

Table 17.3   Mean force coefficients for long hemicylinders [292]

Form	$\bar{C}_D$	$\bar{C}_L$	Ref $\hat{q}$ at top of hemicylinder
Hemicylinder	0.26	0.63	
Flattened section	0.54	0.21	see Figure 17.18(a)
Ridged section	0.20	0.54	see Figure 17.18(b)

Hemicylinders also occur in multi-span form, particularly as horticultural greenhouses. Measurements at full scale [290] indicate that the pressures over the first span are not significantly changed, but are progressively reduced over later spans.

17.2.3 Other curved structures

Curved structures of unusual shapes that are not included in the pool of available data are obvious candidates for *ad-hoc* wind tunnel tests (§13.5.6). Nevertheless, most unusual structures are composed of basic elements similar to the ones already discussed, and components in the flow are recognisable in their characteristics.

The Olympic Coliseum at Calgary has a cable-supported roof which is curved in two opposite directions. Design pressures on this roof were determined by

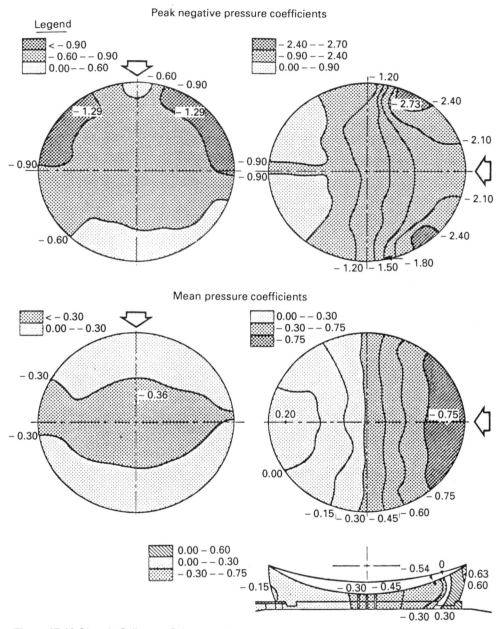

Figure 17.19 Olympic Coliseum, Calgary, design mean and peak pressures (from reference 294)

wind-tunnel tests [294] and examples are given in Figure 17.19. The similarities between these data and the corresponding shallow dome and flat cylinder roof data are clear.

Similarly, the model of the Al Shaheed Monument in the BRE boundary-layer wind tunnel was shown in Figure 13.39 with example data in Figure 13.46. Even this unusual structure shows predictable pressure contours. Wind directly onto the

Values in Pa

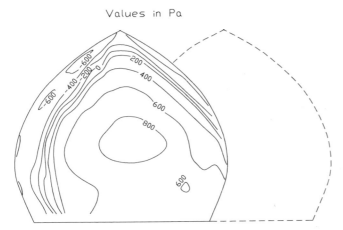

Figure 17.20 Al Shaheed Monument and Museum design pressures for azimuth 180°

convex face (into the page in Figure 13.39) gives the pressure contours shown in Figure 17.20 which are very similar to those on a sphere. Differences are that the stagnation point moves to right of the centre of the shell, towards the centre of the combined pair of shells; and that the pressure contours extend to the ground.

Another study by BRE for the restoration of the Palm House at the Royal Botanic Gardens, Kew, a famous Victorian masterpiece of cast iron and glass, was useful for deducing the effect of domed ends to hemicyliners and reinforcing the observations of the effects of lanterns [293], since the long arms of the Palm House take this form. Figure 17.21 shows the design pressures for winds directly along and normal to the axis of one of these arms.

Accordingly, first-order estimates of loading on even the most unusual curved shape can be deduced from the characteristics described in this section. More accurate estimates can only come from specially commissioned wind-tunnel tests.

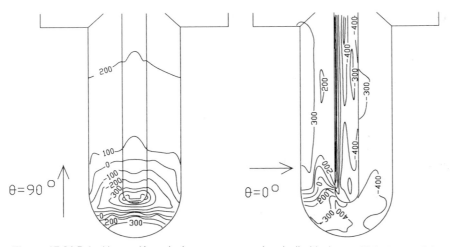

Figure 17.21 Palm House, Kew, design pressures on hemicylindrical arm with lantern and domed end

17.3 Flat-faced structures

17.3.1 Cuboidal buildings

17.3.1.1 Introduction

A simple cuboid with some variations of roof form is the commonest building shape, so is a natural first choice for full-scale studies. These include the BRE experiment at Royex House, London; Confederation Heights[295] and Commerce Court[163], Canada; the Menzies Building[6], Australia; Waseda University[18], Japan; and the mobile-home study of Marshall[165]. All these full-scale studies, except for the mobile homes, are tall: either as a 'tower' in both horizontal axes (Commerce Court[163]) or as a 'slab' in one axis (Menzies Building[6]). They are also a natural starting point for parametric model studies, from which data for the common building forms are derived.

Overall loads on the whole cuboid, base shear forces and moments, are useful design parameters for tall buildings, since these tend to behave monolithically and all the wind, occupancy and dead-weight loads are transmitted through to the foundations at the base. Overall mean loads are also the basis for quasi-steady dynamic response theory, as described in Part 3 'Dynamic structures', and are necessary parameters for design assessments. Accordingly, there have been many studies of overall loads on tall buildings at full and model scale. The same is less true of squat buildings as the load influence functions (§8.6.2.1, §20.2.3) rarely encompass the full width of the building and the structure is usually designed in independent 'bays', as in Figure 8.26(c), requiring distributions of design pressure. Important exceptions to this general trend are mobile homes and other small monolithic structures for which anchorage loads are required and the stability of large, squat, framed buildings where all the lateral resistance is concentrated into wind bracing at one location.

17.3.1.2 Definitions

The convention of the Guide is to denote the longer side dimension as length, L, and the smaller side dimension as width, W, with H as the height. These are body-axis dimensions (§12.3.6), so that the loaded areas for overall forces are $A_x = L\,H$, $A_y = W\,H$ and $A_z = L\,W$. The x-axis is normal to the larger face as shown Figure 12.7, so that at $\theta = 0°$, the long face is normal to the wind. The wind-axes dimensions: cross-wind breadth, B, and inwind depth, D, have already been used in describing the effects of slenderness (§16.1.2) and fineness (§16.1.3) ratios, but these depend on the wind angle. At $\theta = 0°$ or $180°$, $B = L$ and $D = W$ and fineness ratio is smaller than unity; whereas at $\theta = 90°$ or $270°$, $B = W$ and $D = L$ and fineness ratio is greater than unity.

17.3.1.3 Overall forces and moments on tall buildings

In Akins et al.'s study of overall loads on tall buildings[277] measurements were taken of the overall forces and moments on cuboidal buildings over the range of proportions $1 \leqslant H/L \leqslant 8$, $0.5 \leqslant H/W \leqslant 4$ and $1 \leqslant L/W \leqslant 4$. A good collapse of data for all heights, H, in boundary layers varying in power-law exponent from $\alpha = 0.12$ (open country) to $\alpha = 0.38$ (towns) was obtained when the reference dynamic pressure was derived from, V_{ave}, the average wind speed over the height of the building:

$$V_{ave} = \frac{1}{H} \int_0^H V\{z\}\, dz \tag{17.2}$$

This is the same as applying the weighting function $\phi = (z/H)^\alpha$, using the power-law model for velocity profile (§7.2.1.3.1), to the dynamic pressure at the top of the building. In the two cases shown in Figure 17.3, the equivalent weighting functions are $\phi\{z_{ref} = H\} = (z/H)^0 = 1$ and $\phi\{z_{ref} = z\} = (z/H)^{2\alpha}$. As Akins *et al.'s* weighting is halfway between these two cases which straddle the 'perfect collapse' (§17.1.6), it is expected to give a better collapse than either. The collapse obtained for the base shear force, $\overline{C}_{F_x}\{\theta\}$ of square-section cuboids is demonstrated in Figure 17.22. By symmetry, this is the same as $-\overline{C}_{F_y}\{90° - \theta\}$. This compares individual data for ten different combinations of slenderness ratio and velocity profile with their average. The collapse is good when the wind is normal to the face and the load is high, but becomes progressively worse as the wind direction turns. Nevertheless, the collapse is generally good and is typical of all the overall coefficients.

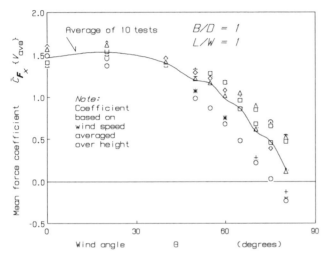

Figure 17.22 Mean x-force coefficient on cuboidal buildings (from reference 277)

The effect of cross-section proportions, L/W, (or fineness ratio, D/B), on the averaged base shear forces, $\overline{C}_{F_x}\{\theta\}$ and $\overline{C}_{F_y}\{\theta\}$, is shown in Figure 17.23. Note that the most elongated cuboid, $L/W = 4$, has a low fineness ($D/B = 0.25$) at $\theta = 0°$ with the longer face normal to the wind, but a high fineness ($D/B = 4$) at $\theta = 90°$. Accordingly, the key for fineness ratio is given The curve for $D/B = 0.25$ compares with the line-like plate data in Figure 16.20, but with the 'wing effect' (§16.2.3.1) missing. Similarly the curve for $D/B = 1$ compares with the square-section data of Figure 16.22, except that the local force reversal (§16.2.3.2) as $\theta = 90°$ is approached is missing in the averaged curve, although some individual cases in Figure 17.22 retain the effect. At the highest fineness, i.e. for the y-axis force at $L/W = 4$, the force reversal is restored. These differences from the line-like cases are due to the effects of the velocity profile and turbulence of the atmospheric boundary layer, as well as the lower slenderness of the buildings. The effect of the cross-section proportions on the height of the centre of force is shown in Figure 17.24. The location remains fairly constant between $0.5 < z/H < 0.6$ for all wind angles, except for the finest sections, where it rises to $z/H = 0.65$. Even so, this variation not large so that the base shear and the corresponding base bending moment remain nearly proportional.

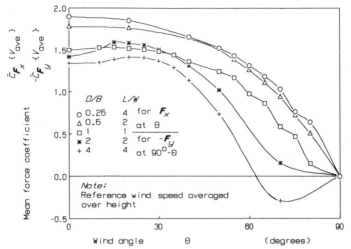

Figure 17.23 Effect on fineness on mean horizontal force coefficient on cuboidal buildings (from reference 277)

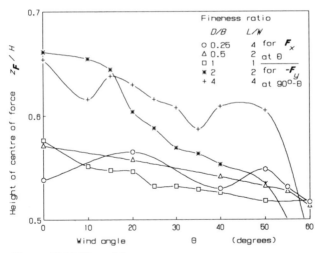

Figure 17.24 Effect of fineness on centre of force on cuboidal buildings (from reference 277)

   The effect of the cross-section proportions on the z-axis loadings: vertical force, $\overline{C}_{F_z}\{\theta\}$, and torque, $\overline{C}_{M_z}\{\theta\}$, are shown in Figures 17.25 and 17.26. The overall vertical force is of little direct interest because it does not reveal much of the complex flow and loading regions over the roof, discussed later (§17.3.3). Figure 17.25 does show that the flat roof always experiences a net uplift, which reduces as the building becomes long in the wind direction ($L/W = 4$ at $\theta = 90°$). The torque, $\overline{C}_{M_z}\{\theta\}$, is of more interest: the form of the curves in Figure 17.26 changes from a $\sin2\theta$ form when the section is elongated to a $\sin4\theta$ form when the section is square. The tendency with the elongated section is always to twist to bring the longer face normal to the wind, and this is unstable when the wind is normal to the shorter face in the manner of flat plates (§8.6.4.2) (but the torsional stiffness of buildings is

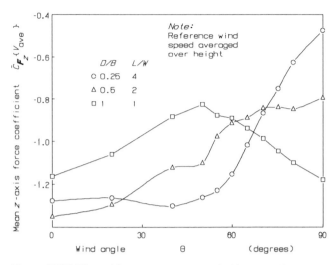

Figure 17.25 Effect of fineness on mean vertical force coefficient on cuboidal buildings (from reference 277)

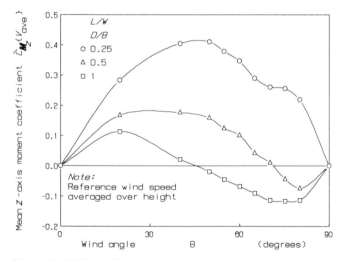

Figure 17.26 Effect of fineness on mean vertical moment coefficient on cuboidal buildings (from reference 277)

usually so high that divergence is impossible). The tendency with the square section is to twist to bring either face normal to the wind, with $\theta = 45°$ as a neutral position by symmetry. Isyumov and Poole[89] compared the distribution of local mean torque, $\bar{c}_{M_z}\{\theta\}$, down a tall square-section building between the reference pressure fixed at $z_{ref} = H$ or varying with position down the building, $z_{ref} = z$, with the equivalent result to Reinhold et al. in Figure 17.3 (§17.1.6). These overall torque characteristics are more relevant to the torsional response of dynamic buildings and are discussed in Part 3.

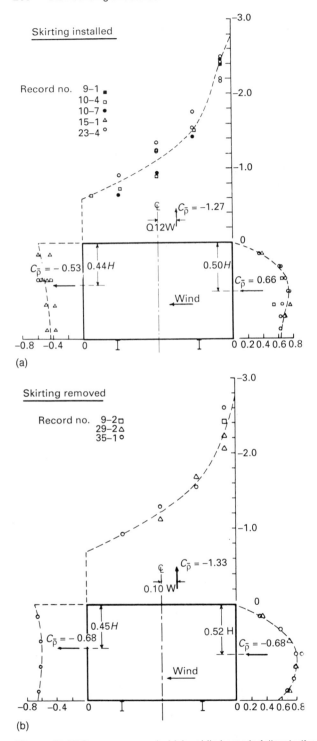

(a)

(b)

Figure 17.27 Pressures on cuboidal mobile home in fullscale (from reference 165)

Akins *et al.'s* complete data set [277] is tabulated in Appendix L, and includes all the necessary loading coefficient definitions for implementation.

17.3.1.4 Overall forces and moments on squat buildings.

Pressures and overall forces on a mobile home, $L = 18.3$ m (60 ft), $W = 3.7$ m (12 ft), $H = 2.3$ m (7 ft 8 in), with the wind normal to the longer side were measured in full scale by Marshall [165]. The mobile home was mounted 1 m clear of the ground to give an eaves height of 3.3 m and measurements were taken with the gap to the ground sealed by a skirt and open. The results, shown in Figure 17.27, give $\overline{C}_{F_x}\{\theta = 0°\} \simeq 1.2$, as is expected for a long low bluff body (§16.1.2). Roy's model comparison at model-scale [96] extends the data to the variation wind direction, giving a similar result to Akins *et al.'s* tall building data, maintaining similar shaped curves but with 35% lower values. (Roy attributes the discrepancy in values to scaling effects but, in the light of the IAWE–ACE study (§13.5.5.3), it is just as likely to be a simple gain error between the reference dynamic pressure and the force balance calibration in one or both experiments.) Earlier work on a typical Australian low-rise house [95] (low-pitch, large eaves overhang) again gives similar curves, but this time with intermediate values.

Figure 17.27 shows that the height of the centre of load remains in the same region as for the taller cuboids. One major difference is that the vertical force from the pressure distribution over the roof makes a bigger contribution to the overturning moment. As this depends strongly on the form of the roof, overall loads and moments for the design of anchorage for mobile homes, shipping containers and other small buildings are better determined by summation of the surface pressures for the particular shape, presented later.

17.3.1.5 Effect of Jensen number

By obtaining a good collapse of data for a wide range of building height and terrain roughness, Akins *et al.* [277], have demonstrated that the effect of Jensen number on overall forces are small enough to be ignored, **provided** Je is in the correct range (§17.1.5).

17.3.1.6 Overall flow characteristics

The overall forces and moments are the integral effect of the whole flow characteristics around the building. Figure 17.28, compiled by Peterka *et al.* [296], gives some impression of the complexity of the flow and may be compared with Figures 17.10(a) and 17.24 for squat cylinders. Each of the coherent flow structures that can form next to the building surface: attached flow, separation bubble, reattachment zone, 'delta-wing' vortex, 'horse-shoe vortex', etc. (see Chapter 8), dominate the surface pressures in the corresponding loaded region. Design pressures for these individual loaded regions are the most general design parameters for buildings of all shapes, since overall loads or loads on any required area can be synthesised from them by summation. Short-duration peak loads are due as much to perturbations of these flow structures under the influence of atmospheric and building-generated turbulence, as to the wind gusts directly. Their characteristics are strongly dependent on the form of the building. The remainder of this chapter is devoted to the discussion of these characteristics on walls and roofs, starting with the simple cuboidal form and extending to cover the range of common building form.

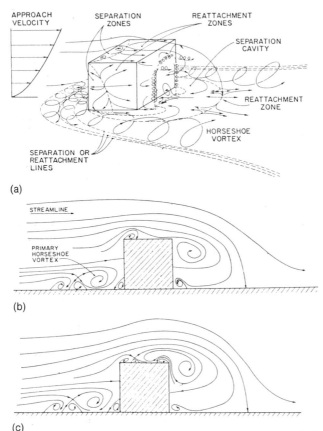

(a)

(b)

(c)

Figure 17.28 Flow around cuboidal buildings (from reference 296): (a) mean streamlines around whole building; and centreline streamlines for (b) flow reattaching to top and (c) flow not reattaching to top

17.3.2 Walls

17.3.2.1 Effect of wind angle

Most of the interesting effects on each face occur in the range $\theta = \pm 90°$ each side of normal to the face. From $\theta = 0°$ to $\pm 45°$, the face is the windward face, experiencing positive pressures. Figure 17.29 shows typical data by Hongo *et al.* [297] for the pressure along the larger face of two cuboids at $z = H/2$. One is a tall square-section tower ($L/W = 1$, $H/L = 3$), so all faces are the same. The other is a rectangular slab ($L/W = 4.6$, $H/L = 0.5$), so is tall when the wind is normal to the smaller face, but is exactly on the tall/squat division (§17.1.4) when the wind is normal to the larger face.

The 'windward face' range, $\theta = 0°$ to $45°$, is shown in (a) and (c). At $\theta = 0°$, the pressure remains fairly constant across the face, but falls at either edge as the flow accelerates around the corners, and the same occurs along the top edge of the face. The stagnation point should be in the centre of the face, with a symmetrical pressure distribution on either side. While this always occurs on a convex face, e.g. cylinder, sphere or dome, the flow normal to a flat face is neutrally stable: only a

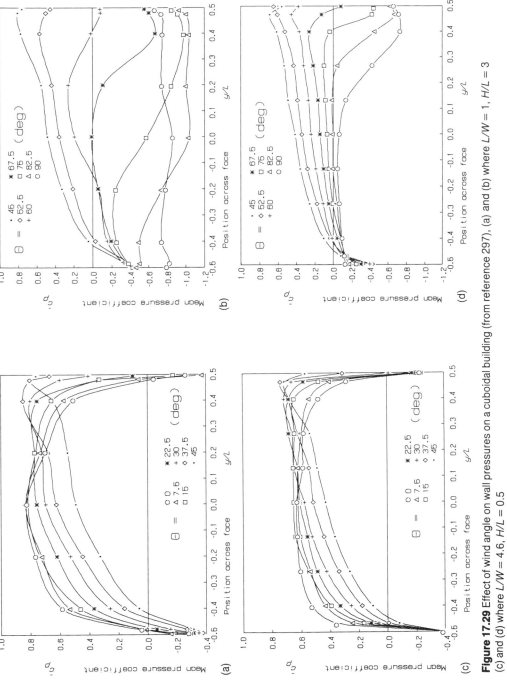

Figure 17.29 Effect of wind angle on wall pressures on a cuboidal building (from reference 297), (a) and (b) where $L/W = 1$, $H/L = 3$ (c) and (d) where $L/W = 4.6$, $H/L = 0.5$

small angular change moves the stagnation point to one side and makes the distribution unsymmetrical. The stagnation point moves further as the angle increases until, at $\theta = 45°$ it reaches the upwind corner.

From $\theta = 45°$ to $90°$, the face becomes a side face, experiencing suctions as shown in Figure 17.29(b) and (d). At $\theta = 45°$ the flow sweeps across the face from the upwind edge to the downwind edge, decreasing almost linearly in pressure. As the face turns further towards parallel, the stagnation point moves onto the next adjacent 'windward wall'. The flow negotiating the sharp corner cannot remain attached and a separation bubble forms with associated suctions. Initially small, this bubble extends further down the wall as the angle increases.

Figure 17.30 shows contours of pressure on two adjacent faces of the rectangular slab [297] for some of the wind angles in Figure 17.29. Note that the wind angle refers to the larger face, as above, so that the corresponding wind angle from normal to the smaller face is $90° - \theta$. Beyond the range of wind angle shown in Figure 17.29 and 17.30, the ranges $\theta = 90°–180°$, the wall becomes a 'leeward face' and experiences the fairly uniform suction of the wake.

17.3.2.2 Effect of slenderness and fineness ratios

For the windward face, slenderness ratio effects whether the wind flows over or around the building, as indicated in Figure 17.2. When slender, as for the smaller face in Figure 17.30, it is seen that the pressure contours are predominantly vertical. There is a small variation with height: as $H/B \rightarrow \infty$ the building becomes line-like and pressure would be proportional to $q\{z\}$ by strip theory (§16.2); however, the three-dimensional flow around bluff bodies tends to reduce the effect by the mechanism of Figure 8.5 (§8.3.2.1.1, §17.1.6). The position of the nearly vertical contours scale to the cross-wind breadth, B (= W for the shorter face). As the building becomes more squat (as for the longer face) the central region expands, the contours move towards the ends of the face, and their position from the edges scales to the height, H, as $H/B \rightarrow 0$. Figure 17.31(a) illustrates this effect for a squat square-section cuboid. This enables loaded regions on a windward face to be defined as vertical strips for design purposes, dimensioned in terms of the smaller of the height or face length (see Chapter 20).

For the side face, the slenderness and fineness ratios affect the size of the separation bubble at the upwind end of the face, and hence the size of the local high suction region. The degree of divergence still depends on the slenderness of the corresponding upwind face, that is still to the cross-wind breadth, B, or height, H. But when B is the scaling length it is always the length of the **other** face: $B = W$ for the longer face, as in Figure 17.29(d), and $B = L$ for the shorter face; so that fineness ratio is also important. Thus for the longer face in Figure 17.29, $B = W$: in (b), because $W = L$, the bubble occupies the entire face at $\theta = 90°$, i.e. there is no reattachment of flow on the face; whereas in (d), $W < L$ so the bubble is smaller relative to the longer face length and the flow reattaches to the face, reducing the suction on the downwind half. Comparing the shapes of the curves in Figure 17.29(b) and (d), it is clear that the relative sizes of the separation bubble at the upwind (right-hand) edge remains scaled by a factor $\simeq 4.6$, corresponding to the fineness ratio. The corresponding contours in Figure 17.30 again remain nearly vertical. For the squat square-section cuboid in Figure 17.31(b), $2H < B$, so the height becomes the scaling length, and the effect is intermediate between Figures 17.29(b) and (d). This enables loaded regions on a side face to be defined as vertical

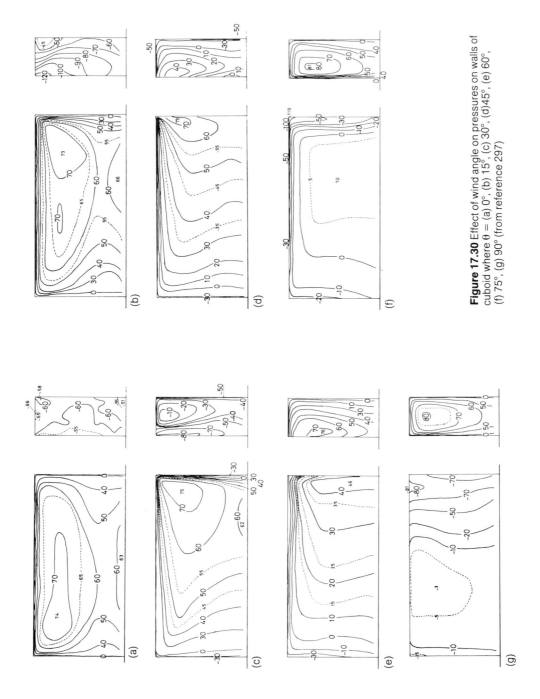

Figure 17.30 Effect of wind angle on pressures on walls of cuboid where θ = (a) 0°, (b) 15°, (c) 30°, (d) 45°, (e) 60° (f) 75°, (g) 90° (from reference 297)

$L/W = 1$ $H/L = 1/3$

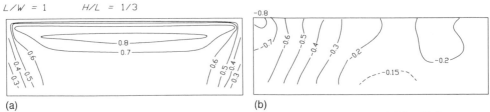

(a)   (b)

Figure 17.31 Mean pressures on walls of squat cuboid, from BRE measurements: (a) θ = 0°, flow normal to face; (b) θ = 90°, flow from left to right

strips for design purposes, dimensioned in terms of the smaller of the height or upwind face length (see Chapter 20).

17.3.2.3 Peak cladding loads

When peak loads are considered, the effects remain similar but are modified by the effects of atmospheric and building-generated turbulence which are usually manifested by expanding the high-load regions. Figure 17.32 gives the contours of the 4 second duration pseudo-steady pressure coefficient, $\bar{c}_p\{t = 4s\}$, on the walls of the squat cuboid corresponding to the mean data in Figure 17.31. Note in (a) the high positive pressure zone in the middle of the face expands, pushing the reducing contours nearer to the edges. This is due to directional smearing (§8.4.2.2) by incident gusts which produces the envelope of peak load over a range of wind direction either side of normal, e.g. the envelope of the curves for 0° ≤ θ ≤ 15° in Figure 17.29(a). Similarly in (b), the contours of peak suction in the separation bubble at the upwind edge expand down the side face. This is due to the additional turbulence in the shear layer which acts on the face in the reattachment region.

Some interesting insights into the behaviour of peak loads have been gained by recent studies. Using a combination of correlation, conditional sampling and signal averaging techniques, Surry and Djakovich [298] found that the peak suctions on

$L/W = 1$ $H/L = 1/3$

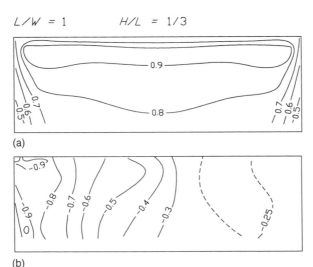

(a)

(b)

Figure 17.32 Peak 4s duration pressures on walls of squat cuboid, from BRE measurements: (a) θ = 0°, flow normal to face; (b) θ = 90°, flow from left to right

the side faces of tall buildings move in a coherent 'wave' from the top upwind corner towards the bottom downwind corner, associated with the weak vortex shedding, and they mapped the envelope of peak suctions for all wind directions. The BRE regression analysis of peak values against mean values, discussed in §15.3.3, compares peak pressures against the quasi-steady model assumption. Values of the regression coefficients, A, B and C in Eqn 15.25 are given in Table 17.4 for durations of $t = 1$ s, 4 s and 16 s. In all cases, the slope is reduced by about the same value as the intercept (allowing for the sign with the minima), which formally confirms the earlier statement that high-load regions expand: mean values near unity take pseudo-steady values near unity, but mean values near zero take pseudo-steady values near the intercept value, B. There is a definite trend with duration of maxima on the windward face, where the slope increases with duration, indicating that the larger gusts have more effect than the smaller gusts on the pressures. This is the same as, but additional to the trend with the admittance function when the pressures are integrated to overall loads. For minima on the side and leeward faces this trend does not occur, probably because the small gusts are augmented by the wake turbulence.

Table 17.4   Regression coefficients in Eqn 15.25 for walls of cuboidal buildings

	Duration t	Slope A	Intercept B	Std dev C
Maxima on	1 s	0.751	0.387	0.093
windward face	4 s	0.809	0.303	0.085
	16 s	0.828	0.204	0.074
Minima on	1 s	0.771	− 0.249	0.173
side and	4 s	0.752	− 0.200	0.123
leeward faces	16 s	0.775	− 0.130	0.089

17.3.2.4 Effect of corner angle

Since the separation bubble at the upwind edge of a side wall is formed by the flow around the corner, the resulting high-suction region will be affected by the angle of the corner, β. This has been studied at BRE at $H/B \simeq 1$ for corner angles in the range $60° \leqslant \beta \leqslant 150°$ and the results for the relevant range of wind angle, θ, are plotted in Figure 17.33. Here, corner angle $\beta = 180°$ is represented by the centre of a straight wall (i.e. no corner). This shows that the cuboidal corner, $\beta = 90°$, represents the worst case. These observations are confirmed as general characteristics for other slenderness ratios by ad-hoc published data for non-cuboidal shapes, e.g. [298].

Chamfering the corners of a rectangular building at 45° to produce an irregular octagon is a viable option for reducing the high suctions in the local corner region and also the wind speeds at ground level near the upwind corners [299]. It follows that there must be a minimum chamfer width to establish flow across the chamfer for it to be effective. Measurements with chamfered roof edges, discussed in §17.3.3.3.4 'Mansard roofs', suggest that a chamfer about $B/6$ wide is sufficient.

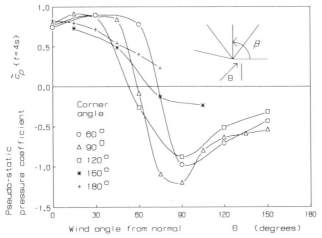

Figure 17.33 Effect of wall corner angle on pressure in edge region, from BRE measurements

17.3.2.5 Irregular-plan and irregular-elevation buildings

17.3.2.5.1 General. Corner angles greater than $\beta = 180°$ represent walls with an internal corner, such as occurs with $\beta = 270°$ where a wing extends from a cuboidal building to form a 'T' shape in plan, or with 'H'- or 'X'- 'Y'-plan buildings. Local high-suction regions do not form on the walls at these corners. Whether the pressure distribution on the faces either side of an internal corner retains the characteristics discussed above depends on the whole plan shape and the slenderness. Knowledge of the complex interactions that can occur between different 'blocks' or 'wings' of an irregular-plan building is almost completely qualitative, since no comprehensive parameteric studies have been performed and the little data that exist come from a few *ad-hoc* design studies.

Irregular elevations are complimentary to irregular plans in the sense that the same changes of shape, cut out corners, etc., can be made to the face in elevation instead of the plan. The general concepts of tall and squat buildings, dependent on slenderness ratio, require modification to account for these changes. The problem breaks down into two complementary parts:

1 Irregular faces – plane faces that are not rectangular, but are usually flush with the external envelope of the building.
2 Inset faces – which are usually rectangular, but are set back from the external envelope of the building.

This section attempts to describe the behaviour of the loading of both categories in terms of the geometry of the building.

17.3.2.5.2 Re-entrant corners. Placing one internal corner between two external corners creates a re-entrant corner and leads to irregular-plan buildings with L-, T-, X- and Y-plan shapes. The principal flow characteristics are indicated in Figure 17.34 for an X-plan building, but apply equally to any plan shape with a re-entrant corner in the corresponding position. Note that the height of the roof on either side of an internal corner need not be equal and the regions of strong suction that occur at the lower roof–wall junction when this is the case are considered in §17.3.2.5.5.

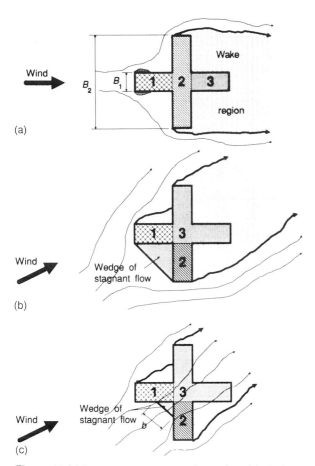

Figure 17.34 Flow around X-plan buildings when (a) wind normal, (b) wind skewed and building is tall, and (c) wind skewed and building is squat

In (a), when the wind blows nearly normally onto the protruding wing '1', the cross-wind breadth for flow around that wing is the width of the wing itself, B_1. The cross-wind breadth for the wide vertical leg '2' is the full breadth of the building, B_2. Accordingly, the implication of this model is that the loading of wing '1' depends on the its proportions, independent of '2'; while the loading of '2' is independent of '1'. This is only approximately true, there being interaction where '1' and '2' join, the positive pressures on the windward faces of '2' acting also on downwind end of the wing '1'; nevertheless this model is a reasonable working assumption. The downwind wing '3' protrudes into the wake of '2', so experiences pressures similar to the rear face of '2'.

When the wind blows at an skew angle into the re-entrant corner, this tends to trap a region of stagnant air in which the pressure rises to the stagnation pressure, i.e. to the face pressure for $\theta_n = 0°$. The size of this region depends on the slenderness. When the building is tall, the region fills the whole re-entrant corner, as in (b), flow choosing to flow around the sides rather than into the corner. When the building is squat, flow tends to enter the corner before rising over the building, forming a stagnant region about equal to the scaling length b ($b = 2H$). The other re-entrant corners, facing downwind into the wake, behave as described above.

17.3.2.5.3 Recessed bays. Two internal corners between external corners produces a recessed bay and leads to irregular-plan buildings with H plan shapes, or more complex plan shapes, e.g. |—|—·|—|—|—|. For all such shapes the simpler H-shape serves as a model for the flow characteristics of recessed bays as indicated in Figure 17.35.

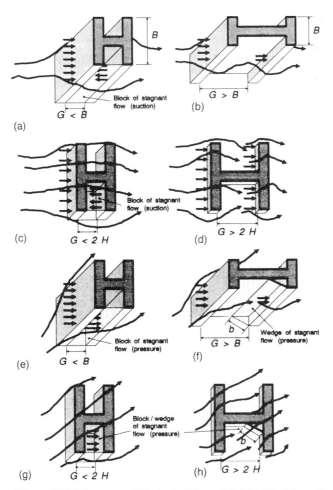

Figure 17.35 Flow around H-plan buildings: (a), (b) tall buildings, flow normal; (c), (d) squat building, flow normal; (e), (f) tall building, flow skewed; (g), (h) squat building, flow skewed

When the width of the gap across the recess, G, is small compared with the scaling length, b, for the building: $G < B$ for tall buildings and $G < 2H$ for squat buildings, the flow tends to skip past the gap, leaving almost stagnant flow in the recess, as in (a), (c), (e) and (g). The pressure in the recess is almost constant at the average of the pressure at the open face. The pressure in the recess is the pressure that would have occurred along the equivalent wall: positive pressure when in the front face, (e) and (g), and suction when in the side or rear faces, (a) and (c). A corollary of this effect is that overall loads tend to be the same as for unrecessed buildings of the same external envelope, so that for tall buildings with cuboidal enevelopes Akins *et al.'s* data are expected to be valid (§17.3.1.3).

When the width of the recess becomes large, the flow tends to enter the recess and act directly on the back face of the recess, as in (b), (d), (f) and (h). This case has flow characteristics very similar to the re-entrant corner: in the normal flow cases (b) and (d), the downwind leg of the H acts like the long arm '2' in Figure 17.34(a); and in the skewed flow cases (f) and (h), a wedge of positive pressure similar to Figures 17.34(c) and (d) occurs in the upwind-facing internal corner.

17.3.2.5.4 Central wells.

More than two internal corners creates a central well open at the roof. Usually four corners are used, giving irregular-plan buildings with an 'O', '8' or more complex plan shape. Flow characteristics in these wells are dominated by the flow over the roof, as shown in Figure 17.36.

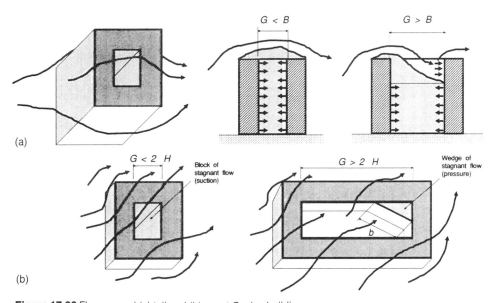

Figure 17.36 Flow around (a) tall and (b) squat O-plan buildings

When the inwind depth of the well, G, is small compared with the scaling length, b, the flow tends to jump over the well, leaving a region of nearly stagnant air in which the pressure is set by the suction expected on the corresponding region on the roof. When the inwind depth of the well is large, flow enters the top of the well and acts directly on the rear face. With squat buildings, this gives flow conditions similar to the corresponding re-entrant corner and recessed bay cases. With tall buildings, this effect occurs only near the top of the well.

17.3.2.5.5 Irregular faces.

Figure 17.37, which is complementary to Figure 17.2, shows the three typical combinations that occur when one or two top corners of a face are removed and the primary effect on the flow. In (a) the face can be

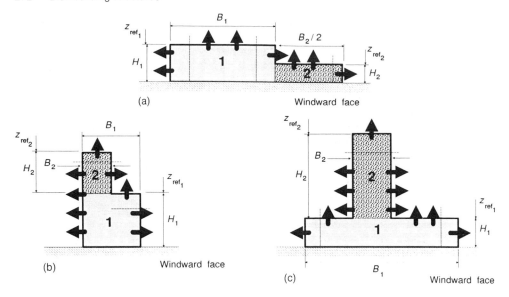

Figure 17.37 Examples of irregular-faced buildings where (a) both parts are squat, (b) both parts are tall and (c) tall part on squat part

considered to be two squat faces, side by side: a higher part '1' and a lower part '2', since the flow predominantly rises over both parts. The junction between the two parts acts like a ground plane for the **lower** part 2 (in a similar manner to changes of diameter of cylinders, §16.2.2.1.4), so that the effective cross-wind breadth of the lower part, B_2, is twice the actual breadth (hence the actual breadth is marked $B_2/2$). Conversely, in (b) the face can be considered to be two tall faces, a wider face '1' underneath a narrower face '2', since the flow predominantly goes around both parts. The junction again acts as a ground plane for the **narrower** part 2, so that the effective height of the part, H_2, is the height from the top of the lower part. The third typical combination in (c) is a tall part '2' on a squat part '1', with flow predominantly over the squat part and around the tall part, with the same consequence on the effective height of 2 as case (b). In all three cases, the effective heights are compared with the respective breadth to determine the appropriate scaling length, b.

The implication of this model is that the loading on an irregular face can be deduced from the data for rectangular faces by dividing the face into these parts. The reference dynamic pressure, $\hat{q}_{ref}$, for each part is taken at the top of that part. *This is not the same as the 'division by parts rule'*, Clause 5.5.2 of the UK code CP3 Chapter V, Part 2[4] which allows buildings to be divided into horizontal strips. Strip theory is only valid when the building is like-like, $H/B > 4$ (§8.2.2, §16.1.3, §16.2.1.1). Here the division is made into logical parts according to the flow characteristics and the particular interaction effects near the junctions of the parts must be separately accounted for.

17.3.2.5.6 Inset faces.

In the examples of irregular-faced buildings in Figure 17.37 the front faces considered are flush in one plane. Flow from right to left in any of the examples would see the higher part as an inset storey, rising from the roof of the lower building.

When the upper storeys are inset on all sides, i.e. if part 2 in Figure 17.37(c) were set back from part 1 in the view shown to produce a tower in the centre of a podium, the inset faces can be regarded as walls of a separate building built on the roof of the lower building as the ground plane. There are no particular problems for these inset faces, but the positive pressures on the windward wall interact with the pressures on the roof of the lower building, as described in §17.3.3.2.8 in the section on roofs.

A particular problem occurs when the edge of the inset face is flush with the side of the buildings, i.e. adjacent to the irregular faces of Figure 17.37. In the region where the flow arrows converge near the internal corner, the interaction of these two crossing flows causes some very high suctions on the roof of the low part and the side wall of the high part. This effect is discussed in §17.3.3.2.8 and is the basis of the enhanced design pressure coefficients for both walls and roof in Chapter 20.

17.3.2.5.7 Comments.   Owing to the qualitative nature of these observations, it is difficult to set hard-and-fast design rules. Nevertheless such rules are necessary if designs for complex façades, such as shown in Figure 17.48, are to be assessed. The corresponding rules in Chapter 20 are the first attempt at formal interpretation of current knowledge.

Figure 17.38 London Borough of Hillingdon Offices, Uxbridge

17.3.2.6 Non-vertical walls

17.3.2.6.1 General.   The concept of non-vertical walls leads to difficulties in dividing the problem between 'walls' and 'roofs'. Figure 17.39(a) shows a building with non-vertical walls (where the balconies form recessed bays, §17.3.2.5.3), while Figure 17.39(b) shows a building with a duopitch roof that extends down to ground level (A-frame building). The pitch of the 'wall' and the 'roof' in these two cases is almost the same, so how are they distinguishable? The answer is that they are not distinguishable at all in terms of the **aerodynamics**, and this is what matters for the loading. Accordingly, this section arbitrarily categorises all non-vertical faces that

(a)

(b)

Figure 17.39 Buildings with non-vertical walls: (a) Shell UK Exploration and Production Office, Aberdeen (courtesy of George Wimpey plc); (b) Ferrybridge Services, West Yorkshire

extend to ground level as 'walls', even when the form of construction and materials clearly indicate that the face is a 'roof', as in Figure 17.39(b).

Figure 17.40 shows four forms of building having non-vertical walls. The A-frame building in (a), of which Figure 17.39(b) is an example, is essentially a duopitch gabled roof without walls. If the vertical gables are also made non-vertical, this leads first to a building which is essentially a hipped roof without walls, then on to the special case of the pyramid in (b). If either the A-frame of (a) or the pyramid of (b) are truncated to form a flat roof, the building forms shown in (c) and (d) result. These forms are appearing in contemporary large plan-area industrial building designs and have been built in the UK and elsewhere.

17.3.2.6.2 A-frame buildings. These are buildings with a rectangular plan with main walls with lean inwards to meet at a ridge and vertical triangular gable end walls, as shown in Figure 17.40(a). This has been studied at BRE for pitches in the range $30° \leq \alpha \leq 75°$, yielding extensive data for the pressures on both the main non-vertical faces and the vertical gable ends.

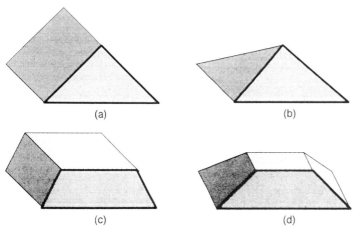

(a)   (b)

(c)   (d)

Figure 17.40 Types of building with non-vertical walls: (a) A-frame building; (b) pyramid; (c) truncated A-frame building; (d) truncated pyramid

Figure 17.41 shows distributions of 4 s-duration pseudo-steady pressure coefficient, $\tilde{c}_p\{t = 4\,s\}$, at a wind angle $\theta = 60°$ from normal to a main face for $H/L = 3$, viewed in plan. On the windward face the pressure contours are similar to the corresponding vertical face in Figure 17.30, except that the values reduce with pitch angle. Figure 17.42 shows that this reduction factor is approximately $\sin\alpha$ for all wind angles giving positive pressures on the windward face. On the leeward face the pressure contours show the formation of a pair of 'delta-wing' vortices,

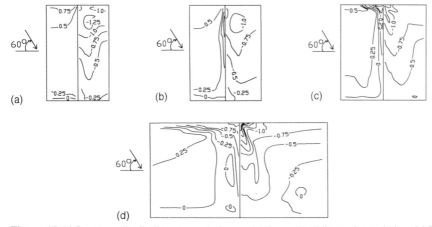

Figure 17.41 Pressure distributions on main faces of A-framed buildings where pitch = (a) 75°, (b) 60°, (c) 45°, (d) 30°

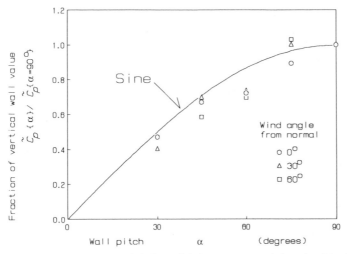

Figure 17.42 Effect of wall pitch on global pressure on windward wall for $H/B = 3$

springing from the upwind corner as in Figure 17.1(b), one behind the ridge and the other behind the gable verge. These are most prominent at the lowest pitch, $\alpha = 30°$, in (d) and are similar in most characteristics to the vortex pair on the leeward half of a duopitch roof (§17.3.3.3.2).

Figure 17.43 shows the pressure distribution on the triangular gable end, pitch $\alpha = 30°$, for a range of wind angle from normal to the face, θ. Again, while the face is to windward giving positive pressures, the contours are similar to Figure 17.30 if allowance is made for the triangular shape. That is, the stagnation point starts at the centre of the face and moves towards the upwind edge as the wind angle

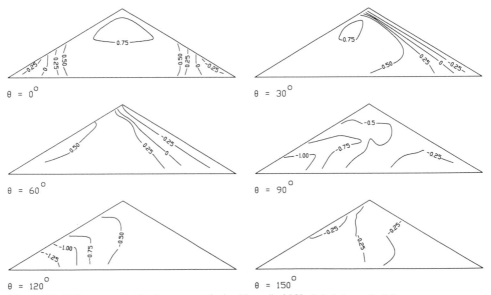

Figure 17.43 Pressure distributions on vertical gable wall of 30° pitch A-frame building

becomes more skewed, but the contours are lower on the face. Other differences occur when the triangular gable becomes a side face, with the highest suctions occurring towards the upwind point at $\theta = 120°$, instead of $\theta = 90°$, caused by a single conical vortex fed by vorticity from the shear layer separating along the edge of the main face. This effect decreases with increasing pitch angle, α.

In summary, the largest overall loads on A-frame buildings occur at the steepest pitches with the wind normal to the main face, while the highest local suctions occur near upwind edges and corners at the lowest pitches at skewed wind angles.

17.3.2.6.3 Pyramids.   Reliable data exist only for the special case of a pyramid formed from four equilateral triangles ($\alpha = 54.7°$), shown in Figure 17.44, taken from a BRE design study. The characteristics are transitional between the two different face types of the A-frame building: i.e. non-vertical **or** triangular, becoming non-vertical **and** triangular. While the face is to windward, $0° \le \theta \le 60°$, the contours are similar to the main face of the A-frame, except that the position of maximum pressure is lower and moves into the bottom corner in skew winds. When the face is the side face, $60° \le \theta \le 120°$, the contours are similar to the triangular gable end of the A-frame, except now that the windward face is also triangular all of the flow passes over the diagonal ridge and around the side face, so that the conical vortex in the windward corner is stronger and the maximum suctions occur at $\theta = 90°$.

17.3.2.6.4 Truncated A-frame buildings and pyramids.   No specific data exist for either form but, fortunately, the loading on the walls can be deduced from the BRE A-frame and pyramid data, and the loading on the flat roofs from the BRE mansard roof data reported later in §17.3.3.3.4. For the truncated A-frame, the rectangular pitched face is expected to behave exactly like the A-frame face when at the front and under positive pressure, and like the walls of cuboidal buildings when at the side or rear, i.e. positive pressures are reduced by the pitch, but

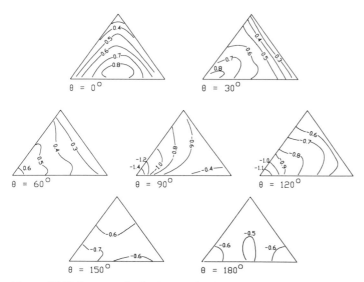

Figure 17.44 Pressure distributions on the face of a pyramid with four equilateral faces

suctions are unchanged. The corresponding gable end face is expected to behave like the A-frame gable face in the upwind triangular corner, but like an ordinary vertical wall over the remainder of the face. For the truncated pyramid, all faces are the same: pitched with triangular tapering corners. These are expected to have characteristics transitional between the two face types of the truncated A-frame. For both forms, the roof pressure characteristics are identical to the mansard roof case of §17.3.3.3.4, and are discussed there.

17.3.2.7 Other aspects of walls

17.3.2.7.1 Effect of Jensen number. The effect of Je on overall forces was demonstrated by Akins *et al.* [277] to be small (§17.3.1.5) within the typical range for buildings, particularly for the base shear forces and overturning moments which depend directly on the wall pressures. Jensen's demonstration in Figure 2.37 shows a small effect on the shape of the wall pressure distribution. This small effect is lost when the loaded areas are defined as vertical strips (§17.3.2.2).

17.3.2.7.2 High-set houses. In tropical regions, such as in northern Australia, there is a tendency to raise houses on stilts so that they are clear of the ground. Studies on this form of 'high-set house' [300] show that the loading is not greater than corresponding conventional housing [301]. Reports of this work may be found in the Australian Housing Research Council publication *Wind pressures and forces on tropical houses* [302], including design loading coefficient values.

17.3.2.7.3 Effect of roof pitch and eaves overhang. The pitch of a roof affects the pressures along to top edge of windward walls. As long as the roof pitch is less than about $\alpha = 35°$, the flow separates at the eave to form a separation bubble or delta-wing vortex along the upwind edge of the roof (see §17.3.3.3.1). In this case the pressure distributions remain similar to those shown in Figure 17.30, reducing rapidly in a narrow band below the eave. With higher pitched roofs, the position of maximum pressure rises towards the eave and the flow remains attached over the windward roof face.

The effect of eaves which project out from the top of the wall, usually to throw rainwater clear of the walls or to provide shade in tropical areas, is to reduce the decrease in pressure along the top edge of the windward wall. As the decrease in pressure below the eave of the windward wall is not exploited in the design data of Chapter 20, there is no need to account for the effect of roof pitch or eaves overhang on the wall itself. However, by trapping the flow, the underside of the eave or 'soffit' is subjected to the positive pressure on the windward wall. With low roof pitches, the combination of high uplift on the top surface at the windward eave with pressure acting on the soffit underneath can be very onerous.

17.3.2.7.4 Friction-induced loads. When buildings are very long, $L >> W$, significant forces parallel to the long faces can be accumulated by the action of friction on the part of the wall downwind of any separation bubble which is swept by the wind. When the surface of the wall is smooth or homogeneously rough the effect can be quantified in terms of a shear stress coefficient, c_τ, as described for canopy roofs in §16.4.4.8. The effect of linear protrusions depends on their alignment: those aligned horizontally, parallel to the wind, will have no significant effect; but those aligned vertically, across the wind, such as mullions, will be loaded like a boundary wall. Multiple mullions will tend to shelter those downwind as described in §18.2.1.

17.3.3 Roofs

17.3.3.1 Introduction

The flow conditions over roofs have similarities with the flow past walls once the differences with wind angle are appreciated (§17.3.3.2.1). With flat roofs, the flow with wind normal to a wall is similar to the flow around the side face, in that the flow separates at the upwind edge and may reattach at some distance downwind to form a separation bubble. One difference in this process on roofs is that the velocity profile and downwards momentum of the Reynolds stress both assist the reattachment to occur sooner (§8.3.2.2.1). The main similarity is that the size of the separation bubble is expected to scale in the same manner to the smaller of the cross-wind breadth, B, or twice the height, $2H$, according to the principles in §17.1.4. The main difference is that the corners of vertical walls remain normal to the flow for all wind angles and the separation bubbles form as cylindrical vortices, while this occurs on roofs only when the wind is normal to the eave. At all other wind angles conical 'delta-wing' vortices form from the upwind corner (§8.3.2.2.2).

The dynamics of these two forms of vortex at the upwind edges of the roof dominate the loading characteristics on roofs. This is described in the remainder of this section in terms of the distribution of peak pressures expressed as pseudo-steady values, $\tilde{c}_p\{t = 4\,\text{s}\}$. Initially the discussion of the principal effects on roof loadings in §17.3.3.2 is conducted in terms of flat roofs only. Other roof forms are individually discussed in the later section, developing the differences from the flat roof case.

It will help clarify the discussion to be familiar with a number of terms used in the Guide to describe forms and parts of roofs:

1 Eave – the horizontal edge of the roof, i.e. both horizontal edges of monopitch and duopitch roofs (ridged or troughed), also the longer edges of flat roofs (assuming that any small pitch to provide a fall for rainwater goes to one long edge) and of hipped roofs.
2 Verge – the non-horizontal edge of the roof, i.e. at the gable edge of monopitch and duopitch roofs, also the shorter edges of flat roofs and of hipped roofs.
3 Hip – triangular pitched face at either end of the main faces of a hipped roof.
4 Ridge – the highest horizontal line formed where the two faces of pitched duopitch roofs meet; the same for hipped roofs, but distinguished as 'main ridge', with 'hip ridges' at the junction of the main and hip faces.
5 Trough – the lowest horizontal line formed where the two faces of troughed duopitch roofs meet.

The dynamics of the vortices at the upwind edges of roofs depend on the main parameters wind angle, slenderness ratio, roof pitch, etc., discussed below, but is also strongly influenced by the incident turbulence. In particular, short duration 'spikes' of very high suctions (Figure 8.20) are found near upwind edges in full-scale measurements and in model simulations where the turbulence of the atmospheric boundary layer has been correctly represented[16] but, significantly, not in 'velocity profile only' simulations (§13.5.3). Melbourne's work[303,304] to elucidate this effect was discussed briefly in §8.4.2.3. Later work by Melbourne and others [305, 306] has added to our knowledge and all seems to confirm this as an instability of the shear layer triggered by the atmospheric turbulence. Melbourne's proposal for a method of suppressing high peripheral loads is discussed in §19.5.3.

By conducting the discussion of load effects in this chapter and presenting the design data of Chapter 20 entirely in terms of peak pressures expressed as pseudo-steady values, this and any other transient effects are automatically included in the data.

17.3.3.2 Flat roofs

17.3.3.2.1 Effect of wind angle on flat roofs. Figure 17.45 shows the effect of wind angle on distribution of pressure coefficient, $\tilde{c}_p\{t = 4\,\text{s}\}$, over the flat roof of a cuboid with proportions $L{:}W{:}H = 3{:}1{:}1$. These are useful proportions for an example because the cuboid is squat, $2H/B = 2H/L = 2/3$, with low fineness, $D/B = W/L = 1/3$ at $\theta = 0°$ and tall, $2H/B = 2H/W = 2$, with high fineness, $D/B = L/W = 3$, at $\theta = 90°$, so shows aspects of all forms.

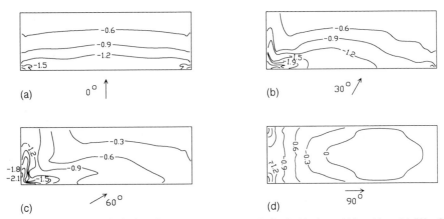

Figure 17.45 Effect of wind angle on pressures on roof of cuboid where $H/L = \frac{1}{3}$ and $L/W = 3$

When the flow is normal, (a) and (d), a separation bubble containing a cylindrical vortex forms at the upwind edge (§8.3.2.2.1), but the variation of wind angle caused by the atmospheric turbulence is sufficient for small conical vortices to form temporarily in either corner. The pressure under the vortex is moderately negative, giving typically $\tilde{c}_p\{t = 4\,\text{s}\} = -1.2$. The rotation of the cylindrical vortex pulls the flow over the roof back down towards the roof surface behind the bubble and the pressure rises towards zero. Whether flow reattachment occurs depends on the fineness ratio (§17.3.3.2.2).

When the wind is skewed more than about $\theta = 10°$ from normal, (b) and (c) a pair of conical 'delta-wing' vortices form permanently from the upwind corner along both upwind edges (§8.3.2.2.2). The strength of each vortex depends on the wind angle and is strongest at about $\theta = 30°$ from normal to the respective edge, i.e. in (b) for the longer eave and (c) for the shorter verge. The pressure along the eave and verge under these vortices is strongly negative, becoming even more negative towards the corners, where values less than $\tilde{c}_p\{t = 4\,\text{s}\} = -2.0$ are consistently obtained. Further from the corners, the negative pressure rises above the typical value of $\tilde{c}_p\{t = 4\,\mathrm{s}\} = -1.2$ for normal flow. The rotation of both vortices pulls the flow down behind them to form a 'V'-shaped wedge of attached flow (see Figure 8.12). In the apex of this 'V' the flow speed is higher than the approach flow and the pressure is still negative, but recovers downwind towards zero as the flow diverges out of the 'V'. Note that the centreline of the 'V', which is the dividing line for the effect of either vortex on the roof surface, remains approximately in the incident wind direction (within 5° or so).

Accordingly, the most negative values occur in the middle of the eave and verge edges when the wind is normal to the edge, but move to the corners and become more negative when the wind is skewed. There has been some recent debate [307] as to how close to the corner it is necessary to measure the pressure in order to obtain representative values. So far, the closer the measurement to the corner, the more negative is the minimum value obtained. This is important for short-duration loads on cladding in the vicinity of the corners. The semi-empirical model described in §17.3.3.2.3, derived from BRE data, predicts that the negative value is unlimited, but this is an asssumption of the model. The theory of vortices in §2.2.8 suggests that viscosity will impose a limit as the radius of the vortex becomes very small. In practice, a limit is set by the load sharing capacity of the cladding. Most codes of practice rely on this to set a limit to the minimum value for short-duration cladding loads on a purely pragmatic basis. For example the latest Swiss code [182] uses $\tilde{c}_p = -2.0$.

17.3.3.2.2 Effect of slenderness and fineness ratios on flat roofs.

Figure 17.46 uses the wind angle $\theta = 30°$ to demonstrate the effect of slenderness ratio on the distribution of pressure coefficient, $\tilde{c}_p\{t = 4\,\mathrm{s}\}$, on flat roofs of cuboids. In (a), (b) and (c) the distributions are given for tall square-section towers of decreasing slenderness, showing that there is little difference for $2H/L > 1$ and $2H/W > 1$, as the flow over the roof is expected to scale to the cross-wind breadth B in this range (§17.1.4). The distribution for $2H/L = 1$ in (d), represents the boundary between tall and squat structures for the longer eave but remains tall for the shorter verge.

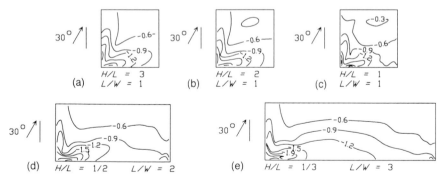

Figure 17.46 Effect of slenderness ratio on pressures on flat roofs

Here the distribution in the upwind corner and along the shorter verge is unchanged, but the contours along the longer eave are stretched out in the middle. Finally, the squat case $2H/L = 2/3$ in (e) shows the upwind corner and verge contours still unchanged and the contours along the eave stretched further in the middle. It is evident that the scaling rules, described in §17.1.4 and demonstrated in

§17.3.2.2 for walls, also work in a very similar manner along the edges of flat roofs, giving distinct regions of different characteristics. There is the additional observation that the delta-wing vortex along the eave appears to be independent of the vortex along the verge, implying that the scaling for slenderness works on the proportions of the respective face rather than the whole building.

The region around the upwind corner is a 'growth region' characterised by the initial conical growth of the vortex, where the circulation (§2.2.8) of the vortex increases almost linearly with distance from the corner fed by the vorticity of the shear layer separating from the eave. The region in the middle of the eave is a 'mature region', where the vortex has grown to fill the available space or, more correctly, where it has reached an equilibrium circulation at which vorticity is shed into the wake at the same rate as it enters from the shear layer. The region around the downwind corner is a 'decay region', where the circulation decreases because the flow tends to spill around the downwind corner of the windward face instead of over the eave, reducing the vorticity supply from the shear layer.

Most studies of fineness have concentrated on flow normal to a face, for example Castro and Dianat [308] demonstrate that the flow characteristics differ between the high-fineness case where the flow reattaches to the roof and the low-fineness case where it does not. This change occurrs at about $D/H = 1$ in the atmospheric boundary layer and about $D/H = 2$ in uniform flow, indicating that the boundary layer promotes reattachment, implying a sensitivity to Jensen number (see §17.3.3.2.3 below) and reinforcing the need for properly scaled boundary-layer simulations. Figure 17.47 demonstrates the effects of fineness ratio, again for the wind angle $\theta = 30°$. Here (a) is the same as in Figure 17.46(e) and (b) extends the length of the verge without changing the eave proportions. The contours of $\tilde{c}_p\{t = 4s\}$ in (a) are superimposed on (b) as the dashed lines showing that the contours along the eave are not much changed, again implying the eave and verge vortices are independent.

When the slenderness is reduced while keeping the fineness constant, i.e. squatter for the same plan shape, the contours contract towards the eave and verge

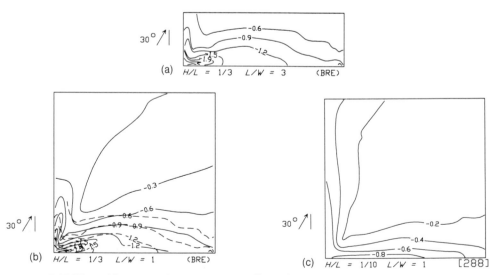

Figure 17.47 Effect of fineness ratio on pressures on flat roofs

edges as shown in (c), this time for the mean pressure coefficient, $\bar{c}_p$. The scaling length for the eave, $b = 2H$, is the same in (a) and (b) and is reduced in (c). This shows that the scaling rules also work along the depth, D, as well as across the breadth of the roof. Note again, that whatever the proportions of the building in Figures 17.46 and 17.47, the centreline of the 'V' shape in the contours, dividing the two vortices, remains approximately in the wind direction.

17.3.3.2.3 Effect of Jensen number on flat roofs.

Jensen's demonstration in Figure 2.37 shows a larger effect on the roof than on the walls, but over a range of Je that is far greater than found in practice. It is very difficult to make direct studies of Je effects in isolation from the other model parameters. Typically, models of different sizes are placed in a number of boundary layers, so that model proportions may also change. Figure 17.48 comes from a study of this kind by

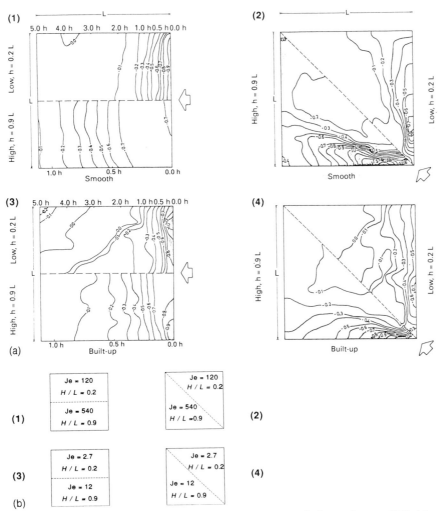

Figure 17.48 Effect of Jensen number on pressures on flat roofs (from reference 309): (a) measured c_p; (b) Key to Je and H/L

Stathopoulos *et al.* [309], showing mean pressures on the roof of a square-plan cuboid for four values of Je and two values of H/L, for wind angles of $\theta = 0°$ and $\theta = 45°$. The authors make use of the symmetry afforded by these two angles to present data for only half the roof. A key to the data is given which shows that a direct comparison for the same proportions, but for Je differing by a factor of 45 can be made by comparing the same half of either model in the top row with the corresponding model on the bottom row. There is some variation, but the form of the contours is similar in each case and the values match to about $\bar{c}_p = 0.15$. Comparing one half of each model with its other half gives a change in Je by a factor of 4.5, one-tenth of that above, together with a change in slenderness ratio by the same 4.5 factor. This time there is a large difference in the contours, confirming that the effects of building proportions are much larger than the effects of variations in Jensen number in the typical range found in practice.

In making the comparisons above, the reader will appreciate the difficulty in making objective assessments of the importance of the various parameters. In essence, reducing Je by reducing building height or making the terrain more rough reduces the size of the vortex at the upwind eave/verge and promotes flow reattachment. This leads to reduced overall uplift forces, reduced suctions over more of the downwind part of the roof but increased suctions under the smaller vortices along eave and verge edges. Nevertheless, these effects do appear secondary to the effects of the principal parameters and good estimates of loading are obtained by standardising on values of Je near the middle of the observed range. This view may change with future improvement in understanding. For example, the new semi-empirical model described below in §17.3.3.2.5 is able to quantify the effect of each parameter and may lead in time to a better understanding of the problem.

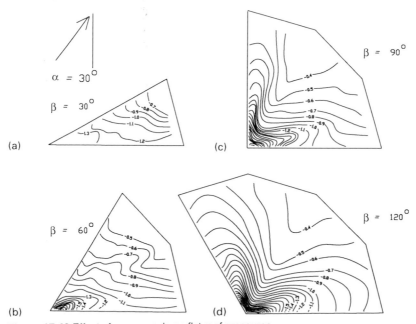

Figure 17.49 Effect of corner angle on flat roof pressures

17.3.3.2.4 Effect of corner angle on flat roofs. It has been suggested several times above that the action of the vortex along the upwind eave appears to be independent of the companion vortex along the upwind verge. This was specifically investigated by Maruta in a comprehensive series of tests conducted at BRE. The conclusions of this study were that the typical corner angle $\beta = 90°$ produced the strongest delta-wing vortices, with more acute or obtuse angles giving less severe suctions, so that the rectangular-plan case used in codes and other design guidance covers the most onerous case.

Figure 17.49 shows some example data for $\tilde{c}_p\{t = 4\,\text{s}\}$ over a range of corner angles with wind at $\theta = 30°$ from normal to the corner, i.e. comparable with Figure 17.46 and 17.47. The most striking difference between the most acute angle, $\beta = 30°$, in (a) and the other data is that the high-suction region close to the corner is absent. Remember that, for the rectangular corners in Figures 17.45–17.47, the centreline of the 'V' dividing the two delta-wing vortices remains close to the wind direction. In (a) the $\beta = 30°$ corner angle is smaller than the expected size of the eave vortex, which is $90° - \theta = 60°$. Instead of the strong eave vortex forming over solid roof, air is directly entrained from the wake by flow up the adjacent leeward face and the high suctions are suppressed. When the corner angle exactly occupies the whole of the expected vortex size, as in (b) $\beta = 60°$, the eave vortex forms normally, but there is no adjacent verge vortex because the wind direction is parallel to the verge, which lies along the centreline of the 'V'. As the corner angle increases, (c) $\beta = 90°$ and (d) $\beta = 120°$, changes in the eaves vortex are small, but now the other verge vortex forms of a size corresponding to the wind angle for the verge.

The variation of pressure with corner angle away from the region near the corner is not large, varying typically by $\tilde{c}_p \simeq \pm 0.1$, with a rectangular corner, $\beta = 90°$, giving the most onerous loading in general. This variation is small compared with the major parameters, such as wind angle, α, and roof pitch, α, that the values for $\beta = 90°$ may be used for design purposes.

Assumption of first-order independence of eave and verge vortices allows loaded regions to be defined behind each upwind eave or verge for flat roofs of any arbitrary polygonal shape. Instead of the scaling length b, based on the cross-wind breadth of the whole building, it is convenient to base the loaded regions on the proportions of the corresponding face. The rules used in Chapter 20 give loaded regions extending away from every upwind external corner, typically only one corner for cuboidal buildings, with special empirical provisions to account for the internal corners of re-entrant corners, recessed bays and central wells (see §17.3.2.5.2–§17.3.2.5.4).

17.3.3.2.5 A semi-empirical model for flat roofs. After completing the study described above, Maruta derived an empirical 'mapping' model for roof pressures, by the same techniques he used to derive the empirical model for wind speed around high-rise buildings [310, 311], described in §18.3.1. This was moderately successful, being limited by the degree of collapse obtained in the data by the empirical method, and by being a 'mapping' rather than a predictive method. It was, however, extremely valuable in pointing the way to the development of the semi-empirical mathematical model for roof pressures described in Appendix M.

A 'mapping' model is one which gives the position of the contour of a given value of pressure on the surface of the roof, whereas a 'predictive' model is one which predicts the value of pressure at a given location. The first is more valuable to the

compiler of codes and design advice, since it allows 'maps' or charts of pressure to
be drawn. The second is more valuable to the designer who needs to predict values
at given locations. The semi-empirical model of Appendix M serves both purposes
because the predictive model equation for c_p in terms of position (Eqn M.10) is also
soluble for the position coordinates in terms of the pressure (Eqns M.13 and M.14).

The model treats the vortex in the 'growth region' near the corner as a conical
vortex, changing to a cylindrical vortex in the 'mature region'. Appendix M gives
the fitted model parameters for a cuboid with $H/L = 1/3$ and $L/W = 1$ for a range of
wind directions. Figure 17.50 shows the distribution of pressure coefficient
$\tilde{c}_p\{t = 4\,\mathrm{s}\}$, synthesised from the model for the wind angle $\theta = 30°$ and is directly
comparable with the experimental data of Figure 17.47(b). The rms standard error
between the model and the experimental data is $\varepsilon\{\tilde{c}_p\} = 0.1$, which is acceptably
small and typically the same size as the variation with secondary parameters (such
as corner angle, β) and assumed to be negligable in the design data of Chapter 20.
Using these model parameters, Figure 17.51 shows the prediction for the case

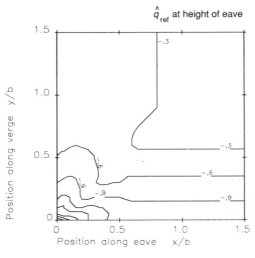

Figure 17.50 Model pressure contours for cuboid
roof with $H/L = 1/3$ and $L/W = 1$ at 30°

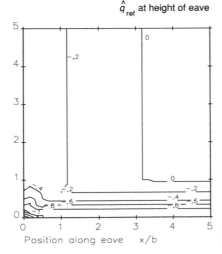

Figure 17.51 Model pressure contours for cuboid
roof with $H/L = 1/10$ and $L/W = 1$ at 30°

H/L = 1/10 and L/W = 1 which is directly comparable with Figure 17.47(c), demonstrating the usefulness of the model in extrapolating to other situations.

Development of this model is incomplete and it will be some time before a sufficient range of model parameter values has been determined to allow its use in design. Until then, the simpler 'loaded region' approach used here and in most codes of practice will have to suffice.

17.3.3.2.6 Effect of parapets on flat roofs.

A comprehensive study of the effect of parapets on mean pressures was made in 1964 by Leuthesser[51] in smooth uniform flow. This indicated that parapets were beneficial in reducing the high suctions at the upwind eave and verge, but made little difference to the overall loads. This problem has been re-investigated recently in terms of peak loads on models in properly scaled boundary layers[52, 312].

Increasing the height of a parapet makes the eave vortex larger but more diffuse, so that the width of the eave high suction region increases in size, changing as indicated in Figures 17.52(a) and (b), but the value of the pressure coefficient becomes less negative. Accordingly, the highest suction coefficients around the periphery are always reduced, but the more moderate coefficients further from the eave may increase. Stathopoulos[312] demonstrates that the suction at locations in the 'V' of attached flow near the corner can almost double with a small parapet. This effect is reduced with large parapets where, in effect, the same total loading is 'smeared' over a much larger area.

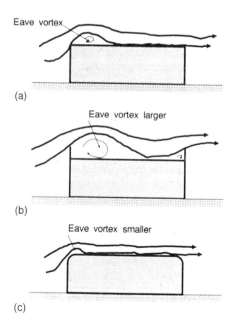

(a)

(b)

(c)

Figure 17.52 Effect of eave profile on the eave vortex: (a) plain eave; (b) parapet; (c) curved eave

Figure 17.53 shows BRE data for $\bar{c}_p\{t = 4\,\text{s}\}$ on the example cuboid, proportions $L{:}W{:}H$ = 3:1:1, at θ = 30° with increasing parapet height, h. The values in (a) for no parapet, h/H = 0, are an independent check on the data presented in Figures 17.45(b), 17.46(e) and 17.47(a). (The different contour values are due to the different scaling factors used in the tests.) Note there is little difference in the

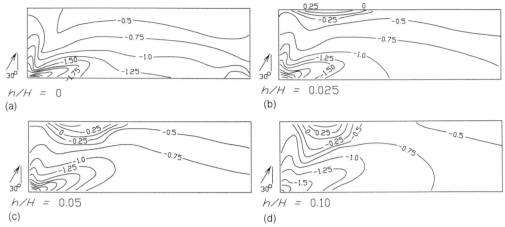

Figure 17.53 Effect of parapet height on pressures on flat roofs

location of the central $\bar{c}_p = -0.75$ contour, but the high local suctions in the corner decrease in value and gradient with increasing parapet height. Note also that the downwind parapet acts as a wall against the attached flow of the 'V'-shaped wedge, creating a region of positive pressure on the roof.

No corresponding detailed studies have been made of the loading on the parapet walls themselves. The suction acting on the rear face of the windward parapet can be expected to be similar to the corresponding roof edge region suctions, while the pressure on the front face is in the top edge region where the pressure is falling. When these expected values are examined, they are found to take values similar to those across boundary walls, including the effect of corners (§16.4.2).

17.3.3.2.7 Effect of curved eaves on flat roofs.
Rounding-off the sharp eave edge with a circular radius, currently a popular design trend in the UK with light industrial buildings clad in profiled steel sheets, has the opposite effect to parapets. With increasing edge radius, the flow attempts to remain attached, decreasing the size of the eaves vortex, as shown in Figure 17.52(c), but the peak suctions also reduce in value. A study of this effect was commissioned by BRE in the BMT wind tunnel. This showed that substantial reductions of suctions near the eave in the 'growth' region close to the upwind corner: suctions were halved with an eave radius $r = 0.1H$ and reduced to a third of the sharp-corner values when the eave radius was increased to $r = 0.4H$. However in the 'mature' region further from the corner, the suctions near the eave initially increased at small radii, before decreasing at larger radii. By radiusing only one eave while keeping the adjacent verge sharp, the study also showed that the eave and verge edge regions are essentially independent.

17.3.3.2.8 Effect of inset storeys on flat roofs.
Here the concern is the effect of the irregular (§17.3.2.5.5) and inset (§17.3.2.5.6) forms of wall faces on the pressures on the adjacent roofs. Apart from a single recent BRE study, reliable data on this subject are very sparse. There are two classes of effect to consider:

(a) the effects on the upper roof level above the inset storey; and
(b) the effects on the lower roof level at the base of the inset storey.

Upper roof: for an eave flush with an irregular-face external wall face, corresponding to the viewed faces in Figure 17.37, the flow over the roof depends on how the face is divided to give the scaling length, b. For an eave to an inset face, corresponding to the either side face in Figure 17.37(c), the scaling height for the eave is the height from the lower roof ($H = H_2$). The flow up this wall faces rises from the lower roof plane, so the pressure distribution on the upper roof behind the eave depends on the lower roof acting as the ground plane. When this scaling height is small, e.g. for a plant-room on a large building, the regions of high pressure scale to this smaller size and not the full building dimensions. The inset face must be set back sufficiently for this effect to be significant, typically by at least $H/2$ or $B/2$.

Lower roof: here there are three main effects:

1 Positive pressures on the windward wall face of the inset storey also act on the adjacent lower roof in the region upwind of the inset storey wall/lower roof junction. The BRE study indicates that the positive pressures extend typically $H/2$ or $B/2$ upwind of the junction. These produce a net **downward** load on this region of the roof which is an unusual loading case for flat roofs that might otherwise be overlooked.
2 Negative pressures (suctions) on the side walls of the inset storey also act on the adjacent lower roof and may be severe near the upwind corner. This is especially the case where the two converging flows meet near the corner of a flush irregular face (§17.3.2.5.6). The BRE study shows that the high suctions in the junction corner are very similar in value to the local roof pressures near a conventional corner.
3 Finally, the pressure in the wake which controls the pressure on the rear wall face of the inset storey also controls the pressure on the part of the roof in the wake. This region is more extensive than the more local effects around the front and sides.

These effects are illustrated in Figure 17.54 by data from Sakamoto and Arie's study [313] of the mean pressure field on a flat ground plane caused by a cube. (Here the depth of the incident boundary layer is only the same as the cube height and in the real situation of inset storeys there would also be a pre-existing pressure field on the 'ground plane' of lower roof, nevertheless the general effects are shown.)

17.3.3.3 Pitched roofs

17.3.3.3.1 Monopitch roofs.   These are roofs formed by one plane face at a pitch angle, α, so that the distinction between eave and verge is no longer arbitrary. The eave remains horizontal, while the verge is at the pitch angle, α, to the horizontal. Recent studies of monopitch roofs have been made by Stathopoulos and Mohammadian [314] and at BRE. Figure 17.55 shows BRE data for pressure coefficient, $\tilde{c}_p\{t = 4\,\mathrm{s}\}$, over monopitch roofs on the example cuboid, proportions $L{:}W{:}H = 3{:}1{:}1$, at $\theta = 30°$ for pitch angles in the range $-45° \leqslant \alpha \leqslant 45°$. In order to make direct comparisons with the earlier example data in Figures 17.45 and 17.53, the reference dynamic pressure is taken at the same height, the height of the low eave. (The design data in Chapter 20 uses the height of the upwind corner as the reference datum to reduce the variability with building proportions.)

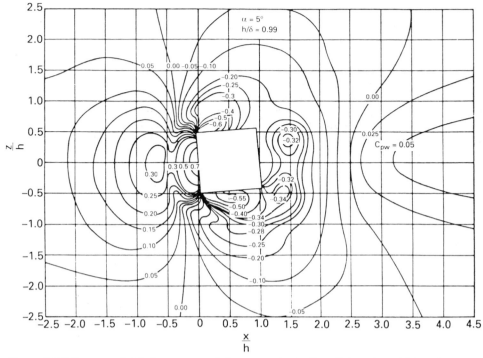

Figure 17.54 Pressure distribution on ground plane around cube (from Sakamoto and Arie, reference 313)

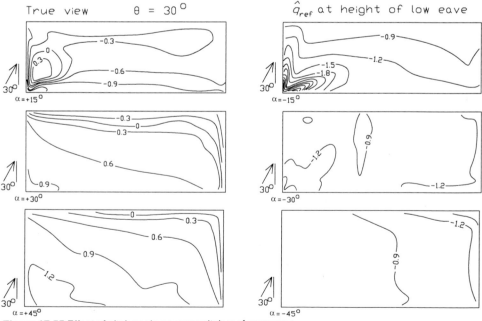

Figure 17.55 Effect of pitch angle on monopitch roof pressures

The left-hand side of Figure 17.55 shows the effect as the pitch angle becomes more positive. Between $\alpha = 0°$ and $15°$ the delta-wing vortices along the upwind eave and verge decrease in strength and size, while the overall pressure rises, reducing the uplift. At $\alpha = 15°$ the delta-wing vortices can be still recognised and the overall pressure remains negative (suction). By $\alpha = 30°$, the delta-wing vortices have disappeared, the flow does not separate from the upwind eave and verge, and the overall pressure is now positive. Increasing the pitch angle to $\alpha = 45°$ increases the positive pressure and values of $\bar{c}_p\{t = 4\,s\}$ in excess of unity are obtained because the roof now rises to $z = 2z_{ref}$ at the downwind high eave where the incident wind speed is higher than the reference value. This effect limits the validity of using one fixed reference dynamic pressure. The upwind low eave height is the best reference only as long as the flow separates from eave and verge. Once the flow remains attached, the downwind high eave height (or ridge for duopitch roofs) becomes the best reference (as used for the A-frame buildings in §17.3.2.6). This is resolved in the design data of Chapter 20 by dividing the low pitch angle 'separating' data from the high pitch angle 'attached' data.

The right-hand side of Figure 17.55 shows the effect as the pitch angle becomes more negative. Initially the delta-wing vortices along the upwind eave and verge increase in strength and size, while the overall pressure falls, increasing the uplift. The delta-wing vortices are at the strongest around $\alpha = -15°$, but by $\alpha = -30°$ they have become diffuse and the pressure distribution is nearly uniform, although the overall uplift remains high. This does not change much at higher pitch angles and is equivalent to the 'stalled' condition of an aircraft wing.

These effects are summarised over the full range of wind angle and pitch angle by Figures 17.56 and 17.57 which show contours of pressure coefficient, $\bar{c}_p\{t = 4\,s\}$, plotted in the $\alpha - \theta$ plane. Figure 17.56 gives $\bar{c}_p\{t = 4\,s\}$ in the 'growth' region of the eave vortex near the upwind corner and illustrates that the highest suctions occur for $\alpha \simeq -13°$ and $\theta \simeq 35°$. The 'growth' region of the verge vortex shows a similar peak of larger value, but the suction remains high for all pitch angles as

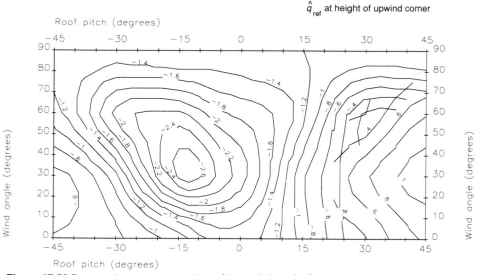

Figure 17.56 Pressure in eave corner region of monopitch roofs (Region A)

Proportions *L:W:H* = 3:1:1

$\hat{q}_{ref}$ at height of upwind corner

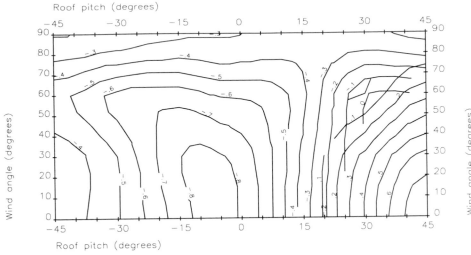

Figure 17.57 Global roof pressure coefficient for monopitch roof on cuboid

$\theta \to 90°$ since the roof appears flat when the wind is normal to the pitch. Figure 17.57 gives the global coefficient, $\tilde{C}_p\{t = 4\,\mathrm{s}\}$, for the whole roof and illustrates that the highest uplift occurs with wind normal to the eave at $\alpha \approx -7°$, but that a lobe of high uplift extends to skew wind angles as α decreases further. In the bottom right-hand corner of both figures is a region where the pressure is positive, corresponding to pitch angles $\alpha < 30°$ facing into the wind.

17.3.3.3.2 Duopitch roofs.   These are roofs formed by two plane faces joined along a common edge to form either a high ridge (positive α), typical of most house roofs, or a low trough (negative α), which is less common. Usually the pitch of both faces is equal, for which many studies have been performed, but sometimes the faces are of unequal pitch, for which there is are almost no data. The BRE full-scale experimental house at Aylesbury was a typical duopitch house, so that the IAWE–ACE comparative data described in §13.5.5.3 are all for this form.

The upwind face behaves almost exactly like a monopitch roof, implying that it is reasonably independent of the downwind face. Figure 17.58 gives the global coefficient, $\tilde{C}_p\{t = 4\,\mathrm{s}\}$, for the whole upwind face and is almost identical to the previous monopitch figure. Slightly smaller values of both positive and negative pressures occur, probably due to the slight influence of fineness ratio which was doubled by the addition of the downwind face, i.e. the supporting cuboid now has proportions $L:W:H = 3:2:1$.

On the other hand, the downwind face has entirely new characteristics. With ridged duopitch roofs a second pair of delta-wing vortices forms along the upwind edges of the face, one behind the ridge and the other along the upwind verge under the existing verge vortex, as shown in Figure 17.59. The plane formed by the downwind face is similar in orientation to a monopitch roof with negative pitch angle (high eave upwind), except that the incident wind flows over the upwind face first. Figure 17.60 gives $\tilde{c}_p\{t = 4\,\mathrm{s}\}$ in the 'growth' region of the ridge vortex near the upwind corner and should be compared with Figure 17.56 for the eaves vortex.

Proportions L:W:H = 3:2:1 $\hat{q}_{\text{ref}}$ at height of upwind corner

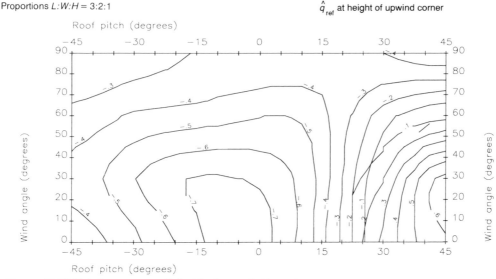

Figure 17.58 Global pressure coefficient for the upwind face of a duopitch roof

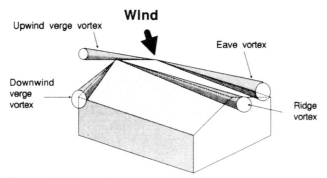

Figure 17.59 Delta-wing vortex pairs on duopitch roof

The above similarity implies an approximate mirror image: although the highest suctions occur at about the mirror position, $\alpha \simeq +15°$ instead of $\alpha \simeq -13°$, their values are reduced to less than half. The highest suctions under the downwind verge vortex are similarly about half the corresponding upwind values. With troughed roofs, α negative, the positive pressure lobe does not occur in the mirror position. Both these effects which make the downwind face less onerously loaded are due to the influence of the upwind face. They are also apparent in Figure 17.61, giving the global coefficient, $\bar{C}_p\{t = 4\,\text{s}\}$, for the whole downwind face. For ridged roofs, the global pressure is about half the corresponding mirror value in Figure 17.57; while for troughed roofs, the positive lobe is suppressed because the downwind face lies in the wake of the upwind face.

Much less is known about unequal pitches, but some deductions can be made. The upwind face is expected to remain effectively independent of the downwind face, so that only the effect on the downwind face of the different upwind pitch angle need be considered. For pitched roofs when the upwind pitch is **steeper**, the

Proportions $L{:}W{:}H = 3{:}2{:}1$ $\hat{q}_{ref}$ at height of upwind corner

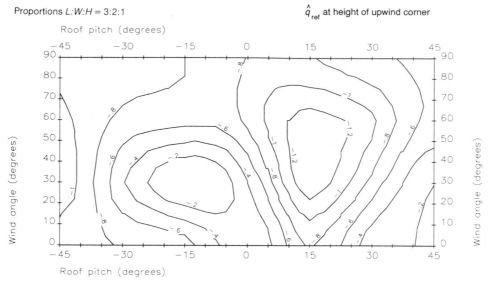

Figure 17.60 Pressure in ridge corner region of downwind face of duopitch roofs (Region A)

Proportions $L{:}W{:}H = 3{:}2{:}1$ $\hat{q}_{ref}$ at height of upwind corner

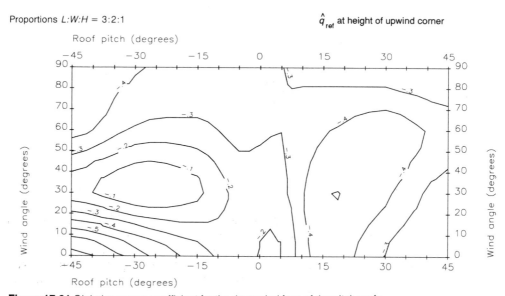

Figure 17.61 Global pressure coefficient for the downwind face of duopitch roofs

loading of the downwind face should be transitional between the equal duopitch and the more onerous negative-monopitch cases, since the negative-monopitch is equivalent to an upwind face pitch of $\alpha = 90°$. For pitched roofs when the upwind pitch is **shallower**, the loading of the downwind face should continue to decrease, so that the equal-pitch case is conservative. For troughed roofs when the upwind pitch is **steeper**, the loading of the downwind face is dominated by the wake of the upwind face, so should take values close to the equal-pitch case for the **upwind** face pitch. For troughed roofs when the upwind pitch is **shallower**, the upwind face wake effect lessens until the upwind face acts like the ground plane, so that the loading of the downwind face is transitional between the equal duopitch and the the front wall face of an A-frame building (§17.3.2.6.2).

17.3.3.3.3 Mansard and multi-pitch roofs.

A mansard roof is a roof where each face has two pitches between eave and ridge, the lower region springing from the eave having a steeper pitch than the upper region rising to the ridge, and this is usually used to enable rooms to be built into the roof space. The opposite is also possible, with the lower eave region at a shallower pitch than the upper region, this sometimes being found when buildings have been extended and in traditional barn designs.

No reliable systematic studies have been performed, except for the special case of a mansard edge to flat roofs. In this study at BRE, chamfering the eaves of a flat roof was found to reduce the size and value of the high-suction regions around the periphery of the roof. The chamfered eave acts in exactly the same manner as the curved eave described in §17.3.3.2.7, but is less effective. Design data for this case are given in Chapter 20.

For the other cases, it is necessary to deduce the loading from the known behaviour of other roof forms. Because we find that the downwind face of duopitch roofs has little effect on the upwind face, we can safely expect the loading of the lower part of the upwind face to be similar to a normal monopitch or duopitch roof in respect of the eave and verge regions as well as the main part of the face. The problem of the upper slope of the upwind face divides into two cases, depending on the difference in pitch angle:

1 When the upper slope is less steep, i.e. a classical mansard, a vortex is expected to form behind the change in pitch if the upper pitch angle is sufficiently low. As the change in pitch is less than would occur if the lower region were a vertical wall, these vortices will be less strong than the eaves vortices of a typical pitched roof. Similarly, a new verge vortex will form at the change in pitch having the characteristics of the new pitch. Accordingly, it is safe to assume the loading equivalent to a single pitched roof having an eave along the change in pitch and a common verge.
2 When the upper slope is steeper, a vortex will not form along the change in pitch because no separation will be induced, but a new verge vortex will form again at the change of pitch. Accordingly it is safe to assume the loading equivalent to a single-pitched roof with a common verge for the upper region, but without the high eave loadings.

A similar argument holds for the two pitches of the downwind face. The ridge vortex depends on the pitch of the upper region since it forms just behind the ridge. The verge vortex forms as normal. Thus the upper region is loaded similarly to a single-pitch roof of the same pitch angle. Being in the wake, the flow is not sufficiently strong or organised for an equivalent vortex to form along the change in slope, but a new verge vortex will form from the change in slope. Accordingly, it is again reasonable to assume the loading equivalent to a single-pitched roof with a common verge, this time for the lower region, but without the high ridge loadings.

17.3.3.3.4 Hipped roofs.

Conventional hipped roofs are formed from duopitch roofs by replacing the vertical gable ends with triangular pitched roofs or 'hips'. This gives four pitched faces for a rectangular-plan building: two trapeziodal main faces and two triangular hips. Data are available only for the case when all four faces have the same positive pitch angle, α. In the special case of a square-plan building, all four faces become indentical and triangular.

With a pitched roof rising from the top edge of each wall, every edge is an 'eave' and there are no 'verges'. Instead there are new small ridges that run from the main ridge to the corners of the building, which are called 'hip-ridges' here. Figure 17.62 shows the typical set of delta-wing vortices that form at skewed wind angles. Note that as the wind angle moves through $\theta = 45°$, the vortices along the hip-ridges move so that they are downwind of the hip-ridge and always start from the upwind end. There are now six vortices in total instead of the four in Figure 17.59. At first sight, it might be thought that these would give additionally severe loadings, but this is not the case. With the earlier duopitch roof, the verge vortices were the strongest whatever the pitch angle, whereas the eave and ridge vortices were consistently less strong. The new hip-ridge vortices are similar in strength to the main ridge vortex so that, having lost the verge vortices, the loading of hipped roofs is much less severe.

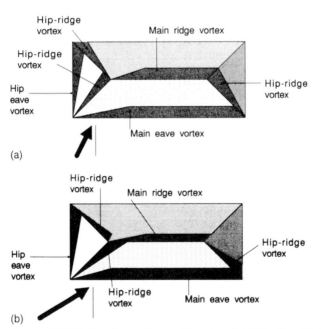

Figure 17.62 Delta-wing vortices on hipped roof: (a) $0° < \theta < 45°$; (b) $45° < \theta < 90°$

17.3.3.3.5 Skew-hipped roofs. If the conventional hip roof reduces edge high suctions by removing the verges, what would be the effect of keeping the verges but removing the eaves? This question was investigated at BRE using the series of models shown in Figure 17.63. These have a square plan and the hipped roof has been skewed by 45° to bring the hip-ridges away from the corners and onto the peak of the gables. The roof/wall joint is not an 'eave', since it is no longer horizontal, nor is it strictly a 'verge', but is transitional between these two cases. This is reflected in the pressures in the corresponding edge regions which are between the

Figure 17.63 Models of skew-hipped roofs in BRE study

duopitch verge and eave values. On the other hand, the pressures in the hip-ridge and interior regions of each face are very similar to the conventional hipped roof values when the wind angle is measured from the normal to the roof face, i.e. from the diagonal of the square plan.

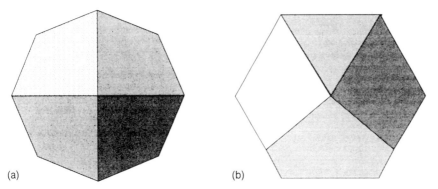

(a)   (b)

Figure 17.64 Examples of other hip forms: (a) hipped roof on regular octagonal-plan prism; (b) unequal-pitch hipped roof on regular hexagonal-plan prism

This not merely an intellectual exercise, since the hip roof is a natural choice for roofing unusual plan shapes. The skew-hip form of the models in Figure 17.63 is found in practice. Figure 17.64 shows two other possibilities, the form of (b) actually existing as shown in Figure 17.65. Empirical rules to cope with these forms are given in Chapter 20.

17.3.3.4 Hyperboloid roofs

There are a number of roof forms that appear to be curved, but are in fact generated from a family of straight lines of differing slope which are used for roofs of shell or tension-web construction. One of the most popular is illustrated by the

Figure 17.65 Unequal-pitch hipped roof on regular hexagonal-plan prism

model in Figure 17.66, which is described here as a 'hyperboloid' roof. This roof is formed by a ridge along the long centreline axis which rises linearly from either end to a peak in the middle. Straight rafters fall from this ridge to the horizontal eave along either long side. When viewed along the long axis, the envelope formed by the family of rafters is the hyperbola which gives this form its name.

Interest was aroused in 1977 when an *ad-hoc* design study of a grandstand with this form of roof for the Bahrain Racing and Equestrian Club gave values of peak suction far in excess of any value previously recorded at BRE. In response, a parametric study was performed using a series of models with ridge pitch angles of $\alpha = 10°$, 20° and 30°. These models were made in sections that could be assembled in various combinations. The 30°-pitch model with $L/W = 3$ is shown in Figure 17.66 with the joints opened so that they can be seen. By removing the middle

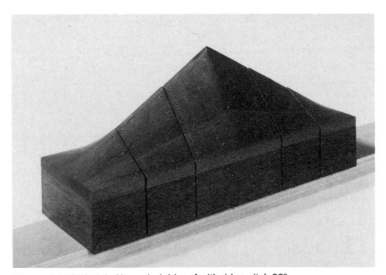

Figure 17.66 Model of hyperboloid roof with ridge pitch 30°

section and closing up the ends, the proportions of $L/W = 2$ and 1 are obtained with a horizontal eave on all sides. By removing the end sections the same proportions are obtained, this time with pitched verges at the gable ends, but tests with this form were performed only for the 20° ridge pitch angle.

$L/W = 3$

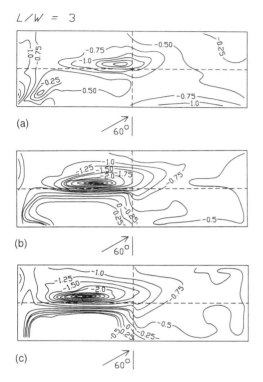

(a)   60°

(b)   60°

(c)   60°

Figure 17.67 Effect of ridge pitch angle on pressures on hyperboloid roofs where ridge pitch = (a) 10°, (b) 20° and (c) 30°

Figure 17.67 shows the effect of ridge pitch angle on the pressure distribution for $L/W = 3$ at a wind angle of $\alpha = 60°$, corresponding to the most onerous loading case. An intense conical vortex grows behind the rising ridge giving very large suctions on the surface. This vortex is exactly the same form as that behind the ridge of the Rock of Gibraltar shown in Figure 7.17(a), the relative wind angles being identical. The hyperboloid roof generates high lift like a sail and the corresponding aerodynamics are much more akin to aircraft aerodynamics than typical building aerodynamics. Perhaps spacecraft aerodynamics is a better analogy because the shape is very similar to lifting-body re-entry vehicles. The design pressure corresponding to the local region under the vortex in (c) is $\tilde{c}_p\{t = 4\,s\} = -3.50$, substantially higher than the next largest case in the verge region of negative-pitch roofs.

The parametric study indicates that the strength of this vortex:

(a) increases with increasing ridge pitch angle, at least to $\alpha = 30°$ which covers the typical practical range;
(b) increases with increasing fineness, i.e. with increasing L/W in this case;
(c) increases for the gabled-end case over the flat-end case for the same overall proportions for the 20° ridge angle tested.

At low ridge pitch angles the dominant high suctions are caused by the eave vortices in the same manner as typical pitched roofs.

17.3.3.5 Barrel-vault roofs

These are essentially curved roofs, but placed on flat-faced walls, giving a hybrid combination which exhibits characteristics of both forms. Only a few studies have been performed, principally by Blessmann[315,316], and then only for squat buildings, cylindrical sections and a limited range of rise ratio. Except at the upper limit of rise ratio when the arch is a hemicylinder, there is a sharp junction betwen the wall and roof at the eave. When the pitch angle at the eave, α_E, is smaller than about 30° the flow separates at the upwind eave to form an eave vortex similar to that on the equivalent pitched roof. Similarly, verge vortices form at the sharp upwind verge at all rise ratios. Eave and verge vortices are the principal characteristics common to flat-faced buildings. On the other hand, the ridge vortex does not form, instead the suction lobe common to curved structures occurs along the crest.

The pressures are very similar to the arched structures of §17.2.2.2.2 when the rise ratio is high, with a lobe of positive pressure at the upwind eave, high suctions along the ridge and uniform moderate suctions in the wake regions. As the rise ratio is reduced, instead of the pressures converging towards zero for a flat ground plane, the pressure distribution converges towards that for flat roofs, i.e. relatively high suctions. This effect is dependent on the eave height for any given rise ratio, and the limit for no walls at all is the arched structure of §17.2.2.2.2.

17.3.3.6 Multi-span roofs

Multi-span roofs are the principal alternative to flat roofs for covering large plan areas and are usually formed of several pitched roofs joined along their eaves. Multi-span monopitch roofs produce a 'saw-tooth' profile and these have been studied at CSIRO by Holmes[317,318]; multi-span equal-pitch duopitch roofs at CSIRO[319] and at BRE, and multi-span unequal-pitch (60°/30°) duopitch roofs at BRE. This form was also included in the NIAE series of full-scale studies using a multi-span greenhouse[320]. These studies reveal some common characteristics:

1 The pressures along the verges are not significantly changed from the single-span values.
2 The pressure along the eaves on the first, upwind span are not significantly changed from the single-span values. 'Eave' regions on central spans become ridge or trough regions.
3 Pressures in the local regions behind the first ridge are not significantly changed from the single-span values, but the values reduce for the second and third ridges, thereafter remaining constant. (Actually the loading of last downwind span is always different because it is directly influenced by the pressures in building wake. However the loading is always slightly less onerous.)
4 With the wind angle predominantly parallel to the ridge line ($\alpha \rightarrow 90°$), interior pressures on all spans are not significantly changed from the single-span values.
5 With the wind angle predominantly normal to the ridge line ($\alpha \rightarrow 0°$), interior pressures on the first span are not significantly changed from the single-span values, but the values reduce for the second and third ridges, thereafter remaining constant. At the steeper pitches where positive pressures would be

expected on the windward-facing faces, this occurs only on the first upwind span, downwind spans being in the wake of the previous span in the manner of troughed roofs (§17.3.3.2).

This enables some general design rules to be formulated in Chapter 20. The tendency to less onerous loading of downwind of the first several spans when the wind is normal to the ridges is caused by the flow skipping from ridge to ridge, leaving separation bubbles in each trough. This effect also occurs with multi-span barrel-vault roofs, as it did with multiple arched buildings in §17.2.2.2, supported by the model study of Blessmann[316] and the NIAE full-scale measurements[290].

An exception to this general rule is multi-span hyperboloid roofs of the form shown in Figure 17.66 (§17.3.3.4). Pressures on second and subsequent spans become progressively more onerous at skew wind angles as the multi-spans act like an aerofoil cascade. Expert advice should always be sought when this form of roof is used.

17.3.3.7 Effect of parapets on pitched roofs

Parapets are commonly used with pitched roofs in order to disguise the roof line, usually by building the parapet up to the ridge line, as in the examples of Figure 17.68, but sometimes raising the parapets substantially above the ridge. Blessmann[321,322] has studied the troughed duopitch form in (c) for a range of parapet heights. The multi-span form shown in (d), where the parapets fill the gable end troughs, has been studied at BRE.

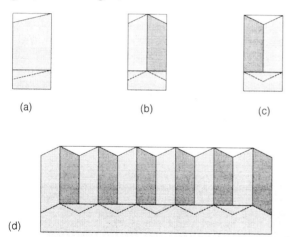

(a)   (b)   (c)

(d)

Figure 17.68 Examples of pitched roofs with parapets: (a) monopitch; (b) ridged duopitch; (c) troughed duopitch; (d) multi-span duopitch

The principal effect of the parapet is to make the flow separation line horizontal along both upwind eave and verge. In the case of the monopitch roof (a) with the low eave with parapet upwind and of the ridged duopitch roof (b), the flow separating from the parapet does not 'see' the pitch of the upwind face of the roof and gives pressures similar to a flat roof with parapets, including the reduced suctions in the upwind eave regions (§17.3.3.2.6). An important result is that the positive pressures expected when the roof pitch exceeds $\alpha \simeq 30°$ are replaced by suctions. This will also occur in the multi-span case (d) if the parapets are extended around the upwind roof face; but as shown in (d) the upwind face is unaffected by the parapet. The BRE study shows that the suctions along the ridge regions of the duopitch (b) and multi-span (d) roofs are not significantly affected.

In the case of the monopitch roof (a) with the high eave without parapet upwind and of the troughed duopitch roof (c), the flow separating from the eave is unaffected. Thus the eave vortex still increases in strength as the pitch becomes more negative to the maximum at $\alpha = -15°$, and the corresponding pressures on the roof are unchanged (§17.3.3.3.1). However, the flow at the verge separates from the parapet so that the verge vortex does not increase in strength and the more onerous verge region suctions are avoided.

In summary: the parapets affect the edge regions immediately downwind, reducing the high suctions; and, where the parapet extends to the ridge level on steep duopitch roofs, the positive pressures on the upwind face are changed to suctions.

17.3.3.8 Friction loads on roofs

When buildings are very squat, significant forces parallel to the wind can be accumulated by the action of friction on the part of the roof in the wedge of attached flow between and downwind of the eave and verge vortices, in a similar manner to walls (§17.3.2.7.4) and canopy roofs (§16.4.4.8), and can be quantified by the use of the shear stress coefficient, c_τ. Rules for this effect are given in Chapter 20.

The action of shear stress on the roof under the strong eave and gable vortices is of no structural signifance, but is the dominant mechanism in the removal of gravel ballast from flat roofs. This mechanism is distinct from that for the removal of unsecured insulation panel systems and paving stones, which is discussed in §18.8.2, later. Gravel scour has been extensively studied by Kind [323, 324, 325] and others [326], leading to procedures for design against this effect [327, 328]. Current BRE advice is given in BRE Digest 311 *Wind scour of gravel ballast on roofs*, based on the full NRCC method of Kind and Warlaw [327].

17.4 Combinations of form and complex bluff structures

17.4.1 Introduction

In the preceding sections of this chapter the loading of the various forms of structure has been discussed. While many buildings are clearly one single part of one distinct form, many others will be constructed by putting together several parts of the same or of a different form. For example, a combination of lattice and line-like forms is obtained when tall cylindrical stacks are supported by a lattice tower or when a lattice gantry is carried on line-like columns. Similarly, combinations of lattice:plate-like, lattice:bluff, line-like:plate-like, line-like:bluff and plate-like:bluff are all found in practice. Particularly among the bluff forms of structure there is the problem of parts of widely different size, e.g. appendages such as chimney stacks on large buildings. No current code of practice or other design guide specifically addresses this problem, yet few buildings and structures are of the 'pure' form for which design data are usually given.

17.4.2 Laws of scale and resonance

In the absence of specific guidance, the designer makes pragmatic decisions based on his experience of how buildings have worked in the past. It is likely that he will make intuitive use of the **law of scale**, but not its counterpart the **law of resonance**. The term 'laws' conveys too great an importance since it implies that, like the law of gravity, they cannot be broken.

In essence, these first of these 'laws' is:

Law of scale: things of a certain size are directly affected by things of a similar size or larger, but are not significantly affected by things that are much smaller. Thus we may expect that a building on a large hill will be affected by the flow over the hill, but that the flow over the hill will not be significantly affected by the building. This seems quite reasonable and is the basis of the Topography Factor in Chapter 9.

On the other hand, the essence of the second 'law' is:

Law of resonance: small things can strongly influence large things if their direct action can be increased by some form of resonant amplification. Thus the provision of small surface roughness on a large cylinder can change the flow regime from subcritical to supercritical, producing the big result of halving the drag coefficient.

At first sight, these two 'laws' seem contradictory, as seem the traditional British maxims: 'Many hands make light work' and 'Too many cooks spoil the broth'. In reality, either of the 'laws' can be true in certain given circumstances, as can either of the maxims. For the many hands to be able to lighten the work-load, the job in hand must be amenable to cooperative effort without overcrowding, i.e. in a parallel process, and one falls foul of 'too many cooks' in the opposite conditions, i.e. in a serial process. So we find that the 'law of scale' works well in most circumstances because the 'law of resonance' requires some pre-existing instability to be triggered in order to amplify the small effect. The large change from subcritical to supercritical conditions is not due to the roughness as such, but occurs because the surface boundary layer is in an unstable condition, ready to be switched from being laminar to being turbulent by roughness or some other small perturbation in just the right position upwind of the separation point.

In summary, the 'law of scale' works for most of the time, but the 'law of resonance' is always waiting to catch the unwary. The purpose of this section is to give guidance on when the 'law of scale' can be safely used.

17.4.3 Combinations of form

17.4.3.1 Combinations with lattices

17.4.3.1.1 General.   In general, in all combinations including lattices, the lattice components are affected by the other forms, while the other forms will experience only the 'wind shadow' effect of the lattice in reducing the incident dynamic pressure. This is true providing the lattice is of sufficiently low solidity to stay in the 'lattice' ranges rather than the 'porous body' ranges (§16.3.1).

17.4.3.1.2 Lattice:line-like.   The combination of a lattice with line-like components usually occurs as the result of a structural lattice supporting non-structural line-like elements like power-lines, pipes, stack flues, cylindrical antennae, *etc*. The shielding effect of the 'wind shadow' of a lattice on downwind line-like components can be exploited. If the line-like component is subject to vortex shedding, an upwind lattice is likely to reduce the effect since the shedding is sensitive to turbulence (§2.2.10.4). The effect of the wakes of the line-like components also tends to reduce the peak loading on a downwind lattice, but most of this reduction is in the mean component and the additional turbulent fluctuations could cause fatigue problems. It is normally safe to treat the lattice and line-like components separately and to ignore any interaction. When many line-like components are used together, they create a lattice in their own right, so that lattice trusses packed with many line-like ancillaries, e.g. pipe-bridges, should be treated in total as one lattice.

17.4.3.1.3 Lattice:plate-like.
This combination occurs when a lattice is used to support the plate and the falsework supporting the bridge deck in Figure 16.50 is a good example. This was discussed earlier in §16.3.6.4 'Partly clad lattice arrays', but in Figure 16.50 the plate is always parallel to the wind. Other combinations include plate-like and dish antennae on lattice towers, where the plates can be normal to the wind, shielding downwind lattice elements as discussed above.

17.4.3.1.4 Lattice:bluff.
Lattices are often used to support bluff components, such as water tanks. Here the higher-drag bluff components will usually dominate the loading, with the lattice giving a small additional load. The 'wind shadow' shielding of the bluff components by the lattice will only be significant if the lattice is relatively dense.

Another common problem is the loading on temporary lattice frameworks, e.g. scaffolding around large bluff buildings. There have been very few relevant studies of this problem. Some measurements of the wind speed close to building façades have been made at TNO in the Netherlands [329, 330] which are relevant to this problem. They are also relevant to the loading of appendages on the face, e.g. mullions, balconies, canopies, sunscreens, etc., as discussed below, as well as for convective heat transfer and rain impingement. Figure 17.69 presents some of the results [329] as contours of wind speed over the face of the building for various wind

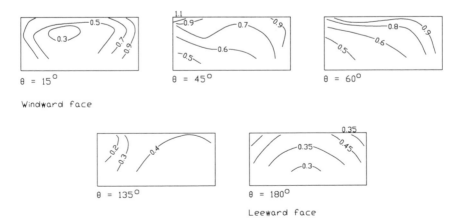

Figure 17.69 Isotachs of wind speed close to building wall (from reference 330)

directions. When these are compared with the pressure contours of Figure 17.30, they appear to mirror the pressure contours in value. If the Bernoulli equation, Eqn 2.6, derived in §2.2.2 held exactly, then the wind speed, V, would be given by:

$$V = (1 - c_p)^{1/2} V_{\text{ref}} \tag{17.3}$$

which is effectively the same as Eqn 14.21 used in the UK slating and tiling code, BS5534, for the local dynamic pressure on roofs (see §14.1.3.2). The top row of Figure 17.69 represents the windward face where the flow is attached, where this equation is the most likely to give a good result, and here the comparison is quite reasonable. However, the wind speeds on the leeward face are much lower than predicted by Eqn 17.3, which is not surprising since the Bernoulli equation is expected to be poor in the turbulent wake. Fortunately, the highest speeds are found when the face is swept by the wind.

An attempt has been made to measure the loads on access scaffolding directly using models, commissioned by the European Council for Standardisation (CEN). Some results from their report [331] are reproduced in Figure 17.70, showing the maximum force coefficient on sections of the scaffolding normal and parallel to the building face, for a range of building face solidity. Clearly the building face shields the scaffolding considerably as the building face becomes more solid (see §16.3.5.2). Unfortunately these results were obtained in smooth uniform flow, so only serve as qualitiative examples. Other measurements have been made in Japan

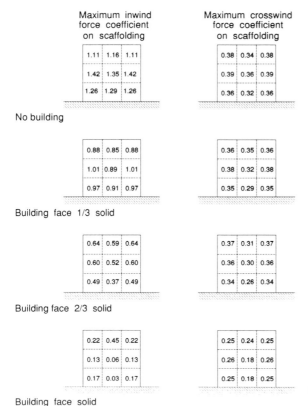

Figure 17.70 Effect of building face solidity on loads on access scaffolding

with the scaffolding clad [332], but these are more relevant to the problem of porous cladding systems discussed in §18.8.2.2.

17.4.3.2 Combinations with line-like components

17.4.3.2.1 Line-like:plate-like. This combination comprises plates supported above the ground by line-like legs or columns, typically signs and hoardings. Canopy roofs are an important member of this class, since the canopy is usually supported on line-like columns. The loading of the plate and line-like elements can always be safely assessed independently. Remember that a plate-like sign supported by a central column is a good candidate for the divergence instability of §8.6.4.2, requiring the test in §10.6.2.2.

17.4.3.2.2 Line-like:bluff. This includes bluff elements supported on line-like columns, such as water tanks on poles, as well as line-like appendages to bluff structures, such as chimney stacks and link bridges between buildings. Here the line-like elements can be taken to have no significant effect on the loading of the bluff elements. Conversely, the bluff elements control the end effects of the line-like elements and hence the effective slenderness ratio, as discussed in §16.1.2 and later.

17.4.3.3 Combinations of plate-like and bluff elements

17.4.3.3.1 Scope. Practical combinations in this category are typically parapets around buildings, which were discussed in §17.3.3.2.6 and §17.3.3.7, canopies attached to buildings and other plate-like appurtenances such as balconies and ribs, which are discussed below.

17.4.3.3.2 Canopies attached to buildings. The loading effects on canopies attached to buildings falls into two classes:

(a) canopies attached near the top, or at least more than half-way up, the building;
(b) canopies attached near the bottom, or at least less than half-way up, the building.

The first class is typically represented by extensions of the roofs of low-rise buildings to provide shelter over loading bays, etc. These are equivalent to freestanding canopies when fully blocked at one edge, so are covered by the discussion in §16.4.4 and the design rules in §20.5.3. The later discussion on open-sided buildings in §18.6 and the corresponding design rules may also be relevant in some cases. Figure 17.71(a) shows that when the canopy is on the windward face of the building, the incident wind is blocked from flowing underneath the canopy, resulting in a large upward force on the canopy.

The second class is typically represented by canopies over entrances of high-rise buildings, for which the flow conditions are very different from the first case. Figure 17.71(b) shows that the flow driven down the windward face by the incident wind profile (§8.3.2) impinges on the top of the canopy, producing a net downward load. When the canopy is on the side face, the accelerated flow in the 'horseshoe vortex' (§8.3.2.1.2) produces high upward loads. When the canopy is on the rear face the loads are much smaller. This has been studied recently by Jancauskas and Eddleston [333] who confirm that the two cases to be considered in the design are a canopy on the windward face for maximum downward load and a canopy on the side face for maximum upward load.

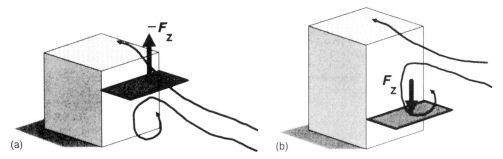

Figure 17.71 Canopy attached (a) high and (b) low on building

17.4.3.3.3 Balconies and ribs. A study of both types of appurtenance, including a review of earlier work, has been recently reported by Stathopoulos and Zhu [334]. Balconies without walls, i.e. horizontal plates, distributed all over the face of a building have little effect on the face pressures, but the addition of parapet walls to the balconies slightly reduces the high suctions in the wall edge regions. Similarly, vertical ribs or mullions make no significant difference to the wall pressures over most of the wall. The exception is the recess caused by the first rib at the upwind corner where the local edge suction is reported to increase substantially in some critical wind directions.

These results imply that the loading of the balconies and ribs is also small, although they were not measured directly. The variation of wind speed over the building façade discussed in §17.4.3.1.4 in combination with the discussion of shelter behind walls in §19.2.2 is relevant to this case, and result in the empirical design rules given in §20.8.2.

17.4.4 Appendages

17.4.4.1.1 General. These are typically small components or services entering or leaving buildings, such as chimney stacks, ventilators, smoke vents, gutters, downpipes, ladders, etc., or minor architectural features. In general, the 'law of scale' applies in the majority of cases, so that the loading of the large bluff structure is not significantly affected by the small appendages. The exceptions, cases where the 'law of resonance' overrides, occur when an instability in the aerodynamics, the structure or both together (aeroelastic instabilities, see §8.6.4) is triggered. The latter two cases are covered in Part 3 of the Guide.

The principal aerodynamic instability is the transition of the flow around curved structures from the laminar separation subcritical flow regime to the turbulent separation supercritical flow regime when at just below the critical Reynolds number (§2.2.10.2). The majority of curved structures will be sufficiently large that the flow is naturally supercritical and, besides, the transition results in a general reduction of loading. This aspect was discussed in the relevant sections of §17.2 'Curved structures'.

Separations on sharp-edged structures are affected by small modifications to the corner detail, so that the form of gutters etc., may affect the loading of roofs. In general, the standard sharp-edged case gives the most onerous loading, so that the data in Chapter 20 are conservative. The effect of parapets, curved eaves and mansard eaves in reducing the loading on roofs was discussed earlier. The use of aerodynamic devices to reduce loading is discussed in §19.5 'Load avoidance and reduction'.

The other side of the problem is the loading of the appendages, and this is now discussed.

17.4.4.2 Small appendages

Where the appendage is much smaller than the main structure, the loading of the appendage depends on the flow around the main structure. The discussion for lattices on bluff structures, §17.4.3.1.4, also applies here. The wind speed close to walls can be determined from Figure 17.69. When the wall is swept by the wind, and on roofs where the dynamics of the delta-wing vortices dominate, estimates of the local wind speed are given by Eqn 17.3. This predicts that wind speeds are less than the reference wind speed where the pressure is positive, typically on windward walls, but greater where there is suction, typically under the delta-wing vortices. This is the approach used by the UK slating and tiling code, BS5534, discussed in §14.1.3.2. Loading of small appendages can therefore be assessed using this local wind speed as if they were attached to a ground-plane. The alternative is to collect and present data for each possible combination. While this is feasable for some simple and common cases, and has been attempted for solar-collector arrays on houses [335], the vast number of possible combinations makes this approach impossible on anything other than an *ad-hoc* basis.

17.4.4.3 Parts of comparable size

The above approach breaks down when the parts are of comparable size, since each can affect the flow around the other. Each combination requires to be specifically addressed. The earlier sections on walls and roofs of inset storeys are examples of this approach. Although the corresponding design rules in Chapter 20 appear to assess the components independently, this is only a design simplification. By careful choice of the scaling length for the loaded regions and reference height for the dynamic pressure, this gives a conservative result in most cases and exceptions may be accommodated by special rules. This approach is shown by measurements, for example studies of lanterns at the ridge of pitched roofs [336,337], to be reasonable.

18 Internal pressure

18.1 Introduction

Internal pressures do not effect the overall wind loads on an enclosed structure because their contributions cancel out over all the internal faces of each room. However they do affect the overall load on an individual building face, which is the algebraic difference of the external pressure, p_e, and the internal pressure, p_i, integrated over the face:

$$F = \int \int (p_e - p_i) \, dy \, dx \tag{18.1}$$

and thereby affect the load paths by which the overall load accumulates through the structure. Differences of internal pressure between adjacent rooms of multi-room buildings generate net loads the internal walls.

The internal pressure of any one room within a building is primarily controlled by the external pressure field around the building, the position and the size of all openings which connect the inside of the building to the outside and which connect the room to every other room. Second-order effects, for example the rate at which the internal pressure changes in response to changes in the external pressure, are also dependent on the volume of the rooms and the stiffness of the walls and roof of the building. Other minor effects that influence the internal pressure are the thermal 'stack effect', and the pressure rise through air-conditioning plant. Stack effect [338, 339] can produce pressure differences up to typically only 100 Pa, whereas inlet and extract fans of air-conditioning plant are usually balanced to give an internal pressure near zero, so that neither effect is significant to the structural design.

Figure 18.1 illustrates three typical situations:

(a) Here there is a large opening in the windward wall, while all other faces are nominally sealed, so that air flows through the opening until the internal pressure equalises to the external pressure on the windward wall. The positive internal pressure acts against the positive external pressure on the windward wall, reducing the net load to zero. However, the positive internal pressure acts with the suctions on the roof and the leeward wall, increasing their loading. This represents a common cause of roof removal after breakage of windows in the windward wall by flying debris.

(b) Here the large opening is in the leeward wall, while all other faces are nominally sealed, so that now air flows through the opening until the internal

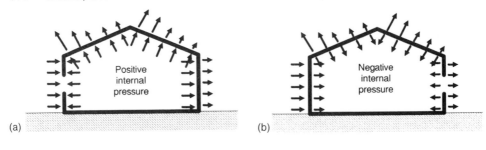

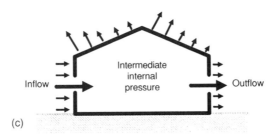

Figure 18.1 Effect of openings in walls where opening in (a) windward wall, (b) leeward wall and (c) windward and leeward walls

pressure equalises to the external suction on the leeward wall. Now the loading of the leeward wall is zero, the loading of the roof is reduced, but the loading on the windward wall is increased. Opening windows in the leeward wall is a recommended action in tropical storms which helps to prevent roof blow-off, provided the windward wall is sufficiently strong.

(c) Here there is an opening of comparable size in both windward and leeward walls, so that now air flows continuously in through the windward opening and out through the leeward opening. The internal pressure is intermediate between the two previous cases, controlled by the balance of the two flows which, by continuity (§2.2.2) must be equal and opposite.

In the first two cases the openings dominate the internal pressure and so are called 'dominant openings'. Dominant openings, such as open windows, doors etc., are typical of service conditions, but can also be critical in the ultimate load condition when the dominant opening is accidental, caused by cladding failure or debris impact, when they can control the mode of more serious structural failure. The third case is relevant to the typical design condition for the ultimate limit state, while the external envelope of the building remains intact, but is also typical of serviceability conditions in buildings without opening windows.

These aspects are discussed in the remainder of this chapter. The discussion starts with consideration of first-order quasi-steady flow conditions on which most codes of practice and design guidance are based, before moving on to more recent research into the response of internal pressure to changes in external pressure, the effects of wall flexibility, sudden breaching of the external envelope, and other second-order effects.

18.2 Quasi-steady conditions

18.2.1 Steady-state flow balance

The flow rate, Q, through an opening is related to the area of the opening, A, and the magnitude of the pressure difference across it, $|p_e - p_i|$ by the proportionality:

$$Q \propto A \, |p_e - p_i|^n \tag{18.2}$$

where the exponent n takes the value $n = 1$ for laminar flow through the opening and $n = \frac{1}{2}$ for turbulent flow. Laminar flow occurs when the opening is long and narrow like a pipe and turbulent flow occurs when the opening is short like a hole in a plate. Later, it will be demonstrated that the value of n determined in buildings varies in the range $0.5 \leqslant n \leqslant 0.7$, because the general porosity is formed by a mixture of laminar and turbulent flow cases. The limit of $n = \frac{1}{2}$ for turbulent flow is the most onerous condition for loading of building faces and internal partitions, so is usually adopted for design.

The best known form of Eqn 18.2 is the orifice-plate meter equation:

$$Q = C_D A_D \left(2 \, |p_e - p_i| \, / \, \rho_a \right)^{1/2} \tag{18.3}$$

where C_D is the discharge coefficient for the orifice which has a standard value of $C_D = 0.61$ for sharp-edged circular orifices, and A_D is the corresponding discharge area. For standard sharp-edged orifices A_D is the actual area, but for more complex or labyrinthine openings A_D can be regarded as the area of an equivalent sharp-edged orifice. The sign of Q is given by the sign of the pressure difference, so that $p_e > p_i$ gives an inflow and $p_e < p_i$ gives an outflow.

Flow continuity requires that the inflow and outflow balance, so that:

$$\sum_{j=1}^{N} Q_j = 0 \tag{18.4}$$

It simplifies the problem if the standard value of discharge coefficient, $C_D = 0.61$, is adopted and any discrepancies compensated by replacing the actual area of the opening by the effective discharge area, A_D. Then Eqn 18.4 becomes:

$$\Sigma \left(A_D \, |p_e - p_i| \right)^{1/2} = 0 \tag{18.5}$$

since the other parameters are constant and cancel out of the equation.

Figure 18.2(a) represents a single-room building with six orifice-like openings on various faces. The quasi-steady internal pressure, p_i, is determined by solving the equation:

$$A_1 \left(p_1 - p_i \right)^{1/2} + A_2 \left(p_2 - p_i \right)^{1/2} = A_3 \left(p_i - p_3 \right)^{1/2} + A_4 \left(p_i - p_4 \right)^{1/2}$$
$$+ A_5 \left(p_i - p_5 \right)^{1/2} + A_6 \left(p_i - p_6 \right)^{1/2} \tag{18.6}$$

This cannot be solved directly, but requires iteration, i.e. an initial value is assumed for p_i and is adjusted until the equation balances. Note that inflow and outflow components have been separated for convenience on either side of the equation, while the order of p_i and the p_e values have been adjusted so that the difference for the square root is always positive. During the iteration process, if the new value of p_i changes an inflow into an outflow, or *vice-versa*, the corresponding term should be moved to the respective side of the equation and the order of p_i and p_e swapped.

Figure 18.2(b) expands the problem to a multi-room situation. In this case there is an Eqn 18.5 for the flow through each room, giving seven simultaneous equations in the example. Connections between rooms means that the internal pressures of

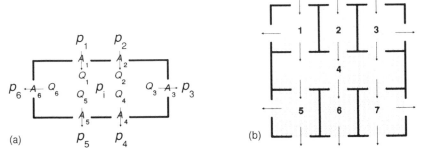

Figure 18.2 Flow continuity through (a) single-room and (b) multi-room building

adjacent rooms appear as extra terms in the equation. In this example, room 1 connects to two external pressures and the internal pressure of the corridor 4. The corridor 4 is unique in that it has no external connections, but connects to the internal pressures of all the other rooms. These internal connections complicate the solution of the seven equations, since each change of internal pressure affects the flows through all connected rooms, requiring solution by simultaneous iterations. This is quite practical, through tedious, to do by hand, but is much more conveniently solved by computer.

18.2.2 Determination of envelope porosity

The porosity of the building envelope is important in the ventilation and spread of fire in buildings, so that most of our knowledge comes from such studies. Estimates of porosity can be made of individual components, such as doors or windows, or else measurements can be made of the overall porosity of rooms or whole buildings. Air-conditioned buildings can be pressurised using their own ventilation plant and the flow rates through the walls measured. Other buildings require special equipment to provide the pressure. Figure 18.3 shows BREFAN, the whole

Figure 18.3 BREFAN system in use on three-storey offices

building system devised by BRE, in use on a three-storey office block. Other, smaller systems are used to test individual rooms.

Research has been carried out in many countries and the results presented in a variety of ways. One way is to combine the area, A_D, with the other constant parameters to give a total parameter, usually denoted by C. Another recognised standard test method for ventilation studies[340] uses a fixed 25 Pa pressure difference (very low for wind loading but typical for natural ventilation) to give the corresponding flow rate, denoted by Q_{25}. For wind loading calculations the equivalent orifice discharge area, A_D, is the most convenient to use, and is adopted for the Guide. This is because the results are also compatible with directly estimated areas of large, possibly dominant, openings like doors and windows.

Table 18.1 gives typical values to the porosity, in terms of the ratio of the effective discharge area of openings, A_D, to the total surface area, A_{Tot}, and values of the exponent, n, both from direct measurements. In 1974, BRE advice in *The wind loading handbook* was that typical porosity is in the range 4×10^{-4} to 10^{-3} and that the exponent should be taken as $n = \frac{1}{2}$. This is shown to be correct by measurements taken in the decade 1976–86. The value of porosity for office buildings in North America, where the cladding and glazing has been engineered, is about 4×10^{-4} and this can be taken as typical because engineered structures are quite similar worldwide. Building practice for traditional housing differs considerably between countries so that, while the porosity of Canadian housing is similar to typical engineered buildings, housing in the UK is three times more porous, while Swedish housing is significantly less porous. Porosity of internal walls is about 50% greater than external walls.

It is clear from these data that it will be very difficult to put accurate absolute values to porosity. Fortunately, only the relative porosity of each of the various walls is required when there are no dominant openings. For buildings that use the same form of construction for all external walls, this simplifies the problem to requiring only the total area of each wall.

Table 18.1   Typical building porosity characteristics from measurements

Office buildings	Porosity A_D/A_{Tot}	Exponent n
Canada [341, 342] (12 buildings)	$3.6 \times 10^{-4} \pm 0.46 \times 10^{-4}$	0.65 ± 0.01
USA [343] (6 buildings)	$3.9 \times 10^{-4} \pm 1.64 \times 10^{-4}$	0.60 ± 0.04

Housing [344]	Porosity A_D/A_{Tot}	
United Kingdom	10.4×10^{-4}	
Canada	3.8×10^{-4}	
Sweden	2.07×10^{-4}	

Partition walls	Porosity A_D/A_{Tot}	Exponent n
Netherlands [344]	$6.7 \times 10^{-4} \pm 1.5 \times 10^{-4}$	0.64 ± 0.13

Single leaf door
United Kingdom (BRE)   Gap width when closed = 1.53 mm/m run $n = 0.59$

18.2.3 Definition and consequence of dominant openings

With typical wall porosities being so small, any large opening such as a door or window would be expected to make a large difference to the loads. If we return to the cases shown in Figure 18.1(a) and (b), where there is a large opening in one face (face 1), but also allow the opposite face (face 2) to have a general porosity, then air will flow in through the opening and out through the general porosity or *vice-versa*. The flow continuity equation for this case is:

$$A_1 (p_1 - p_i)^{1/2} = A_2 (p_i - p_2)^{1/2} \tag{18.7}$$

which, after rearranging, becomes:

$$(p_i - p_2) / (p_1 - p_i) = (A_1 / A_2)^2 \tag{18.8}$$

showing that the pressure drop across either face is in the proportion of the square of the discharge areas.

If the area of the large opening is three times larger than the sum of the distributed porosity of the opposite wall, then the internal pressure will rise so that only about 10% of the load is taken on the wall with the opening, with the remaining about 90% of the load being taken by the opposite wall. This is so close to the original case of the opposite wall solid, that the opening is clearly dominant. On the other hand, if the large opening is equal in area to the sum of the distributed porosity of the opposite wall, the load is shared equally and the opening is not dominant. Somewhere in this range a threshold can be set to define a dominant opening: an opening twice as large as the sum of the distributed porosity of the rest of the building is a good threshold, giving a load ratio of 25%:75%. Openings of this size, or larger can be treated as dominant, so that the internal pressure is set by the pressure at this opening. Openings smaller than this size are not dominant, so that the internal pressure must be estimated from the balance of flow.

18.3 Time-dependent conditions

18.3.1 Compressible flow

Just as the concept of a stiff structure must be abandoned to assess the difference between quasi-steady and dynamic response of the building, so the concept of a 'stiff', i.e. incompressible, gas must be abandoned to assess the difference between quasi-steady and time-dependent internal flow balance. In allowing the density of air to change, the flow continuity equation, Eqn 18.4, must be modified to:

$$\sum_{j=1}^{N} \rho_{a_j}\{t\} \, Q_j\{t\} = 0 \tag{18.9}$$

to conserve mass flow.

The relationship between gas pressure and density is given by the gas equation:

$$p_{abs} / \rho_a = R \, T_{abs} \tag{18.10}$$

where p_{abs} is the absolute atmospheric pressure ($p_{abs} = p_{atm} \simeq 10^5$ Pa), T_{abs} is the absolute temperature in degrees Kelvin (K) and R is the universal gas constant ($R = 287.1$ J/kg K). With the involvement of temperature, the problem becomes one of thermodynamics. When the process is very slow, any change of the temperature is equalised to the constant ambient temperature and the process is **isothermal**, giving:

$$p_{abs} / \rho_a = \text{constant (isothermal)} \tag{18.11}$$

When the process is very fast, there is no time to lose heat and the process is **adiabatic**, giving:

$$p_{abs} / \rho_a^{\gamma_a} = \text{constant (adiabatic)} \tag{18.12}$$

where $\gamma_a = 1.4$ is the ratio of specific heats for dry air. Eqn 18.12 implies that by quickly compressing air its temperature will rise and by quickly expanding air its temperature will fall. Both cases can be covered by using Eqn 18.12 and setting γ_a to unity for the isothermal case.

Unfortunately, allowing for these changes in air density makes the time-dependent flow balance extremely complicated and it has only been determined for the simple cases in Figure 18.1: a single opening, as in (a) and (b); and two openings on opposite faces, as in (c).

18.3.2 Single-orifice case

18.3.2.1 Time-dependent flow balance

For a single orifice as in Figure 18.1(a) and (b) the quasi-steady model predicts that the internal pressure, c_{p_i}, equalises to the external pressure at the opening, c_{p_e}. The time-dependent model predicts that the equalisation process will take a finite time, t_i, to occur, so that $c_{p_i}\{t\} \to c_{p_e}$ as $t \to t_i$.

The single orifice of area A_D is the well-known Helmholtz resonator problem in acoustics [345]. Holmes [346] shows that the time-dependent flow balance is given by a second order differential equation:

$$\frac{\rho_a \, L \, O}{\gamma_a \, p_{abs}} \, d^2 c_{p_i}/dt^2 + \frac{\rho_a \, O^2 \, q}{2 \, C_D^2 \, \gamma_a^2 \, A_D \, p_{abs}^2} \, dc_{p_i}/dt \, |dc_{p_i}/dt| + A_D \, c_{p_i} = A_D \, c_{p_e} \tag{18.13}$$

where O is the volume of the room and L is the depth of the orifice, in effect the length of the neck of the opening. If, as in Figure 18.1, there is virtually no depth to the neck, the effective value of L becomes:

$$L = (\pi \, A_D / 4)^{1/2} \tag{18.14}$$

A similar second-order differential equation describes the dynamic response of buildings:

$$m \, d^2x/dt^2 + d \, dx/dt + k \, x = F\{t\} \tag{18.15}$$

where m is the mass, d is the damping and k is the stiffness of the building for the mode, and $F\{t\}$ is the modal force exciting the building (see Part 3). The terms of Eqn 18.13 have analogous functions. On the left-hand side: the first term represents the inertia of the mass of air moved in and out of the opening; the second represents the damping, the energy lost in this process; and the third term represents the stiffness of the air trapped in the room. The term on the right hand side is the fluctuating external pressure forcing air in and out of the opening.

There are two forms of solution to Eqn 18.13: an oscillatory solution representing Helmholtz resonance when the damping is small; and a gradual decay when the damping is large.

18.3.2.2 Response time

The expectation for conventional buildings with small openings and low overall porosity is for the damped decay solution. This can be described by a characteristic response time for the internal pressures, t_i. Lawson[7] derived an order-of-magnitude estimate for the slower isothermal case from a dimensional analysis, with the result:

$$t_i \propto \frac{\rho_a\, O\, \overline{V}}{A_D\, p_{abs}}\, (\overline{C}_{p_e} - \overline{C}_{p_i})^{1/2} \tag{18.16}$$

For the faster adiabatic solution of Eqn 18.13, the first inertial term can be discarded when the damping term is large, reducing the problem to a first-order differential equation, which Holmes[346], following Liu and Saathoff[347], shows has the solution:

$$t_i = \frac{\rho_a\, O\, \overline{V}}{\gamma_a\, C_D\, A_D\, p_{abs}}\, (\overline{C}_{p_e} - \overline{C}_{p_i})^{1/2} \tag{18.17}$$

This has the same form as the previous equation, except that γ_a from Eqn 18.12 now appears in the denominator, along with the discharge coefficient, C_D. These are both constants which are not parameters in a dimensional analysis, and provide the constant of proportionality required in Eqn 18.16. The difference between isothermal and adiabatic solutions is only the value for γ_a in the denominator of Eqn 18.17.

18.3.2.3 Helmholtz resonance

For large openings, such as a door or broken window, Holmes[346] was the first to show that the damping would not be large enough to prevent Helmholtz oscillations. Discarding the second damping term of Eqn 18.13 gives the undamped resonant frequency:

$$n_i = \frac{1}{2\pi} \sqrt{\left(\frac{\gamma_a\, A_D\, p_{abs}}{\rho_a\, L\, O}\right)} \tag{18.18}$$

If the approximation of Eqn 18.14 for L is used togther with the speed of sound, c_a, given by:

$$c_a^2 = \gamma_a\, p_{abs} / \rho_a \tag{18.19}$$

then Eqn 18.18 reduces to:

$$n_i = \frac{c_a\, A_D^{1/4}}{2^{1/2}\, \pi^{5/4}\, O^{1/2}} \tag{18.20}$$

where the speed of sound in air is $c_a = 340\,\text{m/s}$.

Holmes[346] took measurements on models demonstrating that Helmholtz resonance does occur. Figure 18.4 shows later spectra of internal pressure measured by Liu and Rhee[348] in models, demonstrating the increasing tendency to resonance as the orifice area increases relative to the volume of the room.

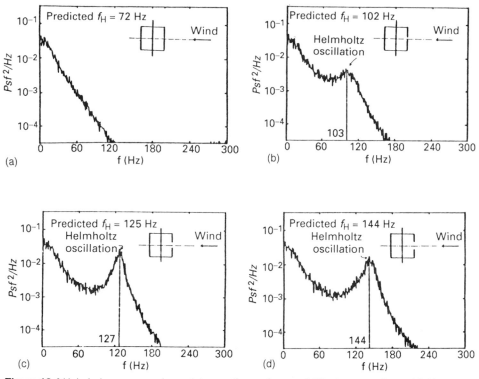

Figure 18.4 Helmholz resonance in model room (from reference 348) where opening = (a) 10 mm × 10 mm, (b) 20 mm × 20 mm, (c) 30 mm × 30 mm, (d) 40 mm × 40 mm

18.3.3 Two-orifice case

A new theory for the two-orifice case of Figure 18.1(c) has been developed for this Guide by Harris in terms of a full non-linear theory and a simplified linearised theory[349]. Even the linearised theory is more complex than the single-orifice case, so only the conclusions will be presented here. Harris confined his attention to the highly damped solution for small-orifice areas on the windward wall, A_W, and leeward wall, A_L. However this has more general applicability since typical distributed porosity can be likened to areas on the windward wall when in a region of positive external pressure, $c_{p_W}\{t\}$, and areas on the leeward wall when in negative external pressure, $c_{p_L}\{t\}$.

The time constant, t_i, for the two-orifice case is given by:

$$t_i = \frac{\rho_a \, O \, \overline{V}}{\gamma_a \, C_D \, p_{abs}} \frac{A_W \, A_L}{(A_W^2 + A_L^2)^{3/2}} \, (\overline{c}_{p_W} - \overline{c}_{p_L})^{1/2} \tag{18.21}$$

The form of Eqn 18.21 is equivalent to Eqn 8.17, except that the pressure coefficient is now the total difference between windward and leeward external mean pressures and that the simple A_D term in the denominator is replaced by the more complex term involving both areas. This equation is valid only when neither A_W or A_L are dominant, which may be taken as the range $0.5 \leqslant A_W / A_L \leqslant 2$.

Harris found that external pressure fluctuations very much slower than this time constant are followed by the internal pressure in a quasi-steady manner. Fluctuations very much faster than this time constant are suppressed in the internal pressure. However a new phenomenon was observed, owing to the non-linearity of the problem, whereby the fast external fluctuations cause a change in the **mean** internal pressure, Δc_{p_i}. The direction of this mean change depends on the relative level of fluctuations on the windward and leeward faces. If the fluctuations on the windward face are larger than those on the leeward face, the internal pressure drops, increasing the pressure difference across the windward face. If the fluctuations on the leeward face are larger, the internal pressure rises, increasing the pressure difference on the leeward face, side faces and roof. The first case occurs for real buildings, because the pressures fluctuate more on the windward face than the leeward face, as shown in Figure 8.17. However, if the leeward face opening is moved to the roof or the side walls, where the fluctuations are greater (see Figure 8.19), the opposite case occurs.

Harris investigated both effects, suppression of fluctuations and the change in mean value of the internal pressure, by superimposing sinusoidal fluctuations of external pressure of period t_e on the mean values and solving the time-dependent flow balance equations for a number of cases. These cases were computed for the following typical conditions:

$$O = 1000\,\text{m}^3,\ A_\text{W} = A_\text{L} = 0.1\,\text{m}^2,\ \bar{c}_{p_\text{w}} = 0.8,\ \bar{c}_{p_\text{L}} = -0.2,\ \bar{V} = 25\,\text{m/s}.$$

Case A corresponds to fluctuations with a peak value half the mean:

$$\hat{c}_{p_\text{w}} = 0.4, \quad \hat{c}_{p_\text{L}} = -0.1,$$

for which the quasi-steady flow balance would give a peak internal pressure fluctuation:

$$\hat{c}_{p_{i,\text{QS}}} = 0.150;$$

Case B corresponds to fluctuations equal to the mean:

$$\hat{c}_{p_\text{w}} = 0.8,\ \hat{c}_{p_\text{L}} = -0.2,$$

for which the quasi-steady flow balance would give a peak internal pressure fluctuation:

$$\hat{c}_{p_{i,\text{QS}}} = 0.300;$$

Table 18.2 lists the computed values of mean change, Δc_{p_i}, and the peak fluctuation, $\hat{c}_{p_i}$, of internal pressure for a range of period ratios, t_i/t_e. Small period ratio corresponds to an internal time constant very much faster than the pressure fluctuations, giving a very small mean change and nearly quasi-steady fluctuations. Large period ratio corresponds to an internal time constant very much slower than the pressure fluctuations, giving a large mean change and suppressing the fluctuations. Note that Δc_{p_i} is always negative in Table 18.2 because $\hat{c}_{p_\text{w}} > -\hat{c}_{p_\text{L}}$ for the typical case computed.

In practice the pressure fluctuations will be random. Those on the windward wall will have a spectrum characteristic of the incident atmospheric boundary layer, which means that more than 90% of the fluctuations will occur for periods much longer than t_i. Those on the leeward wall will have a spectrum characteristic of the building wake and therefore at periods nearer the value of t_i.

Table 18.2   Internal pressure characteristics for Harris' two-orifice case

Period	Case A:		Case B:	
t_i/t_e	Δc_{p_i}	c_{p_i}	Δc_{p_i}	c_{p_i}
0.01	− 0.000	0.150	− 0.000	0.300
0.1	− 0.005	0.129	− 0.019	0.250
0.2	− 0.012	0.097	− 0.061	0.225
0.5	− 0.019	0.048	− 0.108	0.126
1.0	− 0.020	0.025	− 0.120	0.067
2.0	− 0.021	0.013	− 0.122	0.033
5.0	− 0.021	0.005	− 0.124	0.013
10.0	− 0.021	0.003	− 0.124	0.007

18.3.4 Effect of building flexibility

When the building is rigid, a change in internal pressure, Δp, requires a certain volume of air, ΔO, to flow in or out of the room to compress or expand the trapped air in the room volume, O. When a building is flexible, as all are to some degree, the room volume will also expand or contract and that additional change of volume must also be passed through the openings, slowing the response time.

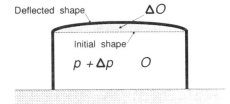

Figure 18.5 Effect of building flexibility

Vickery [350] approached this problem in terms of the bulk modulus of the air, k_a, and of the building, k_b. The bulk modulus is the change in pressure required to make a small volumetric change. Figure 18.5 shows a building of volume, O, with a flexible roof in which a change of internal pressure from p to $p + \Delta p$ displaces the roof, changing the room volume by ΔO. The bulk modulus of the building, k_b is then:

$$k_b = \Delta p\, O / \Delta O \qquad (18.22)$$

The corresponding bulk modulus of air, k_a, is:

$$k_a = \gamma_a p_{atm} \simeq 1.4 \times 10^5 \, \text{Pa} \qquad (18.23)$$

i.e. air is **very** stiff, so that many long-span buildings will be flexible in comparison.

Vickery [350] calculated the damped decay and undamped Helmholtz resonance cases for the single orifice and showed that the result is the same as the rigid building case, except that the effective volume, O_{eff}, increases to:

$$O_{eff} = O\,(1 + k_a / k_b) \qquad (18.24)$$

which replaces the rigid volume, O, in Eqns 18.17 and 18.20. From this it is inferred that O_{eff} should also replace O in Eqn 18.21 for the two-orifice case.

Figure 18.5 represents only the roof as being flexible. In practice, the displacement of all four walls should also be included in the estimation of k_b. For domestic houses of typical stiffness, Vickery suggests that $k_a / k_b \simeq 0.2$, giving an 20% increase in response time over the rigid case. For long-span arenas, where the deflection of the roof dominates, he suggests that $k_a / k_b \simeq 4$, increasing the rigid response time by a factor of five. Clearly, for flexible structures, the response time is dominated by the volumetric change.

Most flexible structures are designed for a limiting deflection-to-span ratio in the range $1/100 < z/L < 1/250$ for a given design pressure. Take the case of a square arena with a span to height ratio of $L/H = 10$. The volume change is related to the midspan deflection by the integral of the deflected shape, $\iint \phi_z\{x,y\} \, dx/L \, dy/W \simeq 0.6$. Assume the design is for a deflection ratio $z/L = 1/150$ for $\Delta p = 2\,\text{kPa}$, the bulk modulus for the building is given from Eqn 18.22 as:

$$k_b = \Delta p \, O \, / \, \Delta O = (2000) \, (L \, W \, H) \, / \, (0.6 \, L \, W \, H \, 10 \, / \, 150) = 50 \times 10^3 \, \text{Pa}$$

so that the response time is increased by the factor:

$$(1 + k_a / k_b) = (1 + 1.4 \times 10^5 / 50 \times 10^3) = (1 + 2.8) = 3.8$$

which is substantial.

18.3.5 Effective averaging time

The effective averaging time to represent the duration of internal pressures depends on the response time of the internal pressure, t_i, and also on the load duration of the external pressures, t, determined from the *TVL*-formula of Eqns 12.37 and 15.27 (§12.4.1.3 and §15.3.5).

Assuming the internal pressure responds in a quasi-steady manner down to the critical period, t_i, but is totally suppressed for shorter periods will give a reasonable model of the actual behaviour. This is analogous to the operation of the *TVL*-formula, as illustrated by the critical frequency in Figure 8.28.

However, the *TVL*-formula also works in parallel with the internal pressure response to give the appropriate external pressure duration, t, to account for the way the external pressure fluctuations are correlated over the areas of the openings. For single openings on faces, as in Figure 18.1, the appropriate length, l, is the diagonal of the opening (§15.3.5). For multiple openings or distributed porosity on a face, such as in Figure 18.2, the appropriate length is the diagonal of the envelope area enclosing the openings.

When $t < t_i$, the internal pressure is unable to respond to the fastest fluctuations of external pressure. When $t > t_i$, there are no external pressure fluctuations faster than t to which the internal pressure can respond. In both cases, the effective averaging time for the internal pressure is the **longer** of the two time parameters, t and t_i. Lawson [7] put this as:

When a window is open in such a room,' this 'gives an averaging time shorter than the averaging time over the area of the window, so that the latter value ought to be used.

The final aspect of effective averaging time is its effect on the design loads. Recalling from §18.1 that the internal pressure controls how the overall external loads are distributed between the faces of the building, the averaging time effects only how this distribution varies. If a slow averaging time decreases the gust loading on one face, it must also increase the loading on other faces by an equal amount. Whether a fast or a slow averaging time for internal pressures tends to be beneficial to a design depends on whether the internal pressure increases or decreases the loading on the most critical face. Suppose in the simple case of Figure 18.1(c) that the quasi-steady balance of flow produces a net positive internal pressure. This reduces the load on the windward wall but increases the load on the roof. Now if the incident wind speed is increased by a gust, but the averaging time for internal pressures is long, the internal pressure will not rise as quickly as the external pressures, so that the benefit to the windward wall is reduced and the penalty to the roof is also reduced. Similarly, if the quasi-steady balance gives a net internal suction, increasing the windward wall load and decreasing the roof loads, the averaging time slows the drop in internal pressure, reducing the penalty to the windward wall and reducing the benefit to the roof. Overall, a slow averaging time for internal pressures tends to reduce problems but also reduce benefits. On the other hand, with a fast averaging time the internal pressure tends to maintain the quasi-steady balance.

18.4 Conventional buildings

In conventional buildings, openings in the external skin tend to be small and distributed over all faces, unless there is a dominant opening. Usually, the ultimate limit state is assessed with controlled dominant openings, like doors or windows, closed. Typical measured values of porosity were given in Table 18.1. Holdø et al. [351] demonstrated that porosities as high as 3×10^{-2} (3%) had no significant effect on the external pressures, provided no individual opening is greater than 75% of the total permeability. This is the reason that it is possible to consider the external pressures to be independent of the internal pressures. Of course, the converse is not true, since the internal pressures are clearly controlled by the external pressures as already discussed.

The quasi-steady flow balance of §18.2.1 using the peak external pressures of the duration given by the effective averaging time, defined above, is appropriate for both single and multi-room cases shown in Figure 18.2. This yields a value of $\bar{c}_{p_i}$ for each internal room, as discussed in §18.7, later. For many typical situations, the standard values of $\bar{c}_{p_i}$ given in Chapter 20 can be used directly. These coefficients are implemented in exactly the same manner as described in §15.3.5 for the external pressures, using a dynamic pressure calculated from the gust speed of a duration equal to the effective averaging time.

In determining the effective averaging time, the response time of the internal pressure, t_i, should be determined from the two orifice model of Eqn 18.21 by taking A_W as the sum of the openings in areas of positive external pressure and A_L as the sum of the openings in areas of suction. Strictly, A_W should be the sum of openings giving inflow and A_L as the sum of the openings giving outflow, dependent on the flow balance itself, but this refinement makes little difference to the value. The constants in Eqn 18.21 can be simplified by using the speed of sound in Eqn 18.19. The potentially important effect of building flexibility can be incorporated by using the effective volume of Eqn 18.24. This gives the equation:

$$t_i = \frac{O \, \bar{V}}{c_a^2 \, C_D} \frac{A_W \, A_L}{(A_W^2 + A_L^2)^{3/2}} (1 + k_a / k_b) \, (\bar{c}_{p_W} - \bar{c}_{p_L})^{1/2} \qquad (18.25)$$

valid in the range $0.5 \leqslant A_W / A_L \leqslant 2$. Standard values for the constants cases are: speed of sound in air, $c_a = 340\,\text{m/s}$; standard-orifice discharge coefficient, $C_D = 0.61$.

The corresponding load duration for the external pressures, t, should be taken as the design value from the TVL-formula as described in §20.2.3, then the effective averaging time for the internal pressures is the larger of t and t_i. This value is used to determine the gust duration for the reference dynamic pressure. The practical limits to the assessment of internal pressure in conventional buildings are, for the fastest response, the quasi-steady peak gust case given by $\hat{q}_i = \hat{q}_{\text{ref}}\{t\}\ \bar{c}_{p_i}$, and, for the slowest response, the mean value given by $\bar{p}_i = \bar{q}_{\text{ref}}\ \bar{c}_{p_i}$.

18.5 Dominant openings

When the single-orifice model, Eqn 18.17, is treated in the same manner as Eqn 18.25 above, it becomes:

$$t_i = \frac{O\ \bar{V}}{c_a^2\ C_D\ A_D}\ (1 + k_a / k_b)\ (\bar{c}_{p_e} - \bar{c}_{p_i})^{1/2} \tag{18.26}$$

The problem here is that the quasi-steady model predicts that $\bar{c}_{p_i}$ equalises to $\bar{c}_{p_e}$. Consider the quasi-steady loading process. The peak external pressure coefficient is $\hat{c}_{p_e} = \bar{c}_{p_e}\ S_G^2\{t\}$. The peak internal pressure coefficient is $\hat{c}_{p_i} = \bar{c}_{p_e}\ S_G^2\{t_i\}$, if $t_i > t$, otherwise it is the same as the external pressure. Accordingly, Eqn 18.26 can be rewritten as:

$$t_i = \frac{O\ \bar{V}}{c_a^2\ C_D\ A_D}(1 + k_a / k_b)\ (\bar{c}_{p_e}\ [S_G^2\{t\} - S_G^2\{t_i\}])^{1/2} \tag{18.27}$$

but this can only be solved iteratively because $S_G^2\{t_i\}$ depends on the result. Eqn 18.27 is valid only for the damped case, where the area of the opening, A_D, is small compared with the volume, O, when it should predict t_i significantly greater than t.

When the opening is large, like an open door or window, the undamped Helmholtz resonance model is more appropriate. Figure 18.4 demonstrates that the resonance frequency is a reasonable approximation to the critical frequency of Figure 8.28, above which fluctuations are suppressed. If the amplification at the resonant frequency is neglected for the moment, the response time for the internal pressure, t_i, can be taken as the reciprocal of n_i. Incorporating the effect of building flexibility gives:

$$t_i = \pi^{5/4}\ (2\ O\ [1 + k_a / k_b])^{1/2} / (c_a\ A_D^{1/4}) \tag{18.28}$$

where the speed of sound in air, $c_a = 340\,\text{m/s}$.

Codes and design guides often simplify the problem by describing the internal pressure as a simple fraction of the external pressure at the dominant opening. This simplification is based on the relative area of the dominant opening to the remaining porosities of the building, which relates to the question of 'dominance' (the distinction between the models for dominant openings and conventional porosity) for the quasi-steady case rather than to the time-dependent case. Nevertheless, this simple pragmatic approach is usually sufficient and is adopted in Chapter 20. The variations in internal pressure caused by dominant openings are greater than for buildings with completely open sides (see §18.6), particularly when the openings are located in positions of high positive pressure or local high suction.

The question of whether a large access door should be assumed open or closed in the design wind conditions is difficult to answer. Assuming the open case makes the estimation of internal pressure easier, but it results in more onerous loads on all faces except the face with the opening. In some circumstances the design would become uneconomic. The key to the answer lies in the intended use of the building. If the probability that the door or window could be open in the design conditions is low, then it would be reasonable to design for the closed case, making additional provisions to ensure compliance by the users. The closed case becomes the ultimate limit state case, while the open case becomes a serviceability limit state case. The difference in these two cases should be examined carefully before they are adopted. A dominant opening in the windward face could double the uplift on a roof, and then a 'closed' design for the ultimate limit state at the standard $T = 50$ year exposure period is equivalent to a Statistical Factor $S_T = 0.71$ in the 'open' serviceablity limit state. The expectation is that this would be exceeded more frequently than once a year (§9.3.2.1), and would not be an acceptable risk. If the 'open' serviceability limit is set at $T = 2$ years, say, then the Statistical Factor for the 'closed' state becomes $S_T = 1.13$, corresponding to an ultimate limit at $T \simeq 1100$ years. Clearly, the serviceability limit state controls the design in this case.

The practical limit to the internal pressure through a dominant opening is the quasi-steady peak gust case given by the peak external pressure at the opening, corresponding to the fastest response.

18.6 Open-sided buildings

18.6.1 Introduction

The problem of open-sided buildings is similar to, but generally simpler than, the problem of dominant openings. It is usually a little less onerous because the wind is able to enter, swirl around inside, then exit again. By not coming to rest the full stagnation pressure is not recovered when the open side faces into the wind. This occurs when one or more adjacent sides are open, particularly the longer side, and this case is discussed next. There are exceptions to this general rule and one in particular, the case of a building with two opposite walls open, is also discussed below.

18.6.2 One or more adjacent open faces

Among the commonest forms of open-sided building are grandstands, where one side is open for spectators to view the action, and some farm and industrial buildings, where one or more sides are open for access. The motion of the wind within the building causes gradients of internal pressure acting on the inside walls and under-surface of the roof. Measurements [352] show these gradients to be much less than those on the external faces and it is often sufficient to assume a uniform internal pressure, or a global value for each internal face.

Figure 18.6 shows the effect of wind direction on the global pressure coefficient for the internal pressure acting on the internal face directly opposite the principal open face, for four geometries:

1 one shorter face open;
2 one longer face open;

3 one longer face and one adjacent shorter face open;
4 one longer face and both adjacent shorter faces open.

The first geometry is typical of farm or industrial buildings for storage, while geometries 3–4 are also typical of grandstands. The consistent trend in all four curves is for the internal pressure to be positive when the open face is facing upwind and negative when facing cross-wind and downwind. The maximum positive internal pressure always occurs at wind angle $\theta = 0°$, when the open face is directly upwind. The maximum negative internal pressure occurs in the ranges $90° \leqslant \theta \leqslant 120°$ and $240° \leqslant \theta \leqslant 270°$, when the open face is on one cross-wind side. The internal suction is always less onerous when the open face is downwind in the wake.

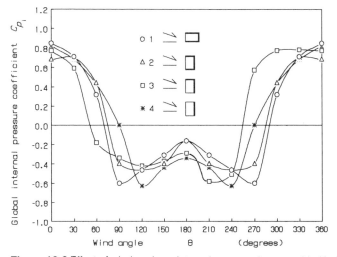

Figure 18.6 Effect of wind angle on internal pressure in open-sided buildings

More detailed inspection of Figure 18.6 reveals the individual differences due to the various geometries. For example, the two cases of one open face 1 and 2 are very similar, except that the suction peak when facing crosswind is sharper and more onerous in the case of the shorter open face. In case 4, the positive lobe is wider and the suction lobe narrower because the missing side faces cannot cast their protective wake over the remaining inside face, but the range of value remains similar. The asymmetrical case 3 follows the weaker suction peak of 2 when the side face is upwind and the stronger suction peak of 4 when the side face is downwind, i.e. the flow quite reasonably chooses between cases 2 and 4 on basis of the upwind geometry, except that the corresponding wind angle is rotated by the flow asymmetry. The effect of the asymmetry on the positive lobe is to offset the centre, extending the lobe for the wind angle that traps the flow in the internal corner.

When the open face is upwind, the positive internal pressure builds up gradually from the open face to the maximum value on the facing internal wall. This build-up is approximately linear for a depth of about $x = H$ or $x = L/2$, whichever is the smaller, and thereafter is constant at the maximum value. Thus with a deep building with a narrow open face, as case 1 of Figure 18.6, the build-up of internal pressure occurs within a short distance of the opening and the majority of the building is at the constant maximum value. However, with a shallow wide building such as a grandstand, the gradient may occupy the whole depth, with the result that the internal pressure on the underside of the roof increases from about 40% of the maximum value at the open front edge linearly to the maximum value at the closed rear edge [352]. This gradient is established by part of the 'horseshoe' vortex (§8.3.2.1.1) which enters the open front of the building. It does not occur when the open face is downwind.

18.6.3 Two opposite open faces

Here there is a different problem in that the wind is able to flow in one end and out the other. When the axis of the building is skewed to the wind, this flow is steered by the side walls, creating a side force. This effect is greatest when the shorter walls are open. Typical buildings with opposite ends open are farm and industrial storage sheds designed for 'drive-through' operations.

This geometry was studied at BRE by measuring the x- and y-axis forces with the upwind short face completely open and the downwind short face of variable solidity. The results are plotted in Figure 18.7, where the upper curves are the

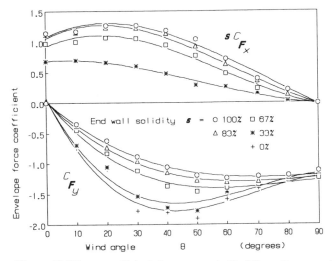

Figure 18.7 Force coefficients for open-ended building with permeable opposite end

x-axis force along the axis of the flow through the building and the lower curves are the corresponding y-axis force. As the x-axis force acts on the porous end face, it is expressed as the envelope coefficient $s\,C_{F_x}$, that is based on the area of the face. As the y-axis force acts on the solid longer faces, it is expressed as the usual force coefficient C_{F_y}, based on the area of one face, A_y. However, C_{F_y} is actually shared between both side faces, but the overall load measurement is unable to show how this is partitioned between them.

The variation of C_{F_x} when the end wall is solid is typical of the load on an enclosed building of the same proportions, but decreases as the solidity of the wall decreases. The corresponding changes in C_{F_y} are more interesting, mirroring C_{F_x} at high solidities, but producing higher loads in skew winds at lower solidities. In the latter case, flow through the building is turned by the side walls and the change of momentum produces a lift force (§2.2.8.5). In this state the building is effectively 'flying' in the same manner as a box-kite. When the wind angle is θ = 90°, the symmetry stops flow through the building and the internal pressure recovers to the suction expected for one open side. Accordingly, the increase in side force at skew wind angles must be accounted for in the stability calculations for the building.

18.7 Loads on internal walls

18.7.1 Effect of internal wall porosity

When a building is divided into rooms the differences of internal pressure between rooms generate loads on the internal walls or partitions. These can be large enough to cause failure, damage to finishes, or at least to cause difficulty in opening internal doors. The largest internal pressure differences are caused when the internal wall forms a barrier to the flow of air from the windward face to the leeward or side faces. Design procedures to cope with this problem were first described in detail in the 1974 BRE *Wind loading handbook* by Newberry and Eaton (now out of print), using the quasi-steady flow balance equations of §18.2.1. In 1979–80, Withey made a parameteric study of the sensitivity of internal pressures using this approach and many of the BRE external pressure measurements from which the design data of Chapter 20 were compiled. This study was started at BRE and completed at Brunel University [353], and the following discussion is based on the results.

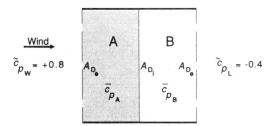

Figure 18.8 Building with single partition

Figure 18.8 represents the simplest case, comprising a single partition spanning between impermeable side faces, dividing the building into two rooms, 'A' and 'B', with internal pressure coefficients c_{p_A} and c_{p_B}. The windward and leeward faces are assigned uniform pressure coefficients, $\tilde{c}_{p_W} = +0.8$ and $\tilde{c}_{p_L} = -0.4$, respectively. The windward and leeward external faces are assumed to have equal discharge areas, A_{D_e}, while the internal partition has the discharge area, A_{D_i}. The quasi-steady flow balance for each room in this example is, from Eqn 18.5:

$$A_{D_e}(0.8 - \tilde{c}_{p_A}) = A_{D_i}(\tilde{c}_{p_A} - \tilde{c}_{p_B}) \tag{18.29A}$$

$$A_{D_i}(\tilde{c}_{p_A} - \tilde{c}_{p_B}) = A_{D_e}(\tilde{c}_{p_B} - (-0.4)) \tag{18.29B}$$

Given a value for the ratio between internal and external porosity, A_D/A_{D_e}, these two simultaneous equations may be solved for the internal pressure coefficients, c_{p_A} and c_{p_B}.

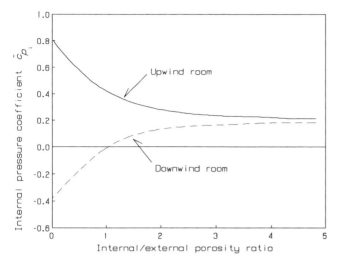

Figure 18.9 Effect of porosity on loading of single internal partition

Figure 18.9 gives the resulting internal pressures for a range of porosity ratio. The pressure across the internal partition is given by the difference between the two curves. The pressure across each external wall is given by the difference between the external pressure and the respective curve. When the partition is impermeable, $A_{D_i}/A_{D_e} = 0$, all the load is taken on the partition and none on the external walls. When the partition and external walls have equal porosity, $A_{D_i}/A_{D_e} = 1$, the load is shared equally. However, for $A_{D_i}/A_{D_e} > 2$, virtually all the load is taken by the external walls and very little by the partition.

Clearly, the way to minimise loads on internal walls is to ensure that the porosity of internal walls and partitions is at least twice that of the external walls. Cubicle partitions that do not reach the ceiling and other internal structures that do not form a barrier to the passage of air take no significant load at all (unless wind is allowed to blow straight through the building as in §18.6.3).

18.7.2 Multi-room, multi-storey buildings

Withey[353] computed solutions for a wide range of building shapes and internal rooms. Figure 18.10 shows the plan for a typical storey which was used as the principal example in the 1974 BRE *Wind loading handbook*. The plan comprises

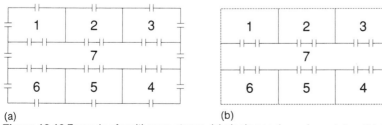

(a)   (b)

Figure 18.10 Example of multi-room storey: (a) single openings of equal size; (b) distributed openings of same net area

three rooms along each long face, separated by a corridor spanning between the short faces. Connections between storeys above and below the one shown can be made in the corridor, representing a stair-well. In (a), discrete discharge areas in each external wall are represented by the single openings, internal walls between rooms and corridor have two openings each of the same area, while the walls between offices are assumed to be impermeable. This is fairly representative of real buildings if the openings in the external walls a regarded as gaps around a window, the openings in the corridor walls as gaps around internal doors (twice as porous) and the walls between offices having no doors. In (b) the model is improved by distributing the external openings along the wall, while maintaining the porosity ratio $A_{D_i}/A_{D_e} = 2$. Withey[353] recomputed these two examples using measured external pressure coefficient data in place of the Code data[4] used in the *Wind loading handbook*.

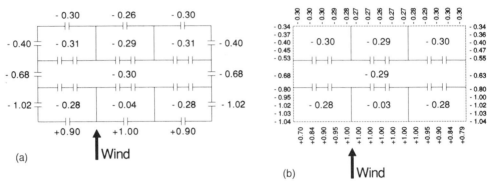

(a)

(b)

Figure 18.11 Example of multi-room storey with wind normal to face: (a) single openings of equal size; (b) distributed openings of same net area

Figure 18.11 shows the external and internal pressure coefficients for the top storey of a three-storey building with the wind normal to a long face. Note that the internal pressure coefficients differ by only 0.01 between (a) and (b), showing that the additional complexity of distributing the external porosity along the faces makes virtually no difference to the result. The internal pressure coefficients in each room are within 0.01 of each other, except for the central room on the windward face, which is more positive by 0.25. If the building is made longer, with more rooms along either long face, the additional rooms on the front face will also have this more positive internal pressure. The internal pressure in corner rooms on the front face does not rise by this value because the high suctions at the upwind part of the side face, reinforced by the corridor suction, overcome the positve pressure from the front face. With suctions on three faces, it is just those rooms that form a barrier to the flow of air from the windward face to the leeward or side faces that have different internal pressures. The net result is that the only internal walls with significant loads are those separating the corner offices from their neighbours on the front face. Results for the other floors were similar, with and without connections between floors in the corridor.

Figure 18.12 shows the same example with flow skewed at $\theta = 60°$ from normal to the long face. Again there is no significant difference between cases (a) and (b). This time both external walls of the room at the windward corner have positive external pressures and its internal pressure becomes positive. Similarly, both

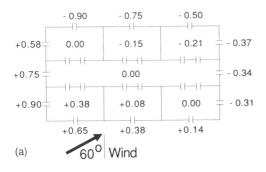
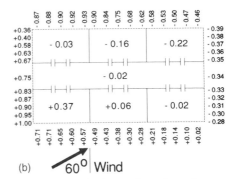

Figure 18.12 Example of multi-room storey with wind skewed to face: (a) single openings of equal size; (b) distributed openings of same net area

external walls of the room at the leeward corner have negative external pressures and its internal pressure becomes negative. Other rooms with one external wall in positive pressure and one in suction, and those with only one external wall, have internal pressures near zero. Again, the biggest internal wall loads occur between corner offices and their neighbours, and this time also on the wall to the corridor. Again, results for the other floors were similar, with and without connections between floors in the corridor.

These same results were obtained with other planform buildings, with buildings from one storey to nine storeys, with and without connections between floors in the corridor. This work shows the following conclusions to be generally valid:

1 All storeys with the same layout of rooms give closely similar results.
2 Connections between storeys make no significant difference.
3 The largest loads occur on internal walls of corner rooms.

The first two conclusions are useful, since they allow the calculations to be reduced to a single floor when a common floor plan is used.

18.7.3 Dominant openings

By ensuring that the porosity of internal walls is twice that of the external walls, the internal wall openings dominate and minimise the internal wall loads. Open doors between rooms provide dominant openings in the **internal** walls, reducing internal wall loads to zero.

An open window, providing a dominant opening through the **external** wall of a room, reverses this effect and the internal pressure becomes a large proportion of the external pressure at the dominant opening. This effectively transfers the external wind loads directly to the internal walls. Whether this should be a design case at the ultimate limit was discussed in §18.5 with respect to the external faces, but this discussion is also relevant to the design of the internal walls. Provided they are not structural, a degree of damage to internal walls may be acceptable in the rare case of an extreme wind coinciding with an accidental dominant opening. In this event, the problem reverts to the serviceability case (b) given in the design cases, below.

18.7.4 Design cases

For typical buildings, the problem of loads on internal walls reverts to two design cases:

(a) all external windows and doors closed and all internal doors closed at the ultimate limit state;
(b) all internal doors closed and any combination of external windows and doors open at the serviceability limit state.

Case (b) may be simplified by considering the practical experience that, for comfort, only windows on the leeward faces are deliberately opened in any wind speeds above moderate. Thus open windows on the windward face are likely to occur singly, through accident or breakage. For buildings without openable windows, case (a) applies at both limit states.

18.8 Multi-layer claddings

18.8.1 Introduction

As long as a multi-layer skin acts structurally as a single unit, the way that wind loads are distributed through the layers of the skin is usually irrelevant. A good example of this is a standard cavity masonry wall, where the internal pressure in the cavity is controlled by the position and size of ventilators. However, provision of wall ties to the specification required by building regulations ensures that the wall acts as a single structural unit. The way the wind load is shared between the two leaves of the wall becomes irrelevant and the problem reverts to considerations of the durability of the ties.

However, there are a number of forms of construction where the way that the wind loads are distributed through the multiple layers is important. This is especially true with the increasing use of cladding systems which are assembled on site from different components by different manufacturers. Each of the components can be separately tested for strength and durability, but the share of the design load that each takes depends on the way they act together as a system. Until recently, system performance could only be assessed by calculation, tests using scale models and by measurements or satisfactory performance at full scale **after** construction. The development of the BRERWULF system[354] at BRE, described later in §19.9.3.3.3, which is able to reproduce measured time-varying pressures on areas of cladding at full scale, now enables the performance of these systems to be analysed in some detail. So far only a few systems have been studied using BRERWULF and the following discussion is naturally biased towards these known cases.

A prominent feature of multi-layer cladding systems where the outer skin is porous or flexible is that the wind loads have two alternative load paths from the outside skin through to the structural members:

(a) **pneumatic load paths,** where the load is shared through the layers by differences in pressure across them; and
(b) **structural load paths,** where the loads pass through the fixings.

The concept of load paths is discussed in a more general context in §19.3.

18.8.2 Permeable outer skin

18.8.2.1 Loose-laid paving and insulation slabs on roofs

Loose-laid slabs are often laid on flat roofs, on top of the weatherproof and structural surface, sometimes using paving stones to provide a suitable surface for access and sometimes using proprietory insulation boards to provide extra insulation (especially to reduce condensation problems in rehabilitation schemes). A typical installation is shown schematically in Figure 18.13(a). The external pressure field acts on the top surface of the boards, varying with position and with time, $p_e\{x,y,t\}$. The pressure under the boards, the 'internal' pressure in this context, also varies with position and time, $p_i\{x,y,t\}$, and is controlled by the leakage of air between the volume of air trapped under the boards and the upper surface through the gaps between the boards.

The stability of a loose-laid board under wind action depends on its effective deadweight remaining greater than the difference between top and bottom surface pressures, $\Delta p = p_e - p_i$. In practice, the volume of air trapped under each board is very small (the gap between the roof surface and the board being typically less than 5 mm) so that the bottom surface pressure around the periphery of each board equalises rapidly to the external pressure through the gaps between boards (typically > 1 mm wide). These characteristics have been studied using model and

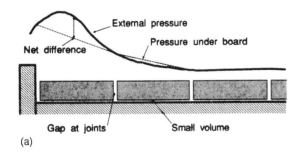

(a)

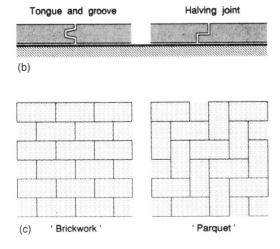

(c)   ' Brickwork '   ' Parquet '

Figure 18.13 Loose-laid insulation boards:
(a) pressure equalisation characteristics;
(b) types of interlock; (c) laying patterns

full-scale boards in wind tunnels in Canada [355, 356] and Japan [357], and by full-scale measurements at BRE and elsewhere. Linear gradients of pressure between the peripheral values are established under each board, so that significant net pressure differences occur across the board only where the instantaneous external pressure field is non-linear. Non-linear gradients of external pressure occur principally along the upwind eave and verge in the direction normal to the edge.

BRE guidance on design values for these systems is given by Digest 295 *Stability under wind load of loose-laid external roof insulation boards* (BRE, 1985), and is included in Appendix K. The pressure equalisation mechanism makes it very unlikely that neighbouring boards will be simultaneously loaded at the maximum pressure difference. In essence, the guidance recommends that the design pressure difference, Δp, on any one board should be taken as one third of the design external pressure (i.e. from Chapter 20), with neighbouring boards along one axis (say x) also loaded to this value but with neighbouring boards along the other axis (y) unloaded.

The term 'effective deadweight' was used above because the actual deadweight of any one board when laid as in Figure 18.13(a) can be improved by interlocking the boards as in (b). If a board attempts to lift, it will also have to lift its unloaded neighbours on either side along one axis by rotating them around their opposite edge (see Digest 295, Appendix K). A full 'tongue and groove' interlock gives half the weight of the board on either side to double the deadweight while a halving joint interlock gives 50% extra deadweight. These values assume a typical 'brickwork' laying pattern of identical boards, as shown in Figure 18.13(c), where the system is weakest along the continuous joints, but much stronger along the staggered joints. The alternative 'parquet' or 'herringbone' pattern, also shown in (c), increases the effect of the typical 'brickwork' interlock by more than a factor of three. A complication of this pattern is that pairs of left-handed and right-handed boards are required, but the advantages are sufficiently great that at least one UK manufacturer has adopted this approach.

There is much more design advantage to be gained by increasing the interlock than by attempting to decrease the external loading by the methods described in §19.5 'Load avoidance and reduction'. However, there is also a potential penalty. As the actual deadweight of the boards is constant while the wind loading is proportional to the square of the wind speed, there will always be a critical wind speed above which the system will fail. Systems without interlocks fail by the loss of the edge boards first, then more boards are lost if the wind speed increases and the failure mode is gradual, analogous to a ductile failure. Interlocks increase the critical wind speed, but failures tend to be sudden and catastrophic, with many boards being simultaneously removed, analogous to brittle failure. As the consequences of the 'brittle' mode are more serious, it is prudent to increase the margin of safety so that the gains of the interlock are reduced, but are still significant.

The effective deadweight can be further increased by adding ballast in the form of heavy paving stones or gravel. This raises the critical wind speed still further, but again there is a potential penalty. The secondary damage that could be caused by a paving stone or a large piece of gravel falling from a tall building is potentially greater than a lighter insulation board. Paving stones follow the loading model given here but, with no interlock, have only the restraint of their own deadweight. Critical wind speeds for gravel scour, introduced in §17.3.3.8, are given by the design procedure in BRE Digest 311 *Wind scour of gravel ballast on roofs* of Appendix K. Other aspects of external roof insulation unconnected with the loading are covered in BRE Digest 312 *Flat roof design: the technical options* and Digest 324 *Flat roof design: thermal insulation.*

18.8.2.2 Porous cladding systems

The use of traditional forms of tiles to clad a boarded roof or a roof with an impermeable sarking membrane might be thought similar to the preceding case, except for a greater volume under each course of tiles. However, the forces caused by the flow of wind over the steps formed by the overlapping tile courses are greater than those caused by the loading mechanism described above. Accordingly, this form of cladding requires the special rules described in §14.1.3.2 'Slating and tiling'.

Porous cladding systems with smooth external surfaces and significant under surface volumes are typically represented by the form of overcladding system shown in Figure 18.14(a), currently popular for the rehabilitation of large panel system buildings built in the 1960s. The usual problems are excessive heat loss and

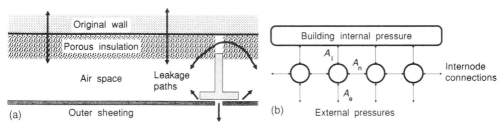

Figure 18.14 Porous overcladding: (a) typical form of overcladding with insulation; (b) representation of flow paths

condensation, requiring additional insulation, and rain penetration through the joints between panels, requiring a new external skin or 'rainscreen'. In this system studied at BRE, structural 'T' rails are fixed at intervals to the existing walls, insulation boards are likewise fixed between each pair of rails, then the outer sheeting panels are screwed or pop-riveted to the 'T' rail. An air space is left between the outer sheeting and the insulation to allow rain penetrating the joints to run down the inside, where it is collected at intervals and routed back to the outer surface. Part of the wind pressure leaks through the gaps between the boards of the outer skin to act directly on the building wall, while the remainder acts directly on the outer sheeting and the loads pass to the building through the 'T' rails and fixings.

The proportion of the load taken by each path determines the required strengths of the outer skin, the 'T' rails and the fixings. This depends on the relative sizes of the air leakage paths and the trapped volume as well as on the distribution of external pressures. The trapped volume is divided into a series of nodes by the fixing rails. The air leakage paths into each node are to the internal volume of the building through the porosity of the original building wall, to the external wind through the porosity of the outer skin and between the nodes on either side through the insulation and under the 'T' rail. A section of these nodes and flow paths is represented in Figure 18.14(b). The discharge area between a node and the inside of the building, A_i, is generally much smaller than that between nodes, A_n, and to the outside, A_e, and will be neglected in the following discussion.

The discharge area to the outside, A_e, for this particular system has special properties. A positive pressure difference tends to press the panels of the external skin against the 'T' rails, decreasing the discharge area, while a negative pressure tends to pull the panels away from the rails, increasing the discharge area. Since the panels are fixed at intervals by screws or rivets, the increase in area by suction is much the larger effect. When the flow-balance equation, Eqn 18.5, is modified to include this effect by varying A_e according to the instantaneous pressure difference, results like those shown in Figure 18.15 are obtained. Other, more typical systems are likely to have constant leakage characteristics.

Figure 18.15 represents the external and cavity node pressure distributions around a square section tower block assuming there are 20 nodes to each face, for four values of the inter-node discharge area, A_n. The solid line, which is the same in each graph, is the external pressure distribution measured on a model of the block with the wind normal to a face. Thus from left to right on each graph, the positive lobe represents the windward face, followed by a side face, the leeward face and the other side face. The stepped broken line represents the corresponding nodal pressure in the cavity between the external skin and the building for each of the 80 nodes. The limits of inter-node area are represented by cases (a) and (d):

(a) Here the inter-node area is small, represented by a 1 mm-wide gap and possible only if a deliberate attempt is made to seal along the lines of the rails. The nodal pressures follow the external pressures closely so that the majority of the

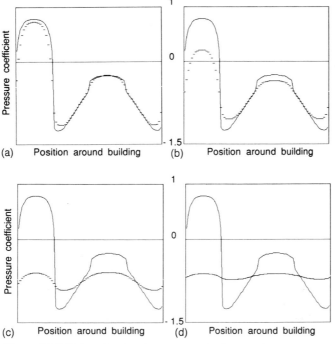

Figure 18.15 External and cavity pressure distributions for porous overcladding where A_n = (a) 0.001 m²/m run, (b) 0.003 m²/m run, (c) 0.01 m²/m run, (d) 0.03 m²/m run

load is transmitted to the building by the pneumatic path and there is little load on the overcladding.

(d) Here the inter-node area is large, represented by a 30 mm-wide gap and typical when no insulation is installed. Now the nodal pressures are almost uniform in the cavity around the building, but the average cavity pressure is much more negative than normal because of the changes in A_e. The openings on the windward face have closed and those on the side faces have opened to create a large bias. The panels on the windward face are strongly pressed against the building and may become overstressed. Panels on the leeward face are also pressed against the building because the cavity suction exceeds the external suction. Panel and rail fixings are in tension only on the side faces. In this state, fixing strengths may not be as important as the bending strength of the panels.

Cases (b) and (c) lie between these extremes, but are still closer to each extreme than each other. This illustrates the dominating effect of $(A_e/A_n)^2$ described by Eqn 18.8 in §18.2.3. The 'real' situation for this practical system is expected to lie between cases (b) and (c).

The large bias of cavity pressure to the side wall suctions and the problem of overstressing the panels on the windward face can be reduced by providing cavity closures at the corners. Perfect closures would make each face act independently of the others, but this is unlikely to be achieved. Figure 18.16 shows two examples where closures equivalent to a 1 mm-wide gap (as Figure 18.15(a)) are installed where the inter-node gap is 10 mm, so should be compared with Figure 18.15(c).

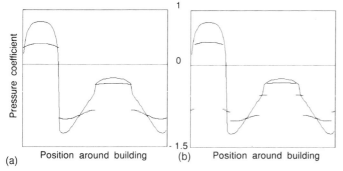

Figure 18.16 Effect of cavity closures on external and cavity pressure distributions for porous overcladding, with closers (a) at corners and (b) in two bays in from corners

These two cases are:

(a) Here the closures are at the corners. Restricting the flow inside the cavity from the windward and leeward faces to the more negative side faces allows the nodal pressures of the windward and leeward faces to rise towards the average external pressure for each face. The nodal pressures for the side faces do not change much because these dominated previously in Figure 18.15(c). One interesting feature on the windward face is that, although the external pressure is always positive, the net pressure difference across the end panels next to the corners is negative, tending to pull the panel away from the face.

(b) It is not always practical to construct a closure exactly at the corner, so here closures have been placed at positions two panels either side of each corner.

The pressures in the cavity compartments made by this action are transitional between the adjacent face values. At the edges of the windward face, the net suction difference of (a) is replaced by a much larger positive pressure, similar to that previously in Figure 18.15(c).

Clearly installing cavity closures at the very corners is the best single strategy for reducing panel loads, but sub-dividing the cavities on each face with more closures makes further reductions. Ganguli and Dalgliesh[358] show from full-scale measurements that cavity compartmentation is very effective in reducing panel loads.

Some overcladding systems have no specific inter-node restriction that can be equated to the discharge area A_n, in which case cross-flows inside the cavity are resisted only by the skin friction, τ. The flow balance equations can be modified to cope with this effect but, unless the cavity is very narrow, the friction is small and the result will be close to the large A_n case of Figure 18.15(d).

This discussion applies to all forms of porous cladding with a cavity of significant volume. Damage frequently occurs to suspended ceilings on external balconies, particularly by uplift on the windward face. The problem often occurs because unsuitable systems, designed only for use **inside** buildings, are sometimes wrongly installed. Compartmentation of the ceiling cavity reduces the problem considerably, but sometimes the addition of deadweight provides sufficient resistance.

18.8.3 Flexible outer skin

18.8.3.1 Unbonded and ballasted membranes

Figure 18.17(a) shows a typical unbonded and ballasted membrane system. If the principal resistance to uplift is provided by the ballast, the weight of ballast must exceed the design uplift load. This gives an inefficient structural design since the roof must always carry this additional weight.

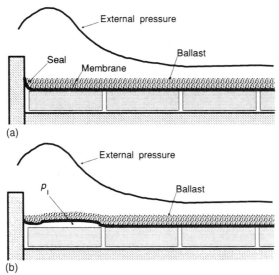

Figure 18.17 Unbonded and ballasted membrane: (a) ballast deadweight exceeds uplift; (b) uplift exceeds ballast deadweight

If the volume between the membrane and the roof deck is completely sealed and both the membrane and the deck are impermeable, then a small volume of air is trapped between the insulation boards which will obey the gas equation, Eqn 18.10, and will also follow the adiabatic equation, Eqn 18.12, for fast changes of pressure. If a gust occurs that causes the external suction to exceed the deadweight of the ballast, the membrane will start to rise, as shown in Figure 18.17(b). The pressure in the trapped volume reduces until it balances the difference between the external suction and the ballast deadweight, then the membrane stops rising. In this state the excess loading over the ballast deadweight is taken entirely by the pneumatic load path.

For this system to work well, both the membrane and the deck must be almost impermeable, the periphery must be well sealed and the trapped volume of air must be small. This last condition may be met by sub-dividing the roof into sealed compartments. These requirements make the system very vulnerable to construction defects. Gravel scour of the ballast, introduced in §17.3.3.8, is also a problem and a design procedure is in BRE Digest 311 *Wind scour of gravel ballast on roofs* of Appendix K.

18.8.3.2 Mechanically-fixed membranes

Figure 18.18 shows a typical mechanically fixed system comprising, from bottom to top:

- purlins or rafters which support the roof deck and transmit the wind loads to the main structural members;
- profiled steel sheet decking, fixed to the purlins at intervals;
- a polythene or similar sheet membrane laid over the decking as a vapour retarder;
- insulation boards, fixed at intervals through the vapour retarder to the decking; and finally
- a flexible weather-proof membrane, fixed at intervals through the insulation boards and the vapour retarder to the decking.

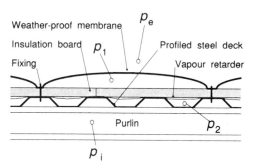

Figure 18.18 Mechanically fixed membrane

The fixings are usually self-tapping screws for the decking and insulation, the latter requiring large load-spreading washers. Penetration of the vapour retarding membrane by the fixing screws gives a small porosity which allows air to pass through to the underside of the main membrane. The main membrane can be fixed in a variety of ways: with washers or long bars fixed with screws which penetrate the membrane, requiring patches to be bonded over to restore the weathertightness; or with hidden fixings systems that are bonded to the underside of the membrane or grip the membrane from underneath.

Whatever the fixing method, as the main weather-proof membrane is unballasted it lifts up under suction until the external pressure, p_e, is balanced by the pressure under the membrane, p_1, and the catenary action of the membrane in tension. At the same time the vapour retarder is pressed up against the underside of the insulation boards by the difference in pressures, $p_2 - p_1$, which draws air through the vapour retarder. Under positive external pressure, the main membrane is pressed down onto the insulation boards and the vapour retarder sags down into the troughs of the decking.

As the porosity of the vapour retarder is very small, the response time for this process can be from several to tens of seconds. Figure 18.19 shows traces of the pressure difference across the main membrane, the vapour retarder and the steel deck in response to a fluctuating suction generated by BRERWULF. This

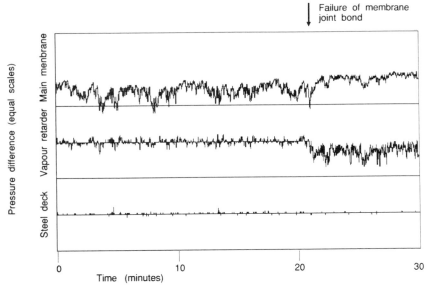

Figure 18.19 Performance of mechanically fixed membrane system at four times design load

fluctuating suction trace is represents a typical wind loading trace with a peak suction of four times the design value. For the first 20 minutes the system is intact and the traces show the balance of loads across the three components. About 80% of the wind load, comprising the mean and the low-frequency components slower than the response time, is taken across the main membrane and transmitted through its fixings. About 20% of the wind load, comprising only the high-frequency components faster than the response time, is taken by the vapour retarder. Note that the vapour retarder trace has flat regions of zero pressure between small positive and negative peaks. The positive peaks occur when the retarder is pressed up against the insulation, the negative peaks when the retarder is pulled taut into the decking troughs, and the flat regions when the retarder is slack between these two states. The loading across the deck is minimal.

After the main membrane was breached (at a safety factor of four), the balance changed dramatically. The hole in the main membrane was only about $A_D = 0.001$ m² in a total area of $A = 25$ m², representing a porosity of 4×10^{-5} which is a tenth of the normal range for building façades given in Table 18.1. Nevertheless, this dominated the porosity of the vapour retarder, changing the balance of the loading to about 30% on the main membrane and 70% on the vapour retarder.

18.8.3.3 Bonded membranes

Figure 18.20 shows a typical fully-bonded membrane system. The load paths through this system depends on whether the volume between the top and bottom membranes, containing the insulation, is vented to the atmosphere or is sealed. In

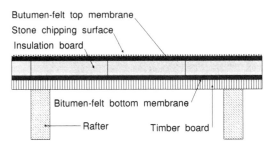

Butumen-felt top membrane
Stone chipping surface
Insulation board

Bitumen-felt bottom membrane
Rafter   Timber board

Figure 18.20 Bonded membrane

the sealed case, the majority of the loading passes from the top membrane to the bottom membrane through the insulation, but a proportion will be transmitted pneumatically. The balance depends on the stiffness of the insulation. However, variation of atmospheric pressure and variation of the vapour pressure of trapped moisture with temperature changes may cause delamination leading to blistering of the membrane. In the vented case, venting to the top surface of the roof through 'mushroom' or similar vents equalises the void pressure to the external pressure, while venting to the inside of the building equalises the void pressure to the internal pressure. The first case reduces the pressure difference across the top weather-proof membrane to near zero, while the second case gives the maximum difference, $p_e - p_i$. In all the above combinations, the bond between the bottom membrane and the roof boards always carries the full wind load.

In practice, damage to fully bonded membrane systems tends to occur in the form of blisters over poorly bonded patches which increase in size with time, or as sudden catastrophic delamination from the upwind edge due to poor edge detail which allows positive pressure from the windward face to enter the insulation cavity.

18.9 Control of internal pressure

Internal pressure can be controlled by the provision of vents in specified locations. This is most often done for other reasons, usually to provide ventilation and to suppress condensation. Roof spaces are usually cross-ventilated from upwind eave to downwind eave, or from eave to ridge, and the sizes and positions of the vents controls the internal pressure of the roof space. Inappropriate positioning of vents may lead to damage to non-structural components such as ceilings.

Control is effected by providing a vent which serves as a dominant opening to an external pressure of the required value. This treatment is especially valuable to reduce the uplift on low-pitch roofs, when this is a critical aspect of the design. To do this, vents are required in a region of consistently high suction and the best compromise position is along the line of the ridge. The discharge area of the ridge ventilators should exceed the accumulated porosity of the remainder of the building envelope by a factor of two. The penalty for such action is an increase in the net loading across the windward wall and this should always be considered.

If a region with consistently the required pressure cannot be found, this can be circumvented by using a series of vents for which the net effect is consistent, or by using vents with one-way action (flap valves, etc.). In the latter case, the reliability of moving components needs to be taken into account. Another approach is to use vents that are sprung-loaded to remain closed until the pressure difference exceeds a pre-set threshold. These can often also double as fire vents. Again the reliability of the components needs to be taken into account.

19 Special considerations

19.1 Scope

In this chapter various special considerations not covered in previous chapters are discussed. These are a mixed bag of considerations not normally covered in codes of practice and other design guidance, but which nevertheless contribute to the level of safety enjoyed by building structures. The subjects discussed are:

1 Groups of buildings – the effects of shelter and negative shelter caused by neighbouring buildings.
2 Load paths in structures – considerations of how the external and internal pressure are taken up and passed to the structural members.
3 Serviceability failure – considerations of the performance of the structure in service when the design is principally for safety at the ultimate limit state.
4 Load avoidance and reduction – techniques by which a design may be optimised to resist the action of wind.
5 Optimal erection sequence – planning the erection sequence so that high wind loads are avoided or reduced while the structure is below its design strength.
6 Variable-geometry structures – structures that change their shape in service and hence change the character of their loading.
7 Air-inflated structures – discussion of pneumatically-stiffened structures, where the wind effects are considered in terms of stiffness rather than strength.
8 Fatigue – discussion of the particular problems caused by the repeated actions of gusts.

Two of the above subjects – shelter of low-rise buildings in an array of many similar buildings, part of the section on groups of buildings, and the section on the experience and design practice in Australia for fatigue – are so specialised that those sections were specially commissioned by BRE from experts in those subjects.

19.2 Groups of buildings

19.2.1 Introduction

Preceding chapters have dealt with the loads on structures in isolation. Where shielding has been discussed, it has been confined to parts of the same lattice structure. In this section the effects of grouping structures together are discussed. Here the principal effects are of one structure on another – the effects of shelter and negative shelter (§16.1.4). The discussion concentrates on building structures, rather than industrial plant and other specialised structures. There are two reasons for this: firstly because there are little enough data on many structural forms when isolated, so that data in combination with other forms are non-existant, and this Guide does not include speculation about the unknown; secondly because many of the excluded non-building forms are dynamic, such as tall cylindrical stacks, and their interaction requires knowledge of the principles of dynamic response and aeroelastic instabilities, so that their discussion is reserved for Part 3 of the Guide.

The scope of this section is therefore principally buildings, which are by far the most common form of structures, but includes boundary walls and fences, since these are often built specifically for the shelter they provide. Even so the range of available data is small, so that the discussion is able to concentrate on only a few key aspects. Before discussing the characteristics of the loading of buildings arranged in groups, it is prudent to discuss the philosophy of their incorporation into the design assessment.

The concept of shelter defined in §16.1.4 implies that the different buildings are independent, which leads to the likelihood that each are built at different times, possibly by different contractors for different owners. In this case there are two classes of problem: the effects of the existing buildings on the new building and the effects of the new building on the existing buildings.

In the first case, because the existing buildings are already in place, the detrimental negative shelter effects can be assessed, ensuring a safe design. However, the beneficial effects of shelter are more likely to occur, so the designer may wish to exploit them in the design. If the owner of the new building has no control over the sheltering buildings, what happens if they are demolished? To avoid this problem, most codes of practice exclude the direct effects of shelter by neighbouring buildings, but do include the general effects of urban development on the incident wind speed profile. On the other hand, where the owner has control of the whole site and this will be maintained over the design life of the buildings, there may be a case to exploit the shelter. At least, the buildings can be arranged to minimise loading and the gains over the isolated state used as an additional factor of safety. Provision of suitable boundary walls or fences can provide shelter at the design stage, or can be used later to mitigate problems of damage to existing buildings.

In the second case, shelter produced by the new building on existing buildings is of no benefit to their owners, except as an additional factor of safety. However, the less likely but still frequent detrimental effects of negative shelter will be of great concern to their owners, since the design safety factors may be so eroded that damage may occur. On several occasions in the UK during the last decade, the construction of a tall building has resulted in damage to nearby low-rise buildings. So far the liability for such damage has been accepted by the developers of the new buildings, who have paid for the necessary repairs and strengthening. As yet no cases have reached the courts, so that no legal precedents have been set. There is as yet in the UK no legal concept of 'ancient wind' to compare with the concept of ancient light. The failure to anticipate the detremental effects of a development on neighbouring existing buildings may yet lead to a serious collapse, loss of life, and the penalty of punative damages against the developer in addition to the direct liabilities.

In summary, while the knowledge of beneficial shelter effects are sufficient to derive appropriate design rules in some common cases, they should be used only after careful consideration of the consequences. On the other hand, action should always be taken to mitigate the enhanced loading caused by negative shelter effects on the new building and on existing neighbours.

19.2.2 Shelter effects

19.2.2.1 Shelter from boundary walls and fences

19.2.2.1.1 Solid boundary walls. The Oxford study of boundary walls and fences for BRE reported in §16.4.2, also included measurements of loads and base moments on pairs of solid walls and slatted fences in various combinations of porosity.

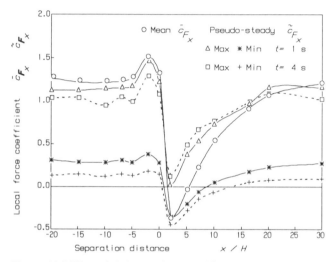

Figure 19.1 Effect of shelter on the normal force on one of a pair of solid walls

Figure 19.1 shows the results for the base shear force on one of a pair of identical long parallel walls with the wind normal to the walls at various separations of the walls, x/H. Positive values of x/H are for the downwind wall and negative values are for the upwind wall. Mean coefficients are compared with pseudo-steady maximum and minimum coefficients on the same axes. For the upwind wall at large spacings the mean, $\bar{c}_{F_x}$, and the maximum $\tilde{c}_{F_x}$ remain close to the design value for an isolated wall of 1.2. The corresponding minimum $\tilde{c}_{F_x}$ remains positive, i.e. there is no reversal of load. At close spacings, the load on the upwind wall increases and this corresponds to an open-roofed building or central well, with a separation bubble trapped between the two walls containing the high suctions of the missing roof regions which act on the rear face of the wall. The corresponding load on the downwind wall at small spacings is negative, i.e. acting upwind, since the high suction in the separation bubble acting on the front face is greater than the wake suctions on the rear face. As the spacing increases, the load on the downwind wall increases towards the isolated value, there being a significant shelter effect up to about 30 wall heights separation.

In comparing the mean and pseudo-steady values, the characteristic noted in §15.3.3 for the silo example of Figure 15.12 is again evident – the maximum values are close to the mean when the mean is positive, the minimum values are close to the mean when the mean is negative, and the maxima and minima have similar absolute values when the mean is near zero.

It is clear that there is a considerable benefit to the shelter offered by walls of similar heights up to quite large spacings. Nothing is known for walls of different heights, but it can be assumed that downwind walls will be sheltered by at least the same amount when shorter, but will be less sheltered when taller. When much taller, typically three times, the 'Wise effect' of §8.3.2.1.1 and §19.3.3.2.1 may slightly increase the loading. This shelter is also available to downwind buildings of similar height to the wall. The detremental effect of increased load on the upwind wall at small spacings is relevant to the case of building façades during renovation when the roof is absent.

19.2.2.1.2 Slatted fences. The study examined the similar effect with slatted fences. Figure 19.2 shows data for pairs of identical fences of various solidities. This time the data are expressed as a shelter factor, η, for the downwind fence, which is identical to the shielding factor for pairs of lattice frames in §16.3.5. Several points are of interest here. Firstly the position of maximum shelter is about five fence heights downwind, this being a well-known effect[359] exploited in the design of shelter-belts for crops, etc. Secondly, the shelter becomes less, but more uniform with separation as the solidity decreases, as expected for lattice frames. Thirdly, an anomaly appears at small spacings depending on whether the downwind fence slats are in line with the wakes of the upwind fence slats, where the increased shelter is equivalent to the wake effect for lattices of Eqn 16.63 in §16.3.6.2.1. However, when the fences are staggered so that the downwind slats are in the jets, the shelter effect is almost completely negated. This effect occurs at low solidities, $s = 0.167$ and 0.33 in Figure 19.2, where the jets are wider than the wakes, but is indistinguishable at $s = 0.67$.

The study also examined these effects with fences of different solidities and concluded that the shelter factor value is set principally by the solidity of the upwind fence and the separation in terms of the upwind fence height, and the solidity of the downwind fence has only secondary effects.

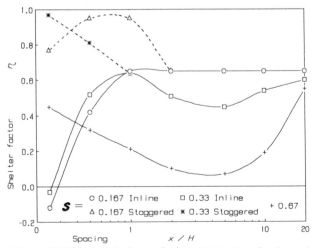

Figure 19.2 Effect of shelter on the base moment on the downwind of a pair of slatted fences

19.2.2.2 Shelter from low-rise buildings

19.2.2.2.1 Introduction. Figure 19.3 shows a typical suburban area of houses in the UK. The mutual shelter of such arrays of buildings was the particular interest of Professor B E Lee during the 1970s, while at Sheffield University, and resulted in a series of papers on wind loading and natural ventilation [360, 361, 362, 363, 364, 365]. This section is an edited version of a summary of this work written for BRE by Professor Lee. Other work on this subject is reported in the published literature [99, 296, 366, 367].

Figure 19.3 Typical suburban housing in the UK (Aerofilms Ltd)

19.2.2.2.2 Defining the geometry of building groups. The plan-area density of buildings, a, defined in Figure 9.2 (§9.2.1.2), has been adopted in the past as a simple and useful single index for defining group form. For most purposes, the shapes of individual buildings can be defined by the heights, H, and their proportions: slenderness ratio, H/B (§16.1.2), and fineness ratio, D/B (§16.1.3). The number of parameters with which to describe the group array is very large, particularly if the type of pattern is irregular, as in Figure 19.3. However, particular regular groups appear in the overall irregular pattern which can be used for the purpose of study and, if the choice of group is restricted to what are called here 'normal' and 'staggered' arrays, then the problem becomes manageable. These

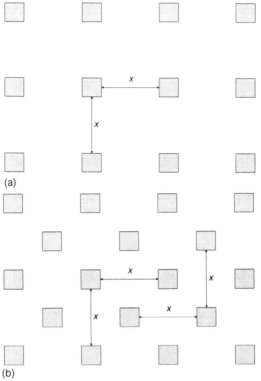

(a)

(b)

Figure 19.4 Types of group: (a) normal array; (b) staggered array

pattern types are illustrated in Figure 19.4. The normal array in (a) is a square matrix of buildings, separated by a constant distance, x. For the staggered array, additional buildings are placed in the centre of each square, so that the density, a, is doubled for the same separation distance, x, as shown in (b).

The range over which the geometrical parameters must vary to encompass most practical examples is significant for the design of experiments as well as for the presentation of design data. A recent survey of 110 housing schemes has shown that almost two out of three schemes have a value of density in the range $0.11 \leqslant a \leqslant 0.20$. Alternatively, this may be expressed as a range of spacing between buildings in the range $1 \leqslant x/H \leqslant 4$, depending on the proportions of the individual buildings.

19.2.2.3 Reference dynamic pressure.   Changes in the density of the buildings changes their mutual shelter, which changes their drag, which in turn changes the wind speed profile above the array. This is accounted for in the design wind speed data of Chapter 9 by the S-Factors, particularly the Fetch Factor, S_X. However, for the purpose of studying the effects of housing density, a fixed reference wind speed giving a constant value of reference dynamic pressure was adopted, to give the changes in loading in absolute terms. The natural choice for this fixed wind speed was the gradient wind speed at the top of the boundary layer, V_g (§5.2.1.3), since this was independent of changes in the housing density, and this was adopted for the studies.

However, since wind loading codes of practice and design guides adopt a reference wind speed and dynamic pressure at some height relative to the building, the relationship between the mean wind speed at the building height, $\overline{V}\{z = H\}$, and the gradient wind speed, V_g, was determined in the study. For the normal array, this relationship was given by the empirical equation:

$$\overline{V}\{z = H\} / V_g = 0.281 + 0.116 \log(x/H) \tag{19.1}$$

in terms of the separation distance, x.

19.2.2.2.4 The three flow regimes.   The study identified three distinct flow regimes characterised by the flow patterns shown in Figure 19.5:

(a) isolated roughness flow regime;
(b) wake-interference flow regime; and
(c) skimming flow regime.

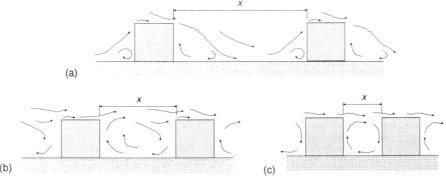

Figure 19.5 The three-flow regimes: (a) isolated roughness flow regime; (b) wake-interference flow regime; (c) skimming flow regime

In the isolated flow regime, Figure 19.5(a), the buildings are sufficiently far apart that each acts in isolation. A 'horshoe vortex', as in Figure 8.7(a) (§8.3.2.1.1), forms around each individual building in the array and the flow reattaches to the ground behind the near-wake circulation bubble shown in Figure 8.13 (§8.3.2.3.1) before the next building is reached. Thus here the separation distance, x, is greater than the sum of the upwind separation and downwind reattachment distances. The shelter in this case is small and forces on each individual building are similar to the values for the building in isolation.

In the skimming flow regime, Figure 19.5(c), the buildings are sufficiently close that a stable vortex can form in the space between them and the flow appears to skim over the roofs. Here the shelter is large and forces on each individual building are very small. The flow around the buildings in this case forms the interfacial layer of Figure 7.2 (§7.1.3) and the zero-plane displacement, d, is close to the building height, H.

The wake-interference flow regime, Figure 19.5(b), represents an intermediate state between isolated and skimming flows, where the near-wake circulation bubble does not have sufficient space to develop fully between the buildings, but where the separation is too large for a stable vortex to exist.

19.2.2.2.5 Effect of plan-area density on loading.

The effect of plan-area density and spacing on the global front and rear face pressures, C_p, and on their sum, the overall drag, C_D, is shown in Figure 19.6 for cuboidal buildings in the normal array. Here the coefficients are based on the gradient wind speed, V_g. The dashed vertical lines mark the boundaries of the three flow regimes, determined from the breaks in slope of the curves, showing that the middle wake-interference regime exists over the range $0.08 \leqslant a \leqslant 0.18$ or $1.4 \leqslant x/H \leqslant 2.5$. The corresponding data for the staggered array are shown in Figure 19.7. These results are very similar, except now the wake-interference regime exists over the range $0.14 \leqslant a \leqslant 0.32$ or $1.5 \leqslant x/H \leqslant 2.7$. It was noted in §19.2.2.2.2 that the density of the staggered array was twice that of the normal array for the same separation.

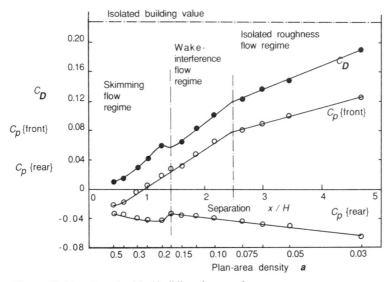

Figure 19.6 Loading of cubical buildings in normal array

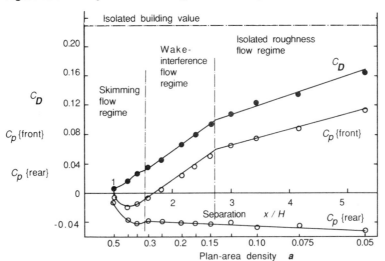

Figure 19.7 Loading of cubical buildings in staggered array

The data of Figures 19.6 and 19.7 collapse quite well in terms of the separation but not in terms of the plan-area density, indicating that the separation is the best index of the shelter effects. Consider replacing the cuboidal buildings with flat hoardings: the plan-area density would be zero, but the drag of the hoardings and their mutual shelter would not be very different, so clearly the plan-area density is a poor index of shelter. Referring back to the study of housing scheme densities indicates that most building layouts will operate aerodynamically in the wake-inteference flow regime and the lower part of the isolated flow regime.

The drag and global pressure data of Figures 19.6 and 19.7 are based on the gradient wind speed. In order to convert these data to a form compatible with the reference used in this Guide and elsewhere, the relationship of Eqn 19.1 must be used. The results of this conversion procedure are shown in the standard form of

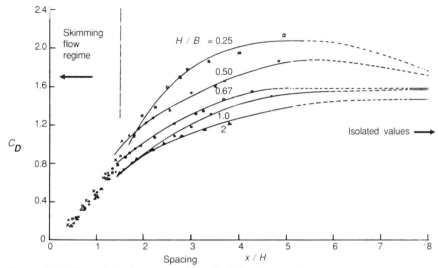

Figure 19.8 Effect of slenderness on drag of buildings in normal array

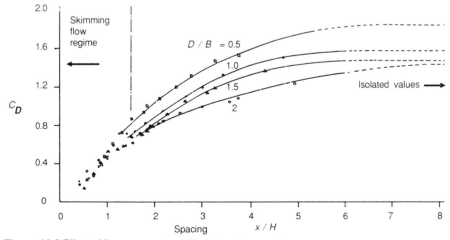

Figure 19.9 Effect of fineness on drag of buildings in normal array

drag coefficient against separation for a range of slenderness ratios in Figure 19.8 and for a range of fineness ratios in Figure 19.9. A number of features are notable in these diagrams. Firstly, they depict results for a wide range of building proportions. Secondly, it can be seen that the clear distinction between flow regimes can no longer be detected, not only because the different slenderness and fineness ratios have different flow regime limits, but also because the normalising velocity, the incident wind speed at the height of the building $\overline{V}\{z = H\}$, depends on the degree of shelter so is a function of the spacing. Thirdly, while the shelter effect is positive and reduces the building drag in the majority of instances, there are some cases of low slenderness and wide spacing where the shelter is negative and drag is increased from the isolated values. Finally, the data at close spacings for the skimming flow regime seem to collapse for all proportions, indicating that the shape of the buildings is unimportant when they are constructed very close together.

19.2.2.2.6 Effect of wind direction. The effect of wind direction was studied for arrays of cubical buildings in terms of the body-axis force coefficient, C_{F_x}. This is the same as the drag, C_D, when the wind direction is normal, $\theta = 0°$. The data are presented in Figure 19.10 and show that the cosine variation applicable to isolated buildings becomes flattened as the group shelter becomes effective. At very close spacings, in the skimming flow regime, no directional influence is apparent over a wide range of direction.

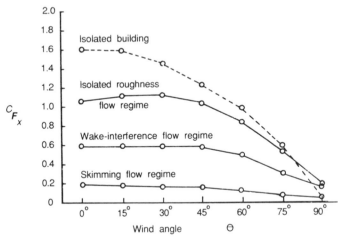

Figure 19.10 Effect of wind direction on shelter

19.2.2.2.7 Influence area. An influence area was defined as that part of the surrounding group around a building where the addition or removal of another building of the group influences the loading on the central building. This concept may be useful in estimating the consequence of changes to buildings around a particular site.

The size and shape of the influence area was determined by removing members of the group and noting whether the force coefficient, C_F, changed by more than 5% or 10%. This was repeated in 15° intervals of wind direction from $\theta = 0°$ to

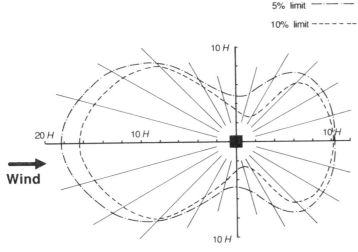

5% limit —·—·—

10% limit ——————

Figure 19.11 Influence area for normal array at 10% density

$\theta = 90°$. When the data for all wind directions were superimposed in plan, the 5% and 10% influence areas shown in Figure 19.11 were obtained.

19.2.2.2.8 Effect of building height relative to the group. The concept of large groups of buildings all of the same height is somewhat idealistic because, even in large estates of nominally identical houses, variations in the site datum levels may make the height of any one building differ from its neighbours. Figure 19.12 shows the effect of building height in the range $0.5 \leqslant H / H_n \leqslant 4$ on the drag of the central building, where H is the height of the central building and H_n is the height of all the other buildings of the array. The data were obtained by varying the height of the central building and hence its slenderness ratio. As the drag coefficient is

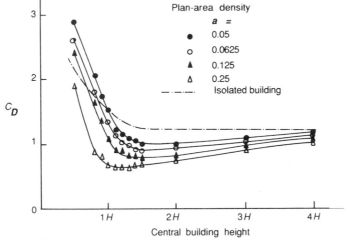

Figure 19.12 Influence of central building height on drag

dependent on slenderness, the corresponding curve for an isolated building is included for comparison.

It is seen that the group always provides positive shelter for $H / H_n \geqslant 1$, but as the height increases a smaller proportion of the building is sheltered and so the benefit reduces. At low densities the drag increases as $H / H_n \rightarrow 0$ above the value for an isolated building, and this is the effect noted earlier in Figure 19.8.

19.2.2.2.9 Consequences for design.
The knowledge of the characteristics of shelter on a low-rise building in an array of similar buildings provided by these studies does not give much direct benefit to the designer unless the spacing is so close that skimming flow is ensured. Figures 19.8 and 19.9 show that the drag reduces dramatically as spacings reduce from $x/H = 1$ towards zero. But in the typical practical range of $1 \leqslant x/H \leqslant 4$ the benefit is not large and the loading can be worse in some cases. Accordingly, it is prudent not to reduce the design loads to account for shelter, but to treat any benefit obtained as an additional factor of safety. This is the approach adopted by most codes of practice. This additional safety factor is probably the main reason for the survival of buildings in the middle of groups where buildings at the edge of the group and nearby isolated buildings have suffered damage in wind storms.

19.2.2.3 Shelter from medium-rise buildings

19.2.2.3.1 Scope.
Between the many low-rise buildings discussed above and the tall city-centre tower is a range of medium-rise cuboidal office and apartment buildings that are tall in respect of their smaller faces, but are squat or only just tall in respect of their larger faces. This form of building is commonly called a 'slab' block, and many of these are found in rows of two, three or more slabs on a common centreline. Figure 8.13 (§8.3.2.3) illustrated the structure of the flow in the wake of a slab block, showing the near-wake circulation bubble in which the momentum of the flow is much reduced.

The shelter afforded by an upwind slab on a downwind slab on the same centreline and aligned parallel has been studied recently by English and Durgin at MIT [368, 369]. In this study, the effects on the base shear and base moment on a slab of proportions $H{:}L{:}W = 8{:}12{:}3$ of the spacing between the slabs, the wind direction and variations in the height, breadth and depth of the upwind sheltering slab were studied. Figure 19.13 shows the principal parameters of the study.

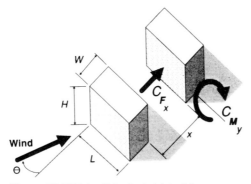

Figure 19.13 Pair of identical slab buildings

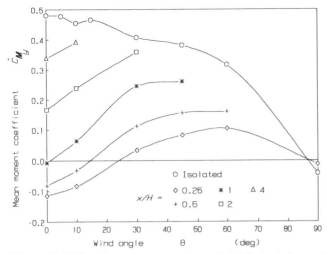

Figure 19.14 Mean base moment on downwind slab of pair (from reference 368)

19.2.2.3.2 Effect of spacing and wind angle. The varation of the mean base moment, $\overline{C}_{M_y}$, with wind angle and spacing between slabs when both slabs are identical is shown in Figure 19.14, compared with the data for an isolated slab. At close spacings, the shelter effect is strong over a wide range of wind angle, giving reversed moments when the wind is normal. This is the same effect as in Figure 19.1 for boundary walls, so it is expected that the corresponding loading of the upwind slab increases, but this was not investigated in the study. Figure 19.15 shows the shelter effect on peak values, this time for the maximum base shear force coefficient, $\hat{C}_{F_x}$. (**Note:** the peak coefficient has not been converted to pseudo-steady values because the corresponding gust factors were not given.) The peak shelter effect is not as large as the mean shelter effect and the maxima do not reverse sign at close spacings. The various combinations of parameters in the study

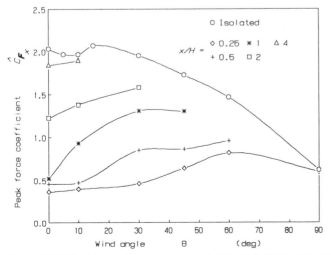

Figure 19.15 Peak base shear on downwind slab of pair (from reference 368)

gave 80 sets of data of the same form as Figures 19.14 and 19.15. However, English and Durgin[368,369] reported only their values for identical pairs of slabs. This restriction is not a great loss to the designer for the reasons discussed below.

While the base shear and moment are always reduced by the action of shelter, the torsional moment, C_{M_z}, is increased at some wind angles. At $\theta = 0°$ the whole of the downwind slab is sheltered by the wake of the upwind slab. As θ increases, part of the downwind slab emerges from the wake and becomes fully loaded, producing an asymmetric load and hence a net torque on the slab. English and Durgin's data indicate that the increase in torque is most onerous at $\theta = 45°$ and a spacing of $x/B = 1$, making the torsional moments some 30% higher than given in Figure 17.26 and Appendix L. Fortunately, torsion is usually not critical in the design of static slab buildings and this effect is not expected to be significant in their design.

19.2.2.3.3 Shelter factors for identical slabs on common centreline.

Figure 19.16 gives the shelter factor, η, for mean base shear and moment for the downwind slab of a pair of identical slabs, as indicated in Figure 19.13, in terms of the wind angle and separation between slabs. Similarly, Figure 19.17 gives values of η for the peak

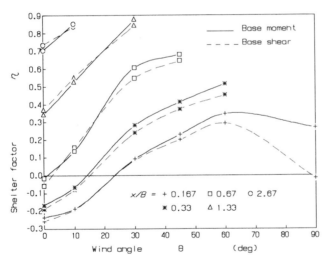

Figure 19.16 Mean shelter factors for downwind slab of pair (from reference 369)

base shear and moment. These curves are equivalent to the curves of Figures 19.14 and 19.15 expressed as fractions of the isolated value, so that the reliability of the shelter factor estimates decreases near $\theta = 90°$ because it is the ratio of two values converging towards zero. The base shear and base moment give similar values of η, indicating that the shelter does not significantly affect the height of the centre of pressure.

19.2.2.3.4 Consequences for design.

For design purposes the curves of Figures 19.16 and 19.17 could reasonably be represented as a family of straight lines of constant slope. Since peak load and moments are required for static structures, fitting the data of Figure 19.17 to this model gives the empirical equation:

$$\hat{\eta} = 0.331 - 0.664 \log(x/L) + 0.0067\,\theta \qquad \leqslant 1 \qquad (19.2)$$

where the peak symbol on $\hat{\eta}$ denotes its use to derive peak loads and the limit indicates a maximum value of unity. Figure 19.13 serves as the key to the dimensions. For design assessments, this shelter factor would be applied to the overall load coefficients given in §20.7.2.1 and Appendix L.

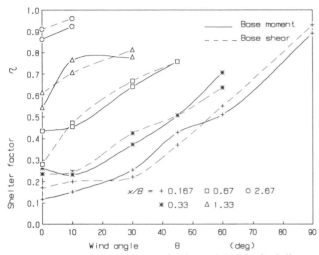

Figure 19.17 Peak shelter factors for downwind slab of pair (from reference 369)

The major question remains: when is it appropriate and safe to use the shelter factor, $\hat{\eta}$, of Eqn 19.2? With a pair of slabs, the downwind slab is sheltered, but the upwind slab is fully exposed. In winds from the opposite direction, the position is reversed. Both slabs are fully exposed at some wind direction, although not simultaneously. Accordingly, the shelter factor is only of use in the design of either individual slab if there is a strong directional bias in the wind climate or exposure. It is much more useful when considering the stability of the pair if on a common foundation, since the foundation needs to support one fully exposed slab and one sheltered slab. The argument extends to the stability of any podium carrying both slabs.

When there are more than two slabs on a common centreline, as with the three shown in Figure 19.18, only the end slabs are ever fully exposed. In this case the shelter factor could be applied to the centre slab, although it is likely that the designer would wish to use the same design specification for all three slabs. The usefulness again comes in assessing the stability of all three slabs on their common podium, since now only the upwind slab is fully exposed and both downwind slabs are sheltered.

19.2.3 Negative shelter effects

19.2.3.1 Introduction

Two ways are used to assess the negative effects of shelter caused by one building on another:

1 The effects of one building on the field of wind speed around it are assessed. The wind speed incident on a second building is taken as the wind speed at the

Figure 19.18 The three slab blocks of DOE headquarters, Marsham St, London

corresponding position in this field. This two step approach requires the second building to be small and the 'law of scale' to apply, so that the field of wind speed is not significantly altered by its presence and it does not alter the loading of the first building. The approach is independent of the shape of the second building, which is a major advantage.

2 The effects of one building on another are assessed directly. This single step approach allows assesment of the mutual effects of each building on the other, so that the buildings may be of comparable size. The disadvantage is that the effects must be determined for each combination of building shape and this is can only be done for a limited range of common shapes. In practice, suitable data exist only for pairs of identical buildings.

19.2.3.2 Surface winds near high-rise buildings

19.2.3.2.1 Published sources. This is the approach of (a), above. The wind speeds near the base of tall buildings were studied extensively in the 1970s for the comfort and safety of people under the classification of 'wind environment studies'. Various guides for designers were produced: BRE published *Wind environment around buildings* by AD Penwarden and AFE Wise in 1976 (London, HMSO) giving design advice based on case studies at full and model-scale and a range of parametric model studies; also in 1976 CTSB in France published *Intégration du phénomème vent dans la conception du milieu bati (Incorporating wind effects in the design of the built environment)* by J Gandemer and A Guyot (Paris, Secretatiat General de Groupe Centrale des Villes Nouvelles) which gives advice on the causes and mitigation of wind effects for use at the planning stage of the development of large sites; TNO in the Netherlands published *Beperken van windhinder om gebouwen (Reducing wind discomfort around buildings)* by WJ Beranek and H van Koten in 1979 (Den Haag, Stichting Bouwresearch) giving ground surface wind speed and direction data for a range of building shapes from sand-erosion and pigment-streak methods with models. These three complementary guides give data and advice which remains current today and more recent publications, such as *The abatement of wind nuisance in the vicinity of tower blocks* by UK Gerry and GP Harvey (London, Greater London Council, 1983) and *Klimatplanering vind (Planning wind climate)* by M Glaumann and U Westerberg (Stockholm, Svensk Byggtjänst, 1988), adopt the same principles.

19.2.3.2.2 Maruta's method. Parametric model studies in contemporary boundary-layer simulations, combined with full and model-scale comparisons of the wind effects around tall buildings in Tokyo, conducted at Nihon University in Japan by Maruta [310, 370] has led to the development of a prototype prediction method. The method is an empirical mapping model which predicts the location of contours of constant wind speed to a claimed standard error of $\pm 10\%$ in the wind direction and $\pm 15\%$ cross-wind. The full method is too complex to reproduce here, but Figure 19.19 compares the prediction of the method with model-scale data for two

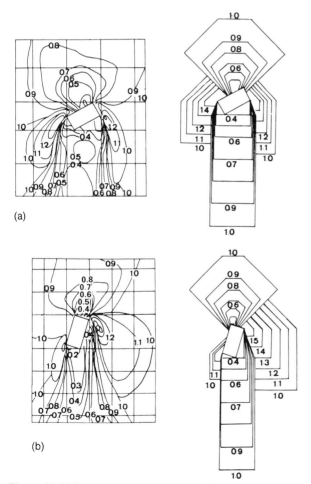

Figure 19.19 Examples of Maruta's method for wind speeds around tall buildings where (a) $\theta = 30°$ and (b) $\theta = 75°$

cases [310]. Fortunately, Maruta also proposed a simpler mapping method which gives the envelope of the highest wind speeds for all directions, which he calls the 'global' method.

Figure 19.20 is the key to Maruta's global method adapted here to determine the reference wind speed for low-rise buildings near a cuboidal tall building. The

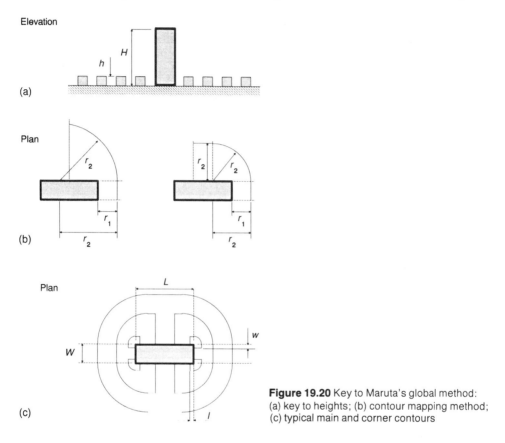

Figure 19.20 Key to Maruta's global method: (a) key to heights; (b) contour mapping method; (c) typical main and corner contours

dimensions of the tall building are the standard dimensions H, L and W used before for cuboidal buildings. The height of the surrounding low-rise buildings is h, and this is assumed also to be the required reference height, z_{ref}. Being purely empirical, the method is valid only over the range of parameters to which the model was fitted and these are:

$$0.5 \leqslant H/L \leqslant 6 \quad 1 \leqslant L/W \leqslant 4 \quad 2 \leqslant H/h \leqslant 32 \quad \text{and} \quad 0.25 \leqslant \alpha \leqslant 0.45$$

The last of these parameters, α, is the power-law exponent of the mean velocity profile for which design values are given in Table 9.3 (§9.4.1), and its range of value confines the validity of the method to urban areas.

The result of the method is a factor on the mean wind speed, equivalent to the S-Factors of Chapter 9, and will be denoted here by S_B using the subscript 'B' for 'building'. Since this is a factor on **mean** wind speed, it is assumed to act on gust speeds in the same manner as the Topography Factor, S_L, in §9.4.3.4, i.e. the mean component is changed but the turbulence component is assumed to remain constant. Thus Eqn 9.39 must be modified to:

$$S_B S_L S_G = S_B S_L + G - 1 \tag{19.3}$$

to represent this effect, and the S-Factor product $S_B S_L S_G$ is included with the other S-Factors in Eqn 9.1 to give the design gust speed, $\hat{V}\{t\}$.

The value of S_B is derived by mapping contours of constant value around the tall building in terms of the dimensions r_1 and r_2, as shown in Figure 19.20(b) and (c). Arcs of radius r_2 are drawn around each corner from a centre on the long face located $r_2 - r_1$ from the corner, as shown in the key. These arcs are extended across the short face and, if necessary, to the centreline of the long face by tangents parallel to the faces. In the corner regions, the tangent extensions are limited to the dimensions l and w, as shown in (c).

The contour for a given value of S_B is mapped by the following steps:

1 Determine the value of $S_{B_{max}}$ from:

$$S_{B_{max}} = 1.7^{0.28 - \alpha} (H/h)^{\alpha - 0.08} \tag{19.4}$$

This represents the maximum value of S_B occurring at the corner.

2 Determine the values of the dimensions r_1 and r_2, from:

$$r_1 = 0.27 (L + W) (H - 1.7 h)^{0.36} (S_{B_{max}} - S_B) / (S_{B_{max}} - 1) \tag{19.5}$$

$$r_2 = (L / W)^{0.34} r_1 \tag{19.6}$$

3 Determine the limits to the tangent extensions, l and w, from:

$$l = 0.8 r_2 \tag{19.7}$$

$$w = 0.8 r_1 \tag{19.8}$$

4 Draw the contour as shown in Figure 19.20(b) and (c).

19.2.3.2.3 ECCS method. A simpler, but cruder, rule-of-thumb method for estimating the effect of a neighbouring tall building is given in the current (second edition) model wind loading code produced by the ECCS [371]. Instead of giving a factor on wind speed, this method defines an effective reference height, z_{ref}. Figure 19.21 is the key: H and h are the heights of the tall and low-rise building,

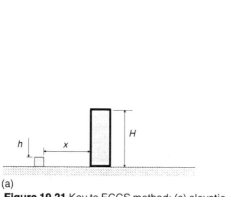

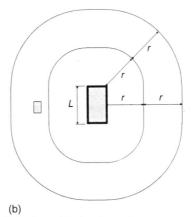

(a)   (b)

Figure 19.21 Key to ECCS method: (a) elevation dimensions; (b) plan dimensions

repectively, x is their separation and L is the length of the longer face. Two bands are defined around the tall building, each of width r given by:

$$r = H \qquad\qquad \text{for } H \leqslant 2\,L$$
$$\;= 2\,L \qquad\qquad \text{for } H > 2\,L \qquad\qquad\qquad (19.9)$$

The reference height of the design dynamic pressure for the low-rise building is taken as:

$$z_{ref} = r\,/\,2 \qquad\qquad \text{in the inner band, } x \leqslant r$$
$$\;= \tfrac{1}{2}\,(r - [1 - 2h/r]\,[x - r]) \quad \text{in the outer band, } r < x < 2\,r$$
$$\;= h \qquad\qquad \text{further from the tall building, } x \geqslant 2\,r$$
$$(19.10)$$

Note: Because the approaches are so different, the ECCS method and Maruta's method do not give the same result. In general, the ECCS method is expecteed to overestimate and so be safe everywhere except near the corners of the tall building.

19.2.3.2.4 'Wise effect'.

It was noted in §8.3.2.1 that placing a low building upwind of a tall slab intensifies the 'horse-shoe' vortex by the 'Wise effect', as shown in Figure 8.7. The effect is caused because the vortex is trapped and stabilised in the space between the two buildings. Wise's paper on the effect, delivered at the Royal Society in 1971 [372] includes empirical expressions for the wind speed ratio $\bar{V}\{z = h\}/\bar{V}\{z = H\}$. Design guidance for the environmental impact of the Wise effect is given in the BRE publications: *Wind environment around buildings* by AD Penwarden and AFE Wise (HMSO, 1975) and BRE Digest 141 *Wind environment around tall buildings* (BRE, 1972). In terms of the building S-Factor, S_B, values as high as $S_B = 1.5$ can be reached in the space between the two buildings and $S_B = 2$ just outside the shear layers from the corners of the tall building when $H \geqslant 4h$. Further studies have been made since this guidance was prepared, in particular the experimental and theoretical work of Britter and Hunt [373]. This indicates that, provided the slab is very much taller than the low building, $H >> b$, an approximate expression for the mean wind speed, $\bar{V}_A$, at a point 'A' near the ground half-way between the two buildings is given by:

$$\bar{V}_A = \bar{V}\{z = h\}\,(3L/x)\,(1 - 1\,/\,[1 + kx/L])^2 \qquad\qquad (19.11)$$

where k is an empirical constant approximately $k = 1.5$, and the dimensions are defined in Figure 19.21. This shows that wind speed scales to the incident wind speed at height of the lower block and is independent of the height of the slab, H. Wise's original study [372] indicated that $\bar{V}_A \propto H^{0.8}$, but Maruta's data given by Eqn 19.4 indicates that the dependence on H varies from the natural exponent of the incident profile by only 0.08, supporting Britter and Hunt's view. Differentiating Eqn 19.11 with respect to the separation ratio x/L yields the maximum value as $\bar{V}_A = 1.125\,\bar{V}\{z = h\}$, occurring when $x/L = 1/k = 0.67$.

19.2.3.3 Funnelling between pairs of buildings

19.2.3.3.1 Introduction.   This is the approach of (b) in §19.2.3.1, where the effects of pairs of identical buildings arranged side-by-side are assessed directly. Flow funnels between the two buildings and most changes occur on the two opposing faces, either side of the gap, tending to increase the suctions and pull the buildings together. In practice, data exist for pairs of conventional duopitch houses, pairs of tall square-section towers and pairs of tall circular cylinders.

19.2.3.3.2 Houses. This problem has been investigated several times over the years. Hamilton's 1962 study [374] in smooth uniform flow indicated that the highest mean local suctions on facing gable walls occurred at very close spacings, typically $y/b \simeq 0.05$ where y is the gap width and b is the scaling length introduced in §17.1.4, and these were made worse by tapering the gap at between 4° and 6° degrees. At BRE, Menzies and Bradley's study in 1971 in the velocity-profile only simulation produced by BRE's first wind tunnel showed that offsetting the two buildings in the wind direction, x, which they referred to as a 'stagger', was also significant. The highest mean local pressure coefficient, $\bar{c}_p = -2.4$, occurred near the windward edge of the gable wall at the very small spacing of $y/b = 0.05$, but no stagger, $x/b = 0$. The highest mean global pressure on the whole gable face, $\bar{C}_p = -1.5$, occurred at a much larger separation, $y/b = 0.5$, with an equal stagger, $x/b = 0.5$.

The problem was again investigated at BRE in 1987 by Blackmore for monopitch and duopitch houses, this time in a full boundary-layer simulation and measuring peak as well as mean pressures. The results were surprising in that the pseudo-steady pressure coefficients obtained from the peak measurements were found to be far less sensitive than the mean pressure coefficients. For the gable walls of both monopitch and duopitch houses with the wind normal to the main face, $\theta = 0°$, the local peak suctions were only 20% higher at small gaps and the global suctions only about 10% higher at gaps near $y/b = 0.5$ than the values for an isolated house. However, these suctions were maintained over a wide range of wind angle, θ, by the funnelling action through the gap, where the corresponding values on an isolated house would be much less. Stagger was not investigated in this study.

It now appears that the probability of high loading, and hence the rate of damage to the gable walls, is increased at spacings in the range $y/b < 1$ because the range of critical wind directions is increased and not because the values of the pressure coefficient are very much more onerous.

19.2.3.3.3 Square-section towers. The effect of the spacing and stagger of a pair of square-section towers on the x- and y-axis mean force coefficients has been studied by Blessmann and co-workers first in smooth uniform flow and later in turbulent boundary-layer simulations [375, 376, 377]. The later work [377] indicates that there are two principal loading cases to consider. Case 1 is when the towers are on a common centreline and one tower is directly behind the other, then the drag of the downwind tower is negative at close spacings, but the corresponding drag of the **upwind** tower is increased by up to 22%. Case 2 is when the towers are aligned corner to corner on a common diagonal and the wind is normal to the diagonal, then the body-axis force coefficient that tends push the towers apart increases by about 30%. All other combinations of spacing and stagger remain within these limits.

19.2.3.3.4 Circular cylinders. Circular-section stacks, silos and cooling towers are often placed in pairs, rows or groups. Interest in this problem was stimulated in the UK by the collapse of three cooling towers at Ferrybridge in 1965 (§8.5.3). Ponsford's measurements on a group of closely spaced silos [378] gave a maximum resultant force coefficient of $C_F = 1.4$, compared with the drag coefficients for sub-critical and super-critical flow of $C_D = 1.2$ and 0.6 respectively. The most comprehensive design data for pairs of long cylinders are currently given by ESDU Data Item 84015 [379].

19.2.3.4 Wake buffeting downwind of high-rise buildings

The general effect expected when one building is in the wake of another is that the reduced momentum in the wake will give a net shelter effect, and this is confirmed by the mean and peak shelter factors given in Figures 19.16 and 19.17 (§19.2.2.3). However the mean shelter factor values are always less than the corresponding peak values. The shelter could be entirely a mean component, with the additional turbulent fluctuating loading unchanged. Figure 19.22 shows the corresponding

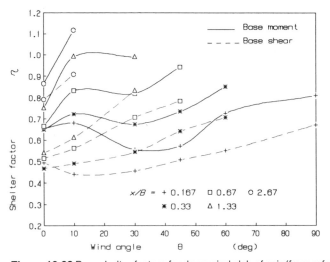

Figure 19.22 Rms shelter factors for downwind slab of pair (from reference 369)

shelter factors for the rms of the fluctuating components of the base shear and moment. Shelter does occur to the fluctuating components at close spacings, but the data indicate that the fluctuating loading increases at spacings ratios greater than 2, that is, for buildings in the far wake, owing to the increased wake turbulence.

This effect increases the dynamic response of buildings in the far wakes of others and is called 'wake buffeting'. Detailed discussion of the effect and its design consequences are reserved for Part 3 of the Guide.

19.3 Load paths in structures

19.3.1 Introduction

The external pressures acting on the outer faces of the building and the internal pressures acting on the inner faces of the building accumulate over the surface of the building to form the wind loads which are passed through the structural members to the foundations. The designer needs to consider the path that these loads take through the structure in order to ensure sufficient capacity in the structural members. These paths depend on several factors: the stiffness of the structure is important; so is the position of dominant openings. Only the net external loads pass through to the foundations. The components of load generated by internal pressures react against each other within the structure of the building.

These key aspects are now discussed briefly. As the wind loads depend only on the shape and porosity of the building, irrespective of the structural load-bearing system, a comprehensive discussion is not possible owing to the vast range of structural solutions to any particular building shape. Instead, the relative importance of the three key aspects discussed below should be assessed in the light of the particular structural form chosen. Although the examples in the discussion are based on a steel-framed building, analogous load paths will occur in other constructional forms to those discussed below.

19.3.2 Stiffness of the structure

The majority of structures have many load paths and the proportion of the loads taken by each path depends on the relative stiffness of each path. In general, the stiffer the path, the greater the proportion of the load taken by that path. Here is a simple example to demonstrate this fact. Consider an extensive steel-framed building. In the UK at present the frame is likely to be designed as if it were pin-jointed, neglecting any moment resistance in the joints, and requiring the provision of 'wind bracing' to resist the lateral loads. Usually this wind bracing is provided in one, or at most two, bays of the frame. The stiffness of the cladding will also be neglected. Unless provision is made so that the joints actually are pin jointed, application of a lateral wind load will be partly resisted by the moment capacity of the joints and not all the load is transmitted through to the wind bracing. If the designer does not ensure that the joint is sufficiently more flexible than the wind bracing, distress may occur in the joints. Similarly, if the cladding is stiffer than the wind bracing, a large proportion of the load may be taken in racking by this 'non-load bearing' component, leading to cracking of cementitious sheeting or distress in the fixings of metal-sheet claddings, which may lead to more serious wind induced failure such as complete cladding removal.

Because of the alternative load paths, the proposition that a strength-based design is all that is required for stiff static structures needs to be modified. If the load paths through 'non-loadbearing' components are stiffer than the design load path through the main structural members, the main load path will not act fully until the 'non-loadbearing' components have failed. Even if the frame joints and the cladding suffer considerable distress, the stability of the building is ensured when the wind bracing is sufficient because, in the words of Professor J W de Courcy lecturing at the Institution of Structural Engineers in 1987, 'A building will never fall down until it has tried every possible way of standing up'. However, this occurrence would be a serious serviceability failure, as discussed later.

In essence, stength-based design of static structures works correctly provided that the main design load path is significantly stiffer than any alternative load path through 'non-loadbearing' components.

19.3.3 Dominant openings

The effects of dominant openings on the pressures on buildings were discussed in Chapter 18. The effects on load paths are exactly complementary. Taking the wind-braced steel framed building again as the example: a dominant opening in the windward face increases the internal pressure, taking load off the windward wall but increasing the loads on the other walls and increasing the uplift on the roof; while a dominant opening on the leeward or a side wall decreases the internal pressure, putting more load on the windward wall while reducing loads on the other walls and the roof uplift. Changes in the roof uplift are reacted down through the structure to the foundations, so sufficient strength is required to resist this action. The effect through the walls is more subtle: the position of application of the horizontal wind load changes from rear wall to front wall with the change of dominant opening, so that the path of this load to the wind bracing also changes. In the case of a pure pin-jointed frame, the load would pass directly through the structural members between the loaded wall and the wind bracing, and the remaining members would play no part. In practice, the additional stiffness of cladding, etc., would mean that other alternative load paths would be established which would pass around the wind bracing, e.g. through the roof purlins, to load the structural members on the other nominally unloaded side.

19.3.4 Internal load paths

When the interior of the building is divided into compartments, the action of the balance of internal flows shares the pressure forces between the external and internal walls (§18.2.1). If the porosity of internal partitions can be maintained more than three times greater than the porosity of the external walls, the loads on the internal partitions will be insignificant. If the reverse is the case, the majority of the load will be taken up by the internal partitions. In practice, the first of these two cases is generally true, unless the outer walls are breached accidentally by the failure of cladding or glazing, or deliberately by opening a door or window. Structural failures of internal partition walls and ceilings of intact buildings are quite rare. Most structural failures are the consequence of an accidental dominant opening, such as is shown in Figure 3.21. Failures of non-structural components are more common, particularly as a result of dominant openings, a good example being uplift of suspended ceilings in entrance foyers

19.4 Serviceability failure

Serviceability failure is failure of components that do not effect the structural integrity of the building and, additionally, do not pose a hazard to the life of occupants or passers-by. This latter criterion is not absolute and requires design judgement. For example, the failure of glazing in a small domestic house would not be regarded as serious of itself, but the dominant opening it creates could lead to structural damage. On the other hand, failure of glazing on the upper storeys of a high-rise block adjoining a public road may give rise to serious injuries. This definition is usually sufficient for static structures. A more general definition of serviceability failure would include any other wind-induced problem that makes the performance of the building unsatisfactory in service, including excessive dynamic response causing motion sickness of occupants or rain penetration of joints.

Treatment of serviceability failures by codes of practice varies considerably. The UK wind code, CP3 ChV Pt 2[4] does not specifically address the problem, whereas the Canadian code[178] uses a 100 year return period dynamic pressure for the design of structural members and only a 10 year return period for cladding and dynamic response. Serviceability failures can be repaired, so that the principal criterion in design is one of economics, and this is why a more frequent occurrence is acceptable. Here is one field where the designer is able to exercise his judgement. What is the economic risk for cracking of plaster or other internal finishes when weighed against the frequency of redecoration and refurbishment? Replacement of roof-top gravel scoured by the wind may be fairly simple and cheap, but what is the risk to the glazing of neighbouring buildings caused by flying gravel and the consequent cost of repair? Saffir [380] concludes that the cost of potential damage to neighbouring buildings far outweighs the advantages of a protective gravel coating on roofs of high-rise buildings in a dense urban area.

The economic balance is not set just by the cost of repair to the damaged building: consequential damage to fittings, furniture and stock through water ingress; consequential damage to other buildings from wind-borne debris; loss of trade while repairs are completed; and, not least, compensation for injuries received by occupants and passers-by are all economic factors to be considered. Given a design judgement on the acceptable risk for service criteria, the Statistical Factor, S_T, of Chapter 9 enables the designer to calculate a reference dynamic pressure with the appropriate risk of exceedence.

19.5 Load avoidance and reduction

19.5.1 Introduction

Provided wind loading is considered early enough in the design stage, the design can be optimised to avoid or reduce wind loads. The term 'avoid' is used to suggest that forms of building known to produce high wind loads are rejected and replaced by less onerous forms. The term 'reduce' is used to suggest that wind loads on a given building form can be reduced by judicious use of local aerodynamic features. Accordingly, high wind loads can be avoided only if considered very early in the design, i.e. at the concept stage when ideas for the size and form of the building are still fairly fluid, although high wind loads may still be reduced after the size and form have been chosen. Once the external details of the building have been decided and modifications can no longer be made, it is too late for either and all the designer can do is provide sufficient strength and stiffness to resist the loads that will occur.

19.5.2 Avoiding high wind loads

19.5.2.1 Squat buildings

With squat buildings, the majority of the flow passes over the roof, so that the choice of the roof form is the more important issue. The various forms of roof discussed in §17.3.3 can be ranked in order of effect, in the same manner as the S-Factors were ranked in §9.5. For conventional squat buildings this yields the following list:

worst
(a) hyperboloid (§17.3.3.4);
(b) duopitch troughed (§17.3.3.3.2);
(c) monopitch (§17.3.3.3.1);
(d) sharp-eaved flat (§17.3.3.2);
(e) duopitch ridged (§17.3.3.3.2);
(f) barrel-vault (§17.3.3.5);
(g) skew-hipped (§17.3.3.3.5);
(h) conventional hipped (§17.3.3.3.4);
best

For large-plan area buildings, where the choice is between forms of flat or multi-span roofs, this yields:

worst
(a) monopitch multi-span (§17.3.3.6);
(b) sharp-eaved flat (§17.3.3.2);
(c) duopitch multi-span (§17.3.3.6);
(d) flat or multi-span with parapets (§17.3.3.2.6 and §17.3.3.7);
(e) mansard-eaved flat (§17.3.3.3.3);
(f) multi-span barrel-vault (§17.3.3.6);
(g) curved-eave flat (§17.3.3.2.7);
best.

Wind loads can be further avoided by minimising vertical walls: either by lowering the eaves or by sloping the walls back from the vertical (§17.3.2.6), or by providing shelter from walls, fences (§19.2.2.1) or neighbouring buildings (§19.2.2.2). The optimum form is a low-rise dome (§17.2.1.2), but this is rarely practical. The only way to avoid wind loads entirely is to build underground!

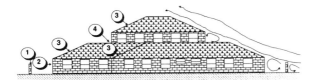

Figure 19.23 Squat building with four load-avoiding features

Figure 19.23 gives an impression of a possible result of optimisation by choice of form. The four load-avoiding features labelled on the figure are:

1 boundary wall or fence of similar height to the eave;
2 a gap spacing, $x < H$, in the skimming flow regime (§19.2.2.2.4);
3 conventional hipped roofs of pitch about $\alpha = 30°$;
4 inset upper storey so that upper and lower roof lines coincide.

Alternatives to the inset upper storey are a continuous roof with flush, inset or 'dormer' windows. In recent years, the UK has seen a resurgence in the use of brick and tile for small office and industrial buildings which would earlier have been built as flat-roofed cuboids with 'system' cladding. Inset terracing of the upper storeys, as shown in Figure 19.23 is also currently in vogue.

19.5.2.2 Tall buildings

With tall buildings, the majority of the flow passes around the sides, so that the forms of the walls becomes the more important issue. Non-vertical walls or inset terracing limits the practical height of the building. An option for avoiding high loads is to replace the rectangular planform with a circular planform, or some intermediate shape such as large-radius curves or wide chamfers on the corners. The directional variation of the wind climate can also be exploited by aligning the long axis in the direction of the strongest winds. In the UK, optimising for wind loads requires the long axis to be east–west, but optimising for daylight requires a

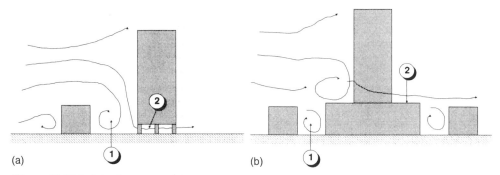

Figure 19.24 Policies for tall slabs: (a) bad neighbour policy; (b) good neighbour policy

north–south alignment and this latter choice is usually made well before the structural engineer has any say in the design process.

Tall buildings which protrude out of the shelter of their low-rise neighbours into the faster wind speeds inevitably experience higher wind loads. It is quite practical and economic to design for these loads otherwise the large population of tall buildings would not exist. However, there remains the effect of the tall building on its neighbours. Construction of town-centre office developments in the UK during the 1960s were often combined with low rise shopping arcades, placing a tall slab on the east side to give the arcade the benefit of the afternoon sun. Sometimes the slab was raised on stilts to give pedestrian access underneath, and such an arrangement is shown in Figure 19.24(a). This form of construction was environmentally disastrous, leading to high wind speeds in the arcade (1) and under the slab (2), generated by the 'Wise effect' (§19.3.3.2.4). These high ground-level wind speeds increased the wind loads on the neighbouring low-rise buildings, often causing damage.

One of the most effective ways of avoiding this problem, a 'good neighbour policy', is shown in Figure 19.24(b). Here the tall building has been provided with a substantial podium at the same height, or just a little higher than the neighbours. The podium is extended to the limits of the site so that at (1) both the podium and the neighbouring buildings enjoy the shelter of the skimming flow regime (§19.2.2.2.4), and at (2) the enhanced wind speeds of the horse-shoe vortex form above the podium roof and pass over the top of the downwind buildings.

19.5.3 Reducing wind loads

So far, all the studies of methods to reduce wind loads have been concerned with the high peripheral suctions on low-pitch and flat roofs. These are all based on modifications to the shape of the eave, from which the flow separates to form the delta-wing vortices (§17.3.3.2.1). The common intent is to explot some 'law of resonance' effect (§17.4.2) to make large changes in the character of the flow over the roof from small modifications to the eave. A number of methods have already been discussed: parapets (§17.3.3.2.6 and §17.3.3.7) are frequently used, but are less effective than previously thought; mansard eaves (§17.3.3.3.3) are effective and can easily be incorporated when the roof is supported by trusses, by taking the cladding line from the lower boom at the eave, along a diagonal bracing member of the truss to the upper boom; curved eaves (§17.3.3.2.7) are even more effective and are a common feature of contemporary steel-framed industrial buildings clad with profiled-steel sheets. The design data for these methods are given in Chapter 20.

Alternative techniques remain experimental, requiring optimisation or confirmation of the resulting benefits by wind-tunnel studies. Considerable advances have been made in recent years at two wind tunnel laboratories: at Monash University, Australia, Professor WH Melbourne has led a research programme devoted to understanding the flow mechanism of separation and reattachment on roofs, resulting in the development of a 'vented eave'; while at the University of Bristol, TV Lawson has successfully used a 'trapped vortex' technique in *ad-hoc* design studies for the National Exhibition Centre, Birmingham, and the Stansted Airport passenger terminal, both on behalf of Ove Arup and Partners.

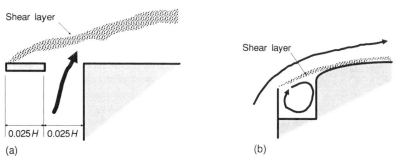

Figure 19.25 Two methods of reducing uplift on roofs: (a) Melbourne's 'vented eave' approach; (b) Lawson's 'trapped vortex' approach

The intent behind Melbourne's 'vented eave' shown in Figure 19.25(a), is to entrain air behind the shear layer to prevent the large suction 'spikes' such as are shown in Figure 8.20 (§8.4.2.3) from forming. The initial research [303, 304] suggested that these 'spikes' were caused by an instability of the shear layer, indicating that a solution exploiting a 'law of resonance' effect (§17.4.2) would be viable. Later work [305, 306] appears to confirm the existence of such an instability. The 'vented eave' successfully inhibits the instability and suppresses the large suction 'spikes'; however it appears to do far more. Cook [352] demonstrated on a model grandstand with a $-7°$ monopitch roof, shown in Figure 19.26, that the 'vented eave' reduces the strength of the eave vortex, lowering the longer-duration peak suctions ($t = 16$s) used for structural design as well as the shorter-duration ($t = 1$s) cladding suctions. In this respect the effect of the entrainment of air through the eave into the eave vortex is similar to that in Figure 17.49(a), where air is entrained into the vortex from behind (§17.3.3.2.4). Accordingly, the 'vented eave' is expected to be a more generally useful mechanism of load reduction than first thought. It may be particularly useful for cantilever grandstands, such as in Figure 19.26, where the open-fronted shape gives high wind loads (§18.6) and the cantilever form of construction is most susceptible to the loads at the eave end of the cantilever. The eave plate and slot need not be exactly of the form shown in Figure 19.25(a). Rotating the plate to the vertical position may still give the desired effect, when the plate may double as the advertising hoardings commonly seen along the eaves of grandstands, but this needs to be confirmed by wind-tunnel tests.

The intent behind Lawson's 'trapped vortex' eave is to exploit the natural features of the eave to promote flow reattachment to the roof. In essence the gutter is enlarged to provide a space in which the eave vortex is trapped. The shear layer

Figure 19.26 Grandstand model with vented eave

separating from the sharp eave reattaches to the specially curved roof edge, immediately behind the gutter, assisted by the force induced by the rotation of the vortex (§2.2.8.4). Because of this reattachment, the supply of vorticity from the shear layer does not accumulate and the vortex does not continue to grow conically. Development of this kind of flow optimisation requires patient trial-and-error experimentation in the wind tunnel, but the benefits can be spectacular. In his optimisation of the roof to Hall 7 of the National Exhibition Centre, Birmingham, Lawson succeeded in reducing the maximum suction in the edge region to $\tilde{c}_p = -0.6$ from the design value of $\tilde{c}_p = -2.0$, and obtained a global value of $\bar{C}_p = -0.23$ over the whole roof.

19.6 Optimal erection sequence

In addition to optimising the completed design to avoid or reduce high wind loads, the designer should also consider the stages of construction. Three aspects are important: the likely wind speeds to be experienced in the construction period; the structural characteristics, strength, stiffness, etc., of the structure while only partially completed; and the loading coefficients appropriate to the phases of construction.

The first of these three aspects is covered by the information on seasonal wind climate in Part 1 of the Guide: the risk of strong winds from depressions in temperate climates being greatest in the winter and the risk from tornadoes in sub-tropical regions and from cyclones in tropical regions being greatest in the summer. Data for the UK are given in the form of the Seasonal Factor, S_S, for various sub-annual periods in Table 9.2 (§9.3.2.4). Use of S_S together with the Statistical Factor, S_T, for an exposure period of one year, $T = 1$ year, gives the design risk of exceedence, P, in the chosen sub-annual period. The designer has two principal decisions to make:

1 What value of design risk, P, is appropriate? If the standard risk of $P = 0.63$ is used with S_S and S_T with $T = 1$, and the standard safety factors are applied, the

risk to the structure during the construction period will be as great as for the whole of the design life. This may not be considered sufficient. The choice of design risk must be made with due regard to the safety of workmen on site and to the public as well as the economics of repair or reconstruction should damage occur during construction.

2 What sub-annual period is appropriate for the construction? In particular, if the lower wind speeds in the calm season are to be exploited, is there a significant risk that problems in construction might delay completion until the windy season? If so, are there contingency plans, such as erecting temporary supporting falsework, that can be implemented?

The second aspect, the structural characteristics of the partially completed building, is in the control of the designer. With careful planning of the erection sequence it will be possible to avoid such problems as unbraced frameworks and unpropped walls, two of the forms most prone to failure during construction [381]. However, situations where major components of structures are in place, but do not yet have their full design strength, are unavoidable.

The third aspect, the loading coefficients during construction, is the principal concern of this Guide. Using the discussion of earlier chapters, particularly Chapters 16, 17 and 18, the designer should be able to identify the appropriate design data in Chapter 20 for any stage in the construction. For example, in the construction stages of a domestic house the walls rise to the first floor level like a rectangular boundary wall; installation of the floor converts this to a cuboid but, as the windows are unlikely to be glazed at this stage, the openings in the walls control the internal pressure and distribution of loading; the walls rise, again unsupported to the final eave level; triangular gable ends are constructed and trussed rafters are installed. The building is now at its most vulnerable, most damage occurring at this stage [382], because the rafters are propped by the gable ends which are free-standing walls, as in Figure 3.16. Progressive collapse of multiple gables can occur in terraces if they are not adequately propped, as shown in Figure 19.27. It is not until the sarking felt has been laid, retained by the tile battens, that the building is aerodynamically closed and the gable ends are propped by the roof structure.

Figure 19.27 Progressive failure of gable walls during construction (courtesy of Oxford Mail)

Addition of the tiles increases the resistance of the roof and, with the windows glazed and the external doors in place, the building is complete. The builder of traditional masonry houses is constrained by this construction sequence and there is very little he can do to reduce the risk. When damage occurs, he must clear the debris and start again. It particularly windy areas it has been known for walls to collapse again after reaching the first-storey height at the second attempt. With timber-framed housing, temporary bracing is usually used to support wall panels and this needs to be adequate, not as in Figure 3.15. Here it can be seen that the walls of one face have been erected to the second-storey level with no cross-walls to provide any shear resistance other than the inadequate temporary bracing. It would not be difficult to devise an erection sequence that provided better resistance to wind than the situation shown in Figure 3.15.

With engineered structures the designer has more control over the construction sequence and is able to minimise the risk to the structure. A sequence in which completed parts shelter vulnerable parts and dominant openings on the windward face are avoided is the best general policy. If this cannot be achieved, temporary structural bracing is required. Large unclad structural frames have been a problem for designers in the UK for some time because the crude lattice frame provisions of the UK wind code, CP3 ChV Pt2[4], does not account for the accumulating shielding of multiple frames and often incorrectly predicts loads on the unclad structural frame to be several times larger than on the completed building. The discussion of lattice frames in Chapter 16 and the design rules in Chapter 20 now dispose of this problem.

19.7 Variable-geometry structures

These are structures that change their shape and hence their wind loads during service. A good example would be a large aircraft hangar which is an enclosed building when the main aircraft access doors are shut, but is virtually an open-sided building when they are fully open. With the advent of larger aircraft, some may be serviced with their tails protruding through partially open doors, giving a dominant opening. Recently, proposals have been made for sports stadia with sliding roofs and other, far more unusual variable geometries are possible for industrial structures.

All is well if the structure is able to resist the wind loads in each of the possible states in the design wind conditions. However, the difference in the loads between hangar doors open and closed, for example, may be so large that it is uneconomic to base the design on the most onerous state. The design is then based on the next less onerous state. There is no intermediate position because a design for loads at some level between these two states will be inadequate for the most onerous state, yet too strong for next less onerous state, and the cost of the extra strength is wasted. This is analogous to the case for cyclone resistant design in regions where cyclones are very rare, discussed in §6.2.3.6. The question becomes one of the risk that the building will be in its most vulnerable state in the design wind conditions. Rules for the use of the structure can be specified, but will they be properly implemented, particularly if there are strong commercial reasons for continuing normal use? For example, opening the large doors of a maintenance hangar at Heathrow to allow an aircraft to be put into service after a major overhaul, while strong winds were blowing directly into the front of the hangar, resulted in failure of the roof glazing. Shards of glass fell onto the aircraft, puncturing the pressure cabin and the wing fuel tanks. A decision had been made for commercial reasons which, in hindsight, was wrong for both commercial and safety reasons. There is also the possibility of mechanical breakdown which might leave the building stuck in its most vulnerable state in the face of an approaching storm.

These are, of course, extreme arguments intended to make the designer consider his options properly. There are a large number of buildings with large doors, etc., which would be vulnerable to wind if they are left open in the design wind conditions, but which are operated satisfactorily. The principle of limit state design is that all these possible states should be considered for the determination of the design limits, even if they are later discarded by a rational decision based on risk.

19.8 Air-supported structures

Air-supported structures are unique in that their stiffness relies on maintaining a difference in pressure between inside and outside that keeps the skin taut. Pressure is maintained inside the structure by means of fans. Typical practical shapes are domes and cylinders with domed ends and these all have a positive external pressure lobe around the windward point. At moderate wind speeds, the internal pressure can be maintained above the maximum external pressure and the skin remains taut, although the whole structure will change shape in response to the external pressure gradients. These changes of shape are also observed on framed structures with flexible skins, such as plastic 'tunnel' greenhouses [289].

The internal pressure rise is fixed by the capacity of the fans, whereas the external pressures depend on the incident wind speed. Where the positive external pressure exceeds the internal pressure, the skin becomes slack and the structure buckles inwards. Damage may occur by the skin snagging on internal structures or by overstressing of the seams when the slack skin flaps in response to atmospheric and building-generated turbulence. Damage also occurs when the structures collapse after interruption of the power supply to the fans.

Air-supported envelopes over sports facilities such as swimming pools and tennis courts can usually be deliberately deflated and safely stowed in response to forecasts of severe winds. Air-supported structures with permanent contents must be provided with sufficient fan capacity to cope with the design wind conditions.

19.9 Fatigue

19.9.1 Introduction

Fatigue is the term used to describe the failure of a material, usually metal, under repeated applications of load below the ultimate static failure load. The fatigue resistance of a structural component is usually defined by determining the number of cycles to failure at various fixed levels of load intensity or stress, to produce a stress-cycle curve of the form shown in Figure 19.28. Here S is the level of stress and N is the number of cycles to failure, so that Figure 19.28 is often called an S–N curve. This curve shows that, as the stress level is reduced, progressively more cycles are required to achieve failure. With ferrous metals the curve eventually levels off after about 10^8 cycles to a constant value of stress, called the fatigue limit, below which fatigue failure does not occur. With non-ferrous metals this limit is never reached and fatigue may eventually cause failure at any stress level.

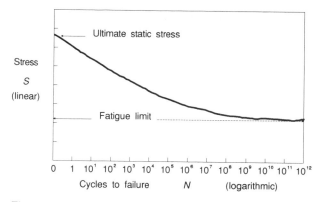

Figure 19.28 Typical stress-cycle fatigue curve

The action of fatigue is believed to be by progressive enlargement of small cracks, that is, by a process of accumulation of damage. Under wind loading the component is subjected to cycles of stress at many different levels. The hypothesis that the amount of accumulated damage at any level of stress is proportional to the number of cycles at that stress, n_i, to the total number for failure, N_i, leads to Miner's rule:

When $n_1/N_1 + n_2/N_2 + n_3/N_3 + \ldots = \Sigma\, n_i/N_i = 1$ failure will occur   (19.12)

Further information on the mechanism of fatigue may be found in most good metallurgy textbooks. In practice, a design limit to the performance of structural components is usually set at below $\Sigma\, n_i/N_i \simeq 0.6$.

Thus a fatigue failure of a component may occur from a very large number of cycles at just above the fatigue limit, sometimes called 'high-cycle' fatigue or a relatively few number of cycles at just below the ultimate static stress, sometimes called 'low-cycle fatigue'. High-cycle fatigue is often caused by a resonant process, typically oscillations driven by vortex shedding over a very long period. This requires the structure to be dynamic and discussion of this aspect of fatigue is reserved for Part 3 of the Guide. Low-cycle fatigue is the principal mechanism of fatigue failure of Static structures, and this aspect is discussed in the remainder of this chapter.

19.9.2 Experience and practice in Australia

19.9.2.1 Preamble

The conditions most conducive to low-cycle fatigue are sustained severe fluctuating loading acting on thin metal components. These are both met in the cyclone-prone regions of Australia, where corrugated-steel sheets are the principal form of roof cladding, so that country has the most extensive experience of fatigue and has evolved the design practice to resist it.   The remainder of this section has been prepared for BRE by Dr G R Walker, at James Cook University of North Queensland, who has been closely concerned with this subject for many years. Dr Walker investigated the damage caused by cyclone Althea to Townsville in 1971 and by cyclone Tracy to Darwin in 1974, and much of §3.3.2.3 and §3.3.2.4 was prepared from his reports. He has also played a major part in the preparation of design regulations in Australia.

19.9.2.2 Failure of roof cladding in Darwin

On Christmas day 1974 the north Australian city of Darwin was devastated by a severe tropical cyclone named Tracy. A large percentage of the buildings were so badly damaged that they had to be demolished and reconstructed and very few escaped with no damage at all [383]. The cause of the damage was the high wind speeds associated with the tropical cyclone. However, with one or two exceptions, it represented a success for the engineering profession as structures fully designed and constructed in accordance with the Australian wind loading code performed remarkably well. The equivalent basic maximum gust wind speed was estimated to be of the order of 65 to 70 m/s, the maximum gust not being recorded owing to instrument failure, which was significantly higher than the equivalent basic design wind speeds which ranged from 49 to 52 m/s.

One of the exceptions was the performance of the metal roof cladding which suffered widespread fastening failures, despite the fixing details having been engineered to the then current codes of practice, including the wind loading code. The failure of this material played a significant rôle in the widespread damage to housing which was the dominant feature of the overall damage pattern.

The principal requirement for metal roof cladding in respect of resistance to wind forces at the time of Tracy was satisfactory performance in a proof test on the roofing system under a static load of 1.8 times the design wind load [384]. This had been introduced following the failure of untested systems in Townsville in 1971 during cyclone Althea which had a maximum basic gust speed of 55 m/s [385]. This should have ensured safety against failure for maximum basic gust speeds up to 75 m/s in most instances, which is significantly higher than the generally accepted maximum in Tracy. It was evident that either the maximum wind speeds had been significantly underestimated (but other observations strongly suggested that this was not so), or the pressure coefficients in the code were much too low, or some other phenomenon was at work. It transpired to be the latter.

After cyclone Althea there had been suggestions that fatigue may be an important factor in the failure of metal roof cladding [385]. Subsequently Beck [386], while testing metal-clad roofing systems in shear, observed significant losses of strength under repeated load conditions from fatigue. In the light of this later work, immediately after Tracy, Morgan and Beck [387] undertook an experimental investigation of the behaviour of the metal-clad roofing systems used in Darwin under repeated loads. The results were astounding. For between 1000 and 3000 load applications they obtained fatigue failures at loads of the order of 15% of the ultimate static load. Furthermore, the failures under repeated load were the result of the formation of distinctive fatigue cracks in the cladding in the vicinity of the fasteners which did not occur under static tests to failure. Subsequent examination of the metal cladding in Darwin after cyclone Tracy revealed evidence of the distinctive fatigue cracking. It was very evident that the major cause of the widespread failure of metal roof cladding was fatigue under the action of the fluctuating wind loads.

As a consequence a major recommendation following cyclone Tracy was that in cyclone areas due account should be taken of the reduction of strength owing to the repeated loading experienced in strong winds and performance tests were developed to account for this [383].

19.9.2.3 General nature of wind-induced fatigue

In this Guide the characteristic features of pressure and forces exerted on buildings have been described. The significant characteristic as far as fatigue is concerned is that they are fluctuating, owing to:

(a) velocity fluctuations in the incident wind arising from atmospheric boundary-layer turbulence;
(b) turbulence induced in the flow past the building due to the interaction with the building, particularly the effects of flow separating from sharp edges and projections;
(c) wind-induced vibration of the building or building components.

Fatigue failures of structural elements generally initiate as small cracks in regions of high-stress concentration following repeated loading. Further load fluctuations cause the cracks gradually to extend in length as a result of successive fatigue failures in the high-stress regions at the crack tips. Structural failure of the elements occurs when the length of the cracks becomes so great that the remaining uncracked sections of the elements can no longer sustain the imposed loads.

Situations in which fatigue failures may occur are therefore generally characterised by:

(a) metallic structural elements;
(b) stress concentrations, generally at connections;
(c) large numbers of repeated moderate to high loads.

Two categories of fatigue problem can be identified in respect of wind loading:

1 Fatigue failure due to fluctuating stress arising when structures or structural elements respond directly to the fluctuating wind pressure without any significant inertial effects, i.e. statically. Such static structures are characterised by high natural frequencies and/or large damping as described in Chapter 10. Light cladding elements, light timber-framed structures and low-rise stiff masonry structures are generally regarded as falling into this category.
2 Fatigue failure due to fluctuating stresses arising from dynamic excitation of structures or structural elements by the fluctuating wind pressure. Such structures are dynamic or aeroelastic depending on the manner in which the excitation occurs.

Estimation of fatigue loads on dynamic and aeroelastic structures requires an investigation of the dynamic response of the structures under wind action. Discussion of this topic will be found in Part 3 of the Guide. The remainder of this section will be concerned with the wind loading and fatigue of static structures.

19.9.2.4 Fatigue loading characteristics of static structures

19.9.2.4.1 Wind climate. In static structures the stresses from wind action are directly proportional to the wind loads, which are in turn proportional to the dynamic pressure or square of the incident wind speed. Design is usually based on an extreme basic wind speed with a very low probability of exceedence. It follows that, in general, commonly occurring wind conditions will not provide fatigue problems as the stresses arising from them are less than the fatigue limit. Therefore attention need be directed only on the likely fatigue under severe wind conditions. For fatigue to be a problem the wind conditions must be sustained for some time. Consequently, if the sources of the extreme wind conditions are thunderstorms, tornadoes or squalls (§5.1.2) fatigue is unlikely to be a problem. However, cyclone Tracy in Darwin demonstrated that fatigue can be a real problem where the severe wind conditions are sustained for a considerable length of time – two to three hours in the case of Tracy.

19.9.2.4.2 Location on the structure.

The frequency content of the pressure fluctuations varies with location on the structure. Regions of positive pressure on the windward faces tend to follow the spectrum of the incident atmospheric boundary-layer turbulence. Regions of suctions on leeward faces and roofs contain additional high-frequency components from building-generated turbulence. The double-peak form of Figure 4.6 appears to be a characteristic feature of leeward surfaces. Spectra obtained for internal pressure fluctuations tend to reflect the characteristics of the location of the dominant opening.

The additional high-frequency components on roofs indicates more rapid fluctuations and hence a higher number of stress cycles during the same event, highlighting the greater susceptibility of roof cladding to fatigue failure over structural elements resisting windward wall pressures.

Location also affects the variation in magnitude of the fluctuations in terms of the probability distribution functions. The fluctuations on the windward walls tend to be normally distributed or Gaussian, whereas the fluctuations in separated regions tend to be exponentially distributed with the standard deviation being higher and giving a larger range of stress. This aspect was discussed in §12.4.3.4.

19.9.2.4.3 Loaded area.

As the loaded area or 'tributary area' for the pressures increases, the influence of small-scale turbulent eddies is reduced by the action of the admittance function (§8.4.2), which effectively damps out the influence of higher-frequency fluctuations. This means that elements with small tributary areas, such as roof cladding connections, are likely to be more at risk from fatigue than those with large tributary areas, such as the connections between roof and walls.

19.9.2.4.4 Simulation of wind fatigue loading.

For fatigue testing of structural components and assemblies to ascertain their adequacy to withstand wind-induced fluctuating loads, a programme of cyclic loading needs to be specified which will reflect the frequency and variation in amplitude of the fluctuations.

Melbourne [388] suggested that two types of loading process be defined: a normally distributed process for elements whose loading is dominated by windward wall pressures or large tributary areas; and an exponentially distributed process for elements in the lee of separation areas and having small tributary areas, such as cladding connections. For each of these processes he suggested a sequence of loading cycles which would be approximately equivalent from a fatigue point of view to the estimated random load fluctuations over an hour at the maximum wind velocities envisaged in design. These were based on an upcrossing analysis (analysis of the number of times any given level of pressure or suction is exceeded) and were defined in terms of the hourly-mean pressure, $\bar{p}$, and the rms of the pressure fluctuations, p', which are in turn defined from the peak value, $\hat{p}$, given a peak

Table 19.1   Proposed load cycles for simulating fatigue in tropical cyclones

Source:	Melbourne [388]	Melbourne [388]	Beck and Stevens [389]
Use:	Positive pressures	Negative pressures	Negative pressures
Rate:	1000 cycles/h	5000 cycles/h	2770 cycles/h
Mean:	$\bar{p} = 0.5\,\hat{p}$	$\bar{p} = 0.25\,\hat{p}$	$\bar{p} = 0.15\,\hat{p}$
Rms:	$p' = 0.125\,\hat{p}$	$p' = 0.094\,\hat{p}$	$p' = 0.115\,\hat{p}$
Composition of cycles in 1 hour:			
	1 @ $\bar{p} \pm 4p'$	5 @ $\bar{p} \pm 8p'$	5 @ $\bar{p} \pm 0.75p'$ to $\bar{p} + 7.25p'$
	10 @ $\bar{p} \pm 3p'$	50 @ $\bar{p} \pm 6p'$	25 @ $\bar{p} + 2.5p'$ to $\bar{p} + 5.5p'$
	100 @ $\bar{p} \pm 2p'$	500 @ $\bar{p} \pm 4p'$	70 @ 0 to $\bar{p} + 5.25p'$
	899 @ $\bar{p} \pm p'$	1000 @ $\bar{p} \pm 2p'$	400 @ $\bar{p} + 0.5p'$ to $\bar{p} + 3.5p'$
		3445 @ $\bar{p} \pm p'$	70 @ 0 to $\bar{p} + 3.25p'$
			1800 @ 0 to $\bar{p} + 1.5p'$
			400 @ 0 to $\bar{p} + 0.5p'$

factor, g, for the process (see §12.4.2). Melbourne's values are given in the first two columns of Table 19.1.

Beck used a modified form of upcrossing analysis, which he described as the 'discriminate range-counting method' [390], to analyse pressure records obtained at a critical roof location during tests in a boundary-layer wind tunnel on a 1:300 scale model of a low-rise building with a 10° pitch roof. On the basis of this analysis Beck and Stevens suggested the values given in the third column of Table 19.1.

The peak pressure, $\hat{p}$, is the estimated maximum peak load which will be imposed on the element during the wind conditions which the element is designed to resist without failure. If these wind conditions correspond with the maximum design wind conditions, the peak pressure corresponds to the design values given by the application of the loading coefficient data of Chapter 20 with the design wind data of Chapter 9. The duration of the wind conditions to be simulated will depend on the nature of the wind climate to be represented. In the case of tropical cyclones, the mean wind speed varies continuously throughout the event. Melbourne suggested that for the extreme load event corresponding to the design ultimate limit state a simulation of three hours at the maximum value would be a reasonable equivalent.

19.9.2.5 Australian design practice

19.9.2.5.1 Darwin reconstruction.   In the light of the findings of Beck and Morgan [391], it was considered essential that any roofing systems approved for reconstruction and new construction in Darwin should be proof tested for fatigue resistance. Because of the urgency of the situation, an interim test was recommended [383] consisting of 10 000 cycles from zero to the design load as defined by the then current Australian wind code [392], followed by a single cycle from zero to 1.8 times the design load, Beck and Morgan [391] having shown that such a test would have indicated the inadequate nature of the fixing systems in use at the time.

This test was incorporated in the Darwin building manual [393], which contained the building regulations for the reconstruction of Darwin. The test was soon made a basic requirement of most building authorities in cyclone-prone areas in Australia. It was a revolutionary change in the design requirements for wind and had a major impact on industrial research and development in the roofing and cladding industry in Australia.

19.9.2.5.2 EBS requirements. In the light of the subsequent work and the suggestions by Melbourne and Beck in Table 19.1, there was a considerable body of opinion that the Darwin test procedure was too severe. The type of test procedures suggested by them were not considered practical but, at a workshop convened by the Experimental Building Station in Sydney, a compromise set of loading cycles was adopted, as given in Table 19.2. The loading cycles should be applied in the sequence given.

Table 19.2 EBS test load cycles for simulating fatigue in tropical cyclones

Composition of cycles in test:	cladding	overall loads
	$8000 @ 0 \text{ to } 0.625\hat{p}$	$800 @ 0 \text{ to } 0.625\hat{p}$
	$2000 @ 0 \text{ to } 0.75\hat{p}$	$200 @ 0 \text{ to } 0.75\hat{p}$
	$200 \ @ 0 \text{ to } \hat{p}$	$20 \ @ 0 \text{ to } \hat{p}$
	$1 \ \ @ 0 \text{ to } \gamma\hat{p}$	$1 \ \ @ 0 \text{ to } \gamma\hat{p}$

where $\hat{p}$ is the unfactored design pressure, γ is the load factor for the ultimate limit state given by: $\gamma = 2.0$ for one test, $\gamma = 1.8$ for two tests and $\gamma = 1.6$ for three or more tests, each test representing about four hours of exposure to severe winds.

These requirements have subsequently been largely adopted in cyclone-prone areas; however some caution needs to be taken in using them. Fatigue testing, by its nature, is an ultimate strength type of test and should be designed to simulate the design ultimate wind conditions. The current Australian test requirements are a curious mix of fatigue tests based on design working loads followed by an ultimate static test.

Fortunately there are some mitigating factors, such as the cyclone factor peculiar to Australian codes, which in effect represents an additional load factor of 1.3, and the apparent conservativeness of the EBS test sequence [389], which should ensure resistance to another event of the magnitude of cyclone Tracy.

Note: since preparing this contribution, Dr Walker has published proposals for a simplified wind code for small buildings in any tropical cyclone-prone region [394]. This proposal presents values of design pressure directly for a small range of critical parameters, e.g. roof slope, with multiplying factors for ground roughness, topography and cyclone intensity. The fatigue test requirements in this proposed code are identical to the EBS test of Table 19.2 with $\gamma = 2$.

19.9.3 Experience and practice in Europe and North America

19.9.3.1 Importance of fatigue in current practice

Perception of the importance of fatigue in current the design practice in the temperate climates of North America and Europe is very different for that in Australia. In these regions, there have been no fatigue failures comparable with those due to Althea and Tracy. The principal cause of wind damage appears to be exceedence of the static ultimate stress.

The discussion of this section compares the experience and practice in Europe and North America with that described above for Australia, concentrating on the differences and trying not to repeat the common aspects already covered. The two cases differ on two major aspects, the wind climate and the structural use of metal claddings.

19.9.3.2 Fatigue characteristics in static structures

19.9.3.2.1 Wind climate.
Although the wind climates differ between temperate and tropical regions, the design wind speed values are not greatly different and, in any case, the design ultimate static stresses are scaled to the design dynamic pressure and so compensate for any difference. The durations of storms are also similar, typically 3–5 hours for the strongest winds. In the storm of 15–16 October 1987 in south-east England, a 250 year return event, the strongest winds lasted for about 3 hours, comparable to cyclone Tracy. The significant difference is in the characteristic product, Π, of the extreme wind climate (see §5.3.1.2) which controls the variability of the values of the extremes. For wind speeds in the UK, Π is relatively high which means that the extremes are not very variable. In cyclone-prone regions, Π is relatively low and the extremes are much more variable. This is illustrated by Figure 6.2, where the slope of the Gumbel plot is greatest for cyclones and smallest for depressions. It means that if a temperate region and a cyclone-prone region have exactly the same design wind speed, the **amount** by which this could be exceeded is much greater in the cyclone-prone region. In the 250 year return storm of 15–16 October 1987 in south-east England, the wind speed was only 9% higher than the design wind speed, giving wind forces about 1.18 times the design values compared with safety factors in the range 1.4 to 1.8. In Tracy, wind speeds were at least 23% greater than the design wind speed and cyclone Althea, giving wind forces between 1.5 and 1.7 times the design values, much closer to the safety factors.

19.9.3.2.2 Structural use of metal claddings.
In the cyclone regions of Australia, where fatigue is clearly a problem, sheet metal roof claddings are prevalent for domestic housing. In North America and Europe their use is principally to clad steel-framed light industrial buildings, using proprietory cladding systems, purlins and fixings. With these systems, the parts of the design are often fixed by factors other than wind loads. For example, the thickness of contemporary profiled steel sheet cladding for roofs in the UK is likely to be set by the imposed load requirements for access. When such sheeting was tested at BRE using the loading sequence given in Table 19.3, below, fatigue failure of the sheeting around the fixings was not induced until the peak pressure reached $-6\,kPa$ on a roof system designed for $-1\,kPa$, representing a safety factor of six. Part of this large safety factor is due to the thickness of the sheeting and part to the relatively high density of fixings in UK practice.

19.9.3.2.3 Glazing.
Glass exhibits characteristics similar to fatigue in metals, in that it undergoes a reduction in strength depending on the level of the stress and also its duration. This latter complicates the reponse to fluctuations by making it dependent on frequency as well as amplitude. This has been considered recently by Holmes[395], who concludes that most of the damage is caused by the largest pressure or suction peaks which occur at infrequent intervals, and this supports the use of a single-peak design value in design. However, Holmes indicates the relevant duration of this loading is very short, $t \simeq 0.2\,s$, i.e. shorter than the standard value of $t = 1\,s$ adopted for cladding in §20.2.3 for which Gust Factors are given in Table 9.13. Current design practice does not adopt gust durations as short as this directly in the loading assessment because the difference from the standard value is effectively included in the strength characteristics published by the manufacturers.

19.9.3.2.4 Assessment of loading cycles. The method for estimating the load cycles on building components in wind storms from the probability distributions of the parent wind climate and the building loading coefficients was given by Davenport[396] more than 20 years ago. The method, which relies on the estimates of upcrossings by Rice[9], is also the basis of dynamic response calculations and is discussed in some detail in Part 3. The method integrates the probability distribution of the wind climate with that of the fluctuations of loading on the structure.

Lynn and Stathopoulos[397] proposed a hybrid Gaussian–Weibull model for the probability distribution function of the fluctuations and computed fatigue lifetimes for a range of wind speeds. Their results highlight the importance of the low-cycle aspect of fatigue in strong winds. For the example they give, an incident mean wind speed of $\overline{V} = 21$ m/s corresponds to the fatigue limit and fatigue will not occur at all. At 10% higher wind speed, $\overline{V} = 23$ m/s, the fatigue life is 20 hours. At 20% higher wind speed, $\overline{V} = 25$ m/s, the fatigue life is only three hours, the typical duration of the strongest winds in a storm. This indicates how abrupt is the threshold between having no possibility of fatigue and having a fatigue failure in a single storm. In analogy to static loading, this is more like a sudden brittle failure than a gradual ductile failure.

Clearly, in view of this apparent 'brittleness' of the fatigue response, it is not wise to be complacent about the possibility of fatigue, especially as the use of metal claddings is a recent innovation in the UK and the population of lightweight all-metal buildings is increasing rapidly. Currently all the national codes of Europe and North America omit low-cycle fatigue for static structures, giving only the ultimate design loads.

19.9.3.3 Simulation of wind fatigue loading

19.9.3.3.1 ECCS recommendations. Conventionally, load cycle sequences are compiled directly from upcrossing analyses of the wind climate and load fluctuations, introduced in §19.9.2.4.4, or are estimated by Davenport's

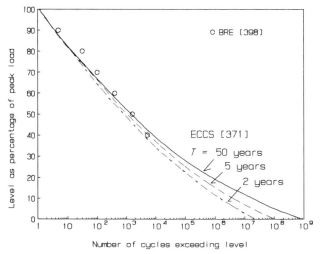

Figure 19.29 ECCS recommendations for simulating fatigue

procedure [396]. In both cases, the number of cycles is effectively counted up from the parent. Although no national codes give any fatigue cycle sequence, the ECCS give recommendations in their model wind loading code [371], derived from Davenport's procedure. These are reproduced in Figure 19.29 in terms of the number of cycles exceeding levels of load expressed as percentage of the peak design load. These recommendations span both the low-cycle and high-cycle ranges.

No advice is given on how a loading sequence should be constructed from these data. It would not be sufficient to start with the low-level cycles and work up to the large because, in nature, the loads cycles will be mixed up randomly and fatigue damage initiated by early cycles of high load is distressed futher by later cycles of low load. Manually controlled cyclic testing machines are unable to cope with a random sequence, so the practical solution is to break up the total number of cycles at each level into suitable batches and reproduce these batches in some random order.

19.9.3.3.2 BRE recommendations.   This 'randomised batch' process was recommended in a paper [398] describing a cyclic pressure test rig developed at BRE. However, in this instance the recommended load cycle sequence was derived by a different synthesis process. Instead of counting up from the parent, the load cycles were derived by counting down from the extreme value distribution of the wind climate and the load fluctuations, using the expressions for Mth highest extremes given in §5.3.1.6. A random number generator was run to give values for the 20 strongest storms in a 50 year period in the UK, then a second random number generator was run to give the Mth highest peak fluctuations in each storm, and the resulting peak loads were counted. This yielded a load sequence comprising cycles of many different levels. This was simplified by rounding up each level to the nearest 10% of the peak load level. The resulting load sequence is given in the first pair of columns of Table 19.3. The load cycles ranges are broken into five batches and given in a mixed order in an attempt to represent the natural mixed occurrence. The cumulative distribution of this sequence is also superimposed on Figure 19.29 and is almost identical to the low-cycle range of the ECCS recommendations.

Table 19.3   BRE test load cycles for simulating fatigue in temperate regions

Mean test sequence for design load		Top-up sequence for 10% load increment	
Apply five times:			
Number of cycles	Percentage of peak load	Number of cycles	Percentage of new peak load
1	90%	1	90%
960	40%	720	40%
60	60%	46	60%
240	50%	180	50%
5	80%	4	80%
14	70%	9	70%
Apply once:			
Number of cycles	Percentage of peak load	Number of cycles	Percentage of peak load
1	100%	1	100%

If a test specimen sustains the full fatigue sequence at the design load without failure, the designer will wish to determine the residual fatigue life by continuing to load. It is not correct merely to repeat the original load sequence because the peak loading for a longer period will be greater. The second pair of columns in Table 19.3 give the loading sequence required to 'top up' the main test sequence at one peak load $\hat{p}$ to give the sequence for a new peak load 10% higher, i.e. at $1.1\hat{p}$. Note that these top-up cycles are given in terms of the new peak load.

The process of incrementing the peak load can be repeated indefinitely. Table 19.4 gives the peak load levels and equivalent exposure periods for subsequent increments, assuming an initial peak design pressure of $\hat{p} = \hat{p}_{50}$ and exposure period $T = 50$ years. The first increment in load is equivalent to a change from $T = 50$ years to $T = 120$ years. With further increments the equivalent return period rises extremely rapidly to very large values that become meaningless. It is far better to use the load factor over the design load as the indicator of safety.

Table 19.4   Increments in the top-up sequence of Table 19.3

Increment number	$n = 1$	2	3	4	5	
New peak pressure	$\hat{p} = 1.10\hat{p}_{50}$	$1.21\hat{p}_{50}$	$1.33\hat{p}_{50}$	$1.46\hat{p}_{50}$	$1.61\hat{p}_{50}$	
Exposure period	$T = 120$	320	900	2800	12 000	(years)

19.9.3.3.3 BRERWULF.

A simple cyclic test rig [398] was constructed at BRE to reproduce the test sequence of Table 19.3. The response of this rig was so rapid that it was thought practical to add some sort of servo-control to follow a preset trace and so reproduce realistic fluctuations of pressure. The result of this development is BRERWULF [354] – the BRE Real-time Wind Uniform Load Follower, principal components of which are represented in Figure 19.30. The test chamber represents the specimen to be loaded and an enclosure to contain the controlled pressure. The pressure control valve was devised and constructed at BRE. Its operating principle cannot be revealed here for commercial reasons, but it is capable of reproducing any pressure in the test chamber in the range ± 8.5 kPa, which represents the

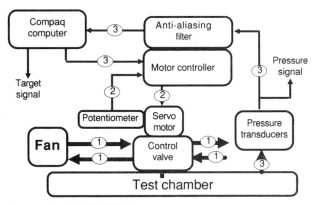

Figure 19.30 BRERWULF system components

pressure rise through the centrifugal fan that continuously circulates air through the control valve from atmosphere in the path marked '1'. The valve position is controlled by a DC servo motor which forms a classical servo-control loop with a feedback potentiometer and motor controller, marked '2'. The command signal to the motor controller is generated from the target pressure trace, which is stored in a microcomputer, by a digital-to-analogue converter (DAC) and is calculated by software, depending on the target and the current pressures in the test chamber. The latter is acquired by two pressure transducers, allowing a choice of position within the chamber for the feedback signal, passed through an anti-aliasing filters and read by a pair of analogue-to-digital converters (ADC). This completes the main feedback loop, marked '3', which includes the servo-motor loop '2' and the air flow loop '1'. Although not strictly part of the servo-control system, a second DAC generates a target pressure signal from the stored pressure trace which can be compared with the achieved pressure to monitor the system performance.

The use of a servo-controlled system like BRERWULF obviates the need to derive equivalent load cycle sequences, like those in Tables 19.1–19.3. BRERWULF is able to reproduce any target trace with a frequency content in the range $0 \leqslant n \leqslant 5\,Hz$ as a uniform fluctuating pressure on the test specimen. The target pressure trace can be obtained in many ways: directly from full-scale measurements in storms; from wind-tunnel simulations; from the anemometer record of a storm, converted to dynamic pressure and multiplied by a suitable pressure coefficient assuming quasi-steady response; or synthesised in some artificial or arbitrary manner. This pressure trace can be up to 53 hours long at the fastest system response, or proportionally longer if it is slowed down. This trace is scaled so that the highest or lowest value corresponds to ± 2048 and stored as a linear array of integer numbers. BRERWULF is able to reproduce this trace with the peak value rescaled to any pressure in the system range $\pm 8.5\,kPa$. Up to 32 768 peak values can be held in another linear array and the trace reproduced for each, so that BRERWULF could repeat a 53 hour long trace at different peak values for 198 years! In practice, tests are much shorter than this because the intervals between storms can be omitted. Typically, testing for a 50 year life can be done by reproducing the 20 highest storms, each of 7 hours duration, taking about 6 days.

The prototype BRERWULF was developed principally as a research tool to investigate the response of multi-layer cladding systems to fluctuations in wind loading, as described in §18.8. However, its range which is greater than three times the maximum pressure predicted by the UK wind loading code of practice [4] and also greater than loads predicted in tropical cyclones, makes it a suitable apparatus for realistic proof testing of structural systems for fatigue. A new application, not previously attempted in this field, is 'forensic' investigations, that is the reconstruction in the laboratory of the effects of past storms, such as the 15–16 October storm in the UK or cyclones such as Althea and Tracy. The nearest analogue to this process is the testing of structural systems for earthquake resistance on shaking tables, using traces of past earthquake events.

20 Design loading coefficient data

20.1 Introduction

In this chapter, the design loading coefficient data for the building forms covered in earlier chapters are presented. The loading coefficient data are used in conjunction with the design wind speed data of Chapter 9 in Part 1 of the Guide, according to the rules set out below. In fact, Chapters 9 and 20 contain all the necessary data for the assessment of wind loads on static structures, but Chapter 10 is also required to confirm the static classification and to give the value of the dynamic amplification factor, γ_{dyn}, if a mildly dynamic structure is to be assessed as if it were static. Nevertheless, the other chapters of the Guide give the background theory and discussion, and it is always worthwhile reviewing the discussions of the appropriate form of the structure, so suitable cross-references have been included in the text. The data given in this chapter may be more detailed than necessary for simple small structures, in which case the simpler codified data of Appendix K can be used.

After using Parts 1 and 2 of the Guide for a little while, the designer is likely to become more familiar with the design wind speed data of Chapter 9 than with the loading coefficient data of this chapter, since most of the design wind data are required for every structure, while only small parts of the loading coefficient data are relevant to any given structure. Anticipating that the experienced user will jump straight to this chapter, the opportunity has been taken in the following section to include a brief review the initial steps that must be taken at each assessment.

20.2 Initial steps

20.2.1 Structural forms

The first, and most important step, is to identify the form of the structure so that the correct loading coefficient data can be located. 'Form of the structure' means the external shape which determines the characteristics of the flow and the consequent wind loads. It does not mean the structural form by which the structure derives its strength and stability, e.g. steel frame, masonry shear walls, reinforced concrete core, etc., which is irrelevant to the size and distribution of the wind loads, although it is relevant to the load paths (§19.3) by which the structure transmits these loads to the foundations.

Figure 20.1 summarises the main categories of form for which data are given in this chapter. There is some overlap between many of these categories: some may be

FORM

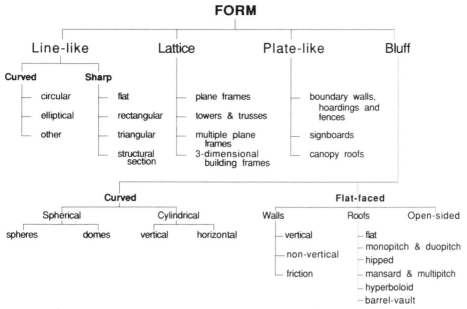

Figure 20.1 Forms of structure for which data are given

included twice – for example, lattices of low solidity are merely a collection of line-like elements and can be treated as such, although this involves much more work for the designer; alternatively, sometimes an arbitrary break is made between categories – for example, the load characteristics of steeply pitched roofs merge with non-vertical walls so an arbitrary break is made at a pitch of 45°.

The user may find that the data do not correspond exactly with the form of his building, in which case he has two choices: he can interpolate or extrapolate the given data to the required form using his experience and commonsense, or he can seek expert advice. In the latter case, it may be that data do exist, but were deemed too specialised for the Guide, or that new data have become available. If the form of the building does not correspond to any of the forms in Figure 20.1, the designer will need to seek expert advice immediately. It might be necessary to resort to wind-tunnel tests, but with most small buildings it will generally be possible to specify upper bounds to the loads which will give a safe design. Another possible choice, that of changing the design to match the available data, could be construed as showing undue caution and a lack of imagination. On the other hand, modifying the form in order to minimise the wind loading, as described in §19.5, is an entirely different matter showing imagination and initiative. Owing to the difficulty in distinguishing between these last two approaches, the designer should always take the opportunity to claim the due credit for the latter.

20.2.2 Orientation of the structure

The next step is to determine the angle between the principal axes of the structure and the direction of North. This is needed because all the assessment of the wind data in Chapter 9 is made in the twelve 30°-wide sectors based on the wind direction, Θ, measured from North, while the loading coefficient data of this chapter are given in terms of the wind angle, θ, relative to the principal body-axis of the structure or the local wind angles, θ_i, relative to the faces of the structure.

The designer will probably find it more convenient to redefine the wind data in terms of the body-axis wind angle, θ. The simplest way to do this is to mark the twelve sectorial wind directions, Θ, onto the site plan of the building, centred on the origin of the body axes.

20.2.3 Influence functions and load duration

20.2.3.1 Influence functions

The influence functions for static structures were defined in §8.6.2.1. They define the way that the wind-induced pressures acting on the surface of the structure accumulate into the forces and moments acting on the structural members. The consideration of load paths in structures in §19.3 is also relevant here. The designer should use these concepts to determine the relevant loaded areas for integration of pressures into forces and moments.

20.2.3.2 Load duration

20.2.3.2.1 *TVL*-formula. The design approach used by the Guide for static structures uses low-pass filtering of the load fluctuations to represent the way that the effect of eddies smaller than the loaded area cancel out. This is called the 'equivalent steady gust model' and was discussed in §8.6.2.3, §12.4.1.3 and §15.3.5.

The simplest implementation of this model is the *TVL*-formula, Eqns 12.37 and 15.27, reproduced here again as:

$$t = 4.5\, l\,/\, \overline{V} \qquad\qquad (20.1)$$

Note that the mean wind speed is used here. The value of load duration, t, is used to determine the corresponding duration of gust wind speed, through the Gust Factor, S_G. (This is why Eqn 20.1 is framed in terms of the mean wind speed and not the gust speed, since determination of the gust speed requires that t be already known.) The resulting gust speed is not very sensitive to the value of the characteristic size, l, and, rather than make use of the complex hierachy of size parameters in §8.6.2.2, the characteristic size is taken as the diagonal of the loaded area through Eqn 15.28. Figure 20.2 gives the key to this procedure in the form of typical examples. This approach has been adopted for the new UK wind loading code, BS6399 Part 2, which is due to replace the current CP3 ChV Pt2 [4] sometime around 1989–90.

It is considered appropriate to adopt $t = 1\,$s as the minimum load duration for design under normal circumstances, even if the *TVL*-formula indicates a shorter duration. The principal reason for this is that durations shorter than $t = 1\,$s correspond to elements smaller than about $l = 5\,$m. These tend to be individual system components, like cladding panels, which have inherent load-sharing capacity within the component. This capacity is usually incorporated into the design strength specifications by the manufacturers in terms of the $t = 1\,$s duration loads specified by codes of practice, such as CP3 ChV pt2 [4] (see §14.1.2.1). Whatever the critical load duration really is, the strength of the element has been assessed *as if it were $t = 1\,$s*, so that the quoted design strength is lower than the actual strength.

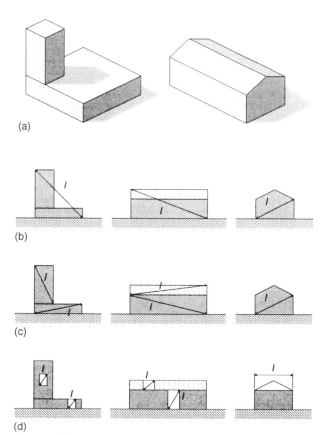

(a)

(b)

(c)

(d)

Figure 20.2 Examples of characteristic size: (a) example buildings; (b) base shear and moments; (c) roof and wall faces; (d) cladding and glazing panels

This form of standardised performance calibration is widespread in UK practice because of the very long period that the UK wind code has been in force. The conseqence is that other codes, such as the code for slating and tiling BS5534, and the masonry code BS5628 have used the $t = 1\,\text{s}$ duration load to calibrate their design procedures. Using the actual critical load duration would require changing all these design procedures to avoid invoking the same load factors twice.

20.2.3.2.2 Standard values of load duration.   For most conventional buildings the determination of the load duration through the TVL-formula can be substituted by three standard values:

1 For cladding and its immediate fixings and individual members of lattice structures; structures and components of structures with $l \leqslant 5\,\text{m} - t = 1\,\text{s}$.
2 For structures or components of structures with $l = \quad 25\,\text{m} - t = \quad 4\,\text{s}$.
3 For structures or components of structures with $l \geqslant 100\,\text{m} - t = 16\,\text{s}$.

These correspond approximately to the Classes A, B and C of static structures in the classification procedure of Chapter 10 and to the classes A, B and C of the UK code, CP3 ChV Pt2[4]. Gust factors for these standard durations are given in

Tables 9.13, 9.14 and 9.15. Interpolation may be used between A and B for the range $1\,\mathrm{s} < t < 4\,\mathrm{s}$, and between B and C for the range $4\,\mathrm{s} < t < 16\,\mathrm{s}$. (**Note**: as the durations increase by factors of four between classes, logarithmic interpolation is the most appropriate.)

20.2.3.2.3 Dynamic amplification factor. For structures in the mildly dynamic class D1, Chapter 10 gives a value of the dynamic amplification factor, γ_{dyn}, in Figure 10.9 to be used on the design loads derived by a static assessment. The recommendation in Chapter 10 was that the boundary between mildly dynamic D1 and fully dynamic D2 should be set at $f\,F = 1$, hence the position of the boundary drawn in Figure 10.9. Subsequent calibration calculations for the incorporation of the classification method into the new UK wind code, BS6399 Pt2, completed after the publication of Part 1, have shown that the boundary can safely be raised to $f\,F = 1.5$.

As the value of the dynamic amplification factor, γ_{dyn}, is derived in terms of the Class A loads in §12.5 the load duration should strictly be taken as $t = 1\,\mathrm{s}$. However, the classification procedure was designed deliberately to be conservative so that no potentially dynamic structures were inadvertantly missed. The calibration calculations for BS6399 Pt2 covered a wide range of typical buildings and showed that the inherent conservatism of the classification method is almost exactly balanced by the variation of load duration with characteristic size, l.

Accordingly, it is now recommended that the advice for mildly dynamic structures given in §10.8.2.1 be modified so that the dynamic amplification factor, γ_{dyn}, is applied to the loads assessed in the same manner as static structures of the same characteristic size, that is with the load duration given by Eqn 20.1.

20.2.3.2.4 Area-averaged loading coefficients from other sources. Other sources of data may use direct area averaging of pressures (see §13.3.3.1) or directly measured peak values of forces and moments. In these cases, the correlation of small eddies over the surface is automatically included in the measurement and the load duration predicted by the *TVL*-formula for the equivalent steady gust model is not appropriate. The maximum response of any structure is produced by a load duration equal to half the natural period of oscillation, i.e. for $t = 1/(2n)$. For static structures, this will be a very short duration and certainly less than $t = 1\,\mathrm{s}$. The argument in §20.2.3.2.1 for a minimum load duration of $t = 1\,\mathrm{s}$ also applies to such data.

20.2.4 Design dynamic pressures

The design dynamic pressure is a peak value calculated for each wind angle, θ, from the gust wind speed of the required duration at the reference height as:

$$\hat{q}_{ref}\{\theta, t, z_{ref}\} = \tfrac{1}{2}\,\rho\,\hat{V}^2\{\theta, t, z = z_{ref}\} \tag{20.2}$$

which is the Bernoulli equation, Eqns 2.6 and 12.1, formulated in terms of the reference values. (**Note**: the assumption has been made that the wind direction relative to North, Θ, has been transformed to the wind angle, θ, in body axes as instructed in §20.2.2.) Care has been taken to define the reference height, z_{ref}, in every case for which data are given, firstly in the main text of the chapter, secondly in the key diagram and thirdly at the head of the data table or graph containing the data.

Values of $\hat{V}^2\{t, z = z_{ref}\}$ are derived from the data of Chapter 9, or from the first two supplements to Part 1: Supplement 1 – 'The assessment of design wind speed data: manual worksheets with ready-reckoner tables' and Supplement 2 – 'BRE program STRONGBLOW' (see front flyleaf for details). Alternatively the simplified procedure set out in Appendix K can be adopted, but will give a conservative result. (This simplified procedure has been adopted for the new UK code, BS6399 Pt2.)

20.2.5 Format of the design data

20.2.5.1 Loading coefficients

The equivalent steady gust model has been adopted as the standard for the design data of the Guide. Wherever possible, the data have been derived from peak values and presented in the pseudo-steady format described in Chapter 15, in terms of the global or local coefficients as a function of wind angle, denoted by $\tilde{C}\{\theta\}$ or $\tilde{c}\{\theta\}$. Where peak data are unavailable, mean values have been used, presented in the quasi-steady format described in §12.4.1, in terms of the coefficients denoted by $\overline{C}\{\theta\}$ or $\overline{c}\{\theta\}$. The user need not be concerned with the two types of derivation because they are implemented identically, the only difference being in the accuracy of the model.

Thus for design pressures:

$$\hat{p}\{\theta\} \equiv \hat{q}_{ref}\{\theta, t, z_{ref}\}\, \tilde{c}_p \qquad (20.3)$$

$$\hat{p}\{\theta\} \simeq \hat{q}_{ref}\{\theta, t, z_{ref}\}\, \overline{c}_p \qquad (20.4)$$

which are Eqns 15.26 (§15.3.5) and 12.40 (§12.4.1.5), respectively, formulated in terms of the reference values. The only difference between these two equations is the use of the pseudo-steady coefficient in Eqn 20.3, resulting in an equivalence ($\equiv$), and the mean coefficient in Eqn 20.4, resulting in an approximation ($\simeq$). This distinction having been formally made, will now be dropped for convenience and later equations will show an equality ($=$). Design pressures act normally to the loaded surface and may be integrated to give overall forces and moments.

Force coefficients are given directly for line-like, lattice and plate-like structures. The definitions are as given in Eqns 12.13, 12.14 and 12.15, where the reference areas are the loaded areas. Thus:

$$\hat{F}_x\{\theta\} = \hat{q}_{ref}\{\theta, t, z_{ref}\}\, \tilde{c}_{F_x} A_x \qquad (20.5)$$

$$\hat{F}_y\{\theta\} = \hat{q}_{ref}\{\theta, t, z_{ref}\}\, \tilde{c}_{F_y} A_y \qquad (20.6)$$

$$\hat{F}_z\{\theta\} = \hat{q}_{ref}\{\theta, t, z_{ref}\}\, \tilde{c}_{F_z} A_z \qquad (20.7)$$

established the general principal for the three orthogonal axes. As the reference areas, A_x, A_y and A_z, are the loaded areas, the defininitions are effectively identical to that of the design pressure. The exception to this rule is the special format used in the 'reference face method' for lattice towers and trusses (§20.4.4.1), when the reference area is the area of the reference face.

Moment coefficients are not used directly. Instead, the position of the force is given for tall buildings to enable the overall base moments to be calculated from the overall base shear. For other forms of structure, moments are calculated from integration of the surface pressures or from summation of the force coefficients for the structure broken down into individual loaded areas.

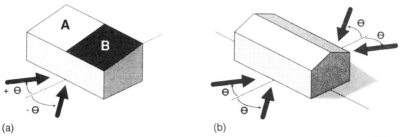

Figure 20.3 Examples of symmetry: (a) one degree of symmetry – monopitch building; (b) two degrees of symmetry – duopitch building

20.2.5.3 Symmetry

Most common forms of structure have at least one, and many have two degrees of symmetry, and this has ben exploited to reduce the amount of data required. Figure 20.3 shows two examples:

(a) A monopitch roof on a rectangular-plan building has one degree of symmetry. The loading on the half of the roof marked 'A' at wind angle $+\theta$ is the same as the loading on the half marked 'B' at wind angle $-\theta$. Only half the full range of data are unique. Contours of pressure on the building for the two wind angles shown are mirror images about the line of symmetry.

(b) A duopitch roof on a rectangular-plan building has two degrees of symmetry. Now only a quarter of the full range of data are unique. Contours of pressure on the building for the four wind angles shown are mirror images about the two lines of symmetry.

Thus, with one degree of symmetry, the amount of data presented can be halved and, with two degrees of symmetry, quartered. Either data are given for a half or a quarter of the building for all wind angles, $0° \leqslant \theta \leqslant 360°$, or data are given for the whole building for half the range of wind angles, $0° \leqslant \theta \leqslant 180°$, or a quarter of the range of wind angles, $0° \leqslant \theta \leqslant 90°$. The former choice is usually made for wind-tunnel studies, since this requires fewer pressure tappings to be installed in the model. For design purposes, it is better to present the data for the whole building face over the range of unique wind angles.

Accordingly the data of this chapter are presented for loaded areas covering the whole building face for one or two quadrants of wind angle. The designer is required to complete the assessment for all other relevant wind angles by employing the mirror image effect of symmetry. Most codes of practice, including the UK code [4], employ this device.

A few curved structures are axisymmetric about the vertical axis; these include vertical circular cylinders, such a stacks or silos, and spherical domes. In these cases all wind directions are effectively the same, and only one set of loading coefficients is required.

20.2.5.4 Loaded areas

In the case of plate-like and bluff structures, the loading coefficients vary with position. Only in the few, simplest axisymmetric cases has it been possible to give contours of loading coefficient over the external surface. In the vast majority of cases the data are broken down into suitable 'loaded areas', with global values ascribed to the whole area. Sometimes the difference between the short-duration cladding loading coefficients and the longer-duration structural loading coefficients is not large, i.e. the quasi-steady model works well, and a single value can be given. At other times it is necessary to give separate values for structral and cladding components.

Coefficients categorised 'structural loads' should be assumed to act uniformly over the whole loaded area for determining loads in structural members. However, in reality the loading will not change abruptly and, if the assumption results in unrealistically high stresses in members, the steps in loading should be smoothed out. Figure 20.4 gives a typical example of this procedure. Overall loads are

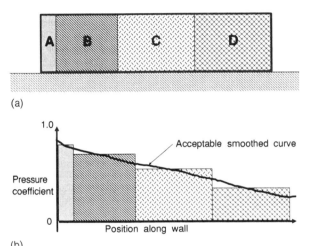

(a)

(b)
Figure 20.4 Example of loaded areas for the wall of a squat building: (a) loaded areas for walls; (b) pressure coefficients for wind angle 45°

obtained by summing the loads on the individual loaded areas. Overall moments may be obtained by assuming the load on each loaded area acts through its centroid of area. (**Note**: This differs from the procedure in the UK code, CP3 ChV Pt2 [4], where local high load areas are excluded from the summation for overall loads. The reason for this change is that the distribution of loading is better represented, leading to better estimates of forces and moments at critical points in the building structure.)

Coefficients categorised 'peak cladding loads' are local values which will act somewhere in the loaded area, but not simultaneously over the whole area, and should be used to determine cladding fixings within the loaded area.

20.2.6 Coefficients from external sources

The data given below are believed to be as comprehensive as is currently possible while still adhering to the required standard of accuracy. The data are also completely self-consistent through the use of the pseudo-steady format (§15.3.2) to adapt both peak and mean data to the equivalent steady gust model. The collection of loading data for new forms and for previously studied forms in better simulations of the atmospheric boundary layer is a continuous process as wind engineers strive to increase the range and accuracy of the available data. The designer may wish to augment the data given below with data from such sources.

The first step is to ensure that the data quality is sufficient by applying the guidelines for commissioning wind-tunnel tests given in Appendix J. Problems may occur when insufficient information is given in the published work to make this judgement, when expert advice should be sought.

The next step is to relate the form of the data to the pseudo-steady format used here. For this it is essential to know the reference height, z_{ref}. If this is close to the height of the structure and the terrain roughness of the experiment is closely similar to the relevant site, the data may be used directly. Otherwise, it is also necessary to know the terrain roughness in terms of the aerodynamic roughness, z_o, or the power-law exponent, α, in order to transform the data. If the data use a reference at the gradient height, z_g (§7.1), large errors can occur unless very great care is taken in the analysis. Figures 7.6 and 7.7 illustrate the large difference in wind speed that occur near the ground for the same gradient wind speed, owing to changes in terrain roughness.

If the designer is extremely fortunate, the new data may already be in the form of pseudo-steady coefficients, $\tilde{c}$. The use of this form as the standard in the Swiss code [182] and the draft for the UK code BS6399 Pt 2 will be more incentive to adopting this standard than just the recommendation of BRE through this Guide. In this case, the data can be used directly in conjunction with the design dynamic pressure at the specified reference height.

More likely, the data will be in the form of the mean, $\bar{c}$, or the peak maximum, $\hat{c}$, or peak minimum, $\check{c}$, coefficients. In the case of the mean, the data can again be used directly, but the resulting estimates of peak design loads will be less accurate. Instead, the peak coefficients should be transformed to pseudo-steady values using Eqn 15.23. This requires that the Gust Factor, S_G, be given in the source data or estimated using the data of Chapter 9. The typical range for Gust Factor is $1.6 \leqslant S_G \leqslant 2$ and this acts on wind speed. The factor on dynamic pressure, relating the peak coefficients to the psuedo-steady coefficients, is therefore the square, in the range $2.5 \leqslant S_G^2 \leqslant 4$.

20.3 Loading data for line-like structures

20.3.1 Scope

This section deals with structures that have their structural and aerodynamic properties concentrated along a line. In practical terms, this is defined as structures whose length, L, is at least eight times greater than their cross-wind breadth, B, or diameter, D. For vertical cantilevers from the ground, the effective length is twice the height, $L = 2H$ (§16.1.2). However, the range of data presented in this section often extends through the two-dimensional line-like range into the three-dimensional bluff range.

Typical line-like structures are industrial chimney stacks, masts, cables, portal frames, portal cranes, and other long structural sections. Many line–like structures, particularly cantilevers or gravity-stiffened structures, will be dynamic and the loading data derived here will be used in the dynamic assessment methods of Part 3.

20.3.2 Definitions

Principal dimensions of line-like structures are:

1 Length, L, or, for vertical structures protruding from the ground plane, height, H, both measured along the long axis.
2 Cross-wind breadth, B, and inwind depth, D, of the cross-section normal to the long axis. In the case of circular cylinders, $D = B$ and is the section diameter.

The aerodynamic effects of these dimensions are described by the two non-dimensional parameters: slenderness ratio, L/B or $2H/B$; and fineness ratio, D/B.

Loading coefficients are defined as acting on the total solid area in projection and data are given in the form of the mean local force coefficients: the drag coefficient, $\overline{c}_D$, for axisymmetric structures (e.g. vertical circular cylinders); and the body-axis force coefficients, $\overline{c}_{F_x}$ and $\overline{c}_{F_y}$, for other structural forms. These are all mean values, applied using the quasi-steady model. The local coordinate convention was defined in Figure 16.6: with the x and y axes as the principal section axes; the z axis as the long axis; the pitch angle, α, as the rotation around the long z axis from parallel to the x axis; and the yaw angle, β, as the angle from normal to the long z axis. The subscript 'o', as in $\overline{c}_{D_o}$, is used to denote the datum value when $\alpha = 0°$ and $\beta = 0°$. In this example $\overline{c}_{D_o}$ is the local drag coefficient with the wind aligned normal to the long z axis and parallel to the x axis. Overall forces and moments are obtained by integration of the local force coefficients through Eqn 16.1 for the mean values and through Eqn 16.2, using the quasi-steady model, for the peak values.

The reference wind speed is the local gust speed, $\hat{V}\{z\}$, acting at the local height, z, above ground. The cumulative effect of the variation of wind speed over the height of a vertical or inclined line-like structure can be accounted for in three ways:

1 Use an influence function to express the weighting on dynamic pressure at any height. This is the influence coefficient, $\phi_q\{z\}$, given by Eqn 16.29.
2 Take the dynamic pressure at the top of the structure to apply to the whole structure. This will always be conservative. The degree of conservatism will be very large for vertical structures, such as industrial chimney stacks, but small for horizontal structures, such as pipe-lines.
3 Divide the structure into a number of small sections, so that the variation of dynamic pressure down each individual section is small, and take the dynamic pressure at the centre of the section. Aggregate the individual section loads to give overall forces and moments. This is the best compromise between simplicity and accuracy.

20.3.3 Curved sections

20.3.3.1 Circular sections

20.3.3.1.1 Smooth and rough cylinders, subcritical flow. Use $\overline{c}_{D_o} = 1.2$ for normal flow ($\beta = 0°$) when $D\hat{V} < 6\,\text{m}^2/\text{s}$ (§16.2.2.1.1), where D is the cylinder diameter and $\hat{V}$ is the design gust wind speed.

20.3.3.1.2 Smooth cylinder, supercritical flow. Use $\overline{c}_{D_o} = 0.6$ for normal flow ($\beta = 0°$) when $D\hat{V} > 6\,\text{m}^2/\text{s}$ (§16.2.2.1.1). The distribution of pressure around the cylinder can be obtained from §20.6.3.2.

20.3.3.1.3 Rough cylinder, supercritical flow. $\bar{c}_{D_o}$ varies with surface roughness as given in Figure 20.5 and empirical equation:

$$\bar{c}_{D_o} = 1.176 - 0.0857 \log(k_s/D) - 0.0902 \, [\log(k_s/D)]^2 \\ - 0.00976 \, [\log(k_s/D)]^3 \tag{20.8}$$

for the range $10^{-5} < k_s/D < 10^{-1}$. The Reynolds number for the onset of supercritical flow, Re_{crit}, reduces with increasing roughness to lower values of $D\hat{V}$ as indicated in Figure 20.5. Use of the criterion $D\hat{V} > 6 \, \text{m}^2/\text{s}$ is always conservative. The equivalent sand grain roughness k_s is given in Figure 20.6 for various surface finishes.

20.3.3.1.4 Effect of length. The overall drag, $\overline{C}_{D_o}$, of circular cylinders in subcritical flow reduces with slenderness ratio (§16.1.2), while the drag in

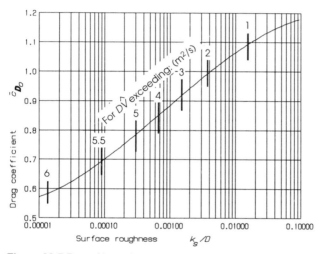

Figure 20.5 Drag of long circular cylinders normal to flow at supercritical Reynolds numbers

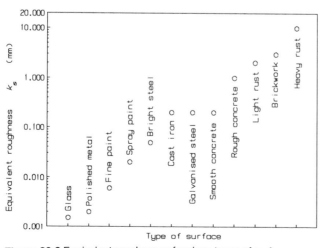

Figure 20.6 Equivalent roughness of various types of surface

supercritical flow remains sensibly constant. The factor to apply to $\bar{c}_{D_0}$ in **subcritical flow only** is given by the empirical equation:

$$\phi\{L/D\} = \bar{C}_{D_0} / \bar{c}_{D_0} \qquad \text{when } D\hat{V} < 6\,\text{m}^2/\text{s (subcritical)}$$
$$= 0.5 \qquad\qquad\qquad \text{for } L/D \leqslant 1.0 \text{ (short cylinders)}$$
$$= 0.5 + 0.3843 \log(L/D) \quad \text{for } 1.0 < L/D < 20 \qquad\qquad (20.9)$$
$$= 1.0 \qquad\qquad\qquad \text{for } L/D \geqslant 20 \text{ (long cylinders)}$$

for $D\hat{V} < 6\,\text{m}^2/\text{s}$, where the length L is the distance between free ends, so that the equivalent length for a cylinder protruding normal from a ground plane is $L = 2H$ (§16.2.2.1.3). The local drag coefficient increases towards a free end or the ground plane as given in Figure 16.10. Rules for changes in diameter, D, are given in §16.2.2.1.4.

20.3.3.1.5 Effect of pitch and yaw angles.   Owing to axisymmetry, pitch angle has no effect on drag, so resolved body axes forces follow simple sine–cosine rules. The resultant force remains normal to the cylinder axis with changes of yaw angle, but the effective fineness ratio (§16.1.3) also changes as discussed in §16.2.4.2. Body-axes forces are given by Eqns 16.25 and 16.26. Care should be taken to account for the additional cross-wind force when sections with a free end are yawed pointing into the wind (§16.2.4.3). With yawed stranded cables check for the effect described in §2.2.10.2 (§16.2.2.1.2).

20.3.3.1.6 Effect of axial protrusions.   Prominent weld lines, ladders, icing and other forms of axial protrusion can produce cross-wind lift forces and changes of drag in the range given in Figure 16.12 (§16.2.2.1.6).

20.3.3.1.7 Effect of ground plane.   The drag and lift coefficients for a long smooth cylinder close and parallel to the ground is given in Table 20.1, based on the data in Figure 16.13 (§16.2.2.1.6), where G is the gap distance between the bottom of the cylinder and the ground and D is its diameter.

Table 20.1   Long smooth cylinder close and parallel to the ground

	Ref $\hat{q}$ at centre of cylinder ($z_{ref} = G + D/2$)			
Gap to ground	Subcritical		Supercritical	
G/D	$\bar{c}_{D_0}$	$\bar{c}_{L_0}$	$\bar{c}_{D_0}$	$\bar{c}_{L_0}$
0	0.8	0.6	0.6	0.84
0.4	1.45	0.15		
1.0	1.2	0.0	0.5	0.0

20.3.3.1.8 Effect of porosity.   Porosity induces flow separation and raises the drag coefficient based on the projected solid area to $\bar{c}_{D_0} = 1.44$ (§16.2.2.1.7). Treat very porous cylinders, $s < 0.3$, as lattice trusses.

20.3.3.2 Elliptical sections

The drag of elliptical cylinders with the long axis normal to the flow ($\beta = 0°$) and either the major ($\alpha = 0°$) or minor ($\alpha = 90°$) cross-section axis normal to the flow varies in the range $0 < \bar{c}_{D_o}\{D/B\} < 2.0$ according to the fineness ratio (§16.1.3) as given by Eqn 16.6 (§16.2.2.2). The effect of pitch angle α on drag and lift at $\beta = 0°$ is given by Eqns 16.7 and 16.9. The cosine model can be assumed for the effect of yaw angle, β, together with the effective fineness ratio.

20.3.3.3 Other curved sections

Values of force coefficients for other curved sections, e.g. square cylinders with rounded corners, are transitional between the elliptical cylinder and the rectangular section of the same fineness ratio (§16.2.2.3). Treat stranded cables as rough circular cylinders. Most available data for other shapes are given in Reference 399.

20.3.4 Sharp-edged sections

20.3.4.1 Flat plates

Pressures acting on a flat plate can only generate a force normal to the plate. This is given for pitch angle, α, in Figure 16.20, and by Eqns 16.12 or 16.13 (§16.2.3.1).

20.3.4.2 Rectangular sections

The drag of rectangular sections with flow normal to one face is given by the upper curve in Figure 16.5 as a function of fineness ratio. The variation with pitch angle is given in Figure 16.22 (§16.2.3.2). The simple design model of §20.3.4.8 is generally appropriate for typical solid and box structural members.

20.3.4.3 Triangular sections

The drag of triangular sections with the apex facing the flow is given in Table 16.1 (§16.2.3.3). Use the simple design model of §20.3.4.8 for triangular sections with a face normal to the flow.

20.3.4.4 Structural sections

Force coefficients for the x and y body axes of various structural sections are given in Figures 16.22 to 16.25 (§16.2.3.4). Owing to the large variation of value with pitch angle exhibited by these figures, the tabular approach started by the 1956 Swiss code [400] and subsequently copied by most codes, including the 1972 UK code [4] and the newest draft code for Australia [401], is not particularly useful. If the detail of Figures 16.22 to 16.25 is not exploited, it is recommended that the simplification of the simple design model in §20.3.4.8 should be used. Even more detailed data are available in ESDU data item 82007 [240].

20.3.4.5 Effect of length

The overall drag, $\bar{C}_{D_o}$, for sharp-edged sections reduces with slenderness ratio in a similar fashion to circular cylinders (§16.2.3.5). The factor, $\phi\{L/B\}$, to apply to $\bar{c}_{D_o}$ is given by the empirical equation:

$$
\begin{aligned}
\phi\{L/B\} &= \bar{C}_{D_o} / \bar{c}_{D_o} \\
&= 0.6 && \text{for } L/B < 1.8 \text{ (short sections)} \\
&= 0.5 + 0.384 \log(L/B) && \text{for } 1.8 < L/B < 20 && (20.10) \\
&= 1.0 && \text{for } L/B \geqslant 20 \text{ (long sections)}
\end{aligned}
$$

20.3.4.6 Effect of yaw angle

The effect of yaw angle on sharp-edged structures is given by the simple cosine model (§16.2.4). The resultant force remains normal to the long axis of the member. Use the simple design model equations, Eqns 20.11 and 20.12 with the actual loading coefficient values for zero yaw in place of the default value of 2.0. Care should be taken to account for the additional cross-wind force when sections with a free end are yawed pointing into the wind (§16.2.4.3).

20.3.4.7 Effect of porosity

Porosity reduces the loading coefficients on typical sharp-edged sections with low fineness ratio by suppressing vortex shedding, but the forces on fine sections may increase (§16.2.3.6). If there is a significant risk that the holes could be blocked by ice in the design wind conditions, no advantage should be taken of the reduced drag. Otherwise the equation for lattice frames, Eqn 16.34 with $\overline{C}_{D_1} = 2.0$, or the corresponding curve in Figure 16.38 can be used (§16.3.3.1.1, §20.4.3).

20.3.4.8 Simple design model

A simple, robust design model is given taking $\overline{C}_{D_0} = 2.0$ as the worst case and the cosine model for pitch and yaw. This leads to:

$$\overline{c}_{F_x}\{\alpha, \beta\} = 2.0\,\phi\{L/B\}\cos\alpha\,\cos\beta \tag{20.11}$$

$$\overline{c}_{F_y}\{\alpha, \beta\} = 2.0\,\phi\{L/B\}\sin\alpha\,\cos\beta \tag{20.12}$$

where $\phi\{L/B\}$ is given by Eqn 20.10. The limits of the term $(2.0\,\phi\{L/B\})$ are 2.0 for long sections and 1.2 for short sections. Care should be taken to account for the additional cross-wind force with angle sections (§16.2.3.4) and when sections with a free end are yawed pointing into the wind (§16.2.4.3).

20.4 Loading data for lattice structures

20.4.1 Scope

This section deals with porous structures composed of a lattice of many structural members, including lattice frames, trusses, towers, cranes, masts and gantries. Important members of this population are building frames while still unclad, during the construction process.

20.4.2 Definitions

Principal dimensions of a lattice are:

1 Length, L, or height, H, of the envelope of the lattice.
2 Cross-wind breadth, B, and inwind depth, D, of the envelope of the lattice.
3 Solidity ratio, s of the lattice, or any part; defined as the ratio of the solid area of all members in projection to the **envelope** area of that part (§13.5.4.2) – (**except** in 'Reference face' approach for towers and trusses, where s_{face} is the ratio of the solid area of the reference face to the envelope area, A_{face}, of the reference face (§16.3.4.3.1)).

Loading coefficients are defined as acting on the total solid area in projection (**except** in the 'Reference face' approach for towers and trusses when the special drag coefficient, $\overline{C}_{D_{face}}$, is taken to act on the solid area of the reference face (§16.3.4.3.1)). The coefficient $\overline{c}_{D_o}$ is the local drag coefficient with $\alpha = 0°$ and $\beta = 0°$ for the members in isolation, taken from the data for line-like structures (§20.3), and is the limit as $s \rightarrow 0$. The coefficient $\overline{C}_{D_1}$ is the overall drag coefficient for a solid structure with the same external envelope, and is the limit as $s \rightarrow 1$. These data are in §20.5, or the factor $\phi\{L/B\}$ of Eqn 20.10 may be applied to the corresponding line-like value from §20.3 when appropriate. These are all mean values, applied using the quasi-steady model.

The reference wind speed is the local gust speed, $\hat{V}\{z\}$, acting at the local height, z, above ground. The cumulative effect of the variation of wind speed over the height of a lattice can be accounted for in three ways:

1 Use an influence function to express the weighting on dynamic pressure at any height. If the solidity of the lattice is uniform, this is the influence coefficient, $\phi_q\{z\}$, given by Eqn 16.29. If the solidity of the lattice is variable, Eqn 16.29 can account for this by also including the weighting $s\{local\}/s\{envelope\}$, but either of the alternative ways is easier.
2 Take the dynamic pressure at the top of the lattice to apply to the whole lattice. This will always be conservative. The degree of conservatism will be very large for tall structures, such as lattice masts, but small for squat structures, such as horizontal gantries.
3 Divide the lattice structure into a number of panels, so that the variation of dynamic pressure down each individual panel is small, and take the dynamic pressure at the centre of the panel. Aggregate the individual panel loads to give overall forces and moments. This is the best compromise between simplicity and accuracy of the three ways, and is recommended by BS8100[189].

20.4.3 Single plane frames

The method described in §16.3.3 is implemented by the following steps:

1 Determine the envelope area for flow normal to the frame, $A_{env} = BL$ if clear of the ground, or $A_{env} = BH$ if built on the ground or a ground plane such as a building roof.
2 Determine the slenderness ratio of the envelope, B/L or $2B/H$ (§16.1.2).
3 Determine the solid area normal to the frame (along x-axis), A_x, and parallel to the frame (along y-axis), A_y. If composed of elements of mixed form: flat, subcritical or supercritical, keep separate sub-totals for each form.
4 Determine the solidity ratios for the frame in the x and y directions, $s_x = A_x / A_{env}$ and $s_y = A_y / A_{env}$.
5 Look up the normal force coefficient for the appropriate form of member in Figures 20.7, 20.8 or 20.9, by interpolating between the individual curves for slenderness ratio. (If of mixed form, a value for each of the forms is required.) These figures are based on Equations 16.34, 16.35 and 20.10. To use the equations directly:
   (a) determine the normal force coefficient for the equivalent solid plate of the same slenderness, $\overline{C}_{D_1} = 2.0\ \phi\{L/B\}$, using Eqn 20.10;
   (b) calculate the normal force coefficient for flat-faced members, $\overline{C}_{F_x}\{flat\}$, using Eqn 16.34 (check value against Fig 20.7);

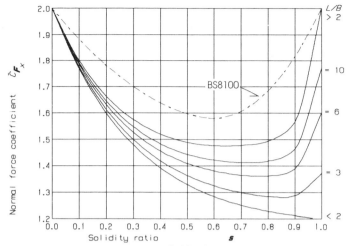

Figure 20.7 Plane frame composed of flat-faced elements

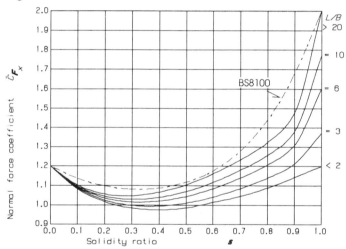

Figure 20.8 Plane frame composed of subcritical circular elements

(c) if circular members, calculate the normal force coefficient, $\overline{C}_{F_x}\{\text{circ}\}$, for circular members from the flat-faced value using Eqn 16.35 and the subcritical or supercritical drag coefficient, $\overline{c}_{D_o}$, in §20.3.3.1. (Check value against Fig 20.8 or 20.9.)

6 If composed of members of mixed form, calculate the normal force coefficient, $\overline{C}_{F_x}\{\text{mixed}\}$, from Eqns 16.38 (§16.3.3.1.3).

7 In the general case, where α is the angle of pitch (around z axis) and β is the angle of yaw (normal to z axis), the resulting force coefficients on a plane frame are:

$$\overline{C}_{F_x}\{\alpha, \beta\} = \overline{C}_{F_x} \cos\alpha \cos\beta \text{ and}$$

$$\overline{C}_{F_y}\{\alpha, \beta\} = \overline{C}_{F_x} \sin\alpha \cos\beta,$$

acting on the solid areas A_x and A_y, respectively. In the case of a vertical lattice, α becomes the wind direction θ, and $\beta = 0$.

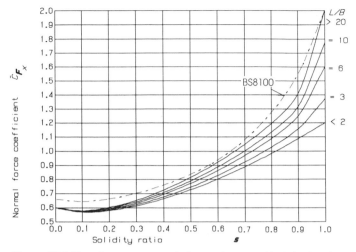

Figure 20.9 Plane frame composed of smooth supercritical circular elements

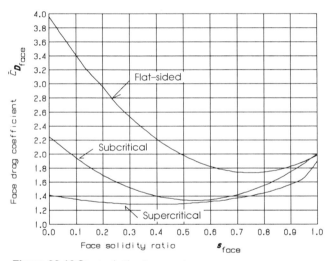

Figure 20.10 Square lattice tower or truss

The largest load will generally occur when the wind is normal to the lattice frame. If the lattice frame is predominantly composed of angle section members aligned in the same direction, account for the additional cross-wind force (§16.3.3.1.4).

20.4.4 Lattice towers and trusses

The designer has the choice of the 'Reference face' approach (§16.3.4.3) or the method of Eden, Butler and Patient (§16.3.4.4). The 'Reference face' approach is restricted to 3- and 4-boom towers and trusses, where each face is similar (i.e. nearly equilateral or square) and accounts only for wind directions normal to the long axis (i.e. accounts for pitch, but not yaw), but is the simpler to apply. It has also been extensively calibrated for lattice towers [190, 196] and is in common use through the British code of practice for lattice towers[189]. The method of Eden, Butler and Patient was developed for use with crane jibs, so can account for the combined effects of skew and yaw.

Rules are given in terms of the overall drag coefficient for trusses of constant section dimensions. The forces and moments on trusses of variable section dimensions are obtained by dividing the truss into suitable panels. The corresponding reference dynamic pressure of the incident wind is taken at the centre of each panel. The drag load on each panel is determined and aggregated to give overall forces and moments on the truss. BS8100 recommends that lattice towers should be divided into between 10 and 20 panels to account for the variation of wind speed with height.

20.4.4.1 Reference face approach

The method described in §16.3.4.3 is implemented by the following steps:

1 Determine the solid area, A_{face}, and solidity ratio, s_{face}, of one 'reference' face of the tower or truss. If there is significant differences between faces, take the largest value of both parameters (which may not correspond to the same face). With horizontal trusses always take a 'side' face as the reference face, not the 'top' or 'bottom' face. If composed of elements of mixed form: flat, subcritical or supercritical, keep separate sub-totals of the areas for each form.

2 Look up the value of $\overline{C}_{D_{face}}$ from Figure 20.10 for a square (4-boom) tower or truss, or from Figure 20.11 for triangular (3-boom) tower or truss.

3 Use this value to assess the following load cases:
   (a) square tower or truss – flow normal to face: take $\overline{C}_{D_{face}}$ acting on reference face area, A_{face} (maximum shear);
   (b) square tower or truss – flow along diagonal (45° to face): take $1.4\,\overline{C}_{D_{face}}$ on reference face area, A_{face}, acting in the wind direction through the diagonal (maximum compression in downwind boom and/or maximum guy tension);
   (c) triangular tower or truss – flow normal to windward face: take $\overline{C}_{D_{face}}$ acting on reference face area, A_{face} (maximum shear and maximum compression in downwind boom);

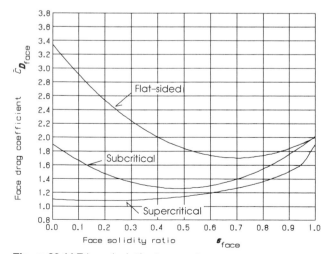

Figure 20.11 Triangular lattice tower or truss

(d) triangular tower or truss – corner into wind (normal to leeward face): take $\overline{C}_{D_{\text{face}}}$ acting on reference face area, A_{face} (maximum tension in guys and/or upwind boom).

These procedures are conservative.

4 If design cases in 3 prove too onerous, better estimates can be obtained by the directional factors of Eqns 16.43 and 16.44 (§16.3.4.3.3). Otherwise, try Eden, Butler and Patient's method (§20.4.4.2).

5 The loading of any discrete or line-like ancillary can be added to the loading of the tower or truss, provided the solidity ratio of the ancillaries is less than that of the reference face. With more solid, line-like ancillaries, such as pipe bridges, use Eden, Butler and Patient's method or treat as porous line-like structure of the corresponding envelope dimensions. For example, treat a dense pipe-bundle almost filling a square truss as a line-like rectangular section, according to the rules in §20.3.4.

20.4.4.2 The method of Eden, Butler and Patient

There are two ways of implementing this method: by considering every wind direction individually, or by considering the three orthogonal directions and using the empirical directional approach of Eqn 16.47. The first option is the more general, since the empirical approach is valid only for crane jibs and similar trusses.

The general method described in §16.3.4.4 is implemented by the following steps:

1 Determine the **total** solid area of all members in the truss, ΣA, and corresponding **total** solidity ratio, $\Sigma s = \Sigma A / A_{\text{proj}}$, for each wind direction of interest.

2 Look up the total drag coefficient, $\overline{C}_{D_{\text{total}}}$, in Figure 20.12. Alternatively, use Eqns 16.45 or 16.46, but keep within the range $0.5 \leqslant \Sigma s \leqslant 10$ for which the original empirical data were valid. The reason for this restriction is that these empirical equations do not converge correctly in the limits $\Sigma s \to 0$ and $\Sigma s \to \infty$.

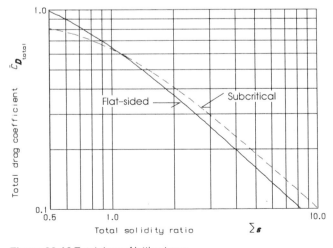

Figure 20.12 Total drag of lattice truss

The experimental scatter was typically as great as the difference between the two curves in Figure 20.12, so it is by no means certain that these differences are real.

3 Apply the total drag coefficient, $\overline{C}_{D_{total}}$, to the total solid area, ΣA, acting in the wind direction.

The empirical directional approach for crane jibs and similar trusses skewed to the wind is implemented as follows:

1 Determine the luff angle, A, and slew angle, B, of §16.3.4.4.3 for each wind direction of interest. Both these terms come from the use of jib cranes. Slew angle, B, is the rotation of the long axis of the truss around the vertical axis, measured from the current wind direction, and comes from the rotation of a crane about its vertical pivot. Luff angle, A, is the angle of the long axis in the vertical plane from the horizontal, and comes from the raising and lowering of a crane jib. Both angles are zero when the truss is pointing directly into the wind.

2 Determine the **total** solid area, ΣA, and corresponding **total** solidity ratio, $\Sigma s = \Sigma A \, / \, A_{proj}$, for the three orthogonal directions:
   (a) A = 0°, B = 0° – flow parallel to long z axis of truss (jib horizontal, pointing into wind);
   (b) A = 90°. B = 0° – flow normal to long z axis of truss, parallel to y axis (jib raised vertical from pointing into wind); and
   (c) A = 90°, B = 90° – flow normal to long z axis of truss, parallel to x axis (actually same value is obtained for any B in this case – jib any angle, normal to wind).

3 Look up the total drag coefficient, $\overline{C}_{D_{total}}$, in Figure 20.12 for the total solidity ratio corresponding to each of the three orthogonal directions. (See earlier comments.)

4 Apply the total drag coefficient, $\overline{C}_{D_{total}}$, to the total solid area, ΣA, to give the drag in each of the three orthogonal directions.

5 Calculate the drag for the required luff and slew angles, A and B, using Eqn 16.47, taking the values of the exponents, n and m, as $n = 1.8$ and $m = 1.4$.

20.4.5 Multiple plane frames and trusses

This section was derived from the model for multiple plane frames (§16.3.6.2), where downwind frames are shielded by upwind frames, but is extended to the more general case of multiple trusses, or other lattice structures. The method described in §16.3.6.2 is implemented as follows:

1 Determine the normal force coefficient, $\overline{C}_{F_x}\{\theta = 0°\}$, of each individual frame or truss. Use steps 1–6 of §20.4.3 for plane frames; steps 1–3 of §20.4.4.1 or steps 1 and 2 of the general method in §20.4.4.2 for trusses.

2 For each wind direction of interest, establish the 'wind shadow' regions as defined in Figure 16.51.

3 The shielding factor for the upwind frame and any unshielded part of downwind frames is $\eta = 1$, so the normal envelope drag for the upwind frame is simply $s \, \overline{C}_{F_x}\{\theta = 0°\}$.

4 Look up the shielding factor, $\eta_{\infty,n}$ for the wind shadow region behind the first frame in Figure 20.13, or calculate using Eqn 16.61 (§16.3.6.2.1).

5 Apply this shielding factor to the wind shadow region of the second frame shielded by the first frame.

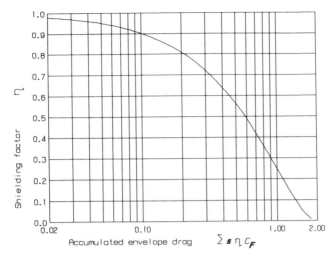

Figure 20.13 Shielding coefficient for lattice frames and trusses

6 Accumulate the envelope drag for the shielded and unshielded regions of the second frame and look up in Figure 20.13 (or Eqn 16.61) the corresponding shielding factors for the wind shadow regions of the third frame shielded by the first and second frame, and shielded by the second frame alone.

7 Progress downwind, accumulating the drag on each wind shadow region of the frames until the shielding factor for every wind shadow region has been determined.

The normal load on any frame in a given wind direction may be taken as:

$$\overline{C}_{F_x}\{\theta\} = \overline{C}_{F_x}\{\theta = 0°\} \cos\theta \, (A_1 + \eta_2 A_2 + \ldots + \eta_n A_n) / A_{env} \qquad (20.13)$$

where η_n is the shielding factor for wind shadow region n, A_n is the envelope area of that region and A_{env} is the total envelope area (and A_1 is assumed to be an unshielded region where $\eta = 1$). Expect the largest normal load to occur when the wind direction is 20°–40° from normal. The corresponding transverse load on any frame in a given wind direction may be taken as:

$$\overline{C}_{F_y}\{\theta\} = \overline{C}_{F_y}\{\theta = 90°\} \sin\theta \, (A_1 + \eta_2 A_2 + \ldots + \eta_n A_n) / A_{env} \qquad (20.14)$$

The above procedure neglects the additional effect of direct wake shielding. This will be significant when the lattice frames are closely spaced and the individual elements are large in breadth or diameter. However, it will only occur for those elements directly in the wakes of other elements, and this may only happen in particular wind directions. Neglecting the effect is recommended unless it is clear that a significant advantage is gained and can be sustained in the design. Clearly, the downwind booms of a square lattice mast will be in the wakes of upwind booms only when the wind is normal to a face and will be exposed in other wind directions, as in the example of Figure 16.41. One example where inclusion of direct wake shielding would be warranted is for a series of closely spaced horizontal trusses with large-diameter top and bottom booms. In this case, the downwind booms will remain in the wakes of upstream booms for a wide range of wind directions, although the downwind bracing members will move in and out of the wakes of the corresponding upstream bracing. Include the effect of direct wake shielding, when justified, as follows:

1 For closely spaced frames, determine the spacing ratio $x/[b \cos\theta]$. For spacing ratios in the range $4 \leqslant x/[b \cos\theta] < 25$, look up the wake shielding function, $\phi\{x/b\}$, in Figure 16.55 or use Eqn 16.63.
2 Include $\phi\{x/b\}$ with the shielding factor η as a factor on the normal drag coefficient *for directly shielded members only*.

20.4.6 Unclad building frames and other three-dimensional rectangular arrays

The method described in §16.3.6.3 is implemented as follows:

1 Divide the three-dimensional frame into individual plane frames in the x and y directions. For building frames, consider each stage of construction (i.e. main structural frame in stages, rafters, purlins, etc.).
2 If the array is much more solid in one of the x or y directions, calculate the normal force in this direction assuming the other orthogonal frames do not exist, using the previous method for multiple frames in §20.4.5. This will be conservative because the shielding effect of the less dense frames will have been ignored.
3 If the array has a similar solidity in x and y directions, calculate the total force coefficients for wind along each axis, $\Sigma \, \overline{C}_F\{\theta = 0°\}$ and $\Sigma \, \overline{C}_{F_y}\{\theta = 90°\}$ using the previous method for multiple frames in §20.4.5.
4 Estimate the total force coefficients for other wind angles, $\Sigma \overline{C}_{F_x}\{\theta\}$ and $\Sigma \, \overline{C}_{F_y}\{\theta\}$ from the empirical equations Eqn 16.66 and 16.67, using values for the envelope breadth, B, and depth, D, for wind along the x axis.

The empirical method used here gives only the total load coefficients, which are required for stability considerations.

Many typical building frames have a stage where floor beams, aligned predominantly in one direction, are inserted into a near-square matrix of structural frames. For this stage, add the loads of the floor beams to the total load for the structural frame in the wind direction normal to the floor beams, using the shielding coefficient corresponding to the wind shadow region cast by the upwind structural frames and including the direct wake shielding function for the breadth and spacing ratio of the floor beams. This action is also appropriate for rafters of roofs.

20.5 Loading data for plate-like structures

20.5.1 Boundary walls, hoardings and fences

20.5.1.1 Solid walls

The variation of load along boundary walls is discussed in §16.4.2. Since most walls and fences are designed in bays between supports, a simple model based on corresponding loaded regions is most convenient. Figure 20.14 gives the key for solid walls. The reference dynamic pressure, $\hat{q}_{ref}$, is defined at the top of the wall, $\hat{z}_{ref} = H$. Loaded regions are defined in terms of the distance in wall heights from a free end or corner. The loading coefficient is defined as a local force coefficient, normal to the wall, which is uniform over the loaded region. Accordingly, the

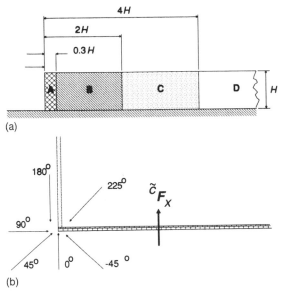

(a)

(b)

Figure 20.14 Key to wall and fence data: (a) key to loaded areas; (b) key to wind angle and force coefficient

height of application of the force (centre of pressure) is half-way up the wall at $z = H/2$. Wind angle is defined from normal to the wall, with the free end or corner pointing upwind for $0° < \theta < 180°$.

Values of local force coefficient for semi-infinite walls, i.e. walls with free ends, are given in Table 20.2. For walls with two free ends, work from both ends. Values of local force coefficient for walls with corners are given in Table 20.3. These are all pseudo-steady values.

Table 20.2   Local force coefficients for semi-infinite solid boundary walls

Key diagram: Figure 20.14	Wind direction measured from normal Ref $\hat{q}$ at $z = H$ (top of wall)								
Wind direction θ	$-30°$	$-15°$	$-7°$	$0°$	$7°$	$15°$	$30°$	$45°$	$60°$
	Pseudo-steady force coefficient, $\tilde{c}_{F_x}\{\theta\}$								
Region A $0 < y/H < 0.3$	0.55	0.83	1.09	1.34	1.54	2.04	3.00	3.41	2.84
Region B $0.3 < y/H < 2$	0.70	0.93	1.04	1.39	1.48	1.83	2.13	2.00	1.82
Region C $2 < y/H < 4$	0.85	1.03	1.18	1.31	1.41	1.66	1.55	1.42	1.40
Region D $y/H > 4$	1.05	1.16	1.19	1.20	1.19	1.16	1.03	0.85	0.60

20.5.1.2 Effect of porosity

With significant porosity, the local high load regions near free ends or corners disappear. Rules for porous fences are therefore:

1 for $s > 0.8$, treat as solid wall;
2 for $s < 0.8$, treat as a plane lattice frame, using Eqn 16.71 for flat-sided elements, Eqn 16.72 for circular elements and the cosine model for wind angle.

The height of application should be taken at the centre of the envelope of the fence, as shown in Figure 16.64, and this is particularly important for horizontally slatted fences with few slats (§16.4.2.4).

Table 20.3   Local force coefficients for corners of solid boundary walls

Key diagram: Figure 20.14	Wind direction measured from normal Ref $\hat{q}$ at $z = H$ (top of wall)			
Wind direction θ	0°	45°	180°	225°
	Pseudo-steady force coefficient, $\bar{c}_{F_x}\{\theta\}$			
Region A $0 < y/H < 0.3$	1.16	2.13	− 1.60	− 0.81
Region B $0.3 < y/H < 2$	1.30	1.80	− 1.64	− 0.81
Region C $2 < y/H < 4$	1.25	1.42	− 1.31	− 0.83
Region D $y/H > 4$	1.20	0.85	− 1.20	− 0.85

20.5.1.3 Shelter from upwind walls and fences

When there are other walls or fences upwind that are equal in height or taller than the wall or fence height, H, an additional shelter factor, η, can be used with the force coefficients for the downwind loaded wall or fence (§19.2.2.1). These rules can also be used for estimating the loads on building façades without roofs, i.e. during construction or renovation, for the design of supporting falsework.

Ref $\hat{q}$ at height of fence or wall

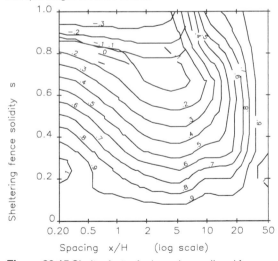

Figure 20.15 Shelter factor for boundary wall and fences

The value of the shelter factor, η, depends on the solidity of the sheltering upwind wall or fence and the separation, x. Design values of η are given in Figure 20.15, derived from measurements on long pairs of walls and fences normal to the wind. The factors may be used for wind directions other than normal using the separation in the wind direction, $x / \cos\theta$. The principle of 'wind shadow' used in §20.4.5 for lattice frames and trusses should be used to determine how much of the length of the downwind wall is sheltered. When there are multiple walls and fences, the separation should be measured to the nearest taller, or as tall, upwind fence or wall casting the 'wind shadow'. This procedure ignores the shelter effect of other lower walls and fences, and of walls and fences further upstream.

At high sheltering solidity and close spacings, Figure 20.15 predicts negative values for the shelter factor, η, indicating that the load on the downwind wall or fence acts in the upwind direction. In this case, the force coefficient for the upwind shielding wall or fence should be increased by applying the factor $\eta = 1.3$ to the force coefficient. Occasions where this is likely to occur are boundary walls and fences either side of narrow paths and building façades without roofs.

20.5.2 Signboards

Loading of signboards is discussed in §16.4.3. Signboards with their bottom edge touching the ground should be treated as a wall or hoarding, as in §20.5.1.1. Figure 20.16 gives the key for signboards held clear of the ground. The reference dynamic pressure, $\hat{q}_{ref}$, is defined at the top of the signboard. Values of the overall normal force coefficient are given in Table 20.4. The height of the normal force should be taken on the centreline of the board, but the horizontal position should be taken to vary between $-0.25 \leqslant y/B \leqslant 0.25$, as defined in Figure 20.16.

At the wind angle of $\theta = 90°$, friction forces from the wind sweeping either face of the board may be neglected, but the forces on any support posts should be considered.

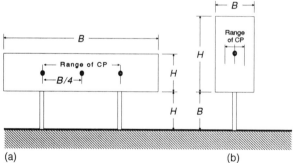

(a)   (b)
Figure 20.16 Key to signboard data where H/B = (a) 0.25, (b) 2

Table 20.4   Overall force coefficients for signboards

Key diagram: Figure 20.16	Wind direction measured from normal Ref $\hat{q}$ at $z = H$ (top of wall)					
Wind direction θ	0°	15°	30°	45°	60°	75°
	Pseudo-steady force coefficient, $\bar{C}_{F_x}\{\theta\}$					
$H/B = 0.25$	1.81			1.78		
0.5	1.80	1.75	1.60	1.55	1.45	0.75
1.0	1.80	1.80	1.75	1.60	1.40	0.70
2.0	1.75	1.75	1.65	1.60	1.15	0.40

20.5.3 Canopy roofs

20.5.3.1 Scope

These data apply only to free-standing canopy roofs that do not have permament walls. Rules for canopies with permanent walls are given in §20.9.2 'Open-sided buildings'. Rules for canopies attached to large buildings are given in §20.8.2 'Canopies attached to tall buildings'. The discussion of free-standing canopy roof loading in §16.4.4 was made with the data presented in the design figures of this section.

In every case, the reference dynamic pressure, $\hat{q}_{ref}$, is defined at the mean height, $\bar{z}_{ref}$, half-way up the roof slope, as shown in the respective key figure.

The data are all pseudo-steady force coefficients. The form of presentation is as contours of value. In most cases, either a positive (downward) or negative (upward) value is given. Where the positive and negative contours overlap represents a case where either is appropriate, depending on the critical direction for the structural form. Overall forces and moments are obtained by summation of the forces acting on the individual local regions.

20.5.3.2 Monopitch canopies

Figure 20.17 gives the key for monopitch canopies. Local force coefficients for each loaded region are given in Figures 20.18–20.21 (§16.4.4.2).

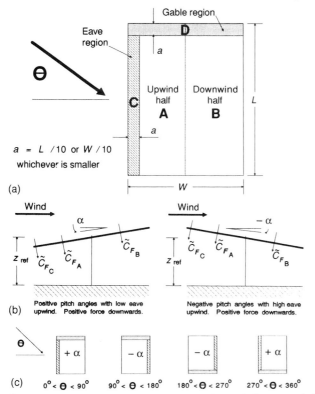

Figure 20.17 Key for monopitch canopies: (a) loaded areas and wind angle; (b) canopy pitch, reference height and force coefficients; (c) local regions and pitch for ranges of wind angle

Key: Figure 20.17   Ref $\hat{q}$ at mean height of canopy

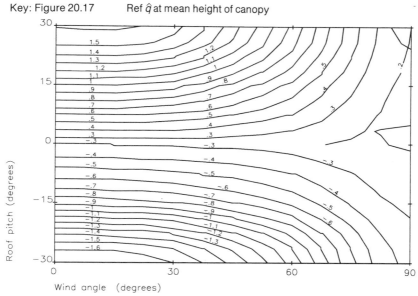

Figure 20.18 Normal force coefficient for upwind half 'A' of a monopitch canopy

Key: Figure 20.17   Ref $\hat{q}$ at mean height of canopy

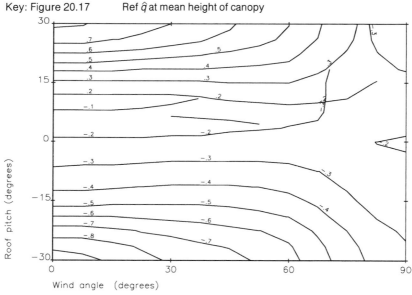

Figure 20.19 Normal force coefficient for downwind half 'B' of a monopitch canopy

20.5.3.3 Duopitch canopies

Figure 20.22 gives the key for duopitch canopies. Local force coefficients for each loaded region are given in Figures 20.23–20.26 (§16.4.4.3). These apply to duopitch canopies of equal pitch. Rules are given in §16.4.4.3 for the case of unequal pitch.

Key: Figure 20.17   Ref $\hat{q}$ at mean height of canopy

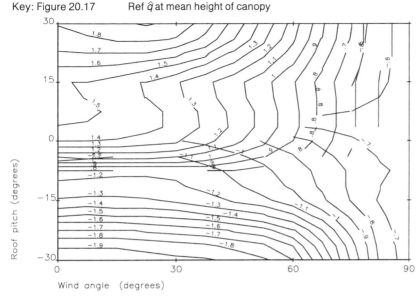

Figure 20.20 Normal force coefficient for upwind eave region 'C' of a monopitch canopy

Key: Figure 20.17   Ref $\hat{q}$ at mean height of canopy

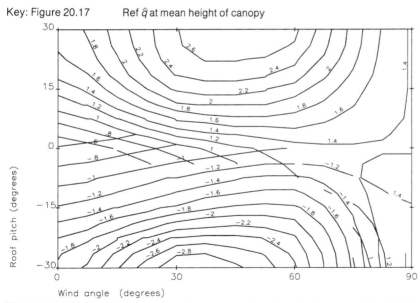

Figure 20.21 Normal force coefficient for upwind gable region 'D' of a monopitch canopy

20.5.3.4 Multi-bay canopies

Figure 20.27 gives the key for multi-bay canopies, which should be treated as 'ridged' or 'troughed' depending on the pitch of the first upstream bay. When the wind is normal to the ridges, $\theta = 0°$ or $180°$, multi-bay canopies are treated as a succession of duopitch canopies, with the reduction factors of Table 16.6 applied to

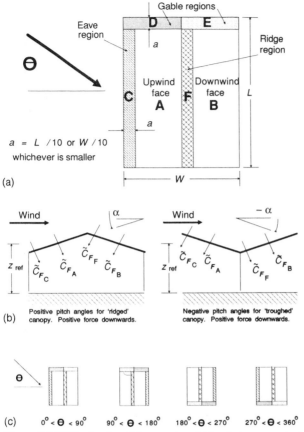

Figure 20.22 Key for duopitch canopies: (a) loaded areas and wind angle; (b) canopy pitch, reference height and force coefficients; (c) local regions for ranges of wind angle

the duopitch data. When the wind is parallel to the ridges, $\theta = 90°$ or $270°$, each bay takes the full duopitch canopy loads (§16.4.4.4).

20.5.3.5 Curved canopies

There are insufficient reliable data to give definitive design guidance. Available data for barrel-vault and domed canopies are given in Figures 16.65 and 16.66, respectively, but their validity is not guaranteed.

20.5.3.6 Effect of under-canopy blockage

Fully blocked canopies should be treated as open-sided buildings as in §20.9.2. In essence, this is done by taking the external pressure coefficients for the equivalent closed building from the data of §20.6 and §20.7 in combination with an internal pressure appropriate to the blockage from Table 16.7.

The loading of partially blocked canopies, where the blockage grows upwards from the ground, can be interpolated from the fully unblocked and fully blocked

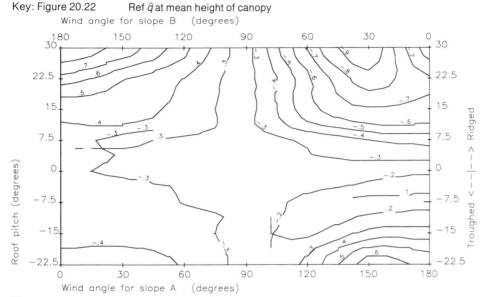

Figure 20.23 Normal force coefficients for each face 'A' and 'B' of a duopitch canopy

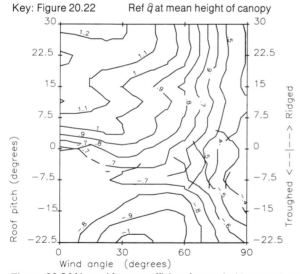

Figure 20.24 Normal force coefficient for upwind eave region 'C' of a duopitch canopy

cases using Eqn 16.77. For blockage solidities of less than 30% the blockage has no significant effect and the data referenced in §20.5.3.2–§20.5.3.5 can be used directly.

20.5.3.7 Fascia and friction loads

The normal force coefficient on a vertical fascia should be taken as:

$$\tilde{c}_{F_x}\{\theta\} = \tilde{c}_{F_x}\{\theta = 0°\} \cos\theta \tag{20.15}$$

Key: Figure 20.22   Ref $\hat{q}$ at mean height of canopy

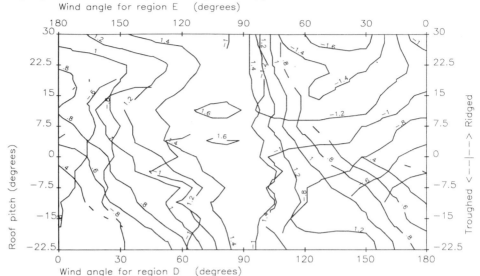

Figure 20.25 Normal force coefficient for upwind gable regions 'D' and 'E' of a duopitch canopy

Key: Figure 20.22   Ref $\hat{q}$ at mean height of canopy

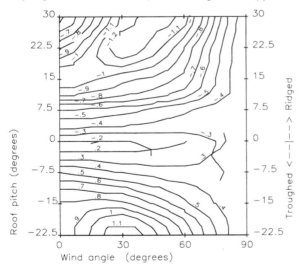

Figure 20.26 Normal force coefficient for downwind ridge region 'F' of a duopitch canopy

where $\tilde{c}_{F_x}\{\theta = 0°\} = 1.2$ for an upwind fascia and $\tilde{c}_{F_x}\{\theta = 0°\} = 0.6$ for a downwind fascia. For 'thick' canopies, such as typical UK petrol-station canopies, take $\tilde{C}_{F_x}\{\theta = 0°\} = 1.2$ as an overall coefficient for both windward and leeward faces.

Friction forces on the canopy are obtained from the shear stress coefficient, c_τ, acting on the area of the canopy swept by the wind. This will be the top and bottom surfaces for unblocked canopies, and the top surface only when the canopy is fully blocked. Values of c_τ are given in Table 16.8. Ribbed or corrugated surfaces should

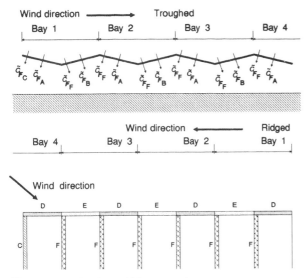

Figure 20.27 Key for multi-bay canopies

be treated as such for the direction normal to the ribs, and as smooth for directions parallel to the ribs. Fascias can be assumed to protect a region $4h$ deep behind the fascia, where h is the height of the fascia above or below the canopy roof skin.

20.6 Loading data for curved bluff structures

20.6.1 Scope

These data apply to bluff structures that are predominantly curved, but may have some flat sections. They include spherical or cylindrical storage tanks, domes, arched structures, cylindrical silos, cooling towers and other similar forms. Barrel-vault roofs on typical rectangular-plan buildings are deferred to §20.7.4.6. The definition of the reference dynamic pressure, $\hat{q}_{ref}$, is given for each form.

The loading coefficients are mean pressure coefficients, applied using the quasi-steady model (except for §20.6.3.1.3 'Flat and monopitch roofs of vertical cylinders'). Curved structures generally have lobes of positive and of negative pressure and regions between where the mean pressure passes through zero. Here the quasi-steady model predicts no load. This is the main application where the quasi-steady to pseudo-steady regression analysis of §15.3.3 would be useful to correct this anomaly and give the range of peak values for cladding. However, this loading is always exceeded in another wind direction and values for the worst cladding suctions are always given.

20.6.2 Spherical structures

20.6.2.1 Spheres

20.6.2.1.1 Definitions.   The diameter of the sphere is D and the gap between the bottom of the sphere and the ground is G. The reference dynamic pressure, $\hat{q}_{ref}$, should be determined at the height of the centre of the sphere, $z_{ref} = G + D/2$, for all cases given below. Owing to the axisymmetry of the sphere, the same data apply to all wind directions. The maximum pressure is experienced at the front stagnation point and the maximum suction is in a vertical ring around the 'equator' (with the front stagnation point as the 'pole'). Every part of the sphere experiences this maximum suction in some wind direction.

20.6.2.1.2 Smooth sphere.

The drag of a smooth sphere clear of the ground may be taken as:

$$\overline{C}_D\{\text{subcritical}\} = 0.47 \text{ when } D\hat{V} < 6\,\text{m}^2/\text{s and}$$

$$\overline{C}_D\{\text{supercritical}\} = 0.19 \text{ when } D\hat{V} > 6\,\text{m}^2/\text{s}$$

(see §17.2.1.1.1 and Table 17.1).

The mean pressure distribution around an isolated sphere is given by Figure 17.4. The range of possible pressure coefficients is $-1.25 < \overline{c}_p < 1.0$. (§17.2.1.1.2).

20.6.2.1.3 Rough sphere.

In the case of subcritical flow, $D\hat{V} < 6\,\text{m}^2/\text{s}$, the drag of rough spheres clear of the ground may still be taken as $\overline{C}_D\{\text{subcritical}\} = 0.47$, provided the height of the roughness is included in the effective diameter of the sphere. Drag of rough spheres clear of the ground in supercritical flow, $D\hat{V} > 6\,\text{m}^2/\text{s}$, can determined from the drag of rough circular cylinders by applying the factor $\overline{C}_D\{\text{sphere}\}/\overline{c}_{D_o}\{\text{cylinder}\} = 0.4$ to the value of $\overline{c}_{D_o}$ for a long cylinder given by Eqn 20.8 or Figure 20.5 for the roughness of various surface finishes given in Figure 20.6 (§17.2.1.1.1).

20.6.2.1.4 Effect of ground plane.

The drag and lift of a sphere close to the ground is given in Table 17.2 (§17.2.1.1.3).

20.6.2.2 Domes

20.6.2.2.1 Definitions.

Figure 20.28 is the key figure: D is the diameter of the dome in plan – the base diameter for hemispherical domes or lower and the diameter of the parent sphere for taller domes; H is the 'rise' or height of the dome; s is the position along the centreline arc of the dome measured from the front edge

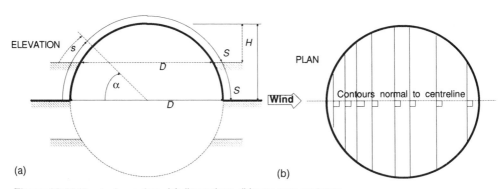

Figure 20.28 Key to dome data: (a) dimensions; (b) pressure contours

and S is the total arc length. The reference dynamic pressure, $\hat{q}_{ref}$, should be determined at the height, $z_{ref} = H$, of the top of the dome for all cases given below. Owing to the axisymmetry of domes, the same data apply to all wind directions. As with the sphere, the maximum suction on a dome is in a vertical arc around the 'equator', and every part of the dome experiences this maximum suction in some wind direction.

20.6.2.2.2 Rise ratio $H/D < 0.5$.   For hemispherical and lower domes, the key Figure 20.28 defines the position along the centreline arc, s, measured from the ground and the total arc length, S. Look up the pressure distribution along the centreline in Figure 20.29, for the rise ratio of the dome. For rise ratios less than $H/D = 0.0625$, interpolate between the values given and zero pressure coefficient at $H/D = 0$. Construct pressure contours normal to the centreline as indicated in key Figure 20.28(b). The maximum pressure is experienced at the front, $s/S = 0$, and the maximum suction around the 'equator', $s/S = 0.5$, depending on wind direction (§17.2.1.1.2).

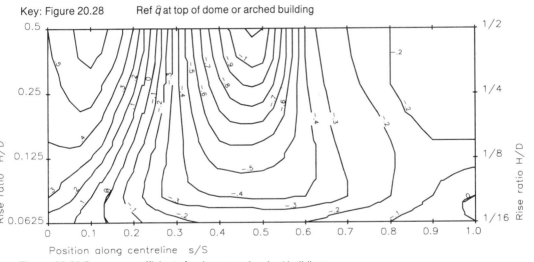

Figure 20.28   Ref $\hat{q}$ at top of dome or arched building

Figure 20.29 Pressure coefficients for domes and arched buildings

20.6.2.2.3 Rise ratio $H/D > 0.5$.   This case is transitional between the hemispherical dome and the sphere close to the ground. Assess both ways and take the most onerous loading. Note the difference in the reference dynamic pressure definitions between dome and sphere: at top of dome and at centre of sphere.

20.6.3 Cylindrical structures

20.6.3.1 Vertical cylinders

20.6.3.1.1 Definitions.   Figure 20.30 is the key figure: in (a) D is the diameter of the cylinder, H is the height to the top of the cylinder walls (at the windward point for monopitch roofs), R is the rise of conical or domed roofs and α is the slope at the roof for conical or monopitch roofs. The wind angle, θ, is measured from the x-axis which passes directly up the slope of monopitch roofs, but is arbitrary for the

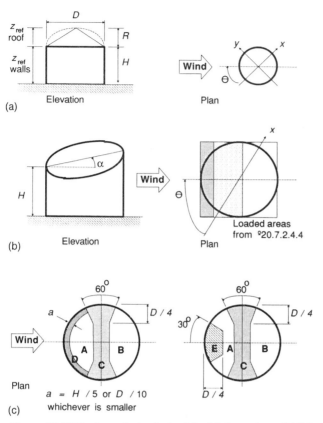

Figure 20.30 Key to vertical cylinder data: (a) dimensions; (b) flat and monopitch roofs; (c) domed and conical roofs

axisymmetric cases of flat, cone or domed roofs. The reference dynamic pressure, $\hat{q}_{ref}$, should be determined at the reference height, z_{ref}, appropriate to the walls or roof as defined in Figure 20.30.

20.6.3.1.2 Walls. The pressure distribution on the walls should be taken as constant with height for $H/D \leqslant 4$. The reference height for the walls is the height of the walls, $z_{ref} = H$. The distribution of mean pressure coefficient around cylinders [283], silos and hyperbolic cooling towers [284] in the range $0.5 < H/D < 2$ is given by the Fourier series:

$$\bar{c}_p = k \Sigma (A_n \cos na) \tag{20.16}$$

where k is a factor and A_n are the Fourier coefficients given in Table 20.5 and a is the angular position around the cylinder measured from the windward point (front stagnation point). Distributions for $H/D = 0.5$, 1 and 2 from Eqn 20.16 are plotted in Figure 17.9. For squatter cylinders, use the distribution for $H/D = 0.5$. For taller cylinders use the distribution for $H/D = 2$. (Cylinders taller than $H/D = 4$ may be treated as line-like in the middle region $2D < z < H - 2D$, and the pressure coefficients applied using the dynamic pressure at the local height.) The maximum

Table 20.5   Parameters for the pressure distribution around circular cylinders

Key diagram: Figure 20.30 and 20.31

Vertical cylinders:	$\hat{q}_{ref}$ at top of cylinder, $z_{ref} = H$
Horizontal cylinders:	$\hat{q}_{ref}$ at axis of cylinder, $z_{ref} = G + D/2$

Cylinders:	$k = 1$ for positive $\bar{c}_p$
	$k = 1 + \log(H/D)$ for negative $\bar{c}_p$ $0.5 \leq H/D \leq 2$
Cooling towers:	$k = 1$ for all $\bar{c}_p$

Harmonic	Cylinders[283] Coefficient	Silos and cooling towers[284] Coefficient
n	A_n	A_n
0	− 0.5	− 0.2636
1	+ 0.4	+ 0.3419
2	+ 0.8	+ 0.5418
3	+ 0.3	+ 0.3872
4	− 0.1	+ 0.0525
5	− 0.05	− 0.0771
6	0	− 0.0039
7	0	+ 0.0341

suction for cladding occurs at the peak of the suction lobe near $\theta = 90°$, but all parts of the walls experience this suction at some wind direction. (§17.2.2.1.2).

20.6.3.1.3 Flat and monopitch roofs.   The pressure distribution over flat or monopitch roofs on cylinders should be deduced from the distribution on the equivalent flat or monopitch roofed cuboid, as indicated in Figure 20.30(b).

For flat roofs, the equivalent square cuboid has the same height, H, and a breadth, B, equal to the cylinder diameter, D, and the reference height is the height of the cylinder, $z_{ref} = H$. For monopitch roofs, the pitch of the equivalent cuboid, α_{equiv}, is the pitch of the roof in the wind direction, so varies with wind direction. This is given from the actual roof pitch, α, by:

$$\alpha_{equiv}\{\alpha\} = \arctan(\cos\theta \tan\alpha) \tag{20.17}$$

The dynamic pressure, $\hat{q}_{ref}$, is given by the rules in §20.7.4.2.2, in terms of the height of the walls at the upwind point, Figure 20.30(b).

The pressures in the loaded areas of the equivalent flat or monopitch cuboid are taken from the pseudo-steady coefficients of Table 20.18 (§20.7.4.2.4), always for the normal flow direction ($\theta = 0°$). The maximum suction for cladding occurs at the windward edge of the roof for flat, low and negative equivalent pitches. This forms an annular region around the periphery when all wind directions are taken into account. For flat-roofed cylinders, the value of the maximum suction will be the value for $\alpha = 0°$ all around the annulus. For monopitch roofs the value of the maximum suction at any location around the annulus will be the value for the effective pitch angle at the local eave, $\alpha_{equiv}\{\theta\}$. For positive effective pitches above about $\alpha_{equiv} = 30°$, the flow remains attached and the pressure is positive over the whole roof, in which case the peripheral region of local suction is not a complete annulus (§17.2.2.1.3).

20.6.3.1.4 Domed and conical roofs.

Loaded areas for domed and conical roofs are defined in Figure 20.30(c). The reference height is the top of the roof, $z_{ref} = H + R$, Figure 20.30(a).

Table 20.6 gives values of mean pressure coefficient for each loaded area for a range of cylinder height, H/D, and roof rise, R/D, or roof pitch angle, α. Unfortunately, insufficient data exist to give reliable values for conical roofs on squat cylinders, $H/D < 0.5$. Note that a value is given for only one of Region D or E, depending whether flow separates at the front edge.

Depending on the rise of the domed or conical roof, the maximum suction for cladding occurs either in the 'equatorial' Region C or the peripheral arc Region D. Considering all wind directions, Region C suctions will apply to the whole dome and Region D suctions to the complete peripheral annulus (§17.2.2.1.4 and §17.2.2.1.5).

Table 20.6   Pressure coefficients for domes and conical roofs on cylinders

Key diagram: Figure 20.30 $\hat{q}_{ref}$ at top of roof, $z_{ref} = H + R$

Domed roofs	Structural loads					Peak cladding loads	
Region	A	B	C	D	E	C	D
$R/D = 1/16$	Mean pressure coefficient, $\bar{c}_p$						
$H/D = 1/16$	0	− 0.1	− 0.2	–	+ 0.1	− 0.25	− 0.2
1/8	− 0.2	− 0.1	− 0.2	–	0	− 0.25	− 0.4
1/4	− 0.4	− 0.2	− 0.3	− 0.9	–	− 0.3	− 0.8
1/2	− 0.6	− 0.2	− 0.4	− 1.1	–	− 0.4	− 1.2
≥ 1	− 0.8	− 0.3	− 0.5	− 1.2	–	− 0.5	− 1.3
$R/D = 1/8$	Mean pressure coefficient, $\bar{c}_p$						
$H/D = 1/16$	− 0.2	− 0.3	− 0.5	0	+ 0.2	− 0.5	–
1/8	− 0.3	− 0.3	− 0.5	− 0.2	–	− 0.5	− 0.4
1/4	− 0.3	− 0.3	− 0.5	− 0.8	–	− 0.5	− 0.9
1/2	− 0.5	− 0.4	− 0.5	− 1.2	–	− 0.5	− 1.3
≥ 1	− 0.7	− 0.4	− 0.5	− 1.2	–	− 0.5	− 1.3
$R/D = 1/4$	Mean pressure coefficient, $\bar{c}_p$						
$H/D = 1/16$	− 0.2	− 0.3	− 0.8	–	+ 0.3	− 0.8	–
1/8	− 0.2	− 0.3	− 0.8	0	–	− 0.8	0
1/4	− 0.3	− 0.4	− 0.8	− 0.1	–	− 0.8	− 0.3
1/2	− 0.4	− 0.4	− 0.8	− 0.3	–	− 0.8	− 0.5
≥ 1	− 0.5	− 0.5	− 0.8	− 0.5	–	− 0.8	− 0.7
$R/D ≥ 1/2$	Mean pressure coefficient, $\bar{c}_p$						
all H/D	0	− 0.2	− 1.2	–	+ 0.8	− 1.25	–

Conical roofs	Structural loads					Peak cladding loads	
Region	A	B	C	D	E	C	D
$H/D ≥ 0.5$	Mean pressure coefficient, $\bar{c}_p$						
Pitch $\alpha = 15°$	− 1.1	− 0.7	− 0.9	− 1.7	–	− 0.9	− 1.7
25°	− 0.8	− 0.7	− 1.4	− 1.1	–	− 1.6	− 1.3
45°	− 0.5	− 0.7	− 1.0	–	+ 0.3	− 1.1	–

20.6.3.1.5 Open roofs.   An open roof forms a dominant opening which controls the internal pressure of the cylinder. With tall cylinders, $0.5 \leqslant H/D \leqslant 4$, the pressure distribution on the outside of the walls is unchanged but the internal pressure becomes negative. Values of external pressure are given by Eqn 20.16 with Table 20.5. Values of internal pressure are given in Table 20.7. For all H/D, the net pressure difference across the walls can be read directly by interpolating between the curves in Figure 17.14 (§17.2.2.1.6).

Table 20.7   Internal pressure coefficients for tall, open-topped cylinders

$\hat{q}_{\text{ref}}$ at top of cylinder, $z_{\text{ref}} = H$

Slenderness	$H/D =$	0.5	1	2
Mean internal pressure coefficient	$\tilde{c}_{p_i} =$	-0.8	-0.9	-0.95

20.6.3.2 Horizontal cylinders

20.6.3.2.1 Definitions.   Figure 20.31 is the key figure: D is the diameter of cylinders, hemicylinders and cylindrical section arched buildings with $H/D > 0.5$, and the base depth for cylindrical section arched buildings with $H/D \leqslant 0.5$ (analogous to the dome rules in §20.6.2.2.1). For cylinders, G is the gap between the bottom of the cylinder and the ground. For arched buildings, H is the height or 'rise' of the top of the arch. The length of the plane cylindrical walls is L and R is the 'rise' of domed ends. The reference dynamic pressure, $\hat{q}_{\text{ref}}$, should be determined at the height of the cylinder axis, $z_{\text{ref}} = G + D/2$, for cylinders clear of the ground ($G \geqslant D/2$) and the top of the arch, $z_{\text{ref}} = H$, for arched buildings.

20.6.3.2.2 Cylinders clear of the ground.   If $G \geqslant D/2$ the ground effect is insignificant. For wind directions within $45°$ of normal to the axis, $\theta \leqslant \pm 45°$, the pressures around the plane cylinder walls are the same as for vertical cylinders (§20.6.3.1.1) and are given by Eqn 20.16 and Table 20.5, using $L/2D$ in place of H/D in determining the coefficient k. Alternatively, interpolate between the curves of Figure 17.9, again using $L/2D$ in place of H/D. Construct pressure contours parallel to the axis in Region A. The maximum suction occurs in a band along the top and bottom of the cylinder. With cylinders with plane ends, continue these contours into Regions C and D. For the plane ends at $\theta = 0°$, take $\tilde{c}_p\{B\} = -0.6$ in the main Region B and $\tilde{c}_p\{E\} = -1.0$ in the upwind local Region E. With cylinders with domed ends, use the rules for spheres in §20.6.2, taking the upwind point of the domed end as the windward 'pole', splitting the equivalent sphere into the two end Regions F and G and constructing pressure contours along lines of 'latitude' (normal to the wind direction) as shown in Figure 20.31(c). The pressure contours will not match exactly at the region boundaries, so adjust them to blend naturally, taking the more onerous value when in doubt. The whole of the domed ends experience the maximum suction in some wind direction.

For wind directions parallel to the axis, $\theta = 90°$ or $270°$, and cylinders with plane ends (Figure 20.31(b)) take $\tilde{c}_p = -0.8$ in whichever of Regions C or D is upwind, $\tilde{c}_p = +0.8$ on the windward end face Regions B and E, and $\tilde{c}_p = -0.2$ on both

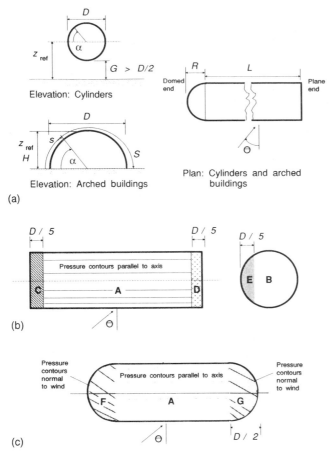

Figure 20.31 Key to horizontal cylinder data: (a) dimensions, (b) loaded areas for plane ends; (c) loaded areas for domed ends

leeward end face regions. For Region A, interpolate linearly from the upwind Region C or D value to zero one diameter downwind and take zero for the remainder of the cylinder (see hemicylinder in Figure 17.17(c)) (§17.2.2.2.1).

20.6.3.2.3 Cylinders resting on the ground.   When $G = 0$, the wind cannot flow under the cylinder and the positive pressure lobe at the front and the wake at the rear extend to the ground. Treat this as a barrel vault roof on a rectangular building with $H/W = 0.5$ and $R/W = 0.5$, using the rules in §20.7.4.6. In essence the region of the cylinder between the height of centre and the ground is assigned the same pressures as vertical walls, using the rules in §20.7.3.1, and the region above the centre is treated as an arched building using the rules in §20.6.3.2.4, below. Force coefficients for this condition were given in Table 20.1 (§20.3.3.1.7).

20.6.3.2.4 Arched buildings.   For the hemicylinder and lower cylindrical section arches, $H/D \leqslant 0.5$, for wind directions within 45° of normal to the axis, $\theta \leqslant \pm 45°$, the pressures around the plane cylinder walls should be taken as the same as the distribution over the centreline of a dome of the same rise ratio, H/D, from Figure 20.29. For the other regions B–E use the same rules in §20.6.3.2.2 as for the cylinder. Use these rules also when the wind is parallel to the axis, $\theta = 90°$ or $270°$, for all regions.

Barrel vault roofs on rectangular-plan buildings become equivalent to arched buildings springing directly from the ground in the limit as the wall height to eaves tends to zero. Hence, the data in §20.7.4.6 for $H/W = 0$ give the same result and may be more convenient to implement. All arched roofs with $H/D > 0.5$ should be treated as equivalent to a barrel vault building with $R/W = 0.5$ and 'eaves' at the height of the centre of the cylinder.

Arched buildings experience reduced pressures when the tops are flattened and increased pressures when the tops are ridged (§17.2.2.2.2).

20.6.2.3.5 Multibay arched buildings.   Limited full-scale data [290] indicate that pressures on arched buildings with multiple bays are the same as on isolated arched buildings for the upwind bay, but reduce for downwind bays in a similar manner to multibay canopy roofs (§20.5.3.4) when the wind is normal to the bay axes, $\theta = 0°$. Reduction factors for this case are given in Table 20.8. Note that the last nth bay is more heavily loaded than the 4th to $(n - 1)$th bays when there are five bays or more. When the wind is parallel to the bay axes, each bay experiences the full pressures of the isolated case. (§17.2.2.2.2).

Table 20.8   Reduction factors for multi-bay arched buildings

Position		Reduction factor, $\phi_{\bar{c}_p}\{n\}$
Upwind	bay 1	1.0
	bay 2	0.85
	bay 3	0.73
	bays 4 to $n - 1$	0.60
Downwind	bay n (> 3)	0.70

Loading data for flat-faced bluff structures

20.7.1 Scope

These data apply to bluff structures that are predominantly flat faced, but may have some curved sections. These include most conventional building shapes: typically cuboidal, or composed of cuboidal elements, with different plane roof forms such as flat, monopitch, duopitch, hipped and mansard. Curved roofs on plane-walled buildings: 'true' curves such as cylindrical section barrel-vault roofs and 'apparent' curves such as hyperboloid roofs; are included in this section because the coherent flow structures, particularly the vortices that form at the junction of the flat and curved faces, have more characteristics common to flat-faced than to curved structures.

The definition of reference dynamic pressure, $\hat{q}_{ref}$, is generally at a fixed reference height, typically the height $z_{ref} = H$ to the eaves of the building or the average height of the roof. An important exception are the data for overall loads on tall buildings, based on the Akins *et al.* data set [277], which use a special reference dynamic pressure.

Loading coefficients are, wherever possible, pseudo-steady values obtained by measurement of peak pressures, but mean values are given when pseudo-steady values are unavailable. Generally, values of pressure coefficient are given for various loaded areas for major structural forms and primary parameters such as wind angle, θ, and roof pitch, α. Factors in the form of influence functions, Φ, or coefficients, $-1 \leqslant \phi \leqslant 1$, are sometimes provided to correct these primary data for significant secondary aspects. Factors to adjust the pseudo-steady coefficients for load duration are only given when it is necessary since, in general, a satisfactory collapse is obtained (§15.3). Typically, the values given for structural loads were derived for duration $t = 4\,\text{s}$, while peak cladding loads were derived for duration $t = 1\,\text{s}$.

20.7.2 Overall loads on cuboidal buildings

20.7.2.1 Tall buildings

20.7.2.1.1 Definitions.   The data in this section enable the horizontal forces and base bending moments to be determined for tall cuboidal buildings. It is assumed that these data will also be valid for rectangular-plan buildings with typical

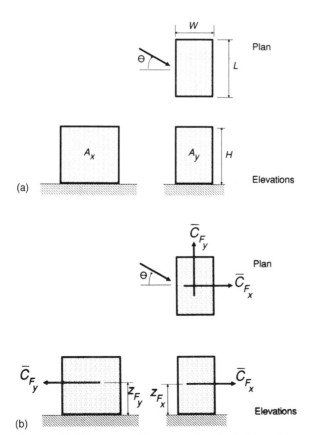

Figure 20.32 Key to overall force data for tall cuboidal buildings: (a) dimensions and reference areas; (b) forces and moment arms

monopitch, duopitch, hipped and mansard roof forms. For the vertical force and torque, and for estimates of reliability refer to Appendix L, where data for all six forces and moments are tabulated.

Figure 20.32 is the key figure: L is the larger horizontal dimension, W is the smaller horizontal dimension and H is the height of the cuboid; the reference areas are $A_x = L\,H$ and $A_y = W\,H$, as defined in (a). For other roof forms take H to the average height of the roof. Loading coefficients are mean values. Force coefficients for the x and y axes, $\overline{C}_{F_x}$ and $\overline{C}_{F_y}$, and their moment arms, z_{F_x} and z_{F_y}, for deriving the base moments are defined in (b) and comply with the convention of Figures 12.7 and 12.9 (§12.3.6).

The data of this section are valid when the proportions of the building are in the range:

$$0.5 \leqslant H/L \leqslant 4 \qquad 0.5 \leqslant H/W \leqslant 4 \qquad 1 \leqslant L/W \leqslant 4 \text{ for all cases.}$$

Taller cuboids, $H/L > 4$ or $H/W > 4$, should be assessed as line-like structures using the data in §20.3.4. The procedure for squatter cuboids is given in §20.7.2.2, below.

20.7.2.1.2 Reference dynamic pressure.
The data of this section are based on a special reference dynamic pressure, $q_{\text{ref}} = \tfrac{1}{2}\,\rho\,V_{\text{ave}}^2$, where the velocity V_{ave} is the average wind speed over the height, H, defined by:

$$V_{\text{ave}} = \frac{1}{H} \int_0^H V\{z\}\,\mathrm{d}z \tag{20.18}$$

For the design of static structures this will normally be the average of the peak gust velocity, $\hat{V}$, obtained from Chapter 9 (§17.3.1.3).

20.7.2.1.3 Overall forces.
The design chart Figure 20.33 gives values of force coefficient, $\overline{C}_F$, for any vertical face in the valid range of L/W for $0° \leqslant \theta \leqslant 90°$. The chart has been prepared by plotting contours of force coefficient with the wind angle from normal to the face as a linear horizontal scale for the range $0° \leqslant \theta \leqslant 90°$

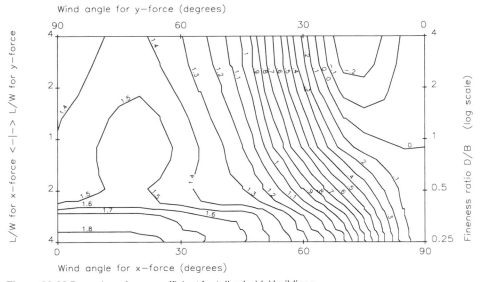

Figure 20.33 Base shear force coefficient for tall cuboidal buildings

(bottom axis), and the fineness ratio with wind normal to the face as a logarithmic vertical scale for the range $0.25 \leqslant D/B \leqslant 4$ (right-hand axis). Since the x-axis of the building is defined normal to the larger face and $L \geqslant W$, the bottom half of the chart gives values for the x-axis force, $\overline{C}_{F_x}\{B \equiv L, D \equiv W\}$, and the top half gives the y-axis force, $\overline{C}_{F_y}\{B \equiv W, D \equiv L\}$. Note the logarithmic scale when interpolating for L/W.

To obtain the x-axis force coefficient, $\overline{C}_{F_x}$, proceed as follows:

1 Locate the required wind angle, θ, along the **bottom** x-axis.
2 Locate the section proportions, L/W, up the **bottom** half of the left-hand y-axis. (Note this axis is logarithmic and reads **downwards** from the middle.)
3 Read off the corresponding value of $\overline{C}_F$ by interpolating between contours.
4 The value of $\overline{C}_{F_x}\{0° \leqslant \theta \leqslant 90°\} = \overline{C}_F$.
5 The value of $\overline{C}_{F_x}\{\theta\}$ for the remaining wind directions is obtained by symmetry:

$$\overline{C}_{F_x}\{\ 90° \leqslant \theta \leqslant 180°\} = -\ \overline{C}_{F_x}\{180° - \theta\}$$
$$\overline{C}_{F_x}\{180° \leqslant \theta \leqslant 270°\} = -\ \overline{C}_{F_x}\{\theta - 180°\}$$
$$\overline{C}_{F_x}\{270° \leqslant \theta \leqslant 360°\} = +\ \overline{C}_{F_x}\{360° - \theta\}$$

To obtain the y-axis force coefficient, $\overline{C}_{F_y}$, proceed as follows:

6 Locate the required wind angle, θ, along the **top** x-axis. (Note this axis is reversed from right to left.)
7 Locate the section proportions, L/W, up the **top** half of the left-hand y-axis. (Note this axis is logarithmic and reads **upwards** from the middle.)
8 Read off the corresponding value of $\overline{C}_F$ by interpolating between contours.
9 The value of $\overline{C}_{F_y} = -\overline{C}_F$. (The sign is negative because the definition of $\overline{C}_{F_y}$ is upwind in the range $0° \leqslant \theta \leqslant 90°$. See Figure 20.32.)
10 The value of $\overline{C}_{F_y}\{\theta\}$ for the remaining wind directions is obtained by symmetry:

$$\overline{C}_{F_y}\{\ 90° \leqslant \theta \leqslant 180°\} = +\ \overline{C}_{F_y}\{180° - \theta\}$$
$$\overline{C}_{F_y}\{180° \leqslant \theta \leqslant 270°\} = -\ \overline{C}_{F_y}\{\theta - 180°\}$$
$$\overline{C}_{F_y}\{270° \leqslant \theta \leqslant 360°\} = -\ \overline{C}_{F_y}\{360° - \theta\}$$

20.7.2.1.4 Overall base moments.   The overall base moments are obtained from the corresponding horizontal forces and the moment arm of their centre of force above the base of the building. The effect of wind angle on the moment arms, z_{F_x} and z_{F_y}, was shown in Figure 17.24. For design purposes the moment arms may be taken as constant:

$$z_{F_x} = 0.56\,H \tag{20.19}$$
$$z_{F_y} = 0.63\,H \tag{20.20}$$

20.7.2.2 Squat buildings

Overall forces and moments on squat buildings should be obtained by integrating the pressure coefficients given for structural loads on all of the loaded regions. This is required because the action of the pressure distribution on the various forms of roof can be significant to the overall forces and moments, particularly when the height to eaves is equal or less than half the height to the ridge. It is therefore not appropriate to give design values of overall loading coefficients for squat buildings (§17.3.1.4).

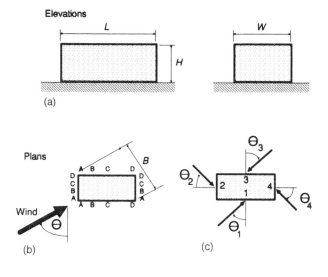

Elevations

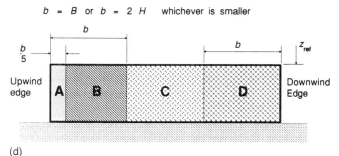

$b = B$ or $b = 2 H$ whichever is smaller

Figure 20.34 Key to wall pressure data: (a) fixed dimensions; (b) variable dimensions; (c) local wind angles for face; (d) key to loaded areas

20.7.3 Pressures on walls

20.7.3.1 Vertical walls

The data in this section enable the loading on vertical walls of rectangular-plan buildings (§20.7.3.1.1) to be determined and also form the basis for empirical rules for polygonal-plan (§20.7.3.1.2) and complex-plan (§20.7.3.1.3) buildings. Rules for non-vertical walls are given in §20.7.3.2.

20.7.3.1.1 Rectangular-plan buildings.   Figure 20.34 is the key figure: L is the larger horizontal dimension, W is the smaller horizontal dimension and H is the height of the cuboid, as defined in (a); B is the cross-wind breadth, which varies with wind direction, as defined in (b). For other roof forms take H to the top of the wall. Wind angle, θ, which is measured from normal to one larger face, is also expressed as local wind angles, θ_n, from normal to each face, where n is an index number for the face as defined in (c). The reference dynamic pressure, $\hat{q}_{\mathrm{ref}}$, is defined at the top of the wall, $z_{\mathrm{ref}} = H$.

Loaded regions are defined in (d) as vertical strips in terms of the scaling length, b, which depends on the slenderness ratio (§17.3.2.2). The value is taken as the **smaller** of $b = B$ or $b = 2H$. Region A is always at the upwind edge of the face as shown in (b), so that the Regions run from left to right for $0° \leqslant \theta_n \leqslant 180°$ as shown in (d), but run right to left for $180° \leqslant \theta_n \leqslant 360°$ (or $0° \geqslant \theta_n \geqslant -180°$), depending on which edge is upwind. Depending on the proportion of the face, not all the regions will exist or be their full defined size. Figure 20.35 shows some typical examples.

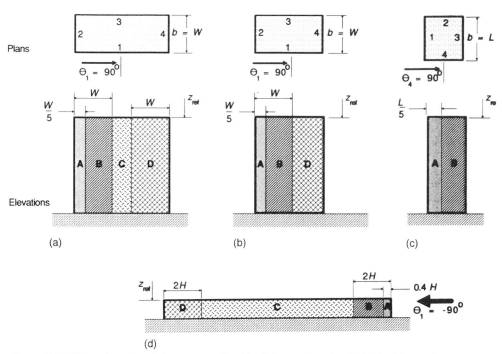

Figure 20.35 Examples of loaded areas on walls: (a) tall, longer face, $L > 2W$; (b) tall longer face, $W < L < 2W$; (c) tall, shorter face $W < L$; (d) squat

To determine the loaded regions proceed as follows:

1 Determine the cross-wind breadth, B, for the current wind direction.
2 Determine the height, H, to the top of the wall.
3 Calculate b as the **smaller** of $b = B$ or $b = 2H$.
4 Define Region A from the upwind edge of the face.
5 Next define Region B, again from the upwind edge. For tall buildings when $2H \geqslant B$, Regions A and B occupy the whole face (as in Figure 20.35(c)) and the procedure is complete.
6 Next define Region D, this time from the downwind edge. If there is insufficient room for the defined size, Region D occupies the remainder of the face after Regions A and B have been defined (as in Figure 20.35(b)).
7 All the remainder of the face after Regions A, B and D have been defined is Region C.

In the special case of wind exactly normal to the face, $\theta_n = 0°$ or $180°$ both edges should be treated as upwind edges and Region D is replaced by Regions A and B. (Values for Region D are correctly given to allow for this case.)

Pressure coefficients from BRE pseudo-steady measurements are given for each region over the range of local wind angle $0° \leqslant \theta_n \leqslant 180°$ in Table 20.9 over the range of slenderness ratio $2H/B \leqslant 8$. (Note that Table 20.9 gives the slenderness in terms of the proportions H/B, without the factor of 2 on H which always appears for surface mounted structures.) The assessment procedure proceeds as follows:

8 Determine the local wind angle, θ_n, for each face, $n = 1$ to 4, from:

$$\theta_1 = \theta, \qquad \theta_2 = 270° + \theta, \qquad \theta_3 = 180° + \theta, \qquad \theta_4 = 90° + \theta \qquad (20.21)$$

as indicated in Figure 20.34(c).
9 Look up values of pressure coefficient for each region the range $0° \leqslant \theta_n \leqslant 180°$ in Table 20.9, interpolating for H/B between columns in the range

Table 20.9 Pressure coefficients for vertical walls

Key diagram: Figure 20.34, 20.35 and 20.36 $\hat{q}_{ref}$ at top of wall, $z_{ref} = H$

	Structural loads ($t = 4$ s)						Peak suctions ($t = 1$ s)	
	$H/B \leqslant 0.5$				$H/B = 1$		$H/B \leqslant 0.5$	$H/B = 1$
Region	A	B	C	D	A	B	A	A
	Pseudo-steady pressure coefficient, $\bar{c}_p\{t\}$							
$\theta_n = 0°$	0.66	0.83	0.86	0.83	0.78	0.89		
15°	0.77	0.88	0.80	0.68	0.91	0.91		
30°	0.80	0.80	0.71	0.49	0.88	0.80		
45°	0.79	0.69	0.54	0.34	0.84	0.68		
60°	0.24	0.51	0.40	0.26	−0.85	0.23		
75°	−0.77	−0.22	0.23	0.08	−1.10	−0.73		
90°	−0.91	−0.68	−0.42	−0.12	−1.21	−0.92	−1.21	−1.27
105°	−0.73	−0.73	−0.48	−0.26	−0.80	−0.80		
120°	−0.63	−0.63	−0.46	−0.29	−0.64	−0.64		
135°	−0.50	−0.50	−0.40	−0.33	−0.59	−0.59		
150°	−0.34	−0.34	−0.26	−0.32	−0.54	−0.54		
165°	−0.30	−0.30	−0.23	−0.35	−0.38	−0.38		
180°	−0.34	−0.34	−0.22	−0.34	−0.24	−0.24		

	Structural loads ($t = 4$ s)				Peak suctions ($t = 1$ s)	
	$H/B = 2$		$H/B \geqslant 3$		$H/B = 2$	$H/B \geqslant 3$
Region	A	B	A	B	A	A
	Pseudo-steady pressure coefficient, $\bar{c}_p\{t\}$					
$\theta_n = 0°$	0.67	0.80	0.62	0.76		
30°	0.83	0.71	0.80	0.68		
60°	0.31	0.31	0.20	0.25		
90°	−1.02	−0.82	−1.09	−0.94	−1.29	−1.32
120°	−0.54	−0.54	−0.67	−0.67		
150°	−0.51	−0.51	−0.51	−0.51		
180°	−0.40	−0.40	−0.54	−0.54		

$0.5 < H/B < 3$. (**Note:** When the result of interpolating between positive and negative values is in the range $-0.2 < \tilde{c}_p < 0.2$, the result should be taken as $\tilde{c}_p = \pm 0.2$ and the design assessed for both possible values.)

10 Values for local wind angles in the range $180° \leqslant \theta_n \leqslant 360°$ (or $0° \geqslant \theta_n \geqslant -180°$) are obtained by symmetry, since:

$$\tilde{c}_p\{\theta_n\} = \tilde{c}_p\{360° - \theta_n\} = \tilde{c}_p\{-\theta_n\} \tag{20.22}$$

Note that in this range the regions are reversed right to left, since the right hand corner is upwind as in Figure 20.35(d).

The worst peak suctions occur in Region A, i.e. near the upwind edge of the face, when the wind is parallel to the face, $\theta_n = 90°$, and values of $\tilde{c}_p\{t = 1\,\text{s}\}$ are given for this condition in Table 20.9 (§17.3.2).

20.7.3.1.2 Polygonal-plan buildings.

The data for vertical walls of rectangular-plan buildings given above should also be used for polygonal-plan buildings, using slightly adapted rules. Figure 20.36 gives new key diagrams to be used with the previous loaded region key of Figure 20.34(d). Most aspects of the problem are

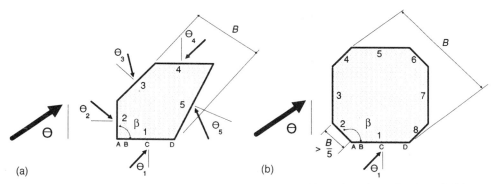

Figure 20.36 Key for walls of polygonal-plan buildings: (a) typical irregular polygon (pentagon); (b) rectangular plan with chamfered corners (octagon)

similar to the rectangular-plan case: the cross-wind breadth, B, is the breadth of the building normal to the wind direction and each face has a local wind angle, θ_n; but now there may be any number of faces greater than three (two faces gives a boundary wall, §20.5.1). The loaded region definitions of Figure 20.34(d) still apply, so that the examples of Figure 20.35 remain valid for the faces shown in elevation, but the plan views may take any polygonal shape. The additional parameter introduced in Figure 20.36 is the corner angle, β, of the **upwind** corner. In both examples (a) and (b), this is shown as the angle between faces '1' and '2', because the left-hand corner of face '1' is upwind for $0° < \theta_1 < 180°$. For the range $180° < \theta_1 < 360°$ (or $0° > \theta_1 > -180°$) the right-hand corner is upwind and the corner angle, β, is between faces '1' and 'n' ($n = 5$ in the case of (a) and $n = 8$ for (b)). The reference dynamic pressure, $\hat{q}_{\text{ref}}$, is taken at the top of the wall, $z_{\text{ref}} = H$.

The assessment for pressures for structural purposes should be made in the same ten steps in §20.7.3.1.1 for rectangular-plan buildings. Provided the length of the adjacent upwind face is greater than $B/5$ (see Figure 20.36(b)), the peak suction coefficient for Region A can be reduced for corner angles $\beta \neq 90°$ by applying the

Table 20.10 Reduction factor for corner angle on worst peak suction coefficients in Region A of vertical walls (derived at $H/B = 1$)

Key diagrams: Figures 20.34 and 20.36	$\hat{q}_{ref}$ at top of wall, $z_{ref} = H$			
Corner angle β	60°	90°	120°	150°
	Reduction factor, $\phi_{\bar{c}_p}\{\beta\}$			
$\theta_n = 90°$ or 270°	0.65	1.00	0.58	0.16

reduction factors in Table 20.10. These have been derived from test values at $H/B = 1$ and have been assumed to apply to other slenderness ratios (§17.3.2.4). **Note:** If the result of applying this factor to Region A gives a less onerous value of suction than on the adjacent Regions, B and C, or at other wind angles, the value can also be taken to apply to these regions and wind angles.

20.7.3.1.3 Re-entrant corners.   This form of corner produces plan forms that include 'L', 'T', 'X' and 'Y' shapes. Figure 20.37 gives the key diagram for adapting the data in Table 20.9 to this case. The X-plan is used in this diagram because it gives all possible orientations of the re-entrant corner in a single diagram. For other plan shapes, apply the rules that give the closest match to each re-entrant corner region.

Follow the ten steps for faces of rectangular buildings in §20.7.3.1.1 with the following changes:

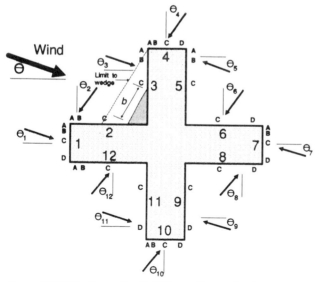

Figure 20.37 Key for walls of buildings with re-entrant corners

1 For the faces of the upwind wing (1, 2 and 12 in Figure 20.37), use the crosswind breadth and the height of the wing to determine the scaling length (b). For all other wings, including any completely within the wake, use the overall building breadth, B, as before.

2 Define the Regions A, B, and C from upwind external corners and Regions C

and D from downwind external corners. In this way, faces with external corners at either end (1, 4, 7, 10 in Figure 20.37) can have all four regions, depending on proportions; while faces with one external and one internal corner lose Region A and B (6, 8, 9 and 11 in Figure 20.37) or Region D (2, 3, 5 and 12 in Figure 20.37), depending on whether the external corner is at upwind or downwind edge.

3 In re-entrant corners that face directly into the wind, define a wedge that extends from the internal corner with the 'face' of the wedge normal to the wind direction, as indicated in Figure 20.37. The size of this wedge is the **smaller** of: cross-wind breadth equal to b, or
    one edge at an external corner ('marked limit to wedge').

4 The pressure coefficient for regions that lie within the defined wedge should be taken for $\theta = 0°$ from Table 20.9, otherwise values are taken for the local wind angle, $\theta = \theta_n$, as before (§17.3.2.5.2).

20.7.3.1.4 Recessed bays.   These occur where there are two internal corners, such as between the two wings of an H-plan building, but also including small recesses such as entrance porches, balconies, etc. Figure 20.38 is the key diagram for adapting the data in Table 20.9 to this case. The local wind angles, which have been omitted for clarity, are defined in exactly the same manner as Figures 20.36 and 20.37.

Proceed in the following steps:

1 First follow the steps 1–3 for faces of rectangular buildings in §20.7.3.1.1 to determine the scaling length, b.

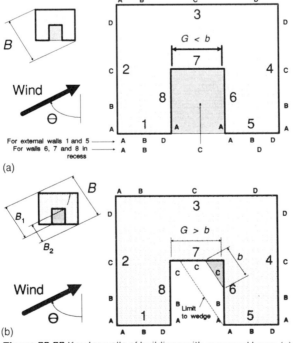

Figure 20.38 Key for walls of buildings with recessed bays: (a) narrow recess; (b) wide recess

2 Next categorise the recess as 'narrow' for $G < b$ or 'wide' for $G > b$, depending on the width of the recess, G.

3 If the recess is narrow refer to Figure 20.38(a) and follow steps 4–6. If the recess is wide refer to Figure 20.38(b) and follow steps 7–8.

4 For a narrow recess, assess all external walls (1, 2, 3, 4 and 5) by completing the ten steps in §20.7.3.1.1.

5 For the internal walls of the narrow recess (6,7,and 8), re-assess the wall in which the recess occurs (1–5) as if the recess did not exist and apply the pressure in the region corresponding to the opening of the recess (C) as the structural loading for all faces in the recess (6, 7 and 8).

6 For peak cladding loads at the mouth of the narrow recess, define local Region A and use the corresponding values from Table 20.9 (faces 6 and 8).

7 For a wide recess, assess all external walls (1, 2, 3, 4 and 5) by completing the ten steps in §20.7.3.1.1.

8 For the walls in the recess, follow the rules for re-entrant corners in §20.7.3.1.3 with the following changes:

   (a) Use the cross-wind breadth of each upwind-facing wing (B_1 and B_2) to determine the scaling length b for the faces of the respective wing.

   (b) Define the regions for the side faces with one internal and one external corner (6 and 8) using the rules in 2 of §20.7.3.1.3, but define the whole of back face with two internal corners (7) as Region C.

   When the recess faces directly into the wind, define the wedge of positive pressure in the downwind corner, exactly as indicated in steps 3 and 4 of §20.7.3.1.3.

Note: In each of the above steps, the face and region indices given in brackets relate to the example in the key Figure 20.38 and will differ in practical cases.

Where the recessed bay is limited in height by a floor or soffit, for example a balcony, the pressure in the recess should be taken to act on the floor and soffit in addition to the walls. For tall buildings with a rectangular envelope, where all recesses are narrow, the overall force and moment data of §20.7.2.1 will be valid (§17.3.2.5.3).

20.7.3.1.5 Central wells.

Here the concern is only for the pressures acting on the walls of the well, since the external walls of the building are unaffected by the well and are assessed as if the well were not there. Pressure in the well is dominated by the flow over the roof.

When the gap across the well, G, is smaller than the scaling length, b, the flow skips over the well and the pressure in the well may be assumed to be uniform and equal to the average pressure appropriate to a roof over the well. This is analogous to step 5 for the narrow recessed bay, except that here the roof pressure is used instead of the wall pressure.

When the gap across the well, G, is larger than the scaling length, b, follow the rules for re-entrant corners, with the exception that the scaling length, b, is always the smaller of the breadth, B, or height, H, for the part of the building containing the well. Some allowance should also be made for the case shown in Figure 17.36(b), where $G < b$ but the well is tall and the flow over the roof acts only on the upper part, but the effect has not been quantified and ignoring it and assuming the whole wall is similarly loaded is conservative.

If access to a narrow central well is provided by a tunnel through an external side, flow will occur through the tunnel driven by the pressure difference between side face and roof. If the area of the central well is greater than three times the cross-sectional area of the tunnel, it may be assumed that the tunnel has no effect on the pressures in the well and pressures on the walls of the tunnel may be assumed to change linearly with position between the well wall and external wall pressures (§17.3.2.5.4).

20.7.3.1.6 Irregular faces.   Figure 20.39 is the key figure for faces of buildings with an rectangular area cut out of one or more upper corner in elevation. This includes any buildings with a lower wing or extension built flush with the main building, as in (a), and buildings with inset upper storeys that are flush with a main face. For the inset faces, see §20.7.3.1.7 'Inset storeys', below.

Proceed for the flush irregular face as follows:

(a) The face should be divided into parts as indicated in Figure 20.39(a).
   (i)   If the tallest part is squat ($H/B < 0.5$) divide from parts either side.
   (ii) If the tallest part is tall ($H/B \geqslant 0.5$) divide from part underneath. Check that it is still tall after this division and if not, treat it as squat as in (i).

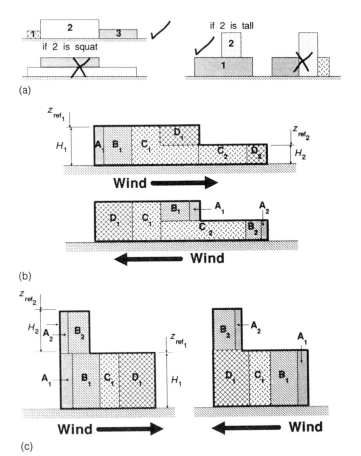

Figure 20.39 Key to flush irregular walls: (a) key to parts; (b) both parts squat; (c) top part tall, bottom part tall or squat

(b) Next divide the parts into Regions A–D, as shown in Figure 20.39(b) and (c). Note in the case of (b) 'Both parts squat', Regions A, B and D reach from the top of the taller part to the top of the lower part if adjacent to that edge, or to the ground if not adjacent. (Also in this case, the division between Regions C_1 and C_2 is important since, although both regions take the same values of pressure coefficient from Table 20.9, the reference dynamic pressures differ.)

(c) Now complete the assessment with steps 8–10 of §20.7.3.1.1 for rectangular-plan buildings.

The reference dynamic pressure, $\hat{q}_{ref}$, is taken at the top of each part determined above for the pressures on that part (§17.3.2.5.5).

20.7.3.1.7 Inset storeys.   Figure 20.40 is the key figure for walls of inset storeys, that is, walls that rise from the roof of a larger-plan lower storey. When the upwind edge of the wall is also set back, as in (a), the wall should be treated in exactly the same manner as a normal building wall of the same proportions assuming the lower roof to be the ground plane. When the upwind edge of the wall is flush with the edge of the lower storey, or is within $b_L/5$ of the edge, define the extra Region E at the bottom of Region A as defined in (b) and use the pressure coefficients given in Table 20.11 (§17.3.2.5.6).

Table 20.11   Pressure coefficients for Region E on walls of inset storeys

Key diagrams: Figures 20.34 and 20.40	$\bar{q}_{ref}$ at top of wall
Region E $\theta_n = 90°$	Pseudo-steady pressure coefficient, $\bar{c}_p\{t\}$ $\bar{c}_p\{t = 4\,s\} = -1.70$ $\bar{c}_p\{t = 1\,s\} = -2.00$

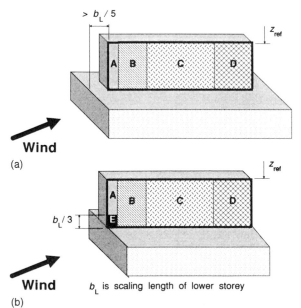

Wind

(a)

Wind b_L is scaling length of lower storey

(b)

Figure 20.40 Key for walls of inset storeys: (a) edge of face inset from edge of lower storey; (b) edge of face flush with edge of lower storey

Table 20.12 Pressure coefficients for non-vertical main walls of A-frame buildings

Key diagram: Figure 20.41(b) and 20.34 $\hat{q}_{ref}$ at ridge. $z_{ref} = H$

Wall pitch $\alpha = 30°$

Structural loads ($t = 4$ s) — Pseudo-steady coefficient, $\bar{c}_p\{t\}$

Region	A	B	C	D	E	F	G
$\theta_1 = 0°$	+0.31	+0.46	+0.51	+0.46			
or 30°	+0.42	+0.42	+0.39	+0.20			
θ_3 60°	-0.88	+0.07	+0.16	+0.07			
90°	-1.07	-0.55	-0.05	-0.09	-0.69	-0.30	-0.19
120°	-1.04	-0.71	-0.37	-0.24	-1.62	-0.75	-0.44
150°	-0.96	-0.65	-0.51	-0.42	-1.26	-0.70	-0.57
180°	-0.45	-0.45	-0.41	-0.45	-0.45	-0.38	-0.45

Peak suctions ($t = 1$ s)

Region	A	E	F	G
$\theta_1 = 0°$				
or 30°				
θ_3 60°				
90°	-1.26			
120°		-1.61	-0.74	-0.53
150°		-1.30	-0.91	-0.74
180°				

Wall pitch $\alpha = 45°$

Structural loads ($t = 4$ s) — Pseudo-steady coefficient, $\bar{c}_p\{t\}$

Region	A	B	C	D	E	F	G
$\theta_1 = 0°$	+0.68	+0.68	+0.68	+0.68			
or 30°	+0.65	+0.63	+0.54	+0.44			
θ_3 60°	+0.20	+0.38	+0.30	+0.22			
90°	-1.00	-0.57	-0.05	-0.14	-0.45	-0.40	-0.45
120°	-0.91	-0.86	-0.62	-0.44	-0.62	-0.58	-0.54
150°	-0.61	-0.65	-0.58	-0.48	-1.15	-0.63	-0.35
180°	-0.49	-0.50	-0.44	-0.53	-0.76	-0.32	-0.18

Peak suctions ($t = 1$ s)

Region	A	E	F	G
$\theta_1 = 0°$				
or 30°				
θ_3 60°				
90°	-1.27			
120°		-0.67	-0.88	-0.88
150°		-1.23	-0.68	-0.55
180°				

Wall pitch $\alpha = 60°$

	Structural loads (t = 4 s)							Peak suctions (t = 1 s)			
Region	A	B	C	D	E	F	G	A	E	F	G
	Pseudo-steady coefficient, $\bar{c}_p\{t\}$										
$\theta_1 = 0°$	+0.50	+0.57	+0.80	+0.57							
or 30°	+0.77	+0.59	+0.62	+0.39							
θ_3 60°	+0.59	+0.37	+0.35	+0.19							
90°	−0.89	−0.44	−0.04	−0.13	−0.72	−0.24	−0.05				
120°	−0.91	−0.74	−0.63	−0.41	−0.95	−0.54	−0.27	−1.21	−1.00	−0.60	−0.42
150°	−0.67	−0.53	−0.60	−0.33	−0.64	−0.54	−0.46		−0.63	−0.71	−0.69
180°	−0.57	−0.41	−0.45	−0.41	−0.49	−0.40	−0.49				

Wall pitch $\alpha = 75°$

	Structural loads (t = 4 s)							Peak suctions (t = 1 s)			
Region	A	B	C	D	E	F	G	A	E	F	G
	Pseudo-steady coefficient, $\bar{c}_p\{t\}$										
$\theta_1 = 0°$	+0.58	+0.81	+0.81	+0.81							
or 30°	+0.85	+0.83	+0.73	+0.55							
θ_3 60°	+0.78	+0.55	+0.41	+0.32							
90°	−0.81	−0.43	−0.07	−0.12	−0.79	−0.42	−0.21				
120°	−0.97	−1.15	−0.61	−0.39	−1.13	−0.67	−0.27	−1.21	−1.13	−0.66	−0.31
150°	−0.70	−0.75	−0.62	−0.46	−0.71	−0.58	−0.45		−0.67	−0.64	−0.63
180°	−0.58	−0.54	−0.45	−0.54	−0.54	−0.43	−0.54				

20.7.3.2 Non-vertical walls

20.7.3.2.1 A-frame buildings. These are buildings of the form shown in Figure 17.39(b), with two main pitched opposite faces which meet at the top edge to form a ridge, and vertical triangular gable end faces. Figure 20.41 is the key figure: showing in (a) the principal dimensions; in (b) the standard wall Regions A–D of Figure 20.34 with additional local Regions E, F and G along the top edge of the main faces to account for the ridge vortex when the face is downwind; and in (c), Regions H–K for the triangular gable. The local wind angles for each face are

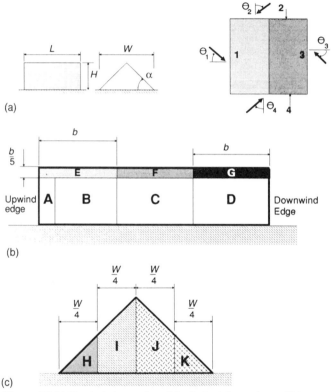

Figure 20.41 Key to A-frame data: (a) principal dimensions; (b) key to additional loaded areas on leeward main face; (c) key to loaded areas on gable face

defined with faces 1 and 3 always the main pitched faces, and with faces 2 and 4 always the triangular gables. The pitch angle of the wall is α, with $\alpha = 90°$ representing a vertical wall. The reference dynamic pressure, $\hat{q}_{ref}$, is taken at the height of the ridge, $z_{ref} = H$.

The assessment of loading proceeds in the same ten steps as for rectangular-plan buildings in §20.7.3.1.1, except that values of pseudo-steady pressure coefficients are taken from Table 20.12 for main faces 1 and 3 and Table 20.13 for gable faces 2 and 4. Note that values for the local ridge Regions E, F and G are given only when the ridge vortex exists, and values from the adjacent Regions B, C and D should be used when there is no given value. Maximum 1 s-duration peak suction coefficients are given for the wind angle at which they occur (§17.3.2.6).

Table 20.13   Pressure coefficients for vertical gable walls of A-frame buildings

Key diagram: Figure 20.41(c) $\hat{q}_{ref}$ at ridge, $z_{ref} = H$

Wall pitch $\alpha = 30°$

	Structural loads ($t = 4$ s)				Peak suctions ($t = 1$ s)
Region	H	I	J	K	H
	Pseudo-steady pressure coefficient, $\bar{c}_p\{t\}$				
$\theta_2 = 0°$	+ 0.25	+ 0.80	+ 0.80	+ 0.25	
or 30°	+ 0.60	+ 0.75	+ 0.50	± 0.2	
θ_4 60°	+ 0.50	+ 0.40	+ 0.20	− 0.20	
90°	− 1.00	− 0.75	− 0.50	− 0.25	
120°	− 1.25	− 0.75	− 0.40	− 0.30	− 1.30
150°	− 0.30	− 0.25	− 0.25	− 0.25	
180°	− 0.25	− 0.25	− 0.25	− 0.25	

Wall pitch $\alpha = 45°$

	Structural loads ($t = 4$ s)				Peak suctions ($t = 1$ s)
Region	H	I	J	K	H
	Pseudo-steady pressure coefficient, $\bar{c}_p\{t\}$				
$\theta_2 = 0°$	+ 0.25	+ 0.80	+ 0.80	+ 0.25	
or 30°	+ 0.60	+ 0.75	+ 0.50	± 0.2	
θ_4 60°	+ 0.50	+ 0.25	± 0.2	− 0.25	
90°	− 0.20	− 0.80	− 0.65	− 0.40	
120°	− 1.00	− 0.75	− 0.65	− 0.60	− 1.25
150°	− 0.03	− 0.25	− 0.25	− 0.25	
180°	− 0.25	− 0.25	− 0.25	− 0.25	

Wall pitch $60° \le \alpha \le 75°$

	Structural loads ($t = 4$ s)				Peak suctions ($t = 1$ s)
Region	H	I	J	K	H
	Pseudo-steady pressure coefficient, $\bar{c}_p\{t\}$				
$\theta_2 = 0°$	+ 0.25	+ 0.80	+ 0.80	+ 0.25	
or 30°	+ 0.70	+ 0.75	+ 0.50	± 0.2	
θ_4 60°	+ 0.40	+ 0.15	± 0.2	− 0.25	
90°	− 1.10	− 0.80	− 0.70	− 0.60	−1.25
120°	− 0.85	− 0.70	− 0.60	− 0.50	
150°	− 0.30	− 0.25	− 0.25	− 0.25	
180°	− 0.25	− 0.25	− 0.25	− 0.25	

20.7.3.2.2 Pyramids.   Design pressure coefficients for the special case of a pyramid formed from four equilateral triangle faces, $\alpha = 54.7°$, are given in Figure 17.44 (§17.3.2.6.3).

20.7.3.2.3 Non-vertical walls of flat-roofed buildings.   Figure 17.39(a) shows a typical building of this type. The problem breaks down into the following face types:

(a) Rectangular non-vertical faces:
    1 Windward face, $\theta \le \pm 60°$: treat as main face of A-frame building using data in Table 20.12.

    2 Side and leeward faces, $60° < \theta < 300°$: treat as typical vertical wall using data of Table 20.9.

(b) Vertical faces with tapered ends:
    1 Divide face into rectangular central region and triangular end regions.
    2 Treat upwind triangular end region as upwind half of gable end face of A-frame building, using data from Table 20.13.
    3 Treat remainder of face as typical rectangular wall using data of Table 20.9.

(c) Non-vertical faces with tapered ends:
    1 Divide face into rectangular central region and triangular end regions;
    2 Windward face, $\theta \leqslant \pm 60°$: treat whole face as main face of A-frame building using data in Table 20.12.
    3 Side and leeward faces, $60° < \theta < 300°$: treat upwind triangular end region as upwind half of gable end face of A-frame building, using data from Table 20.13; treat remainder of face as typical rectangular wall using data of Table 20.9.

Assess pressures on the flat roof using the mansard eave rules in §20.7.4.1.6 (§17.3.2.6.4).

20.7.3.3 Friction-induced loads

Friction forces may be significant on long walls when the wind is parallel to the wall, $\theta_n = 90°$ or $270°$, and may be required for stability calculations when the normal pressure loads on the front and back faces are small, i.e. when $L >> W$. Friction forces act **upwind** in the recirculating flow of the separation bubble at the upwind end of the wall, but act downwind over the parts of the wall swept by the wind after flow reattachment. A reasonable estimate is obtained by treating Regions C and D as being swept by the wind. Values of shear stress coefficient, c_τ, are given in Table 16.8. Taking the reference dynamic pressure, $\hat{q}_{ref}$, at the top of the wall is conservative.

    Loads on individual vertical ribs or mullions in Regions C and D should be estimated by treating them as multiple boundary walls, using the rules in §20.5.1. See also §20.8.3 (§17.3.2.7.4).

20.7.4 Pressures on roofs

20.7.4.1 Flat roofs

20.7.4.1.1 Scope and definitions. The data in this section should be used for all roofs of pitch $-5° < \alpha < 5°$. Figure 20.42 is the key figure defining the loaded regions behind each upwind eave/verge: E is the length of the upwind eave/verge measured from upwind external corner to downwind external corner or **upwind-facing** internal corner, but ignoring any **downwind-facing** internal corners and H is the height to eaves of the wall. The scaling length e is taken as $e = E$ or $e = 2H$, whichever is the smaller. The wind angle, θ, is expressed as the local wind angle, θ_n, from normal to the eave/verge.

    The reference dynamic pressure, $\hat{q}_{ref}$, is defined at the height of the eave, $z_{ref} = H$. Accordingly, the definitions of θ and $\hat{q}_{ref}$ are identical to the corresponding wall definitions (§20.7.3).

20.7.4.1.2 Loaded regions. Loaded regions are defined in strips parallel to the eave/verge which are further divided downwind from the upwind corner. The

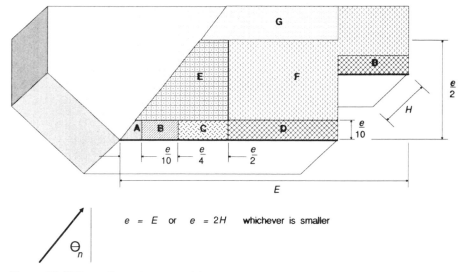

$$e = E \quad \text{or} \quad e = 2H \quad \text{whichever is smaller}$$

Figure 20.42 Key to flat roof pressure data

regions at the corner are divided from the regions along the adjacent eave/verge by the line through the corner in the direction of the wind, allowing the loaded regions to be defined for any corner angle. When the loaded regions behind every upwind eave/verge have been defined, the whole roof is covered.

Figure 20.43 gives examples that cover most of the common cases. Other special cases are covered in §20.7.4.9 and §20.7.4.10. Consider the various upwind corners in (a):

'6' is a typical upwind corner with Regions A–D being defined down either eave/verge '6 → 5' and '6 → 7'.

'7' is the upwind corner for eave/verge '7 → 1', for which Regions A–D are defined, but it is the downwind corner for eave/verge '6 → 7'.

'4' is an internal corner of a recessed bay, but as it faces downwind it is ignored. Accordingly eave/verges '6 → 5' and '4 → 3' are treated as a single eave for the value of E and there are no Regions A–C along '4 → 3'. Note that since the regions are defined from the eave/verge, Regions D and F are set back.

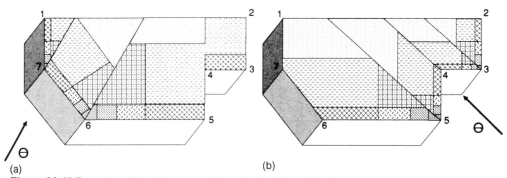

(a)   (b)

Figure 20.43 Examples of loaded areas on flat roofs of arbitrary plan shapes

Now consider the upwind corners for another wind direction in (b):

'5' and '3' are typical upwind corners, like '6' in (a).
'4' is now an upwind-facing internal corner, so terminates the eave/verge '5 → 4' and '3 → 4'.

Determine the position of every upwind eave/verge and define the regions behind each as follows:

1 Determine the length, E_n, of the eave.
2 Determine the height, H_n, of the corresponding wall (i.e. to the ground for simple rectangular-plan buildings or to the lower roof level for inset storeys (§17.3.3.10).
3 Calculate e as the **smaller** of $e = E$ or $e = 2H$.
4 Draw the boundary line through the upwind corner in the wind direction.
5 If the downwind corner of the eave/verge is an upwind-facing internal corner (as corner '4' in Figure 20.43(b)), draw the second boundary line through the downwind corner in the wind direction.
6 Mark out the depth of the edge regions parallel to the eave and define edge Regions A–D. When the wind is exactly normal to an eave/verge with external corners at both ends (e.g. cuboid), define the regions inwards from both corners. (See the special empirical rules for re-entrant corners, recessed bays and central wells in 20.7.4.9.)
7 Mark out the depth of the central regions parallel to the eave and define Regions E and F.
8 All the remainder of the roof downwind of Regions E and F is Region G.

20.7.4.1.3 Flat roofs with sharp eaves.   Pressure coefficients for each region of flat roofs with sharp eaves are given in Table 20.14 in terms of the local wind angle, θ_n, either side from normal to the eave in steps of 15°. Interpolation should be used to give intermediate values. Owing to the way the vortex dynamics dominate the pressures on the roof, the values of pseudo-steady pressure coefficients for the two durations adopted for this Guide ($t = 4\,s$ for structural loads and $t = 1\,s$ for peak cladding loads) remain very similar so that only one value is tabulated for both purposes. Note that the values are mostly high or moderate suctions, but in the far downwind regions where the flow has become reattached behind the 'delta-wing'

Table 20.14   Pressure coefficients for flat roofs with sharp eaves

Key diagram: Figure 20.42 $\hat{q}_{ref}$ at height of eave, $z_{ref} = H$

Region	Structural loads ($t = 4$ s) and peak suctions ($t = 1$ s)						
	A	B	C	D	E	F	G
	Pseudo-steady pressure coefficient, $\tilde{c}_p\{t\}$						
$\theta_n = 0°$	− 1.47	− 1.25	− 1.15	− 1.15	− 0.69	− 0.71	± 0.20
± 15°	− 1.68	− 1.47	− 1.24	− 1.14	− 0.61	− 0.70	± 0.20
± 30°	− 2.00	− 1.70	− 1.38	− 1.03	− 0.66	− 0.67	± 0.20
± 45°	− 1.90	− 1.49	− 1.18	− 0.86	− 0.59	− 0.54	± 0.20
± 60°	− 1.70	− 1.24	− 1.10	− 0.64	− 0.61	− 0.42	± 0.20
± 75°	− 1.45	− 0.85	− 0.69	− 0.35	− 0.61	− 0.24	± 0.20
± 90°	− 1.20	− 0.75	− 0.52	− 0.24	− 0.62	± 0.20	± 0.20

vortices the values range either side of zero and here the value $\tilde{c}_p\{t\} = \pm 0.20$ is given. The designer should account for the possibility of both the negative and the positive value (§17.3.3).

20.7.4.1.4 Flat roofs with parapets.

A parapet along any eave or verge is taken to reduce the suctions in the edge Regions A–D, but not the other Regions E–G. Figure 20.44(a) defines the parapet dimensions. Accordingly, Table 20.15 gives values of the reduction factor, $\phi\{h\}$, in terms of the parapet height, h, to be applied to the sharp-eave pressure coefficients of Table 20.14 to give $\tilde{c}_p\{h\} = \phi\{h\}\,\tilde{c}_p$. These values are dependent on local wind angle, θ_n, and are tabulated in steps of 30°. Interpolation should be used to give intermediate values.

Loading of the parapet walls should be determined using the boundary-wall rules of §20.5.1.1, including the effect near corners. Loads on the downwind parapet may be reduced by the shelter offered by the upwind parapet and any plant room or other inset storey. The shelter factors of §20.5.1.3 may be used, taking the sheltering wall height and spacing from the upwind parapet dimensions or the inset storey dimensions as appropriate (§17.3.3.2.6).

Table 20.15   Reduction factors for effect of parapets on flat roofs

Key diagram: Figure 20.42 and 20.44(a)	$\hat{q}_{ref}$ at height of eave, $z_{ref} = H$		
Eave edge Regions A, B, C and D	Structural loads ($t = 4$ s) and peak suctions ($t = 1$ s)		
Parapet height $h/H =$	0.0125	0.025	0.10
	Reduction factor, $\phi_{\tilde{c}_p}\{h/H\}$		
$\theta_n = 0°$	0.76	0.67	0.56
30°	0.90	0.88	0.70
60°	1.00	0.97	0.74
90°	0.84	0.60	0.60
Interior Regions, E, F and G	Structural loads ($t = 4$ s) and peak suctions ($t = 1$ s)		
Parapet height $h/H =$	0.0125	0.025	0.10
	Reduction factor, $\phi_{\tilde{c}_p}\{h/H\}$		
All θ	1.00	1.00	1.00

20.7.4.1.5 Flat roofs with curved eaves.

Figure 20.44(b) is the key to the curved eave details. The loaded Regions A–D start from the edge of the flat part of the roof. Pressure coefficients for each region of flat roofs with curved eaves are given in Table 20.16 in terms of the eave corner radius, r, and the local wind angle, θ_n, either side from normal to the eave in steps of 15°. Interpolation should be used to give intermediate values (using the values in Table 20.14 for $r/b = 0$). Pressures on the curved part of the wall/eave junction should be assessed using the barrel-vault rules in §20.7.4.6 (§17.3.3.2.7).

20.7.4.1.6 Flat roofs with mansard eaves.

Figure 20.44(c) is the key to the mansard eave details. The loaded Regions A–D start from the edge of the flat part of the roof. Pressure coefficients for each region of flat roofs with mansard eaves are given in Table 20.17 in terms of the mansard pitch in the range, $30° \geqslant \alpha \geqslant 45°$,

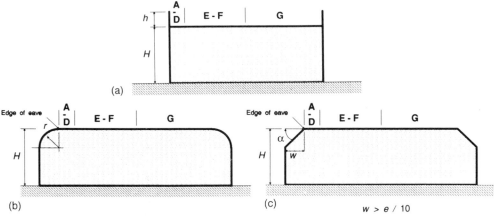

Figure 20.44 Key to eave details for flat roofs: (a) parapets; (b) curved eaves; (c) mansard eaves

Table 20.16   Pressure coefficients for flat roofs with curved eaves

Key diagram: Figure 20.42 and 20.44(b) $\hat{q}_{ref}$ at height of eave, $z_{ref} = H$

Eave radius	$r/b = 0.05$		Structural loads ($t = 4$ s) and peak suctions ($t = 1$ s)				
Region	A	B	C	D	E	F	G
	Pseudo-steady pressure coefficient, $\bar{c}_p\{t\}$						
$\theta_n = 0°$	− 0.81	− 1.00	− 1.15	− 1.26	− 0.39	± 0.20	± 0.20
± 15°	− 0.79	− 1.06	− 1.16	− 1.09	− 0.37	− 0.22	± 0.20
± 30°	− 0.66	− 0.97	− 1.07	− 1.06	− 0.35	− 0.29	± 0.20
± 45°	− 0.61	− 0.80	− 0.90	− 0.92	− 0.35	− 0.35	± 0.20
± 60°	− 0.66	− 0.64	− 0.69	− 0.62	− 0.35	− 0.38	± 0.20
± 75°	− 0.79	− 0.48	− 0.53	− 0.48	− 0.37	− 0.40	± 0.20
± 90°	− 0.81	− 0.48	− 0.39	− 0.29	− 0.39	− 0.43	± 0.20

Eave radius	$r/b = 0.10$		Structural loads ($t = 4$ s) and peak suctions ($t = 1$ s)				
Region	A	B	C	D	E	F	G
	Pseudo-steady pressure coefficient, $\bar{c}_p\{t\}$						
$\theta_n = 0°$	− 0.77	− 0.73	− 0.79	− 0.79	− 0.30	− 0.21	± 0.20
± 15°	− 0.64	− 0.65	− 0.70	− 0.69	− 0.29	− 0.22	± 0.20
± 30°	− 0.56	− 0.60	− 0.62	− 0.63	− 0.29	− 0.25	± 0.20
± 45°	− 0.49	− 0.51	− 0.56	− 0.58	− 0.28	− 0.28	± 0.20
± 60°	− 0.56	− 0.40	− 0.43	− 0.46	− 0.29	− 0.30	± 0.20
± 75°	− 0.64	− 0.39	− 0.36	− 0.36	− 0.29	− 0.30	± 0.20
± 90°	− 0.77	− 0.43	− 0.37	− 0.25	− 0.30	− 0.30	± 0.20

Eave radius	$r/b \geq 0.20$		Structural loads ($t = 4$ s) and peak suctions ($t = 1$ s)				
Region	A	B	C	D	E	F	G
	Pseudo-steady pressure coefficient, $\bar{c}_p\{t\}$						
$\theta_n = 0°$	− 0.51	− 0.54	− 0.54	− 0.56	− 0.30	− 0.21	± 0.20
± 15°	− 0.46	− 0.49	− 0.52	− 0.53	− 0.28	− 0.22	± 0.20
± 30°	− 0.40	− 0.43	− 0.47	− 0.51	− 0.26	− 0.25	± 0.20
± 45°	− 0.38	− 0.41	− 0.43	− 0.43	− 0.26	− 0.27	± 0.20
± 60°	− 0.40	− 0.38	− 0.40	− 0.38	− 0.26	− 0.29	± 0.20
± 75°	− 0.46	− 0.35	− 0.31	− 0.28	− 0.28	− 0.29	± 0.20
± 90°	− 0.51	− 0.40	− 0.36	− 0.23	− 0.30	− 0.30	± 0.20

and the local wind angle, θ_n, either side from normal to the eave in steps of 15°. Interpolation should be used to give intermediate values (using the values in Table 20.14 for $\alpha = 90°$). The data are not valid for mansard pitch angles less than 30°. Pressures on the mansard slope of the wall/eave junction should be assessed using the monopitch roof rules in §20.7.4.2 or the hipped roof rules in §20.7.4.3, as appropriate to the form of the mansard eave (§17.3.3.14).

Table 20.17   Pressure coefficients for flat roofs with mansard eaves

Key diagram: Figure 20.42 and 20.44(c) $\hat{q}_{ref}$ at height of eave, $z_{ref} = H$

Mansard pitch	$\alpha = 30°$	Structural loads ($t = 4$ s) and peak suctions ($t = 1$ s)					
Region	A	B	C	D	E	F	G
	Pseudo-steady pressure coefficient, $\tilde{c}_p\{t\}$						
$\theta_n = 0°$	− 0.93	− 0.98	− 0.98	− 0.98	− 0.27	± 0.20	± 0.20
± 15°	− 0.76	− 0.85	− 0.91	− 0.94	− 0.22	± 0.20	± 0.20
± 30°	− 0.66	− 0.73	− 0.75	− 0.88	− 0.20	± 0.20	± 0.20
± 45°	− 0.60	− 0.59	− 0.63	− 0.66	− 0.21	− 0.25	± 0.20
± 60°	− 0.66	− 0.40	− 0.42	− 0.36	− 0.20	− 0.30	± 0.20
± 75°	− 0.76	− 0.34	− 0.30	− 0.23	− 0.22	− 0.30	± 0.20
± 90°	− 0.93	− 0.39	− 0.30	− 0.22	− 0.27	− 0.26	± 0.20

Mansard pitch	$\alpha = 45°$	Structural loads ($t = 4$ s) and peak suctions ($t = 1$ s)					
Region	A	B	C	D	E	F	G
	Pseudo-steady pressure coefficient, $\tilde{c}_p\{t\}$						
$\theta_n = 0°$	− 1.19	− 1.24	− 1.29	− 1.34	− 0.44	± 0.20	± 0.20
± 15°	− 1.10	− 1.22	− 1.22	− 1.24	− 0.39	± 0.20	± 0.20
± 30°	− 0.98	− 1.06	− 1.06	− 1.05	− 0.35	− 0.24	± 0.20
± 45°	− 0.87	− 0.89	− 0.88	− 0.80	− 0.35	− 0.36	± 0.20
± 60°	− 0.98	− 0.62	− 0.64	− 0.34	− 0.35	− 0.46	± 0.20
± 75°	− 1.10	− 0.50	− 0.45	− 0.24	− 0.39	− 0.48	± 0.20
± 90°	− 1.19	− 0.56	− 0.41	− 0.21	− 0.44	− 0.46	± 0.20

Mansard pitch	$\alpha = 60°$	Structural loads ($t = 4$ s) and peak suctions ($t = 1$ s)					
Region	A	B	C	D	E	F	G
	Pseudo-steady pressure coefficient, $\tilde{c}_p\{t\}$						
$\theta_n = 0°$	− 1.27	− 1.27	− 1.27	− 1.23	− 0.59	± 0.20	± 0.20
± 15°	− 1.37	− 1.25	− 1.27	− 1.17	− 0.54	± 0.20	± 0.20
± 30°	− 1.32	− 1.22	− 1.08	− 1.02	− 0.49	− 0.26	± 0.20
± 45°	− 1.21	− 1.11	− 0.97	− 0.77	− 0.45	− 0.45	± 0.20
± 60°	− 1.32	− 0.81	− 0.73	− 0.35	− 0.49	− 0.60	± 0.20
± 75°	− 1.37	− 0.70	− 0.54	− 0.23	− 0.54	− 0.66	± 0.20
± 90°	− 1.27	− 0.69	− 0.48	− 0.21	− 0.59	− 0.66	± 0.20

20.7.4.1.7 Flat roofs with inset storeys.   Figure 20.45 is the key for fully-inset storeys, such as plant rooms. For the upper roof:

1 Define the Regions A−G on the upper roof by following steps 1–8 of §20.7.4.1.2, taking H as the height from the eave to the lower roof level as instructed in step

Plan

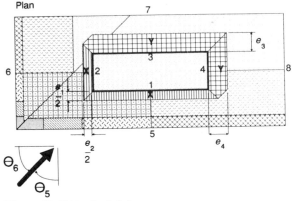

Figure 20.45 Key for fully inset storey

1. (**Note:** the Regions A–G correspond to the eaves marked '1' and '2' in the key figure on the **upper** roof, but have been omitted from the figure for clarity.)

2 Look up the pressure coefficients for each Region A–G from Tables 20.14, 20.15, 20.16 or 20.17, depending on the form of the eaves and the local wind angles, θ_n.

For the lower roof:

3 Define the Regions A–G on the lower roof by following steps 1 to 8 of §20.7.4.1.2, assuming that the fully inset storeys did not exist. (**Note:** the Regions A–G correspond to the eaves marked '5' and '6' in the key figure on the **lower** roof.)

4 Define the $e_n/2$-wide Regions X adjacent to each upwind-facing wall of the inset storey, i.e. walls for which $0° \leqslant \theta_n \leqslant 90°$ or $270° \leqslant \theta_n \leqslant 360°$. (**Note:** where e_n corresponds to the eaves marked '1' and '2' on the key figure.)

5 Define the e_n-wide Regions Y adjacent to each downwind-facing wall of the inset storey, i.e. walls for which $90° < \theta_n < 270°$. (**Note:** where e_n corresponds to the eaves marked '3' and '4' on the key figure.)

6 Look up the pressure coefficients for each Region A–G from Tables 20.14, 20.15, 20.16 or 20.17, depending on the form of the lower roof eaves and the local wind angles, θ_n.

7 Extend the **wall** Regions A–D on each adjacent inset storey wall normal to the wall/roof junction to fill the new regions X and Y, and take the pressure coefficients determined in §20.7.3.1.7 to act on the corresponding areas within regions X and Y.

Figure 20.46 is the key for inset storeys flush with an irregular face. For the upper roof:

1 Define the Regions A–G on the upper roof by following steps 1–8 of §20.7.4.1.2, taking H as the height of the corresponding eave as instructed in step 1. (**Note:** the Regions A–G correspond to the eaves marked '1' and '2' in the key figure on the **upper** roof, and are shown in (a) of the figure, together with the relevant values of H_n.)

2 Look up the pressure coefficients for each Region A–G from Tables 20.14, 20.15, 20.16 or 20.17, depending on the form of the eaves and the local wind angles, θ_n.

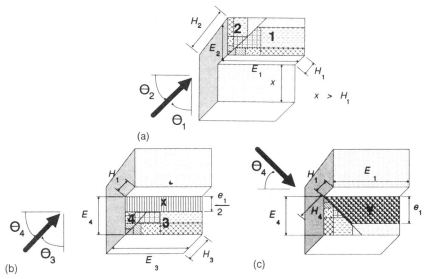

Figure 20.46 Key for inset due to irregular face: (a) upper roof; (b) lower roof upwind; (c) lower roof downwind

For the lower roof with an external corner upwind as in (b) of the key figure:

3 Define the Regions A–G on the lower roof by following steps 1–8 of §20.7.4.1.2, assuming that the fully-inset storeys did not exist. (**Note**: the Regions A–G correspond to the eaves marked '3' and '4' on the **lower** roof in (b) of the key figure.)

4 Define the $e_n/2$-wide Region X adjacent to each upwind-facing wall of the inset storey, i.e. walls for which $0° \leqslant \theta_n \leqslant 90°$ or $270° \leqslant \theta_n \leqslant 360°$. (**Note**: where e_n for Region H in (b) of the key figure corresponds to the eave marked '1' in (a).)

Or for the lower roof with an internal corner of the flush irregular face upwind, as in (c) of the key figure:

5 Define the Regions A–G on the lower roof from the internal corner behind the corresponding eave by following steps 1–8 of §20.7.4.1.2. (**Note**: the Regions A–G correspond to the eave marked '4' in (c) of the key figure.)

6 Define the e_n-wide Region Y adjacent to each downwind-facing wall of the inset storey, i.e. walls for which $0° \leqslant \theta_n \leqslant 90°$ or $270° \leqslant \theta_n \leqslant 360°$. (**Note**: where e_n corresponds to the eave marked '1' on the key figure.)

7 Define the corresponding Regions E, F or G on any remaining part of the roof downwind of the new Region Y.

Then for either case:

8 Look up the pressure coefficients for each Region A–G from Tables 20.14, 20.15, 20.16 or 20.17, depending on the form of the lower roof eaves and the local wind angles, θ_n.

9 Extend the **wall** Regions A–D on each adjacent inset storey wall normal to the wall/roof junction to fill the new Regions X and Y, and take the pressure coefficients determined in §20.7.3.1.7 to act on the corresponding areas within Regions X and Y.

(§17.3.3.10).

20.7.4.2 Monopitch and duopitch roofs

20.7.4.2.1 Scope and definitions.
Data for monopitch and duopitch roofs break into two categories:

1 Low pitch angles, where the flow predominantly separates from the upwind eave or verge, and
2 High pitch angles, where the flow remains attached to the upwind face and separates from the ridge (or high eave),

with an overlap region between the two ranges where either effect may dominate depending on wind direction.

The data in this section are for category 1 and should be used for conventional monopitch and duopitch roofs with gables on rectangular-plan buildings and pitch angles of $\alpha \leqslant 30°$. They can also be used in the overlap region, for pitch angles in the range $30° \leqslant \alpha \leqslant 45°$. Roofs in category 2, with pitch angles $\alpha > 45°$, should be assessed using the rules for A-frame buildings in §20.7.3.2.1, and these rules can also be used in the overlap region $30° \leqslant \alpha \leqslant 45°$. In the overlap region the two approaches will give slightly different results: roofs of tall buildings, $2H > B$, are better assessed by the data in this section, while roofs of squat buildings $2H < B$ are better assessed by the A-frame rules of §20.7.3.2.1. Data for hipped roof forms are presented separately in §20.7.4.3, where advice is also given for adapting the data to non-rectangular-plan buildings.

The eave and verge delta-wing vortices develop different characteristics, so it is no longer possible to define a universal set of data in terms of the local wind angle, θ_n, from normal to eave or verge as given for flat roofs in Tables 20.14 to 20.17. Because separate data must be given for eave and verge regions, it is convenient to define one wind angle, θ, *from normal to the upwind eave* and amalgamate the eave and verge data into a single table for each face of the roof.

Figures 20.47 and 20.48 are the key figures defining the loaded regions on each face: L is the length of the upwind eave and W is the width of the upwind verge. Note that W for duopitch roofs is the total width of both faces as defined in Figure 20.48. The scaling length, e, for the eave regions is taken as the smaller of $e = L$ or $e = 2H$ as before, but now the scaling length for the new verge regions, v, is similarly taken as the smaller of $v = W$ or $v = 2H$. For monopitch roofs the pitch angle, α, is defined as positive when the low eave upwind and negative with the high eave upwind, as in Figure 20.47(a). For duopitch roofs the pitch angle, α, is defined as positive when the roof has a central ridge and negative when the roof has a central trough, as in Figure 20.48. The wind angle, θ is defined from normal to the upwind **eave**.

20.7.4.2.2 Reference dynamic pressures.
The reference dynamic pressure for each plane face is always taken at the height of the upwind corner. This always gives two possible values. For the monopitch roof these are the two cases shown in Figure 20.47(a) – with corner of the low eave upwind (as in Figure 20.47(b)) and with a corner of the high eave upwind. For the ridged duopitch roof these are for the two faces in Figure 20.48(a) – the low eave for the upwind face and the ridge for the downwind face. For the troughed duopitch roof these are for the two faces in Figure 20.48(b) – the high eave for the upwind face and the trough for the downwind face.

There is a good reason for using these two different values instead of a single reference dynamic pressure. When the building is very squat, the high corner may

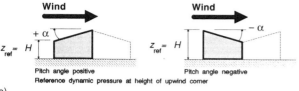

(a)

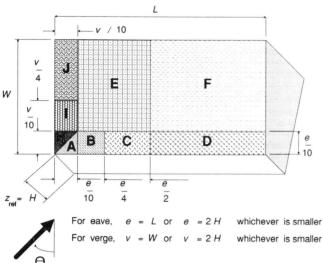

For eave, $e = L$ or $e = 2H$ whichever is smaller

For verge, $v = W$ or $v = 2H$ whichever is smaller

(b)

Figure 20.47 Key for monopitch and duopitch roofs: (a) roof pitch and reference height; (b) key to loaded areas

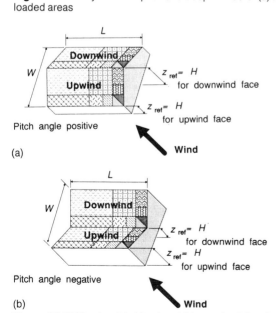

(a)

Pitch angle positive

(b)

Pitch angle negative

Figure 20.48 Key for: (a) ridged and (b) troughed duopitch roofs (loaded areas defined in Figure 20.47)

be several times higher than the low corner, in which case the local pressures along the high eave or ridge may be much more onerous than along the low eave or trough. Whereas when the building is tall, there is little difference between these two cases. Use of the two reference dynamic pressure values enables these effects to be represented by a single table of values.

20.7.4.2.3 Loaded regions. Loaded regions are defined for one plane face in Figure 20.47(b) from the upwind corner of the face. Unlike the data for flat roofs, these regions are the same for all wind directions for which the corner is upwind. Regions A–F correspond to the same regions as on flat roofs (§20.7.4.1.2). Regions H, I and J are the new verge edge regions. Strictly there should also be a Region G further downwind, similar to region G for flat roofs (§20.7.4.1.2), however there are no supporting data available at present. Fortunately, the pitch angle limits the practical size of pitched roofs and the region G rarely exists, except for large-span buildings with very low eaves where the omission is not serious but may lead to overestimates of loading.

20.7.4.2.4 Monopitch roofs. These are roofs with a single pitched plane face, as in Figure 20.47(a). For a given roof pitch, α, data are required for the range of wind

Table 20.18   Pressure coefficients for monopitch roofs and upwind faces of duopitch roofs

Key diagram: Figure 20.47 and 20.48   for $\hat{q}_{ref}$ see §20.7.4.2.2

| **Negative pitch angles:** | Monopitch roof – high eave upwind |
| | Troughed duopitch roof – upwind face |

Roof pitch $\alpha = -45°$ Structural loads ($t = 4$ s) and peak suctions ($t = 1$ s)

Region	A	B	C	D	E	F	H	I	J
	Pseudo-steady pressure coefficient, $\bar{c}_p\{t\}$				($\hat{q}_{ref}$ at high eave height)				
$\theta = 0°$	− 0.61	− 0.58	− 0.56	− 0.41	− 0.76	− 0.78	− 0.62	− 0.79	− 0.94
30°	− 0.53	− 0.50	− 0.49	− 0.55	− 0.55	− 0.81	− 0.52	− 0.58	− 0.58
60°	− 1.11	− 1.29	− 1.36	− 0.96	− 0.97	− 0.91	− 1.05	− 0.97	− 1.17
90°	− 1.25	− 0.81	− 0.62	− 0.42	− 0.77	± 0.20	− 1.48	− 1.05	− 0.97

Roof pitch $\alpha = -30°$ Structural loads ($t = 4$ s) and peak suctions ($t = 1$ s)

Region	A	B	C	D	E	F	H	I	J
	Pseudo-steady pressure coefficient, $\bar{c}_p\{t\}$				($\hat{q}_{ref}$ at high eave height)				
$\theta = 0°$	− 0.76	− 0.68	− 0.60	− 0.50	− 0.76	− 0.63	− 0.76	− 0.85	− 0.90
30°	− 1.13	− 1.02	− 0.89	− 0.79	− 0.84	− 0.76	− 1.17	− 0.87	− 0.73
60°	− 2.06	− 2.33	− 2.17	− 1.22	− 1.03	− 0.80	− 1.69	− 1.18	− 1.21
90°	− 1.28	− 0.94	− 0.70	− 0.37	− 0.70	± 0.20	− 1.54	− 1.10	− 1.01

Roof pitch $\alpha = -15°$ Structural loads ($t = 4$ s) and peak suctions ($t = 1$ s)

Region	A	B	C	D	E	F	H	I	J
	Pseudo-steady pressure coefficient, $\bar{c}_p\{t\}$				($\hat{q}_{ref}$ at high eave height)				
$\theta = 0°$	− 1.08	− 1.05	− 0.97	− 0.92	− 0.88	− 0.82	− 1.10	− 0.96	− 0.83
30°	− 2.64	− 2.37	− 1.71	− 1.00	− 0.93	− 0.85	− 2.75	− 1.66	− 1.11
60°	− 2.25	− 2.15	− 1.85	− 1.02	− 0.76	− 0.72	− 2.44	− 1.60	− 1.07
90°	− 1.22	− 0.79	− 0.58	− 0.31	− 0.60	± 0.20	− 1.51	− 1.12	− 1.13

Table 20.18 continued

Roof pitch $\alpha = -5°$ Structural loads ($t = 4$ s) and peak suctions ($t = 1$ s)

Region	A	B	C	D	E	F	H	I	J
	Pseudo-steady pressure coefficient, $\bar{c}_p\{t\}$				($\hat{q}_{ref}$ at high eave height)				
$\theta = 0°$	− 1.49	− 1.34	− 1.19	− 1.12	− 0.83	− 0.82	− 1.47	− 0.91	− 0.67
30°	− 2.36	− 2.21	− 1.63	− 1.04	− 0.82	− 0.77	− 2.24	− 1.30	− 0.91
60°	− 1.85	− 1.57	− 1.28	− 0.77	− 0.65	− 0.54	− 2.10	− 1.67	− 1.09
90°	− 1.30	− 0.79	− 0.58	− 0.27	− 0.59	± 0.20	− 1.65	− 1.13	− 1.20

Positive pitch angles: Monopitch roof – low eave upwind
Ridged duopitch roof – upwind face

Roof pitch $\alpha = 5°$ Structural loads ($t = 4$ s) and peak suctions ($t = 1$ s)

Region	A	B	C	D	E	F	H	I	J
	Pseudo-steady pressure coefficient, $\bar{c}_p\{t\}$				($\hat{q}_{ref}$ at low eave height)				
$\theta = 0°$	− 1.39	− 1.24	− 1.11	− 1.19	− 0.56	− 0.59	− 1.39	− 0.69	− 0.43
30°	− 1.78	− 1.64	− 1.34	− 1.09	− 0.62	− 0.60	− 1.75	− 1.02	− 0.76
60°	− 1.67	− 1.33	− 1.12	− 0.71	− 0.64	− 0.42	− 2.05	− 1.51	− 1.05
90°	− 1.21	− 0.83	− 0.55	− 0.25	− 0.61	± 0.20	− 1.48	− 1.12	− 1.30

Roof pitch $\alpha = 15°$ Structural loads ($t = 4$ s) and peak suctions ($t = 1$ s)

Region	A	B	C	D	E	F	H	I	J
	Pseudo-steady pressure coefficient, $\bar{c}_p\{t\}$				($\hat{q}_{ref}$ at low eave height)				
$\theta = 0°$	− 0.91	− 0.83	− 0.78	− 0.81	− 0.21	− 0.31	− 0.90	− 0.36	− 0.30
30°	− 0.84	− 0.88	− 0.82	− 0.83	− 0.21	− 0.37	− 0.63	± 0.20	− 0.32
60°	− 1.27	− 0.86	− 0.70	− 0.61	− 0.54	− 0.33	− 1.57	− 1.21	− 0.93
90°	− 1.20	− 0.84	− 0.58	− 0.27	− 0.64	± 0.20	− 1.42	− 1.10	− 1.30

Roof pitch $\alpha = 30°$ Structural loads ($t = 4$ s) and peak suctions ($t = 1$ s)

Region	A	B	C	D	E	F	H	I	J
	Pseudo-steady pressure coefficient, $\bar{c}_p\{t\}$				($\hat{q}_{ref}$ at low eave height)				
$\theta = 0°$	+ 0.38	+ 0.69	+ 0.73	+ 0.77	+ 0.39	+ 0.40	± 0.20	± 0.20	± 0.20
30°	+ 0.75	+ 0.74	+ 0.72	+ 0.59	+ 0.41	+ 0.26	+ 0.78	+ 0.69	+ 0.47
60°	− 0.14	+ 0.43	+ 0.39	+ 0.33	± 0.20	± 0.20	− 0.80	− 0.89	− 0.83
90°	− 1.13	− 0.94	− 0.77	− 0.19	− 0.78	± 0.20	− 1.25	− 1.06	− 1.36

Roof pitch $\alpha = 45°$ Structural loads ($t = 4$ s) and peak suctions ($t = 1$ s)

Region	A	B	C	D	E	F	H	I	J
	Pseudo-steady pressure coefficient, $\bar{c}_p\{t\}$				($\hat{q}_{ref}$ at low eave height)				
$\theta = 0°$	+ 0.82	+ 1.02	+ 1.11	+ 1.13	+ 0.75	+ 0.74	+ 0.69	+ 0.56	+ 0.43
30°	+ 1.11	+ 1.09	+ 1.03	+ 0.88	+ 0.77	+ 0.55	+ 1.12	+ 1.00	+ 0.85
60°	+ 0.79	+ 0.69	+ 0.62	+ 0.46	+ 0.38	+ 0.21	+ 0.84	+ 0.82	+ 0.54
90°	− 1.17	− 0.96	− 0.86	− 0.33	− 0.88	− 0.28	− 1.25	− 1.08	− 1.36

direction for which the low eave is upwind (pitch angle positive) and also for the range of wind direction for which the high eave is upwind (pitch angle negative). Pressure coefficients for each of the regions are given in Table 20.18. Note that at $\theta = 90°$, when the wind is normal to the verge, the values are different for positive and negative pitch angles of the same value, especially the highest pitch of $\alpha = \pm 45°$. This is partly because the reference dynamic pressure is at the low eave height for $+ \alpha$ and at the high eave height for $- \alpha$, and partly a real effect of pitch angle in the regions near the corners. Interpolate for intermediate wind angles. (§17.3.3.3.1). (**Note**: when the result of interpolating between positive and negative values is in the range $- 0.2 < \tilde{c}_p < 0.2$, the result should be taken as $\tilde{c}_p = \pm 0.2$ and the design assessed for both possible values.)

20.7.4.2.5 Duopitch roofs.   These are roofs with two pitched plane faces joined at a common ridge or trough as in Figure 20.48.

The pressure on the upwind face is effectively the same as on a monopitch roof of the same pitch, hence the pressure coefficients for each of the regions are also given by Table 20.18. The pressure coefficients for each of the regions of the downwind face are given in Table 20.19. Values for positive pitch angles should be used for ridged roofs, as in Figure 20.48(a). Values for negative pitch angles should be used for troughed roofs, as in Figure 20.48(b). Note that the reference dynamic pressure for the upwind face it is at the height of the upwind corner of the building and for the downwind face is at the height of the ridge/trough.

These data were derived from model tests with both duopitch faces of equal pitch. They are expected to be reasonable for duopitch roofs of unequal pitch, α_U for the upwind face and α_D for the downwind face, providing the following rules are observed:

Ridged duopitch

1 Upwind faces are assessed without modifications, for $\alpha = \alpha_U$.
2 For downwind faces that are less steep than the upwind face, $\alpha_D < \alpha_U$
   (a) look up the pressure coefficients for the downwind face from the equal-pitch duopitch data in Table 20.19, corresponding to the actual pitch of the downwind face, $\alpha = \alpha_D = \alpha_U$;
   (b) look up the pressure coefficients for a monopitch roof, corresponding to a **negative** pitch angle of the same value as the downwind face, $\alpha = - \alpha_D$, for which $\alpha_U = 90°$;
   (c) interpolate between these two sets of coefficients for the value of the upwind face pitch, α_U, between the coefficients for the equal-pitch data ($\alpha_U = \alpha_D$) and the coefficients for the monopitch data ($\alpha_U = 90°$).
3 For downwind faces that are steeper than the upwind face, $\alpha_D > \alpha_U$, use the equal-pitch coefficients of Table 20.19 for the actual downwind face pitch, $\alpha = \alpha_D$, which will give a conservative result.

Troughed duopitch

1 Upwind faces are assessed without modifications, for $\alpha = \alpha_U$.
2 For downwind faces that are less steep than the upwind face, $- \alpha_D < - \alpha_U$, use the equal-pitch coefficients of Table 20.19 for the upwind face pitch, $\alpha = \alpha_U$, which will give a conservative result.

Table 20.19   Pressure coefficients for downwind face of duopitch roofs

Key diagram: Figure 20.47 and 20.48   for $\hat{q}_{ref}$ see §20.7.4.2.2

Negative pitch angles: Troughed duopitch roof – downwind face

Roof pitch	$\alpha = -45°$		Structural loads ($t = 4$ s) and peak suctions ($t = 1$ s)			
Region	A–B–C		D–F	E		H–I–J
	Pseudo-steady pressure coefficient, $\tilde{c}_p\{t\}$			($\hat{q}_{ref}$ at trough height)		
$\theta = 0°$	− 0.92		− 0.75	− 0.75	− 0.63	
30°	− 1.12		− 0.52	± 0.20	− 0.32	
60°	− 1.04		− 0.24	− 0.73	− 1.05	
90°	Use values in Table 20.18 for $\alpha = +45°$			($\hat{q}_{ref}$ at trough height)		

Roof pitch	$\alpha = -30°$		Structural loads ($t = 4$ s) and peak suctions ($t = 1$ s)			
Region	A–B–C		D–F	E		H–I–J
	Pseudo-steady pressure coefficient, $\tilde{c}_p\{t\}$			($\hat{q}_{ref}$ at trough height)		
$\theta = 0°$	− 0.78		− 0.66	− 0.47	− 0.40	
30°	− 0.44		− 0.52	± 0.20	± 0.20	
60°	− 0.74		− 0.27	− 0.62	− 1.01	
90°	Use values in Table 20.18 for $\alpha = +30°$			($\hat{q}_{ref}$ at trough height)		

Roof pitch	$\alpha = -15°$		Structural loads ($t = 4$ s) and peak suctions ($t = 1$ s)			
Region	A–B–C		D–F	E		H–I–J
	Pseudo-steady pressure coefficient, $\tilde{c}_p\{t\}$			($\hat{q}_{ref}$ at trough height)		
$\theta = 0°$	− 0.69		− 0.52	− 0.26	− 0.21	
30°	± 0.2		± 0.2	± 0.2	− 0.55	
60°	− 0.67		± 0.2	− 0.65	− 1.03	
90°	Use values in Table 20.18 for $\alpha = +15°$			($\hat{q}_{ref}$ at trough height)		

Roof pitch	$\alpha = -5°$		Structural loads ($t = 4$ s) and peak suctions ($t = 1$ s)			
Region	A–B–C		D–F	E		H–I–J
	Pseudo-steady pressure coefficient, $\tilde{c}_p\{t\}$			($\hat{q}_{ref}$ at trough height)		
$\theta = 0°$	− 0.34		− 0.25	− 0.25	− 0.28	
30°	± 0.20		± 0.20	− 0.26	− 0.48	
60°	− 0.69		± 0.20	− 0.66	− 0.88	
90°	Use values in Table 20.18 for $\alpha = +5°$			($\hat{q}_{ref}$ at trough height)		

Positive pitch angles: Ridged duopitch roof – downwind face

Roof pitch	$\alpha = 5°$			Structural loads ($t = 4$ s) and peak suctions ($t = 1$ s)					
Region	A	B	C	D	E	F	H	I	J
	Pseudo-steady pressure coefficient, $\tilde{c}_p\{t\}$				($\hat{q}_{ref}$ at ridge height)				
$\theta = 0°$	− 0.32	− 0.27	− 0.28	− 0.28	± 0.20	± 0.20	− 0.36	− 0.30	− 0.24
30°	− 0.70	− 0.46	− 0.30	− 0.23	− 0.31	± 0.20	− 0.71	− 0.59	− 0.46
60°	− 1.04	− 0.90	− 0.52	± 0.20	− 0.56	± 0.20	− 0.97	− 0.83	− 0.73
90°	− 0.90	− 0.83	− 0.58	± 0.20	− 0.60	± 0.20	− 0.89	− 0.89	− 1.09

Table 20.19 continued

Roof pitch	$\alpha = 15°$		Structural loads ($t = 4$ s) and peak suctions ($t = 1$ s)						
Region	A	B	C	D	E	F	H	I	J
	Pseudo-steady pressure coefficient, $\bar{c}_p\{t\}$				($\hat{q}_{ref}$ at ridge height)				
$\theta = 0°$	− 0.83	− 0.81	− 0.80	− 0.78	− 0.39	− 0.40	− 0.85	− 0.55	− 0.39
30°	− 1.32	− 1.14	− 1.11	− 0.88	− 0.46	− 0.34	− 1.47	− 1.25	− 0.81
60°	− 1.31	− 0.92	− 0.72	− 0.58	− 0.57	− 0.23	− 1.45	− 1.08	− 0.75
90°	− 0.81	− 0.74	− 0.54	± 0.20	− 0.58	± 0.20	− 0.83	− 0.77	− 0.92

Roof pitch	$\alpha = 30°$		Structural loads ($t = 4$ s) and peak suctions ($t = 1$ s)						
Region	A	B	C	D	E	F	H	I	J
	Pseudo-steady pressure coefficient, $\bar{c}_p\{t\}$				($\hat{q}_{ref}$ at ridge height)				
$\theta = 0°$	− 0.29	− 0.26	− 0.25	− 0.30	− 0.30	− 0.30	− 0.31	− 0.32	− 0.33
30°	− 0.74	− 0.63	− 0.52	− 0.43	− 0.39	− 0.43	− 0.76	− 0.51	− 0.40
60°	− 1.04	− 1.05	− 0.98	− 0.64	− 0.58	− 0.47	− 1.02	− 0.67	− 0.64
90°	− 0.66	− 0.61	− 0.49	− 0.21	− 0.49	± 0.20	− 0.67	− 0.58	− 0.69

Roof pitch	$\alpha = 45°$		Structural loads ($t = 4$ s) and peak suctions ($t = 1$ s)						
Region	A	B	C	D	E	F	H	I	J
	Pseudo-steady pressure coefficient, $\bar{c}_p\{t\}$				($\hat{q}_{ref}$ at ridge height)				
$\theta = 0°$	− 0.21	− 0.21	− 0.21	− 0.20	− 0.23	− 0.23	− 0.21	− 0.24	− 0.26
30°	− 0.21	− 0.20	− 0.20	− 0.27	− 0.23	− 0.26	− 0.20	− 0.21	− 0.22
60°	− 0.54	− 0.54	− 0.51	− 0.41	− 0.44	− 0.38	− 0.55	− 0.47	− 0.50
90°	− 0.55	− 0.46	− 0.38	− 0.20	− 0.40	± 0.20	− 0.60	− 0.45	− 0.47

3 For downwind faces that are steeper than the upwind face, $-\alpha_D > -\alpha_U$,
   (a) look up the pressure coefficients for the downwind face from the equal-pitch duopitch data in Table 20.19, corresponding to the actual pitch of the downwind face, $\alpha = \alpha_D = \alpha_U$, and calculate the resulting design pressures;
   (b) look up the pressure coefficients for the front face of an A-frame building using the rules in §20.7.3.2.1, for which $\alpha_U = 0°$, and calculate the resulting design pressures;
   (c) interpolate between these two sets of design pressures for the value of the upwind face pitch, α_U, between the pressures for the equal-pitch data ($\alpha_U = \alpha_D$) and the pressures for the A-frame data ($\alpha_U = 0°$). (**Note**: it is necessary to interpolate design pressures, not pressure coefficients, because the reference dynamic pressure is defined differently in either case and the loaded regions do not match.)

(§17.3.3.3.2). (**Note**: when the result of interpolating between positive and negative values is in the range $- 0.2 < \tilde{c}_p < 0.2$, the result should be taken as $\tilde{c}_p = \pm 0.2$ and the design assessed for both possible values.)

20.7.4.2.6 Pitched roofs with inset storeys.   The situation with inset storeys on pitched roofs is almost exactly the same as for flat roofs. The key figures for flat roofs, Figures 20.45 and 20.46, can be made to serve the pitched-roof case also. Follow the procedure given in §20.7.4.1.7 'Flat roofs' with inset storeys, **except** that the regions defined on the roofs should be the appropriate pitched-roof regions from Figures 20.47–20.50 with their corresponding pressure coefficients. The additional Regions X and Y in the definition Figures 20.45 and 20.46 should be mapped onto the pitched roofs in plan view and the inset storey height, H, is always measured at the upwind corner. Figure 20.56 gives several examples equivalent to Figure 20.46.

20.7.4.3 Hipped roofs

20.7.4.3.1 Scope and definitions. The data in this section are for conventional hipped roofs on rectangular-plan buildings, where all faces of the roof have the same pitch angle in the range $5° \leq \alpha \leq 45°$ (§17.3.3.3.4) and are valid for the trapezoidal main faces and the triangular hip faces. They can also be used for faces that are hipped at the upwind end, but have a conventional vertical gable at the downwind edge. If the hipped roof butts up against an inset storey, the data are still valid but the rules for inset storeys of pitched roofs given in §20.7.4.2.7 should also be followed. Empirical rules for skew-hipped roofs (§17.3.3.3.5) and other hipped roof forms are given at the end of this section.

Figure 20.49 is the key figure: L and W are the lengths of the main face and hip faces eaves, respectively. The height, H, and reference height, z_{ref}, for the dynamic

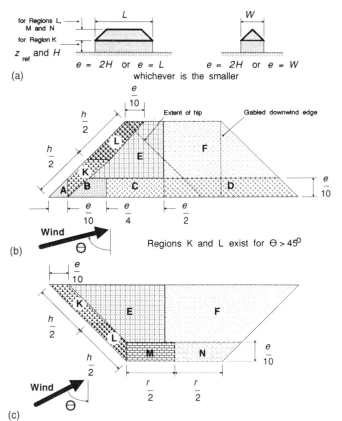

Figure 20.49 Key for hipped roofs: (a) dimensions and reference heights; (b) upwind face; (c) downwind face

pressure are either the height to eaves or to the ridge, depending on the loaded region, as explained in §20.7.4.2.2. The scaling length, e, for determining the sizes of the regions is then the smaller of $e = 2H$ or $e = L$ for the main faces, and of $e = 2H$ or $e = W$ for the hip faces.

As all faces of the roof are treated identically, the wind angle for any particular face of the roof is measured from normal to the eave in the manner of the flat roof data (§20.7.4.1.1). Thus to obtain the loading for all four faces of a hipped roof, four orthogonal wind angles should be considered, but the definition of upwind and downwind faces reduces this to two: θ_1 for the trapezoidal main faces and θ_2 for the triangular hip faces. Thus if the main wind angle, θ, is measured from normal to the eave of a main face, $\theta_1 = \theta$ and $\theta_2 = 90° - \theta$.

20.7.4.3.2 Reference dynamic pressures. In order to minimise the effect of building proportions, two reference heights are used in the same manner as for duopitch roofs (§20.7.4.2.2), as defined in Figure 20.49(a).

20.7.4.3.3 Loaded regions. Loaded regions are shown on a trapeziodal main face in Figure 20.49(b) when an upwind face and in (c) when a downwind face. Regions are the same on the triangular hip faces, but are limited by the extent of the hip face as shown in (b), with the result that the ridge Regions M and N do not exist on the triangular hip face. In the special case of a square-plan building, all faces are identical and triangular.

On the upwind face, Regions A to F are equivalent to the corresponding regions on the upwind face of a duopitch roof. The verge Regions H, I and J are replaced with hip-ridge regions K and L. Note that hip-ridge Regions K and L do not exist on the upwind face until the wind angle has turned to more than $\alpha = 45°$, so that flow separates from the hip ridge. Accordingly, the Regions K and L exist either on the trapeziodal main face or the adjacent triangular face, but never simultaneously on both.

On the downwind face, Regions E and F are equivalent to the corresponding regions on the downwind face of a duopitch roof. The ridge Regions A–D on the main trapezoidal face are replaced by new ridge Regions M and N and the verge Regions H, I and J are again replaced with hip-ridge Regions K and L.

20.7.4.3.4 Pressure coefficients. The pressure coefficients on the upwind faces for Regions A–F should be taken for the corresponding pitch angle from Table 20.18 (only positive pitch angles are valid) and for the new Regions K and L from Table 20.20. Note that the reference dynamic pressure, $\hat{q}_{ref}$, for Regions A–F and K are at the height of the eave, while $\hat{q}_{ref}$ for Region L is at the height of the ridge. In the corner, where Regions A, B and K appear to overlap, the more onerous values should be taken.

Similarly, the pressure coefficients on the downwind faces for Regions E and F should be taken for the corresponding pitch angle from Table 20.18 and for the new Regions K–N from Table 20.20. Note that now the reference dynamic pressure, $\hat{q}_{ref}$, for Regions E, F, L, M and N are at the height of the ridge, while $\hat{q}_{ref}$ for Region K is at the height of the eave (§17.3.3.3.4).

20.7.4.3.5 Skew-hipped roofs. Figure 20.50 is the key figure for skew-hipped and other forms of hipped roofs. The pitch of the roof is always measured down the 'fall line', i.e. the line of maximum pitch angle down which a ball would roll, and the

Table 20.20 Pressure coefficients for hipped roofs

Key diagram: Figure 20.49 $\hat{q}_{ref}$ depends on region

Region K   Structural loads ($t = 4$ s) and peak suctions ($t = 1$ s)

	Upwind face				Downwind face			
Pitch α =	5°	15°	30°	45°	5°	15°	30°	45°
	Pseudo-steady pressure coefficient, $\tilde{c}_p\{t\}$				$\hat{q}_{ref}$ at EAVE height			
θ = 0°					− 0.31	− 0.44	− 0.53	− 0.65
30°					− 0.60	− 1.00	− 0.74	− 0.52
60°	− 1.13	− 0.94	− 0.99	− 1.11	− 0.76	− 1.43	− 1.25	− 0.67
90°	− 1.19	− 1.09	− 1.10	− 1.22	− 0.89	− 0.97	− 1.40	− 1.35

Region L   Structural loads ($t = 4$ s) and peak suctions ($t = 1$ s)

	Upwind face				Downwind face			
Pitch α =	5°	15°	30°	45°	5°	15°	30°	45°
	Pseudo-steady pressure coefficient, $\tilde{c}_p\{t\}$				$\hat{q}_{ref}$ at RIDGE height			
θ = 0°					− 0.45	− 0.83	− 0.33	− 0.24
30°					− 0.46	− 0.99	− 0.55	− 0.22
60°	− 0.63	− 0.52	− 0.47	− 0.33	− 0.51	− 0.71	− 0.82	− 0.35
90°	− 0.76	− 0.77	− 1.01	− 0.71	− 0.50	− 0.59	− 0.62	− 0.43

Region M   Structural loads ($t = 4$ s) and peak suctions ($t = 1$ s)

		Downwind face			
Pitch α =		5°	15°	30°	45°
	Pseudo-steady pressure coefficient, $\tilde{c}_p\{t\}$	$\hat{q}_{ref}$ at RIDGE height			
θ = 0°		− 0.58	− 1.17	− 0.28	− 0.20
30°		− 0.47	− 1.31	− 0.51	− 0.22
60°		− 0.38	− 0.78	− 0.77	− 0.32

Region N   Structural loads ($t = 4$ s) and peak suctions ($t = 1$ s)

		Downwind face			
Pitch α =		5°	15°	30°	45°
	Pseudo-steady pressure coefficient, $\tilde{c}_p\{t\}$	$\hat{q}_{ref}$ at RIDGE height			
θ = 0°		− 0.58	− 1.17	− 0.28	− 0.20
30°		− 0.54	− 1.13	− 0.50	− 0.28
60°		− 0.36	− 0.80	− 0.49	− 0.41

---

wind angle is also measured from this line. For the skew-hipped roofs of Figure 17.63, this is the diagonal of the square building plan.

The loaded Regions E and F, and the hip-ridge Regions K and L are defined exactly as before, Regions K and L existing for θ > 45°, as described in §20.7.4.3.3. The loading coefficients for these regions can be taken directly from the hipped roof data in Table 20.20, in terms of α and θ, without further modification.

The edge regions differ because they are now transitional in form between the eave Regions A–D and the verge Regions H–J. These are now labelled Regions A–H, B–I, C–J and D–J in Figure 20.50(b) to denote this changed status and are

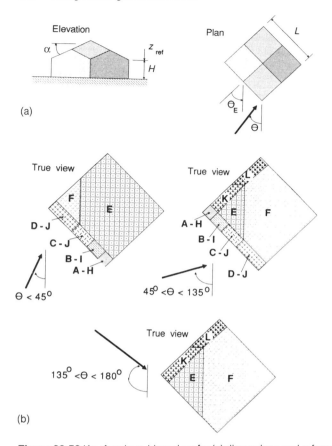

Figure 20.50 Key for skew-hipped roofs: (a) dimensions and reference height; (b) key to loaded areas

defined in a strip parallel to the eave-verge edge. The normal to the edge from which the corresponding local 'edge' wind angle would be defined is at $\theta = \theta_E$, which for the square-plan skew-hipped roof is $\theta_E = 45°$. The loading coefficients in the edge regions take values somewhere between the values for the corresponding eave and verge regions, i.e.:

$$\tilde{c}_p\{A, \theta-\theta_E\} \geqslant \tilde{c}_p\{A - H, \theta\} \geqslant \tilde{c}_p\{H, 90° + \theta - \theta_E\} \tag{20.23}$$

Note that since the local wind angle for the eave Region A is $\theta - \theta_E$, the equivalent wind angle for the verge Region H is $90° + \theta - \theta_E$, since when flow is normal to the edge it is equivalent to an eave at $\theta = 0°$ or a verge at $\theta = 90°$. For the square-plan, skew-hipped roofs $\theta_E = 45°$, half the standard $\beta = 90°$ corner angle, so that the values should be about half-way between the limits, e.g.:

$$\tilde{c}_p\{A - H, \theta\} \simeq \tfrac{1}{2}\,(\tilde{c}_p\{A, \theta - 45°\} + \tilde{c}_p\{H, 45° + \theta\}) \tag{20.24}$$

(§17.3.3.3.5).

20.7.4.3.6 Other hipped roof forms.

Other wall corner and roof skew angles can also be accommodated, e.g. the octagonal-plan prism of Figure 17.64, by assuming a linear transition between 'eave' and 'verge' characteristics with corner angle. As

$\theta_E \to 0°$ the edge becomes more like an eave, while as $\theta_E \to 90°$ the edge becomes more like a verge, so that values of loading coefficient could be obtained by interpolation between the corresponding eave region and verge region in Table 20.18, using:

$$\tilde{c}_p\{A - H, \theta\} \simeq ([90° - \theta_E] \, \tilde{c}_p\{A, \theta - \theta_E\} \\ + \theta_E \, \tilde{c}_p\{H, 90° + \theta - \theta_E\}) / 90° \tag{20.25}$$

but the accuracy of this assumption is entirely unknown. A safe alternative is to use the most onerous of the eave and verge region values (§17.3.3.3.5).

20.7.4.4 Mansard and multi-pitch roofs

These are roofs where the main faces are composed of two or more planes at different pitch angles. In classical mansard roofs, the lower plane to the eaves has a steeper pitch angle than the upper plane to the ridge. Rules for the special case of a pitched mansard edge to a flat roof were given in §20.7.4.1.6, earlier.

Consider each plane face of the roof separately and use the data in §20.7.4.2.5 'Duopitch roofs' for the loaded regions defined in Figure 20.47 for roofs with gable verges, or the data in §20.7.4.3 'Hipped roofs' for the loaded regions defined in Figure 20.49 for roofs with hips, with the following modifications:

1 Include the eave Regions A–D on the lowest plane along the actual eave of the upwind face.
2 Include the eave Regions A–D along the bottom edge of other planes of the upwind face when the pitch of the plane is less steep than the plane below forming an external corner, i.e. classical mansard; but **exclude** the eave Regions A–D when the pitch of the plane is steeper than the plane below forming an internal corner.
3 Include the ridge Regions A–D for buildings with square gables or ridge Regions M and N for buildings with hips, as appropriate, on the highest plane along the actual ridge of the downwind face.
4 **Exclude** ridge Regions A–D on all lower planes of the downwind face, whatever the change in pitch angle.
5 Verge Regions H, I and J or hip-ridge Regions K and L, as appropriate, should always be included, together with the interior Regions E and F

(§17.3.3.3.3).

20.7.4.5 Hyperboloid roofs

20.7.4.5.1 Scope and definitions.
Data from the BRE study on the form of hyperboloid roof shown in Figure 17.66 are presented in this section. They are included because of the extremely high suctions generated by this form, even though its use is still relatively rare. The range of data is restricted by the range of the study, so that the most onerous combination of parameters may not have been found.

Figure 20.51 is the key figure: L and W are the longer and shorter dimensions and H is the height to the eaves. Here α is pitch angle of the long ridge, and not the pitch angle of the actual roof surface which varies with position. The wind angle is defined from normal to the longer eave. The reference dynamic pressure, $\hat{q}_{ref}$, is taken at $z_{ref} = H$.

Table 20.21 Pressure coefficients for hyperboloid roofs, proportions: $L/W = 3$ – no gable

Key diagram: Figure 20.51(a) and 20.51(b) $\hat{q}_{ref}$ at height of upwind corner, $z_{ref} = H$

Ridge pitch	$\alpha = 10°$	Structural loads ($t = 4$ s) and peak suctions ($t = 1$ s)										
Region	A	B	C	D	E	F	G	H	I	J	K	L
	Pseudo-steady pressure coefficient, $\bar{c}_p\{t\}$											
$\theta = 0°$	−0.95	−0.88	−0.58	−0.58	−0.88	−0.95	−0.45	−0.60	−0.58	−0.58	−0.60	−0.45
30°	−0.91	−0.76	−0.67	−0.73	−0.88	−0.79	−0.75	−0.84	−1.05	−0.82	−0.72	−0.43
60°	−0.78	−0.49	−0.53	−0.72	−0.73	−0.57	−0.83	−0.76	−0.98	−0.67	−0.41	−0.25
90°	−0.83	−0.34	−0.36	−0.29	−0.26	−0.22	−0.83	−0.34	−0.36	−0.29	−0.26	−0.22
Region	a	b	c	d	e	f	g	h	i	j	k	l
	Pseudo-steady pressure coefficient, $\bar{c}_p\{t\}$											
$\theta = 0°$	−0.59	−0.54	−0.32	−0.32	−0.54	−0.59	−0.58	−0.78	−0.55	−0.55	−0.78	−0.58
30°	−0.61	−0.65	−0.60	−0.59	−0.65	−0.54	−0.83	−1.23	−1.46	−0.92	−0.77	−0.52
60°	−0.72	−0.43	−0.59	−0.74	−0.56	−0.40	−0.87	−1.04	−1.81	−1.03	−0.48	−0.29
90°	−0.80	−0.31	−0.45	−0.36	±0.2	±0.2	−0.80	−0.31	−0.45	−0.36	±0.2	±0.2

Ridge pitch	$\alpha = 20°$	Structural loads ($t = 4$ s) and peak suctions ($t = 1$ s)										
Region	A	B	C	D	E	F	G	H	I	J	K	L
	Pseudo-steady pressure coefficient, $\bar{c}_p\{t\}$											
$\theta = 0°$	−0.88	±0.2	+0.56	+0.56	±0.2	−0.88	−0.58	−0.61	−0.45	−0.45	−0.61	−0.58
30°	−0.52	+0.21	+0.51	±0.2	−0.67	−0.83	−0.74	−0.76	−0.68	−0.78	−1.09	−0.72
60°	−0.25	±0.2	±0.2	−0.60	−0.62	−0.64	−0.96	−1.36	−1.51	−0.98	−0.63	−0.47
90°	−0.72	−0.51	−0.60	−0.41	−0.28	−0.23	−0.72	−0.51	−0.60	−0.41	−0.27	−0.23
Region	a	b	c	d	e	f	g	h	i	j	k	l
	Pseudo-steady pressure coefficient, $\bar{c}_p\{t\}$											
$\theta = 0°$	−0.63	−0.40	±0.2	±0.2	−0.40	−0.63	−0.61	−0.55	−0.41	−0.41	−0.55	−0.61
30°	−0.53	−0.40	±0.2	−0.50	−0.78	−0.77	−0.80	−0.91	−0.65	−0.67	−1.04	−0.71
60°	−0.56	−0.55	−0.42	−0.84	−0.59	−0.57	−0.99	−2.48	−2.44	−1.20	−0.52	−0.40
90°	−0.66	−0.54	−0.92	−0.56	±0.2	±0.2	−0.66	−0.54	−0.92	−0.56	±0.2	±0.2

Ridge pitch	α = 30°	Structural loads (t = 4 s) and peak suctions (t = 1 s)										
Region	A	B	C	D	E	F	G	H	I	J	K	L
Pseudo-steady pressure coefficient, $\bar{c}_p\{t\}$												
θ = 0°	−0.71	+0.31	+0.84	+0.84	+0.31	−0.71	−0.69	−0.62	−0.49	−0.49	−0.62	−0.69
30°	±0.2	+0.69	+0.80	+0.29	−0.38	−0.92	−0.67	−0.68	−0.65	−0.64	−1.05	−1.00
60°	±0.2	+0.47	+0.34	−0.51	−0.62	−0.71	−1.08	−1.85	−1.46	−0.98	−0.68	−0.63
90°	−0.61	−0.67	−0.65	−0.47	−0.28	−0.25	−0.61	−0.67	−0.65	−0.47	−0.28	−0.25

Region	a	b	c	d	e	f	g	h	i	j	k	l
Pseudo-steady pressure coefficient, $\bar{c}_p\{t\}$												
θ = 0°	−0.62	−0.38	+0.46	+0.46	−0.38	−0.62	−0.68	−0.52	−0.43	−0.43	−0.52	−0.68
30°	−0.42	+0.28	+0.54	−0.43	−0.72	−0.90	−0.71	−0.64	−0.61	−0.57	−0.89	−0.72
60°	−0.52	±0.2	±0.2	−0.86	−0.68	−0.70	−1.16	−3.50	−1.92	−1.16	−0.58	−0.68
90°	±0.2	−0.91	−0.95	−0.64	±0.2	±0.2	±0.2	−0.91	−0.95	−0.64	±0.2	±0.2

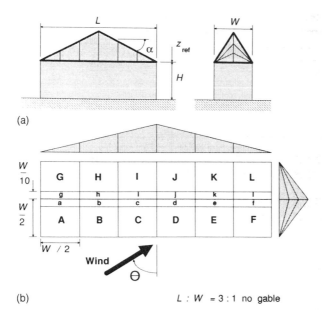

(a)

(b) $L : W$ = 3 : 1 no gable

Figure 20.51 Key for hyperboloid roofs: (a) dimensions and reference height; (b) main key to loaded areas where $L:W = 3{:}1$, no gable

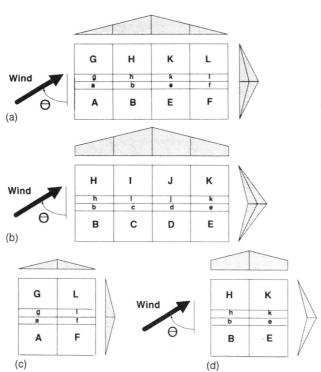

(c)   (d)

Figure 20.52 Examples of hyperboloid roofs: (a) $L:W = 2{:}1$, no gable; (b) $L:W = 2{:}1$, with gable; (c) $L:W = 1{:}1$, no gable; (d) $L:W = 1{:}1$, with gable

Table 20.22   Pressure coefficients for hyperboloid roofs, proportions: $L/W = 2 - no\ gable$

Key diagram: Figure 20.51(a) and 20.52(a) $\hat{q}_{ref}$ at height of upwind corner, $z_{ref} = H$

Ridge pitch $\alpha = 10°$ Structural loads ($t = 4$ s) and peak suctions ($t = 1$ s)

Region	A	B	E	F	G	H	K	L
	Pseudo-steady pressure coefficient, $\bar{c}_p\{t\}$							
$\theta = 0°$	− 1.01	− 1.01	− 1.01	.− 1.01	− 0.44	− 0.70	− 0.70	− 0.44
30°	− 1.00	− 0.84	− 0.97	− 0.87	− 0.65	− 0.74	− 0.63	− 0.43
60°	− 0.93	− 0.54	− 0.77	− 0.59	− 0.80	− 0.68	− 0.46	− 0.27
90°	− 0.87	− 0.40	− 0.33	− 0.24	− 0.87	− 0.40	− 0.33	− 0.24

Region	a	b	e	f	g	h	k	l
	Pseudo-steady pressure coefficient, $\bar{c}_p\{t\}$							
$\theta = 0°$	− 0.65	− 0.67	− 0.67	− 0.65	− 0.67	− 1.07	− 1.07	− 0.67
30°	− 0.63	− 0.69	− 0.86	− 0.65	− 0.77	− 1.23	− 1.14	− 0.59
60°	− 0.77	− 0.43	− 0.70	− 0.46	− 0.88	− 0.96	− 0.77	− 0.35
90°	− 0.88	− 0.38	− 0.34	± 0.20	− 0.88	− 0.38	− 0.34	± 0.20

Ridge pitch $\alpha = 20°$ Structural loads ($t = 4$ s) and peak suctions ($t = 1$ s)

Region	A	B	E	F	G	H	K	L
	Pseudo-steady pressure coefficient, $\bar{c}_p\{t\}$							
$\theta = 0°$	− 0.88	− 0.34	− 0.34	− 0.88	− 0.54	− 0.60	− 0.60	− 0.54
30°	− 0.69	− 0.21	− 0.67	− 0.82	− 0.73	− 1.04	− 0.94	− 0.68
60°	− 0.33	− 0.20	− 0.80	− 0.66	− 0.86	− 1.14	− 0.78	− 0.46
90°	− 0.80	− 0.58	− 0.52	− 0.28	− 0.80	− 0.58	− 0.52	− 0.28

Region	a	b	e	f	g	h	k	l
	Pseudo-steady pressure coefficient, $\bar{c}_p\{t\}$							
$\theta = 0°$	− 0.59	− 0.38	− 0.38	− 0.59	− 0.62	− 0.58	− 0.58	− 0.62
30°	− 0.57	− 0.56	− 0.69	− 0.74	− 0.87	− 1.56	− 0.96	− 0.70
60°	− 0.60	− 0.54	− 1.03	− 0.57	− 1.00	− 2.06	− 1.08	− 0.44
90°	− 0.74	− 0.68	− 0.64	− 0.23	− 0.74	− 0.68	− 0.64	− 0.23

Ridge pitch $\alpha = 30°$ Structural loads ($t = 4$ s) and peak suctions ($t = 1$ s)

Region	A	B	E	F	G	H	K	L
	Pseudo-steady pressure coefficient, $\bar{c}_p\{t\}$							
$\theta = 0°$	− 0.79	+ 0.38	+ 0.38	− 0.79	− 0.64	− 0.60	− 0.60	− 0.64
30°	± 0.20	+ 0.56	− 0.34	− 0.85	− 0.74	− 0.89	− 0.87	− 0.81
60°	± 0.20	± 0.20	− 0.78	− 0.65	− 0.94	− 1.44	− 0.86	− 0.54
90°	− 0.84	− 0.74	− 0.67	− 0.36	− 0.84	− 0.74	− 0.67	− 0.36

Region	a	b	e	f	g	h	k	l
	Pseudo-steady pressure coefficient, $\bar{c}_p\{t\}$							
$\theta = 0°$	− 0.62	± 0.20	± 0.20	− 0.62	− 0.63	− 0.49	− 0.49	− 0.63
30°	− 0.47	± 0.20	− 0.64	− 0.80	− 0.95	− 1.00	− 0.82	− 0.73
60°	− 0.52	− 0.36	− 1.10	− 0.54	− 1.15	− 2.68	− 1.05	− 0.45
90°	− 0.60	− 0.78	− 0.76	± 0.20	− 0.60	− 0.78	− 0.76	± 0.20

20.7.4.5.2 Loaded regions. Main regions are defined from A to L and local regions either side of the long ridge from a–l in Figure 20.51(b). Each letter pair, e.g. A–a, corresponds to a physical section of the model. Figure 20.51(b) shows the regions for $L/W = 3$, for all ridge pitch angles, corresponding to the complete model. Figure 20.52 shows the other combinations for which data are presented.

The key figure does not include any local eave or verge edge regions, which are important near the corners. Accordingly, the eave edge Regions A–D of Figure 20.47, and the verge edge Regions H–J if the ends are gabled, are also required.

20.7.4.5.3 Pressure coefficients. The design pressure coefficients are given in Tables 20.21–20.25, each corresponding to a particular combination. Values may be interpolated for other proportions in the range $1 \leqslant L/W \leqslant 3$ and other ridge pitch angles less than the maximum, $\alpha \leqslant 30°$. For $\alpha < 10°$, interpolate between the $\alpha = 10°$ values and the flat roof data of §20.7.4.1.3 ($\alpha = 0°$). The gabled-end data are only available for $\alpha = 20°$. Expert advice should always be sought if $L/W > 3$ or $\alpha > 30°$. (**Note**: when the result of interpolating between positive and negative values is in the range $-0.2 < \bar{c}_p < 0.2$, the result should be taken as $\bar{c}_p = \pm 0.2$ and the design assessed for both possible values.)

In addition to the specific hyperboloid data, values for the additional eave and verge edge regions are required. Along the long eave the effective roof pitch changes from the angle at the end gables, which may be flat, to a maximum pitch opposite the central peak. Use the monopitch/duopitch data for eave Regions A–D in Table 20.18 corresponding to the local roof slope. If the verges are horizontal (Figures 20.51(b), 20.52(a) and (c)), use the flat roof data for eave Regions A–D as instructed in §20.7.4.1.3. If the verges are gabled, use the duopitch roof for verge Regions H–J data as instructed in §20.7.4.2.5 (§17.3.3.4).

20.7.4.6 Barrel-vault roofs

Barrel-vault roofs are essentially arched roofs (§20.6.3.2.4) springing from walls instead of directly from the ground. Data are only available for squat

Table 20.23   Pressure coefficients for hyperboloid roofs, proportions: $L/W = 2$ – with gable

Key diagram: Figure 20.51(a) and 20.52(b) $\hat{q}_{ref}$ at height of upwind corner, $z_{ref} = H$

Ridge pitch	$\alpha = 20°$		Structural loads ($t = 4$ s) and peak suctions ($t = 1$ s)					
Region	B	C	D	E	H	I	J	K
		Pseudo-steady pressure coefficient, $\bar{c}_p\{t\}$						
$\theta = 0°$	± 0.20	+ 0.53	+ 0.53	± 0.20	− 0.68	− 0.55	− 0.55	− 0.68
30°	+ 0.42	+ 0.56	− 0.20	− 0.58	− 0.99	− 0.81	− 0.80	− 0.91
60°	± 0.20	± 0.20	− 0.60	− 0.50	− 1.16	− 1.53	− 1.04	− 0.61
90°	− 0.97	− 0.69	− 0.45	− 0.23	− 0.97	− 0.69	− 0.45	− 0.23
Region	b	c	d	e	h	i	j	k
		Pseudo-steady pressure coefficient, $\bar{c}_p\{t\}$						
$\theta = 0°$	− 0.48	± 0.20	± 0.20	− 0.48	− 0.63	− 0.46	− 0.46	− 0.63
30°	− 0.28	± 0.20	− 0.48	− 0.70	− 1.53	− 0.90	− 0.77	− 0.89
60°	− 0.59	− 0.40	− 0.85	− 0.53	− 2.14	− 2.96	− 1.27	− 0.52
90°	− 0.95	− 0.89	− 0.60	− 0.21	− 0.95	− 0.89	− 0.60	− 0.21

Table 20.24   Pressure coefficients for hyperboloid roofs, proportions: $L/W = 1 -$ no gable

Key diagram: Figure 20.51(a) and 20.52(c) $\hat{q}_{ref}$ at height of upwind corner, $z_{ref} = H$

Ridge pitch	$\alpha = 10°$		Structural loads ($t = 4$ s) and peak suctions ($t = 1$ s)					
Region	A	F	G	L	a	f	g	l
	Pseudo-steady pressure coefficient, $\bar{c}_p\{t\}$							
$\theta = 0°$	− 0.97	− 0.97	− 0.36	− 0.36	− 0.57	− 0.57	− 0.52	− 0.52
30°	− 0.93	− 0.80	− 0.55	− 0.43	− 0.57	− 0.58	− 0.66	− 0.60
60°	− 0.94	− 0.59	− 0.75	− 0.42	− 0.81	− 0.51	− 0.85	− 0.47
90°	− 0.96	− 0.36	− 0.96	− 0.36	− 0.96	− 0.36	− 0.96	− 0.36

Ridge pitch	$\alpha = 20°$		Structural loads ($t = 4$ s) and peak suctions ($t = 1$ s)					
Region	A	F	G	L	a	f	g	l
	Pseudo-steady pressure coefficient, $\bar{c}_p\{t\}$							
$\theta = 0°$	− 0.65	− 0.65	− 0.41	− 0.41	− 0.55	− 0.55	− 0.72	− 0.72
30°	− 0.57	− 0.71	− 0.50	− 0.40	− 0.48	− 0.75	− 0.68	− 0.67
60°	− 0.50	− 0.60	− 0.63	− 0.42	− 0.51	− 0.69	− 0.68	− 0.55
90°	− 0.64	− 0.44	− 0.64	− 0.44	− 0.64	− 0.51	− 0.64	− 0.51

Ridge pitch	$\alpha = 30°$		Structural loads ($t = 4$ s) and peak suctions ($t = 1$ s)					
Region	A	F	G	L	a	f	g	l
	Pseudo-steady pressure coefficient, $\bar{c}_p\{t\}$							
$\theta = 0°$	− 0.76	− 0.76	− 0.66	− 0.66	− 0.64	− 0.64	− 0.82	− 0.82
30°	− 0.31	− 0.83	− 0.84	− 0.69	− 0.53	− 0.91	− 1.19	− 0.91
60°	− 0.31	− 0.85	− 0.83	− 0.70	− 0.49	− 0.95	− 0.95	− 0.83
90°	− 0.78	− 0.66	− 0.78	− 0.66	− 0.64	− 0.77	− 0.64	− 0.77

Table 20.25   Pressure coefficients for hyperboloid roofs, proportions: $L/W = 1 -$ with gable

Key diagram: Figure 20.51(a) and 20.52(d) $\hat{q}_{ref}$ at height of upwind corner, $z_{ref} = H$

Ridge pitch	$\alpha = 20°$		Structural loads ($t = 4$ s) and peak suctions ($t = 1$ s)					
Region	B	E	H	K	b	e	h	k
	Pseudo-steady pressure coefficient, $\bar{c}_p\{t\}$							
$\theta = 0°$	+ 0.48	+ 0.48	− 0.62	− 0.62	± 0.20	− 0.50	± 0.20	− 0.50
30°	+ 0.68	± 0.20	− 0.76	− 0.80	+ 0.38	− 0.45	− 0.74	− 0.81
60°	± 0.20	− 0.42	− 1.08	− 0.83	− 0.56	− 0.62	− 1.51	− 0.95
90°	− 1.12	− 0.56	− 1.12	− 0.56	− 1.05	− 0.59	− 1.05	− 0.59

cylindrical-section barrels for the range of rise ratio $0.1 \leqslant R/W \leqslant 0.3$ and are mean values from Blessmann's study [315, 316]. Pragmatic guidance is given for rise ratios outside this range.

Figure 20.53 is the key figure: L is the length parallel to the axis of the barrel and W is the width normal to the axis of the barrel, thus L is not necessarily larger than W. The height of the walls to the eave is H and the rise of the roof above the eave is R. The reference dynamic pressure, $\hat{q}_{ref}$, is taken at the height of the roof crest, $z_{ref} = H + R$.

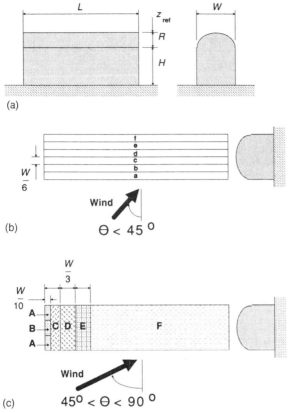

Figure 20.53 Key for barrel-vault roofs: (a) dimensions and reference height; (b) key to loaded areas for wind angles near normal to axis; (c) key to loaded areas for wind angles near parrallel to axis

Owing to the restrictions in the data, loaded regions are defined on longitudinal strips a–f for wind angles near normal to the axis in Figure 20.53(b), and in lateral strips A–F for wind angles near parallel to the axis in Figure 20.53(c). Values of pressure coefficient for each region are given in Table 20.26 for three rise ratios, $R/W = 0.1$, 0.2 and 0.3, and three proportions, $H/W = 0$, 0.25 and 0.5. Values for $H/W = 0$ correspond to arched buildings without walls (§20.6.3.2.4). Interpolation should be used to obtain intermediate values. (**Note**: when the result of interpolating between positive and negative values is in the range $-0.2 < \tilde{c}_p < 0.2$, the result should be taken as $\tilde{c}_p = \pm 0.2$ and the design assessed for both possible values.)

For rise ratios $R/W < 0.1$, the roof should be assessed as if it was flat using the rules in §20.7.4.1.3 'Flat roofs with sharp eaves'. For rise ratios between $R/W = 0.3$ and $R/W = 0.5$ (hemicylinder), the rules in §20.6.3.2.4 'Arched buildings' should be followed (§17.3.3.5).

20.7.4.7 Multi-span roofs

This section gives empirical rules for reducing the loading on downwind spans of multi-span monopitch (sawtooth), duopitch and barrel-vault roofs. A safe result is

Table 20.26   Pressure coefficients for barrel-vault roofs

Key diagram: Figure 20.53 $\hat{q}_{ref}$ at crest of roof, $z_{ref} = H + R$

Rise ratio $R/W = 0.1$ *Structural loads ($t = 4$ s) and peak suctions ($t = 1$ s)*

Region	a	b	c	d	e	f
	Mean pressure coefficient, $\bar{c}_p$					
$H/W = 0$	+ 0.30	− 0.20	− 0.40	− 0.40	− 0.20	− 0.20
0.25	− 0.80	− 0.65	− 0.70	− 0.70	− 0.55	− 0.40
0.50	− 1.27	− 0.85	− 0.85	− 0.75	− 0.60	− 0.45
Region	A	B	C	D	E	F
	Mean pressure coefficient, $\bar{c}_p$					
$H/W = 0$	− 1.00	− 1.50	− 0.70	± 0.20	± 0.20	± 0.20
0.25	− 1.10	− 1.60	− 0.75	− 0.20	± 0.20	± 0.20
0.50	− 1.30	− 1.50	− 0.95	− 0.25	− 0.20	± 0.20

Rise ratio $R/W = 0.2$ *Structural loads ($t = 4$ s) and peak suctions ($t = 1$ s)*

Region	a	b	c	d	e	f
	Mean pressure coefficient, $\bar{c}_p$					
$H/W = 0$	+ 0.45	− 0.10	− 0.60	− 0.40	− 0.20	− 0.20
0.25	− 0.20	− 0.70	− 0.95	− 0.80	− 0.55	− 0.55
0.50	− 0.40	− 0.80	− 0.95	− 0.80	− 0.55	− 0.55
Region	A	B	C	D	E	F
	Mean pressure coefficient, $\bar{c}_p$					
$H/W = 0$	− 0.10	− 0.30	− 0.70	± 0.20	± 0.20	± 0.20
0.25	− 0.20	− 1.70	− 0.80	− 0.20	± 0.20	± 0.20
0.50	− 1.40	− 2.00	− 1.00	− 0.35	− 0.20	± 0.20

Rise ratio $R/W = 0.3$ *Structural loads ($t = 4$ s) and peak suctions ($t = 1$ s)*

Region	a	b	c	d	e	f
	Mean pressure coefficient, $\bar{c}_p$					
$H/W = 0$	+ 0.55	± 0.20	− 0.75	− 0.50	− 0.20	− 0.20
0.25	+ 0.20	− 0.55	− 0.85	− 0.70	− 0.50	− 0.50
0.50	± 0.20	− 0.30	− 0.95	− 0.75	− 0.55	− 0.55
Region	A	B	C	D	E	F
	Mean pressure coefficient, $\bar{c}_p$					
$H/W = 0$	− 0.10	− 1.10	− 0.70	± 0.20	± 0.20	± 0.20
0.25	− 1.30	− 1.30	− 0.70	− 0.20	± 0.20	± 0.20
0.50	− 1.40	− 2.00	− 0.80	− 0.40	− 0.20	± 0.20

Table 20.27   Reduction factors for multi-span roofs

Key diagram: Figure 20.54 $\hat{q}_{ref}$ as defined for single-span roof

Region	A	B	C
Factor ϕ	1.0	0.8	0.6

obtained if these reduction factors are ignored. An exception is multi-span hyperboloid roofs, for which expert advice should always be sought.

Figure 20.54 is the key figure: (a), (b) and (c) denote the valid multi-span roof types. Note that all central spans of duopitch roofs are treated as troughed (negative pitch angle), even when the first span is ridged as shown in (b). Reduction factor Regions A, B and C are defined in (d), for wind angles in the range $0° < \theta < 90°$. When $\theta = 0°$, Region A corresponds to the first span, Region B to the second span and Region C to all remaining spans. When $\theta = 90°$, all spans are in Region A and are thus fully loaded.

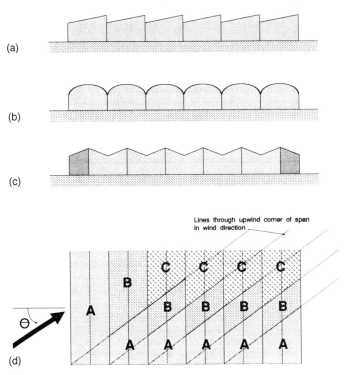

Figure 20.54 Key for multi-span roofs: (a) multi-span monopitch; (b) multi-span barrel-vault; (c) multi-span ridged, ridged or troughed; (d) key to reduction factor regions

Table 20.27 gives values for the reduction factors, ϕ, to be applied to the pressure coefficients for a single span, subject to the following rules to account for the wake shielding effect on upwind-facing steep faces where positive pressure coefficients occur on the first span:

1 **Monopitch and barrel-vault roofs:** any positive pressure coefficient predicted in Regions B and C should be replaced by $\tilde{c}_p = -0.4$.
2 **Duopitch roofs:** treat all central spans downwind of the first span as being troughed (negative pitch angle), even when the upwind span is ridged, as shown in in Figure 20.54(d). This eliminates positive pressure coefficients in Regions B and C.

20.7.4.8 Effect of parapets on pitched roofs

Owing to the way that parapets around roofs change the positive pressures expected on upwind faces with large positive pitch angles to suctions, neglecting their effect is not always conservative. Figure 17.68 should be taken as the key figure:

(a) **Monopitch roofs:**
  (i)   Low eave with parapet upwind – for the part of the roof below the top of the parapet, follow the rules of §20.7.4.1.4 for flat roofs with parapets. For any part of the roof that is above the top of the parapet, i.e. if the top of the parapet is below the level of the high eave, follow the rules given in §20.7.4.2.4 for monopitch roofs.
  (ii)   High eave upwind – follow the rules of §20.7.4.2.4 for monopitch roofs. The reduction factors of Table 20.15 may be used for upwind eave and verge Regions A–D and H–J, with the parapet height h determined at the upwind corner of each respective region. Thus for parapets level with the high eave, as in Figure 17.68(a), $h = 0$ for Regions A–D and H, so that the reduction factor is less than unity only for Regions I and J.

(b) **Ridged duopitch roofs:**
  (i)   Upwind face – follow the same rules as for the monopitch roof with low eave upwind in a(i), above.
  (ii)   Downwind face – follow the rules of §20.4.2.5 for the downwind face of duopitch roofs. The reduction factors of Table 20.15 may be used for the verge Regions H–J, with the parapet height h determined at the upwind corner of each respective region.

(c) **Troughed duopitch roofs:**
  Follow the rules for troughed duopitch roofs in §20.7.4.2.5. The reduction factors of Table 20.15 may be used for the verge Regions H–J of both faces, with the parapet height h determined at the upwind corner of each respective region.

(d) **Multi-span duopitch roofs:**
  (i)   First span ridged with no eave parapet – as in Figure 17.68(d). For the first upwind face, follow the rules for the upwind face of ridged duopitch roofs in §20.7.4.2.5 without modification.
  (ii)   First span ridged with eave parapet – If there is a parapet along the upwind eave of the first span, follow the same rules as for the monopitch roof with low eave upwind in a(i), above.
  (iii)   First span troughed – for the first span, follow the same rules as for the troughed roof in (c), above.
  (iv)   Subsequent spans – all spans downwind of the first ridge should be treated as multi-span roofs using the rules in §20.7.4.7 and the reduction factors of Table 20.27. Additionally, the reduction factors of Table 20.15 may be used for the verge Regions H–J of all faces, with the parapet height h determined at the upwind corner of each respective region.

In addition to the examples in Figure 17.68:

(e) **Barrel-vault roofs**
  (i)   Wind normal to axis – for the part of the roof below the top of the parapet in Regions a, b and c of Figure 20.53(b), follow the rules of

§20.7.4.1.4 for flat roofs with parapets. For any part of the roof in Regions a, b and c that is above the top of the parapet, i.e. if the top of the parapet is below the level of the crest, and for Regions d, e and f downwind of the crest follow the rules of §20.7.4.6 for barrel-vault roofs.

Wind parallel to axis – follow the rules of §20.7.4.6 for barrel-vault roofs. The reduction factors of Table 20.15 may be used for the verge Regions A and B, with the parapet height h determined as the minimum in each respective region. Thus for parapets level with the crest, $h = 0$ for Region B, so that the reduction factor is less than unity only for Region A.

(f) **Multi-span barrel vault roofs**:

(i)   First span – follow the same rules as for a single barrel-vault roof in (e), above.

(ii)   Subsequent spans – follow the rules for multi-span barrel-vault roofs in §20.7.4.7. The reduction factors of Table 20.15 may be used for the verge Regions A and B, with the parapet height h determined as the minimum in each respective Region, for wind parallel to the axis.

20.7.4.9 Re-entrant corners, recessed bays and central wells

Specific rules were required in §20.7.3 'Pressures on walls' to deal with each of these problems. The corresponding position with roofs is, fortunately, much simpler. The rules for each roof face define the loaded regions from the upwind external corner, and the implication for re-entrant corners on flat roofs is that no new regions are defined at internal corners, as in the key figure, Figure 20.42, and the examples in Figure 20.43. Rules are required to cope with the pitched roofs of complex-plan buildings, including 'L', 'T', 'H' or 'O' shapes, when the roof of each wing is at the same level. Rules for roofs at different levels are given in the following section.

Figure 20.55 is the key for roofs at the same level. A typical corner is represented which may be the corner of an 'L' shape building as shown, or the corresponding corner in a more complex shape. The faces labelled 1, 2 and 3 are covered by the earlier rules:

1 One end of face 1 has a vertical gable, so is covered by the rules in §20.7.4.2.5 Duopitch roofs when this is upwind, and one hipped end, so is covered by the rules in §20.7.4.3 'Hipped roofs' when upwind.
2 Face 2 is a main face of a hipped roof, covered by the rules in §20.7.4.3.
3 Face 3 is a hip face of a hipped roof, also covered by the rules in §20.7.4.3.

The faces labelled 4 and 5 are adjacent to the internal corner:

4 Face 4 has one hipped end and the other end is joined to Face 5 at the internal corner equivalent of a 'hip'.
5 Face 5 has one gabled end and the other end is similarly joined to Face 4.

Three flow cases need to be considered, wind into the internal corner as shown in (a), wind out from the internal corner as shown in (b) and wind across the corner as shown in (c). Each of the four corners of the faces is made the key upwind corner in one of the cases and governs the choice of the reference dynamic pressure, $\hat{q}_{ref}$.

In the first case, (a), the faces are both upwind faces and regions should be defined from the upwind external corner as for the appropriate type, gable or hip.

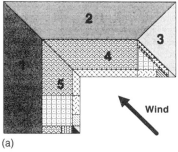

(a)

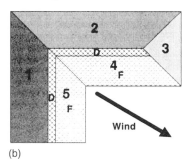

(b)

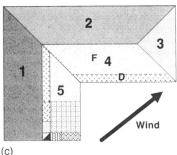

(c)

Figure 20.55 Key for pitched roofs with internal corners: (a) wind into internal corner; (b) wind out from internal corner; (c) wind across an internal corner

The regions terminate at the join with the adjacent face. There are no additional regions, nor are any regions lost. This implies that the design loading is no different from a typical roof, even at the internal corner. This is not exactly true: the corner region is subject to the switching flow mechanism described in §8.4.2.4 in the area where the vortex along the eave of each face cross, as illustrated in Figure 8.22. The values of the peak suction in this region is not more onerous than the design values, so that the appropriate values from Tables 20.18 and 20.20 should be used. It is the tendency for this value to switch to and from a lower value that makes the region prone to fatigue related problems.

In the second case, both faces are downwind faces and only the Regions D and F should be defined from the line of the ridge as shown in the key, (b). This implies that loading of this type of face is the least onerous case, less even than the normal hip.

In the third case, Face 4 is an upwind face and face 5 is a downwind face, although this position is reversed when the angle is rotated by 180°. Now face 5 has the gable end upwind, so the regions are defined from the upwind ridge corner for the appropriate type, in this case as in Figure 20.47 for a gabled face. In constrast, Face 4 has the internal corner upwind, and in this case only Regions D and F should be defined from the eave, as shown in the key (c).

20.7.4.10 Friction-induced loads

Friction forces accumulate on those parts of roofs that are swept by the wind, that is, away from the delta-wing vortices at the upwind edges of the roof, and may be significant when the building is very squat. For flat roofs, a reasonable estimate is obtained by treating Region G of Figure 20.42 as being swept by the wind, in all wind directions. For pitched roofs, the principal horizontal force normal to the eave/ridge comes from the resolved component of the normal pressures. Friction

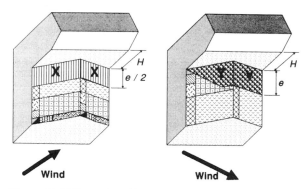

Figure 20.56 Examples of inset storeys with duopitch roofs

forces are important when the wind is parallel to the ridge and the building is very long. In this case a reasonable estimate is obtained by treating Region F of Figure 20.47 or 20.49 as being swept by the wind. Values of shear stress coefficient, c_τ, are given in Table 16.8.

20.8 Rules for combinations of form

20.8.1 Scope

This section gives design rules for those combinations of form for which such rules are needed and available. The more general discussion in §17.4 'Combinations of form' may prove useful in the design context for combinations not included here.

20.8.2 Canopies attached to tall buildings

The scope here is confined to canopies attached below half-way up the building, $h/H < 0.5$. The data are derived from tests on flat canopies, but are expected to be reasonable for pitched canopies. Canopies attached higher than halfway up the building should be assessed using the rules in §20.5.3 for free-standing canopies, fully-blocked at one edge.

Figure 20.57 is the key figure: H is the height of the wall on which the canopy is attached at height h. The reference dynamic pressure, $\hat{q}_{\text{ref}}$, is taken at the top of the building wall, $z_{\text{ref}} = H$. Table 20.28 gives global force coefficients for the normal force, positive downwards as shown in the key, for the two design cases: canopy on windward wall, $\theta = 0°$, and canopy on side wall, $\theta = 90°$.

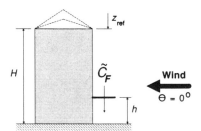

Figure 20.57 Key to canopies attached to buildings

Table 20.28   Global vertical force coefficients for canopies attached to tall buildings

Key diagram: Figure 20.57 $\hat{q}_{ref}$ as height of building, $z_{ref} = H$

$H/h =$	2	3	6	12	18	24	30	36
			Pseudo-steady normal force coefficient, $\bar{C}_F\{\theta\}$					
$\theta = 0°$	+ 0.30	+ 0.40	+ 0.69	+ 0.87	+ 0.92	+ 0.93	+ 0.93	+ 0.91
$\theta = 90°$	− 0.24	− 0.70	− 0.95	− 1.04	− 1.16	− 1.30	− 1.29	− 1.16

20.8.3 Balconies, ribs and mullions

Conservative loads on individual balconies, ribs and mullions should be estimated when necessary by treating them as multiple boundary walls, using the rules in §20.5.1 with the full reference dynamic pressure for the wall.

Better, lower estimates may be obtained using the isotachs of wind speed given in Figure 17.69 to derive a reduction factor for dynamic pressure, ϕ_q, and a direction for the flow. Since the isotachs in Figure 17.69 represent V/V_{ref}, the reduction factor ϕ_q is given by the square of their values. The direction of the flow may be taken as normal to the isotach contours. Thus the maximum loading of vertical ribs and mullions will occur at the ends of the face, while and the maximum loading of horizontal balconies will occur along the top edge.

The first rib or mullion at the windward corner is an exception, since in some critical wind directions the separation point may move from the corner to the end of the rib, exposing the rib to the difference in pressure between the front face and the local edge region of the side face. A normal force coefficient of $\bar{c}_F = 2.0$ should be assumed for this first rib. The maximum local wall suction between the end rib and the corner should be taken as $\bar{c}_p = -2.0$.

With many ribs or mullions on long walls, the loading accumulated on all the ribs should not be determined by summing individual rib loads, but should be estimated using the rules for friction-induced loads on walls given in §20.7.3.3, taking the value of $c_\tau = 0.4$ for ribs from Table 16.8.

20.9 Internal pressures

20.9.1 Scope

This section gives design rules for the pressures inside buildings, which should be used in conjunction with the external pressures to give the net loads on individual faces of the buildings. Loads on internal walls of multi-room buildings should be determined from the difference in internal pressure between adjacent rooms. With enclosed buildings, internal pressures may be neglected for the purpose of stability calculations, because their effect on opposite walls of each room cancel out. With open-sided buildings, such as grandstands, the internal pressure acting on the side opposite an open side does contribute to the net load and must be taken into account for stability.

Internal pressures depend on the position and size of openings in the outer skin of the building, as well as on the position and porosity of internal walls. In the general case, the internal pressures should be calculated using the flow balance equations given in Chaper 18. However, for some specific building forms more general guidance is appropriate and simplified rules for three typical forms are given below.

In addition to the load duration given by the rules in §20.2.3.2, the effective load duration for internal pressures also depends on the response time of the volume of the building and the size of the openings as discussed in Chapter 18 (§18.3). Rules for the effective load duration for internal pressures are given for each typical building form below.

20.9.2 Open-sided buildings

20.9.2.1 Definitions

Figure 20.58 is the key figure. The effective load duration, t, should be determined from the *TVL*-formula, Eqn 20.1, using the diagonal of the largest open face as the reference length, l, **not** the diagonal of the loaded face which is used for the external pressures. The reference dynamic pressure, $\hat{q}_{ref}\{t\}$, for the load duration, t, should be taken at the height of the top of the opening. Wind direction is measured from normal to the principal open face as defined in Figure 20.58. The internal pressure is assumed to be uniform in each region, but there are actually small gradients when the internal pressure is positive as described in §18.6.

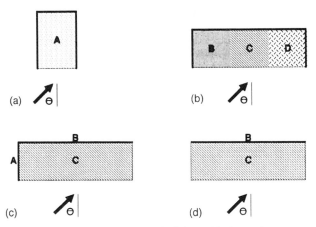

Figure 20.58 Key to open-sided buildings: (a) shorter face open and (b) longer face open (see Table 20.29); (c) longer and shorter face open and (d) longer and two shorter faces open (see Table 20.30)

20.9.2.2 One open face

Internal pressure coefficients for one open face are given in Table 20.29. These should be taken to act uniformly on the inner surfaces of the walls and the undersurface of the roof for the regions shown in Figure 20.58(a) and (b). In the case of the shorter open face there is only one region and the internal pressure is assumed to be uniform. In the case of the longer open face there are three regions, but the internal pressure differs between regions only for the wind angles $\theta = 30°$ and $60°$ when the wind blows into the skewed open face.

Table 20.29   Internal pressure coefficients for buildings with one open face

Key diagram: Figure 20.58(a) and (b) $\hat{q}_{ref}$ as height of top of opening

Open face	Shorter face	Longer face		
Region	A	B	C	D
	Pseudo-steady internal pressure coefficient, $\tilde{c}_{p_i}\{\theta\}$			
$\theta = 0°$	+ 0.85	$\rightarrow$	+ 0.68	$\leftarrow$
30°	+ 0.71	+ 0.54	+ 0.70	+ 0.80
60°	+ 0.32	+ 0.38	+ 0.44	+ 0.54
90°	− 0.60	$\rightarrow$	− 0.40	$\leftarrow$
120°	− 0.46	$\rightarrow$	− 0.46	$\leftarrow$
150°	− 0.31	$\rightarrow$	− 0.40	$\leftarrow$
180°	− 0.16	$\rightarrow$	− 0.16	$\leftarrow$

Note: assume $\tilde{c}_{p_i}\{\theta\}$ to act on inner surface of walls and undersurface of roof.

20.9.2.3 Two or more adjacent open faces

Internal pressure coefficients for two or more adjacent open faces are given in Table 20.30. These should be taken to act uniformly on the inner surfaces of the walls for Regions A and B, and the undersurface of the roof for Regions C, shown in Figure 20.58(c) and (d). The internal pressure is assumed to be uniform.

20.9.2.4 Two opposite open faces

This is the special case where wind can blow through the building, considered in §18.6.3. In skewed winds, the flow through the building is steered by the side walls,

Table 20.30   Internal pressure coefficients for buildings with two or more adjacent open faces

Key diagram: Figure 20.58(c) and (d) $\hat{q}_{ref}$ as height of top of opening

Open faces	Longer and one shorter face			Longer and both shorter faces	
Region	A	B	C	B	C
	Pseudo-steady internal pressure coefficient, $\tilde{c}_{p_i}\{\theta\}$				
$\theta = 0°$	+ 0.76	+ 0.77	+ 0.63	+ 0.82	+ 0.59
30°	+ 0.51	+ 0.59	+ 0.48	+ 0.68	+ 0.52
60°	− 0.35	− 0.18	− 0.43	+ 0.43	+ 0.33
90°	− 0.26	− 0.34	− 0.38	0	0
120°	− 0.36	− 0.42	− 0.47	− 0.63	− 0.61
150°	− 0.31	− 0.37	− 0.39	− 0.44	− 0.49
180°	− 0.26	− 0.29	− 0.33	− 0.34	− 0.39
210°	− 0.43	− 0.58	− 0.64		
240°	− 0.18	− 0.51	− 0.53		
270°	+ 0.68	+ 0.77	+ 0.65		
300°	+ 0.74	+ 0.77	+ 0.65		
330°	+ 0.78	+ 0.78	+ 0.64		

Note: assume $\tilde{c}_{p_i}\{\theta\}$ to act on inner surface of walls for Regions A and B and undersurface of roof for Region C.

creating an additional side force. This effect may be critical in the design of portal frame buildings with open gable ends. The data of Figure 18.7 may be used directly. Alternatively, for buildings with doors at both ends, for which having both doors simultaneously open will be a serviceability limit state, multiplying the forces on the side walls determined for the closed state by a factor of 1.8 is a reasonable working approximation. (In this case, one door open with the other closed is another serviceability limit state.)

20.9.3 Dominant openings

20.9.3.1 Effective load duration

An opening will be dominant if its area is greater or equal to twice the sum of the remaining openings in the building. In this case, the load duration, t, should be determined from the TVL-formula, Eqn 20.1, using the diagonal of the dominant opening as the reference length, l.

The response time of the building, t_i, should also be determined from Eqn 18.28 of §18.5, reproduced again here as:

$$t_i = \pi^{5/4} (2\,O\,[1 + k_a / k_b])^{1/2} / (c_a\,A_D^{1/4}) \tag{20.26}$$

where O is the volume of the building, A_D is the area of the dominant opening and $c_a = 340$ m/s is the speed of sound in air. Eqn 20.26 also involves determining the ratio of the bulk modulus of the air and the building, k_a / k_b, as described in §18.3.4. However, this is likely to be significant only for low buildings with long span roofs, when:

$$k_a / k_b \simeq \frac{84\,000\,L_c}{N\,H\,\Delta p} \tag{20.27}$$

where L_c is the span between columns, H is the height, N is the design span-to-deflection ratio (typically $150 < N < 250$) and Δp is the pressure difference across the roof causing the deflection. (Strictly, Δp is the result of the design assessment, but a reasonable value can be assumed and adjusted later if necessary.) Typical values range from $k_a / k_b \simeq 0.2$ for domestic houses to $k_a / k_b \simeq 5$ for long-span roofs.

The **effective** load duration for internal pressure should be taken as the **larger** of t and t_i.

20.9.3.2 Reference dynamic pressure

The reference dynamic pressure, $\hat{q}_{ref}\{t\}$, for calculating the internal pressure, p_i, from the internal pressure coefficient, $\tilde{c}_{p_i}$, should be taken at the height of the top of the face containing the dominant opening and for the effective load duration, the larger of t and t_i.

20.9.3.3 Internal pressure coefficient

The internal pressure coefficient should be taken as:

$$\tilde{c}_{p_i} = [1 - 1 / (A_{dom} / \Sigma A_D)^2]\,\tilde{c}_{p_e} \tag{20.28}$$

where A_{dom} is the area of the dominant opening, ΣA_D is the sum of the remaining openings in the building and $\tilde{c}_{p_e}$ is the average external pressure at the dominant opening.

Typically, with

$$A_{dom} / \Sigma A_D = 2, \qquad \tilde{c}_{p_i} = 0.75 \, \tilde{c}_{p_e}$$

and with

$$A_{dom} / \Sigma A_D = 3, \qquad \tilde{c}_{p_i} = 0.90 \, \tilde{c}_{p_e}.$$

With the dominant opening in a wall, the likely range of internal pressure is $-1.2 < \tilde{c}_{p_i} < +1.0$. With a dominant ridge ventilator in the roof, $\tilde{c}_{p_i} \simeq -0.8$. A dominant opening in the peripheral region of a low-pitch roof could produce an internal pressure coefficient as low as $\tilde{c}_{p_i} = -2$.

20.9.4 Conventional buildings

20.9.4.1 Scope

These are buildings where the porosity of the external skin is distributed as many small openings over several faces of the building. Most conventional buildings are in this category when all large doors and windows are closed.

20.9.4.2 Effective load duration

The load duration, t, should be determined from the *TVL*-formula, Eqn 20.1, using the diagonal of the largest envelope area enclosing the distributed openings on each face as the reference length, l. With a fully glazed façade, or a building clad in unsealed sheeting, the reference length will be the same as that for the whole face used for the external pressures.

The response time of the building, t_i, should also be determined from Eqn 18.25 of §18.5, reproduced again here as:

$$t_i = \frac{O \, V}{c_a^2 \, C_D} \frac{A_W \, A_L}{(A_W^2 + A_L^2)^{3/2}} \, (1 + k_a / k_b) \, (\bar{c}_{p_W} - \bar{c}_{p_L})^{1/2} \tag{20.29}$$

valid in the range $0.5 \leqslant A_W / A_L \leqslant 2$. The parameters of this equation are explained in §18.3 and §18.4. Standard values for the constants cases are: speed of sound in air, $c_a = 340 \, \text{m/s}$; standard orifice discharge coefficient, $C_D = 0.61$.

The effective load duration for internal pressure should be taken as the **larger** of t and t_i. With most typical buildings the load duration, t, from the *TVL*-formula will be the controlling load duration.

20.9.4.3 Reference dynamic pressure

The reference dynamic pressure, $\hat{q}_{ref}$, for calculating the internal pressures, p_i, from the internal pressure coefficients, $\tilde{c}_{p_i}$, should be taken at the height of the top of the envelope of the openings and for the effective load duration.

20.9.4.4 Internal pressure coefficient

The internal pressure coefficients for single and multi-room buildings should be determined by the quasi-steady balance of flow through Eqn 18.5, considering such combinations of open windows and doors in external and internal walls as are required by the design limit states. Typical values of porosity are given in Table 18.1. Areas of discrete openings should be measured directly.

Table 20.31   Internal pressure coefficients for typical conventional buildings

$\hat{q}_{ref}$ as height of eave, $z_{ref} = H$.

Pseudo-steady internal pressure coefficient,	$\tilde{c}_{p_i}$
Two opposite walls equally permeable; other faces impermeable	
Wind normal to permeable face	+ 0.2
Wind normal to impermeable face	− 0.3
Four walls equally permeable; roof impermeable	− 0.3

Fortunately, experience shows that typical ranges of internal pressure coefficients can be defined for the commonest forms of building porosity and these are given in Table 20.31. In this table, an 'impermeable' face can be taken as any face with permeability less than one third of the 'permeable' face.

From Table 20.31, it can be seen that the internal pressure coefficient for typical conventional buildings is confined to the range $+ 0.2 \leqslant \tilde{c}_{p_i} \leqslant - 0.3$. The most onerous net load on faces with positive external pressure coefficients is obtained at the limit $\tilde{c}_{p_i} = - 0.3$, while the most onerous net load on faces with negative external pressure coefficients is obtained at the limit $\tilde{c}_{p_i} = + 0.2$. Similarly, the most onerous net load on internal walls is expected to be $\Delta\tilde{c}_{p_i} = 0.5$.

Appendix A   Nomenclature

A.1 Symbols

This list contains those symbols in general use in Parts 1 and 2 of the Guide.

Symbol	Units	Description	Equation where defined	Section where introduced
A	m	altitude of site above mean sea level (AMSL)		§9.2.2.1
A_o	m	altitude of average ground plane AMSL		§9.2.2.1
A	m^2	area of structure		§8.3.1
A_D	m^2	orifice discharge area	(17.6)	§17.5.1
a	—	plan-area density of buildings		§9.2.1.2
a	units^{-1}	dispersion^{-1} in FT1 distribution	(5.12)	§5.3.1.2
B	m	crosswind breadth of structure		§8.4.1
b	m	effective breadth parameter		§8.6.2.2.3
b	m	structural breadth parameter		§8.6.2.2.2
C	—	(global) non–dimensional coefficient	(12.4)	§2.2.10.1
C_D	—	drag coefficient		§2.2.10.1
C_D	—	orifice discharge coefficient	(17.6)	§17.5.1
C_F	—	force coefficient	(2.25)	§2.2.10.1
C_L	—	lift coefficient		§2.2.10.1
C_p	—	pressure coefficient	(2.24)	§2.2.5
C_V	—	velocity coefficient	(2.26)	§2.2.10.1
c_a	m/s	speed of sound in air ($c_a = 340\,\text{m/s}$)	(18.19)	§18.3.2.3
c	—	local non-dimensional coefficient	(12.3)	§12.2.3.4
D	N	drag – force parallel to flow		§2.2.10.1
D	m	size, diameter, inwind depth of structure		§2.2.5
d	m	zero–plane displacement	(9.3)	§7.1.3
E	Pa	elastic modulus of structure		§2.4.2
F	N	force		§2.2.10.1
F_S	N	shear force		§8.6.2.1
F_T	N	tension force		§8.6.2.1
F	units^2/s	spectral density function	(2.10)	§2.2.3

Symbol	Units	Description	Equation where defined	Section where introduced
F	—	structural size function	(10.1)	§10.3
f	rad/s	Coriolis parameter	(5.3)	§5.2.1.1
f	—	structural frequency function	(10.1)	§10.3
Fr	—	Froude number		§13.5.1
G	—	gust factor (flat terrain)	(7.22)	§7.4.1
g	m/s^2	acceleration due to gravity		§2.2.1
H	m	height of structure		§8.6.2.1
h	m	effective height parameter		§8.6.2.2
J_{x_i}	—	joint acceptance function of ith mode in x axis	(8.1.3)	§8.6.3.2
Je	—	Jensen number (Je = H/z_o)		§13.5.1
K	m^2/s	circulation of vortex	(2.16)	§2.2.8.1
K		resistance coefficient of lattice	(16.30)	§16.3.2
k	N m/rad	torsional stiffness		§8.6.4.2
k	Pa	bulk modulus of elasticity	(18.22)	§18.3.4
k	—	exponent in Weibull distribution	(5.10)	§5.2.3
k_s	—	equivalent sand grain roughness		§2.2.7.2
L	N	lift – force normal to flow	(2.22)	§2.2.8.5
L	m	effective upwind slope length	(9.26)	§9.4.1.7.4
L	m	length of building (longer axis)		§16.1.2
L	m	depth of orifice	(18.14)	§18.3.2.1
$L_{x,1}$	m	horizontal length of upwind slope of topography		§9.2.2.2
$L_{x,2}$	m	horizontal length of plateau of topography		§9.2.2.2
$L_{x,3}$	m	horizontal length of downwind slope of topography		§9.2.2.2
L_y	m	horizontal length of crosswind slope of topography		§9.2.2.3
L_z	m	vertical height of topography		§9.2.2.1
$\mathscr{L}$	—	length scale factor		§2.4.4
l	m	structural size parameter	(10.2)	§10.5.1
M	N m	moment		§8.6.2.1
M	—	number of exceedences		§5.3.1.5
$\mathcal{M}$	—	mass scale factor		§2.4.4
m	kg/m	mass per unit length	(10.7)	§10.6.2.1.1
m	—	rank of value from smallest $(m = N + 1 - M)$		§5.3.2.1
N	—	number of values in population		§5.3.1.1
n	Hz	frequency		§2.2.3
n_s	Hz	vortex–shedding frequency		§2.2.10.4
n_{y_i}	Hz	frequency of ith mode in y axis		§8.6.4.1
n_θ	Hz	frequency of torsional mode in θ axis		§8.6.4.2
O	m^3	volume of room or building		§18.3.2.1

Symbol	Units	Description	Equation where defined	Section where introduced
P	—	cumulative probability distribution function (CDF)	(B.2)	§5.2.3
p	Pa	pressure	(2.2)	§2.2.1
p_S	Pa	static pressure	(2.6)	§2.2.2
p_T	Pa	total pressure	(2.6)	§2.2.2
Q	—	risk of exceedence ($Q = 1 - P$)	(5.8)	§5.2.3
Q	m³/s	flow rate through opening	(18.2)	§18.2.1
q	Pa	dynamic pressure	(2.6) (12.1)	§2.2.2
$\hat{q}_{ref}$	Pa	reference design dynamic pressure	(20.2)	§20.2.4
R	—	structural response parameter	(10.1)	§10.3
R	years	return (mean recurrence) period	(5.20)	§5.3.1.5
R	J/kg K	universal gas constant (R = 287.1 J/kg K)	(18.10)	§18.3.1
Re	—	Reynolds number		§2.2.6
Ri	—	Richardson number		§13.5.1
Ro	—	Rossby number		§13.5.1
r	m	radial distance		§2.2.8.1
r	years⁻¹	annual rate of occurrences	(5.32)	§5.3.2.2
S_A	—	Altitude Factor	(9.6)	§9.3.2.2
S_B	—	Building Factor (for negative shelter)	(19.3)	§19.2.3.2.2
S_E	—	Exposure Factor	(9.10)	§9.4.1.2
S_G	—	Gust Factor	(9.37)	§9.4.3.1
S_L	—	Topography Factor	(9.23) (9.25)	§9.4.1.7
S_T	—	Statistical Factor	(9.5)	§9.3.2.1
S_S	—	Seasonal Factor		§9.3.2.4
S_u	—	Inwind Turbulence Intensity Factor	(9.31)	§9.4.2.1
S_v	—	Crosswind Turbulence Intensity Factor	(9.33)	§9.4.2.2
S_w	—	Vertical Turbulence Intensity Factor	(9.34)	§9.4.2.2
S_X	—	Fetch Factor	(9.20)	§9.4.1.6
S_Z	—	Height Factor	(9.12)	§9.4.1.3
S_Θ	—	Directional Factor		§9.3.2.3
Sc	—	Scruton number		§13.5.1
St	—	Strouhal number (reduced frequency)		§2.2.5
s	—	solidity ratio	(13.17)	§13.5.4.2
s	—	blockage ratio		§16.4.4.6
s_L	—	speed increment coefficient	(9.25)	§9.4.1.7.1
T	K	temperature		§2.3.1
T	s	time ($T \gg t$)		§2.2.3
T	years	observation or exposure period		§5.3.1.1
$\mathcal{T}$	—	time scale factor		§2.4.4
t	s	time ($t \ll T$)		§2.2.3
U	m/s	mean velocity in x direction		§2.2.1

Symbol	Units	Description	Equation where defined	Section where introduced
U	units	mode (most likely value) of variable		§5.3.1.2
u	m/s	turbulent velocity in x direction		§2.2.3
u_*	m/s	friction velocity	(2.12)	§2.2.7.1
V	m/s	mean velocity in y direction		§2.2.3
V	m/s	wind speed, general flow velocity		§2.4.2
$\overline{V}$	m/s	hourly–mean wind speed		§7.2.1
$\overline{V}_B$	m/s	basic hourly–mean wind speed	(5.6)	§5.2.2
$\hat{V}$	m/s	gust wind speed (of duration t)		§7.4.1
$\hat{V}_B$	m/s	basic gust wind speed	(5.7)	§5.2.2
V_G	m/s	geostrophic wind speed	(5.2)	§5.2.1.1
V_g	m/s	gradient wind speed	(5.4)	§5.2.1.2
V_I	m/s	wind speed at interface of internal layer		§7.2.1.2
V_{max}	m/s	maximum wind speed in period T		§5.3.1.1
$\mathcal{V}$	—	velocity scale factor		§2.4.4
v	m/s	turbulent velocity in y direction		§2.2.3
W	m	width of building (shorter axis)		§16.4.4
W	m/s	mean velocity in z direction		§2.2.3
w	m/s	turbulent velocity in z direction		§2.2.3
X_i	N	modal force of ith mode in x axis	(8.13)	§8.6.3.1
X	km	fetch of ground surface roughness		§9.2.1.3
x_i	m	modal deflection of ith mode in x axis		§8.6.3.1
x	m	horizontal dimension along the flow		§2.2.7.1
y_i	m	modal deflection of ith mode in y axis		§8.6.4.1
y	m	horizontal dimension across the flow		§2.2.1
y	—	reduced variate in FT1 distribution	(5.12)	§5.3.1.2
Z	m	effective height of topography	(9.27)	§9.4.1.7.5
z	m	vertical dimension		§2.2.1
z_g	m	gradient height	(7.14)	§7.1.2
z_I	m	height of interface of internal layer		§7.2.1.2
z_o	m	aerodynamic roughness	(7.9)	§2.2.7.3
α	—	power–law exponent	(7.2)	§7.2.1.3.1
α	°,rad	azimuth angle		§12.4.1.1
α	°,rad	pitch angle		§16.2.1.2
β	°,rad	elevation angle		§12.4.1.1
β	°,rad	yaw angle		§16.2.1.2
γ_a	—	ratio of specific heats of dry air		§18.3.1
γ_{dyn}	—	dynamic amplification factor	(10.11)	§10.8.2.1
δ	—	logarithmic decrement	(10.3)	§10.5.1
κ	—	von Karman's constant (= 0.40)	(7.9)	§7.2.1.3.2
Λ	m, s	turbulence integral length and time parameters		§7.3.1.5
λ	K/km	lapse rate (rate of temperature drop with height)		§2.3.1
λ_A	K/km	adiabatic lapse rate	(2.31)	§2.3.1

Symbol	Units	Description	Equation where defined	Section where introduced
η	—	shielding and shelter factors	(16.50)	§16.3.5.1
μ	Pa s	dynamic viscosity of fluid	(2.1)	§2.2.1
υ	/h	mean crossing rate		§12.4.2.1
Ω	rad/s	angular velocity of Earth's rotation		§2.3.2
ω	rad/s	angular velocity		§2.2.8.2
Φ	°N,°S	latitude of site		§2.3.2
Φ	—	influence function		§8.6.2.1 & §16.3.1
ϕ	—	influence coefficient	(8.5)	§8.6.2.1 & §16.3.1
ϕ_{x_i}	—	mode shape of ith mode in x axis		§8.6.3.1
Π	—	characteristic product of FT1 distribution	(5.14)	§5.3.1.2
Ψ	—	effective upwind slope of topography (Z/L)	(9.25)	§9.4.1.7.1
ψ	—	actual upwind slope of topography $(L_z/L_{x,1})$		§9.2.2.3
ρ	kg/m^3	density		§2.2.1
ρ_a	kg/m^3	density of air (1.225 kg/m^3 in UK)		§2.2.1
ρ_s	kg/m^3	density of structure		§2.4.2
ϱ	—	density scale factor		§2.4.4
τ	Pa	shear stress	(2.1)	§2.2.1
τ_o	Pa	surface shear stress	(7.1)	§7.1.3
Θ	°	mean wind direction from north (in wind axes)		§5.3.4.2
Θ_g	°	gradient wind direction – surface wind direction	(7.16)	§7.2.1.3.3
θ	°,rad	mean wind angle, azimuth (in body axes)		§2.2.10.2
X	—	aerodynamic admittance	(8.4)	§8.4.1
χ	—	impulse response function	(8.3)	§8.4.1
ξ	m/s	vorticity	(2.23)	§2.2.9.1
ζ	—	structural damping ratio	(10.3)	§2.4.2

Notes

1 Units
    '—' in units column indicates a dimensionless number. 'units' in units column indicates dimensions of the variable of which the symbol is a property (see Note 2).
2 Subscripts
    The general convention is that a subscript indicates the variable of which the symbol is a property. Thus C_p is the coefficient, C, of the pressure, p, and P_V is the cumulative distribution function (CDF), P, of the wind speed, V.

Subscripted symbols with particular names which obey the convention are included in the list of symbols, e.g. pressure coefficient, C_p.

Exceptions to the general convention occur when it is necessary to indicate a particular attribute of a variable and these are given in the list of symbols above. Typical examples of exceptions are static pressure, p_S, and basic wind speed, V_B. When a symbol is a property of two variables, or has two attributes, these are separated by commas at the same level; thus $\Lambda_{x,u}$ is the integral length parameter, Λ, of the u turbulence component in the x direction.

When the subscript is a property of another variable or has an attribute so that it has its own subscript, this last subscript appears at the next lower level; thus P_{V_B} is the CDF, P, of the basic wind speed, V_B.

3 Superscripts

The general convention is that a superscript, when it is a value or another symbol, indicates the power by which the symbol is raised (the normal mathematical convention); thus V^2 is the square of the wind speed, V. When the symbol representing a power is also raised to a power, this last power is superscripted at the next higher level; thus e^{-V^2} is e raised to the power of $-V^2$.

Exceptions to this general convention are the mathematical notation for mean, root-mean-square and peak values given in §A.2.1, below.

4 Brackets

In equations the curved, square and angle brackets, () [] <>, are used as separators in the conventional sense.

The curly brackets, {}, are exclusively used to indicate functional dependence; thus $F_u\{n\}$ indicates that the spectral density function, F, of the u-component of turbulence (subscript u) is a function of the frequency, n. An equivalence in these brackets indicates the specific value of the function at that equivalence; thus $\overline{V}\{z = 10\,\text{m}\}$ indicates the value of mean wind speed, $\overline{V}$, at the height $z = 10\,\text{m}$ above ground.

A.2 Mathematical notation

A.2.1 Standard notation

Symbol	Description
$\overline{X}$	mean of X
$|x|$	absolute magnitude of x irrespective of sign ($|x|=|-x|$)
$\hat{X}$	maximum or peak value of X
$\check{X}$	minimum or negative peak value of X
$\tilde{X}$	pseudo-steady value of X (see Chapter 15)
x'	root-mean-square of x ($\overline{x} = 0$)
$N!$	factorial of $N = (N)(N - 1)(N - 2)\ldots(3)(2)(1)$
$\sin X$	sine of X
$\cos X$	cosine of X
$\tan X$	tangent of X
$\arcsin Y$	arc-sine, or inverse sine of Y (angle whose sine is Y)
$\arccos Y$	arc-cosine, or inverse cosine of Y (angle whose cosine is Y)
	(The forms $\sin^{-1} Y$ and $\cos^{-1} Y$ for arc-sine and arc-cosine have been avoided to prevent any confusion with the cosecant and secant, the reciprocals of the sine and cosine, $\sin Y^{-1}$ and $\cos Y^{-1}$.)
e	the base of natural logarithms, $e = 2.7183$
$\exp X$	e raised to power X (e^X)
$\ln X$	natural logarithm (logarithm to base e) of X
$\log X$	logarithm to base 10 of X
Π_a^b	product between limits a and b
Σ_a^b	sum between limits a and b
$\int_a^b$	integral between limits a and b
dX/dY	differential of X with respect to Y
δX	small change in X

A.2.2 Special notation

Symbol	Description
$\Rightarrow$	'becomes'; used to specify changes of ground surface roughness (§9.2.1.3). Thus $z_0 = 0.003 \Rightarrow 0.03$ m indicates that the ground roughness $z_0 = 0.003$ m (Category 0) 'becomes' $z_0 = 0.03$ m (Category 2).
$\downarrow$	'smaller of'; used at changes of roughness (§9.4.1.6) to make a choice between two possible values. Thus $a \downarrow b$ means the smaller value of 'a' or 'b'.
$\uparrow$	'larger of'; used at changes of roughness (§9.4.1.6) to make a choice between two possible values. Thus $a \uparrow b$ means the larger value of 'a' or 'b'.
*	denotes value derived through convolution of Cook–Mayne method (§15.2).

A.3 Units

The International System of Units (SI) is used as standard in this publication.

A.3.1 SI base units

Symbol	Name	Quantity
m	metre	length
kg	kilogram	mass
s	second	time
K	kelvin	temperature

A.3.2 SI supplementary unit

Symbol	Name	Quantity
rad	radian	plane angle

A.3.3 SI derived units with special symbols

			Expression in terms of:	
Symbol	Name	Quantity	Other units	Base units
Hz	hertz	frequency		s^{-1}
mb	millibar	atmospheric pressure	100 Pa	$100\,m^{-1}\,kg\,s^{-2}$
N	newton	force		$m\,kg\,s^{-2}$
Pa	pascal	pressure, stress	N/m^2	$m^{-1}\,kg\,s^{-2}$
°	degree	plane angle	(value/57.296) rad	
°C	degree Celcius	temperature		$(value + 273.15)K$

Appendix I   Bibliography of modelling accuracy comparisons for static structures

Including static loading (e.g. cladding loads) of dynamic structures

I.1 Full scale to model scale comparisons

(may also include model scale to model scale comparisons)

I.1.1 High-rise buildings

Dalgliesh W A. Comparison of model/full-scale wind pressures on a high-rise building. *Journal of Industrial Aerodynamics*, 1975, **1** 55–66.

Dalgliesh W A, Templin J T and Cooper K R. Comparisons of wind tunnel and full-scale building surface pressures with emphasis on peaks. *Wind engineering: proceedings of the fifth international conference, Fort Collins, Colorado, USA, July 1979* (Ed: Cermak JE), Vol 1, pp 553–565. Oxford, Pergamon Press, 1980.

Dalgliesh W A. Comparison of model and full-scale tests of the Commerce Court building in Toronto. *Wind tunnel modeling for civil engineering applications* (Ed: Reinhold T A), pp 575–589. Cambridge, Cambridge University Press, 1982.

Lee B E. Model and full-scale tests of the Arts Tower at Sheffield University. *Wind tunnel modeling for civil engineering applications* (Ed: Reinhold T A), pp 590–604. Cambridge, Cambridge University Press, 1982.

I.1.2 Low-rise buildings

Marshall R D. A study of wind pressures on a single-family dwelling in model and full scale. *Journal of Industrial Aerodynamics*, 1975, **1** 177–199.

Apperley L, Surry D, Stathopoulos T and Davenport A G. Comparative measurements of wind pressures of the full-scale experimental house at Aylesbury, England. *Journal of Industrial Aerodynamics*, 1979, **4** 207–228.

Holmes J D. Discussion: Comparative measurements of wind pressures of the full-scale experimental house at Aylesbury, England. *Journal of Wind Engineering and Industrial Aerodynamics*, 1980, **6** 181–182.

Tieleman H W, Akins R E and Sparks P R. A comparison of wind-tunnel and full-scale wind pressure measurements on low-rise structures. *Journal of Wind Engineering and Industrial Aerodynamics*, 1981, **8** 3–19.

Tieleman H W, Akins R E and Sparks P R. Model/model and full-scale/model comparisons of wind pressures on low-rise structures. *Designing with the wind* (Eds: Bietry J, Duchene-Marullaz P and Gandemer J), Paper IV–5. Nantes, Centre Scientifique et Technique du Bâtiment, 1981.

Holmes J D. Comparison of model and full-scale tests of the Aylesbury house. *Wind tunnel modeling for civil engineering applications* (Ed: Reinhold T A), pp 605–618. Cambridge, Cambridge University Press, 1982.

Holdø A E. Some measurements of the surface pressure fluctuations on wind-tunnel models of a low-rise building. *Journal of Wind Engineering and Industrial Aerodynamics,* 1982, **10** 361–372.

Macha J M, Sevier J A and Bertin J J. Comparison of wind pressures on a mobile home in model and full scale. *Journal of Wind Engineering and Industrial Aerodynamics,* 1983, **12** 109–124.

Roy R J. Wind tunnel measurements of total loads on a mobile home. *Journal of Wind Engineering and Industrial Aerodynamics,* 1983, **13** 327–338.

Vickery P J. Wind loads on the Aylesbury experimental house: a comparison between full scale and two different scale models. Faculty of Engineering Science, MESc Thesis. London, Ontario, University of Western Ontario, 1984.

Stathopoulos T. Discussion: Comparison of wind pressures on a mobile home in model and full scale. *Journal of Wind Engineering and Industrial Aerodynamics,* 1985, **21** 343–344.

Hansen S O and Sørensen E G. The Aylesbury experiment. Comparison of model and full-scale tests. *Journal of Wind Engineering and Industrial Aerodynamics,* 1986, **22** 1–22.

Vickery P J, Surry D and Davenport A G. Aylesbury and ACE: some interesting findings. *Journal of Wind Engineering and Industrial Aerodynamics,* 1986, **23** 1–18.

Mousset S. The international Aylesbury collaborative experiment in CSTB. *Journal of Wind Engineering and Industrial Aerodynamics,* 1986, **23** 19–36.

Apperley L W and Pitsis N G. Model/full-scale pressure measurement on a grandstand. *Journal of Wind Engineering and Industrial Aerodynamics,* 1986, **23** 99–111.

Robertson A P and Moran P. Comparisons of full-scale and wind-tunnel measurements of wind loads on a free-standing canopy roof structure. *Journal of Wind Engineering and Industrial Aerodynamics,* 1986, **23** 113–125.

Surry D and Johnson G L. Comparisons between wind tunnel and full scale estimates of wind loads on a mobile home. *Journal of Wind Engineering and Industrial Aerodynamics,* 1986, **23** 165–180.

I.1.3 Curved structures

Schnabel W. Field and wind-tunnel measurements of wind pressures acting on a tower. *Journal of Wind Engineering and Industrial Aerodynamics,* 1981, **8** 73–91.

Pirner M. Wind pressure fluctuations on a cooling tower. *Journal of Wind Engineering and Industrial Aerodynamics,* 1982, **10** 343–360.

Batham J P. Wind tunnel tests on scale models of a large power station chimney. *Journal of Wind Engineering and Industrial Aerodynamics,* 1985, **18** 75–90.

I.2 Model scale to model scale comparisons

I.2.1 Effect of building model and boundary-layer simulation linear scale factors

Hunt A. Scale effects on wind tunnel measurements of wind effects on prismatic buildings. College of Aeronautics, PhD Thesis. Cranfield, Cranfield Institute of Technology, 1981.

Hunt A. Scale effects on wind tunnel measurements of surface pressures on model buildings. *Designing with the wind,* (Eds: Bietry J, Duchene-Marullaz P and Gandemer J), Paper VIII–8. Nantes, Centre Scientifique et Technique du Bâtiment, 1981.

Holdø A E, Houghton E L and Bhinder F S. Some effects due to variations in turbulence integral scales on the pressure distribution on wind-tunnel models of low-rise buildings. *Journal of Wind Engineering and Industrial Aerodynamics,* 1982, **10** 103–115.

Hunt A. Wind-tunnel measurements of surface pressures on cubic building models at several scales. *Journal of Wind Engineering and Industrial Aerodynamics,* 1982, **10** 137–163.

Bächlin W, Plate E J and Kamarga A. Influence of the ratio of the building height to boundary-layer thickness and of the approach flow velocity profile on the roof pressure distributions of cubical buildings. *Journal of Wind Engineering and Industrial Aerodynamics,* 1983, **11** 63–74.

Stathopoulos T and Surry D. Scale effects in wind tunnel testing of low buildings. *Journal of Wind Engineering and Industrial Aerodynamics,* 1983, **13** 313–326.

Robins A G. Discussion: Influence of the ratio of the building height to boundary-layer thickness and of the approach flow velocity profile on the roof pressure distributions of cubical buildings. *Journal of Wind Engineering and Industrial Aerodynamics,* 1984, **17** 159–160.

Bächlin W, Plate E J and Kamarga A. Discussion: Influence of the ratio of the building height to boundary-layer thickness and of the approach flow velocity profile on the roof pressure distributions of cubical buildings. *Journal of Wind Engineering and Industrial Aerodynamics,* 1984, **17** 161–162.

I.2.2 Comparisons between different wind tunnel facilities using the same model

Melbourne W H. Comparison of measurements on the CAARC standard tall building model in simulated model wind flows. *Journal of Wind Engineering and Industrial Aerodynamics,* 1980, **6** 73–88.

Vickery P J and Surry D. The Aylesbury experiments revisited – further wind tunnel tests and comparisons. *Journal of Wind Engineering and Industrial Aerodynamics,* 1983, **11** 39–62.

Sill B L, Cook N J and Blackmore P A. IAWE Aylesbury Comparative Experiment–preliminary results of wind tunnel comparisons. *Journal of Wind Engineering and Industrial Aerodynamics,* 1989, **32** 285–302.

Appendix J Guidelines for *ad-hoc* model-scale tests

J.1 Introduction

The advice given in this appendix is fairly basic, since most of the detail about wind tunnels, measurement techniques, etc., are given in the main chapters, particularly Chapter 13. This appendix is useful as a checklist of decisions the designer should take and the questions he should ask of the wind engineer undertaking the tests. The intention is to put you, the designer in the driving seat so that you may steer the testing programme. Appendix G of Part 1 gave the necessary provisions for wind-tunnel tests.

J.2 Advice on defining the test programme

Before even approaching a wind engineering laboratory, you need to know why you are commissioning tests and the range and sophistication of the data you expect from them. Ask yourself the following questions:

Why do I need to commission wind-tunnel tests?
Am I concerned with stability?
Am I concerned with structural loads?
Am I concerned with cladding and glazing loads?
Am I concerned with deflections or accelerations?
Am I concerned with ventilation, smoke emission or wind environment?
Am I concerned about effects **of** neighbouring buildings?
Am I concerned about effects **on** neighbouring buildings?
Am I concerned about the site wind climate?
Am I concerned about the site exposure?
Am I concerned about the site topography?
Is the form of the building structure unusual?
Is the external shape fixed?
Is the structural form fixed?

The reasons for asking these questions and the consequences of the answers are all covered somewhere in the Guide. If they are not apparent, then perhaps it is time to read through the Guide again! There is one exception: the sixth question regarding ventilation, smoke emission and wind environment is not covered by the Guide at all. The reason for asking this question is that, having gone to the expense of constructing a wind-tunnel model and commissioning tests for structural design, it is sensible to get as much useful information on the other aspects of wind engineering as you can.

From the answers to the above questions you should now be able to define the ultimate limit state criteria and the serviceability limit state criteria for the building structure. You need the data required to satisfy these design criteria.

J.3 Advice on selecting the contractor

Wind-tunnel testing is not yet such a standardarised exercise that you can expect an 'off the peg' service. The form of buildings and their environment is sufficiently varied that the wind-tunnel simulation and measurement techniques must often be tailored to suit the problem at hand, as part of a more 'bespoke' service. There should be a partnership formed between the designer and the wind engineer with the common aim of satisfying the design criteria.

In the past, when there were relatively few boundary-layer wind tunnels, the designer's choice was restricted and sometimes he had little control over the form and quality of the testing. The range and quality of the facilities offered to the designer varies considerably – from expert wind engineers with boundary-layer wind-tunnel facilities, the necessary acquisition equipment and considerable expertise in the interface between research and design – to the lone consultant with an interest in the subject and access to an aeronautical wind tunnel. The organisations with the necessary facilities, expertise and experience that are listed in Appendix G represent the best of the first category. This list should not be taken as being completely comprehensive. It should not be forgotten that many of these expert organisations started in the second category and developed their expertise over many years, and that members of the second category require support and encouragement to achieve this status. This can be done quite rapidly, given the determination to learn from the several decades of experience of others given in the published literature. However, serious problems can arise when this necessary preliminary is neglected or skimped.

Now that more wind-tunnel facilities offer testing services to the building designer, a recently increasing trend is for the designer to invite several to tender for the work on the basis of a technical specification. Good technical specifications are the ones that are written specifically for the building structure in question, that specify the form and extent of the data required to satisfy the ultimate limit state and serviceability criteria and request information on how the tests will be conducted as well as their cost. To these the wind engineer should reply with the proposals for testing and their cost. Technical specifications that define a rigid test programme (sometimes copied from a particular organisation's response to a previous tender and not related to the current structure at all) may enable the designer's accountant to make direct cost comparisons, but they tend to prevent the wind engineer from exploiting his equipment and expertise to your best advantage.

J.4 Advice on defining the data format

With an *ad-hoc* test the designer has the opportunity to tailor the data format to the problem at hand. The format used in Chapter 20 which is best for the general case may not be the best for a particular structure. Pressure tappings should be located to give the required information for the structural design. If the structural form is fixed, the designer has the opportunity to define loaded areas for pneumatic averaging (§13.3.3.1) which will considerably reduce the number of measurements and the cost, but at the expense of losing information on the distribution of the pressure. The tappings in these loaded areas can also be area-weighted to represent the influence functions (§8.6.2.1) for the loading. The choice here is the designer's and is a balance between detail of information and cost.

One aspect of the data format which may not be in the designer's control is the form of the analysis. Most of the expert organisations have established analysis routines of which only some may be sufficiently flexible in their approach to permit variations. Many of the organisations listed in Appendix G use the fully probabilistic design method of §15.2, usually by the simplified method of §15.2.4, including BRE, BMT and Oxford University in the UK, Colorado State University in the USA and CSIRO in Australia. Others use a different methodology to achieve an equivalent result, for example Bristol University uses Lawson's quantile-level approach [19] (§15.2.5.3), while the University of Western Ontario integrates peak pressure measurements with the parent wind climate using the method of Davenport [396] (§19.9.3.2.4).

The design data of Chapter 20 are in coefficient form to enable them to be applied to any situation of wind climate and site exposure. In *ad-hoc* tests the data are specific to the structure and this general format is not required. Instead it is simpler to simulate a direct model of the site and structure at chosen length and time scales (see §2.4). Design data are given directly in the required engineering units from the results, by applying the corresponding pressure and force scale factors, without the need of the intermediate pressure coefficients.

In general, the site will be sufficiently heterogeneous for the exposure to be different in all directions, especially if there are close neighbouring buildings. Simulation of the exact site surroundings tends to inhibit exploitation of any symmetry in the model (§20.2.5.3) to reduce the amount of data required. However, if the site is reasonably homogeneous in exposure and the shelter of neighbouring buildings is not to be exploited, representative uniformly rough surroundings can be substituted for the exact site surroundings, halving the data for a structure with one degree of symmetry, quartering for two degrees of symmetry, and reducing the tests for an axisymmetric building to one wind direction.

J.5 Advice on quality assurance

J.5.1 General

There are still sufficient differences between the wind-tunnel facilities, methods of atmospheric boundary-layer simulation and techniques of measurement for you to need to check the quality of the design data. Some of the organisations may have formal quality-assurance procedures. In any event, you should make your own assessment and advice to assist you is given below.

J.5.2 The atmospheric boundary-layer simulation

Ask for information on the boundary-layer simulation to ensure that the necessary provisions for wind-tunnel tests listed in Appendix G of Part 1 are actually met. Consider the provisions in §G.1 for testing static structures: a figure of the form of

Figure 13.24 will show that the first provision for the mean wind speed profile has been met when the data match the design data of Chapter 9 for the site at the required scale factor. Similarly, figures of the form of Figures 13.25 and 13.26 will show whether the second and third provisions have been met.

In practice, a natural or accelerated-growth boundary-layer simulation is essential to meet these provisions. Roughness, barrier and mixing device methods are the most appropriate contemporary method (§13.5.3.2). These require a wind-tunnel section with a roughened floor at least six times longer that the height of the building for an urban simulation, and longer still for rural simulations. You should be able to recognise simulation hardware similar to that shown in the photographs in §13.5.

Velocity-profile-only simulations, using graded grids or curved gauzes, may give representative mean pressures on single structures, but peak values will not be represented and multiple buildings will be poorly represented. Uniform flow is acceptable only for lattice structures.

J.5.3 Proximity modelling

Unless the symmetry of the structure is to be exploited, it is usual to model the exact surroundings of the structure for a distance around the site, usually called the 'proximity model' (§13.5.3.3). The extent of the proximity model is usually taken as the size of the wind-tunnel turntable. However, depending on the size of the turntable and the linear scale factor of the model, this may prove to be too large or too small. As the proximity model is usually a large proportion of the cost of the model, the first case is a waste of the designer's money, while the second case means the site is less well represented.

An indication of the extent of influence by neighbouring buildings on the site wind characteristics is given by Pasquill's 'roughness footprint' concept [149] as well as the work of Lee and others, reported in §19.2.2.2 and illustrated by the influence area in Figure 19.11. Studies at Bristol University with arrays of cubes [150] and tall buildings [151] suggested that five rows of similar-sized elements upwind of the site were sufficient to mask the individual effects of the upwind row. Taken together, these studies indicate that the proximity model in urban areas should include at least five city blocks, but that more than about 15 city blocks would be wasteful. A separate problem occurs when the building under study is tall in an area of generally low-rise buildings, but there are other tall buildings some distance away. Wake effects from these buildings when they are upwind will be detectable to distances of up to $20B$ downwind, where B is the crosswind breadth of the upwind building. It is recommended that all such buildings should be represented upwind of the site within a distance of $10B$, and others within $20B$ should be included if practicable, even if they lie upwind of the turntable (in which case they must be relocated by hand as the wind angle is changed). The example photographs in §13.5 show typical proximity models.

J.5.4 The reference flow parameters

Accuracy in the reference parameters: the static pressure, p_s, which is the base from which pressures are measured; and the reference dynamic pressure, $\hat{q}_{ref}$; is essential. The IAWE–ACE model-scale comparisons of §13.5.5.3 show that variation in these reference quantities is the principal source of variation in the results.

Variation in the measured static pressure is caused principally by changes in the zero of the pressure transducer, which is due to changes of temperature and other environmental parameters and is called 'drift'. Another source is changes of wind speed in the wind tunnel and in the blockage caused by the model, but these should happen only when the wind angle is changed if the static pressure is acquired in the same cross-section as the model. Drift is monitored by taking a zero reading from a static pressure tube at each wind direction before and after the measurements of surface pressure are made. The drift during a measurement sequence is the difference between these values and should be less than 1% of the reference dynamic pressure. Drift should be logged during the test and should be available for inspection.

Variation of the reference dynamic pressure may be due to several causes. As the reference dynamic pressure is determined from the difference between the total and static pressures (Eqn 2.6, §2.2.2), static pressure drift causes a corresponding reference dynamic pressure error. Changes in wind-tunnel speed during measurements also produces an error. Changes in wind-tunnel speed and model blockage due to changes of wind angle are accounted for by measuring the reference dynamic pressure in each wind direction. A further perceived source of variation is the use of a reference height too far above the model building structure, when errors in the boundary-layer calibration become included. This source of error is worst when gradient wind speed is used as the reference. The reference dynamic pressure should be taken at, or just above, the height of the model building. Variations in value should be less than 1%, should be logged during the test and should be available for inspection.

J.5.5 Measurement accuracy

The fourth necessary provision in Appendix G requires 'the response characteristics of the wind-tunnel instrumentation are consistent with the measurements to be made'. This implies that the measurement instrumentation be accurate in terms of calibration, range and frequency response. Instrumentation was reviewed in §13.3.

Pressure transducers (§13.3.2.4), force balances (§13.3.2.2) and laser anemometers (§13.3.1.2) do not generally change their calibrations and only need to be checked occasionally. Hot-wire anemometers (§13.3.1.2) generally require to be recalibrated each time they are used, but some of the later shielded and temperature-compensated probes have fixed calibrations. Use of tranducers of an appropriate range or sensitivity is required to keep the signal-to-noise ratio as high as possible. Acoustic noise in the wind-tunnel generated by the fan or by organ-pipe resonances in the wind tunnel working section will appear as unwanted 'noise' on pressure tranducers signals in addition to any electronic noise of the pressure tranducer systems. This can be eliminated by measurement, using a static pressure probe, and simultaneous subtraction from the pressure signals. The total rms noise levels should be less than 1% of the reference dynamic pressure.

Pressure tappings should be sufficiently densely located to define the pressure distribution over the model. This requires them to be closer together near the edges of the model than in the centre of the faces. The exception to this rule occurs when pneumatic averaging (§13.3.3.1) is used, when the positions are determined by the weighting function for the averaging, but the density must still be sufficient. Tappings should be square to the surface and should not protrude (see Figure 13.9(b)). They should also be checked to ensure that they are not blocked (both before and after testing).

Evidence of the frequency response of the measurement equipment should also be sought. There are no frequency response problems with hot-wire and laser anemometers, but the former will have problems in resolving very high intensities of turbulence (§13.3.1.2). Hot-wire 'X-probes' should be monitored for clipping of the signal at turbulence intensities above about 12% and should not be used within the depth of the interfacial layer around the buildings. Single-wire probes will overestimate the mean, underestimate the rms in the interfacial layer, but will still give reasonable measurements of peak value (irrespective of direction). Most pressure transducers also a have good frequency range, but the problem here is the response of the connecting tubing (§13.3.2.5). Make sure that the pressure tubing is fitted with restrictors like those shown in Figure 13.12, or that compensation is applied in the analysis process. The overall length of the tubing should not be much in excess of 500 mm and, if the model is larger in extent, the tappings must be connected in smaller batches. This means that the transducer and scanning switch must be in, or directly under, the model building structure. Connections from the model to tranducers by the side of the wind tunnel, requiring several metres of tubing, will not give a sufficient frequency response. Force and moment balances should be stiff enough that their first natural frequency is well above the frequency of interest, or else compensation is required in the analysis process.

J.5.6 Peak value measurements

With static structures, the main concern is the assessment of peak values. The various methods were reviewed in §12.4. They are ranked here in order from the best, downwards in accuracy.

(a) peaks from extreme-value analysis or quantile-level analysis;
(b) peaks estimated from mean and rms by peak-factor method;
(c) peaks from single or average peak measurements;
(d) mean values using quasi-steady model.

There may be some debate as to the correct order of items b and c in the list, particularly by those who use the single peak measurements of c as a fast, hence cheap, method of obtaining peak data. It is now generally agreed that the absence of all knowledge about the probability of the directly measured single peak makes these data less reliable for design than data derived by the indirect estimation of the peak-factor method.

In order to make a fully probabilistic assessment comparable with the design data of this Guide, the use of an extreme-value or quantile-level analysis is essential.

Appendix K A model code of practice for wind loads

K.1 Introduction

Reproduced in this Appendix are the BRE Digests concerned with the assessment of wind loading of building structures current in 1989 or issued during that year. These have been compiled from the data in Parts 1 and 2 of the Guide, but have been presented in a much simplified format. This results in the loss of much of the detail offered by the Guide, but still provides a suitable basis for design wind load assessments of typical building structures. Taken together these Digests form a model code of practice which follows the procedures and data proposed for the latest UK code, BS6399 Part 2, expected to replace current code CP3 ChV Pt2 [4] sometime after 1989. (The draft Eurocode for actions on building structures has incorporated many of the the loading coefficient data, but has not adopted the Guide approach for the wind data.) They have been compiled from the design wind data of Chapter 9, the classification procedure of Chapter 10 and the loading coefficients of Chapter 20, reduced to simple design rules. It has always been BRE policy to reflect the advances in code development in its series of Digests, and amending or issuing new Digests slightly in advance of code changes has assisted designers to adjust to the transition as well as making the latest data available.

BRE Digest 119 *The assessment of wind loads* has mirrored the current UK code and has been amended and augmented by other Digests, e.g. Digest 283 *The assessment of wind speed over topography*, as the code was amended. Digests 119 and 283 were replaced during 1989 by Digest 346 in a series of parts which cover and extend their range. It is anticipated that major amendments of the other existing Digests and all new Digests will be issued as new parts to Digest 346. It must also be expected that the Digests reproduced here will be amended from time to time, and the current versions may be obtained from the BRE Publications Sales Office or any HMSO bookshop.

It must be stressed that the provisions of these Digests are a simplified **alternative** to the design data of the Guide, i.e. the two approaches are exclusive. The simplifications of the Digests cause bias errors which have been calibrated to balance out partially in the complete assessment, but to retain a degree of conservatism on average. On the other hand, the full procedures of the Guide are intended to be unbiased. Accordingly the two approaches should not be mixed. It is recommended that the Digest method is used for a quick, safe first assessment and that the Guide is used to refine that assessment.

K.2 Contents of the Digests

K.2.1 Nomenclature

The nomenclature adopted in the Digests is that proposed for BS6399 Part 2 and differs from the standard nomenclature of the Guide. It has been derived by adapting the nomenclature of CP3 ChV Pt2 [4], with the number subscripts to the S-factors now replaced by letters which relate to the function of the factor, e.g. S_{TB} is the 'terrain and building factor'. While this conforms to the proposed ISO model wind code specification, it does not conform to BSI standard practice and may very well be changed again in the published form of BS6399 Part 2.

K.2.2 Digest 346 The assessment of wind loads

K.2.2.1 Part 1, Background and method

This Part of Digest 346 provides an outline in principle of the procedures to be used in assessing wind loads, referring to the other Parts which contain the specific rules and data.

K.2.2.2 Part 2, Classification of structures

This Part of Digest 346 gives a simplified method of classifying structures as static or dynamic and provides a value for the dynamic amplification factor, γ_{dyn}, for mildly dynamic structures (called the 'dynamic magnification factor' and denoted by C_R in the Digest for compatibility with BS6399 Pt2).

K.2.2.3 Part 3, Wind climate in the United Kingdom

This Part of Digest 346 gives the extreme wind speed data for the UK for the basic terrain in terms of the geographical location, altitude, wind direction and, for temporary structures, the sub-annual period of exposure. Note that the design risk is maintained at 0.02 annually (or in the sub-annual period). Variation from this standard value of risk is expected to be made using appropriate partial factors on the wind loads.

K.2.2.4 Part 4, Terrain and building factors and gust peak factors

This Part of Digest 346 gives the procedure for adjusting the wind speeds obtained from Part 3 for the basic terrain to apply to the exposure, height and gust duration appropriate for the structure to give the design dynamic pressure.

K.2.2.5 Part 5, The assessment of wind speed over topography

This Part of Digest 346 gives the procedure to be adopted when the influence of topography is significant at the site. It supersedes Digest 283 as the approach of the Guide and BS6399 Part 2 is based on basic hourly-mean wind speeds rather than the basic gust speeds used by CP3 ChV Pt2, but is based on the same approach.

K.2.2.6 Part 6, Loading coefficients for typical buildings

This Part of Digest 346 gives the design pressure coefficients for a range of typical building forms selected from the fuller range of data in Chapter 20.

K.2.2.7 Part 7, Wind speeds for serviceability and fatigue assessments

This Part of Digest 346 gives a procedure for estimating more frequent parent wind speeds (§5.2) from the design extreme wind speeds (§5.3). This is based on the approach used in BS8100 *Lattice towers and masts* (§14.1.3.3) which has been augmented with new data and further refined since publication of Part 1 of the Guide and BS8100. Load cycles for calculating resistance to low-cycle fatigue are also given.

K.2.3 Digest 284, Wind loads on canopy roofs

This is the 1986 edition of the Digest which gives loading coefficients for free-standing canopy roofs, including the effect of blockage. The source data are the same as used in the Guide, except that the worst loadings irrespective of wind direction are given. The Digest complies with Amendment 4 to CP3 ChV Pt, published in September 1988. See **Note** below.

K.2.4 Digest 295, Stability under wind load of loose-laid external roof insulation boards

This is the Digest published in 1985 which effectively summarises the topic covered in §18.8.2.1 as it was understood at that time. See **Note** below.

K.2.5 Digest 311, Wind scour of gravel ballast on roofs

This is a simplified implementation of the full design method of Kind and Wardlaw [327], referred to in §17.3.3.8. See **Note** below.

Note: Digests 284, 295 and 311 refer to Digests 119 and 283 for the design wind speeds. The design wind speeds should now be obtained from Digest 346.

Concise reviews of building technology

CI/SfB (J4)

The assessment of wind loads
Part 1: Background and method

This is the principal Digest in a series which is compatible with the
forthcoming British Standard BS 6399:Part 2. As this new Standard
incorporates several changes from the previous CP3 Chapter V:Part 2:
1972, it is considered appropriate to introduce this series of Digests
by providing some background and guidance to the new provisions.

This Digest considers the assessment of wind loads on domestic,
commercial and industrial buildings and their associated ancillary
constructions. It describes:

- the procedures used in assessing wind loads;

- the principal changes in practice between the old BS and its replacement;

- the response to wind effects of different structures;

- the wind climate and the derivation of wind speeds to be used in design

- load assessment and pressure coefficients.

This Digest supersedes Digest 119 which is now withdrawn.
The other parts to this Digest series are:

Part 2   Classification of structures

Part 3   Wind climate in the UK

Part 4   Terrain and building factors and gust peak factors

Part 5   Assessment of wind speed over topography

Part 6   Loading coefficients for typical buildings

Part 7   Wind speeds for serviceability and fatigue assessments

 **Building Research Establishment**
DEPARTMENT OF THE ENVIRONMENT

Technical enquiries to:
Building Research Establishment
Garston, Watford, WD2 7JR
Telex 923220 Fax 664010

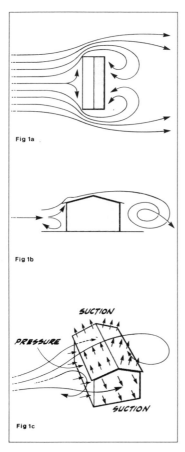

Fig 1a

Fig 1b

SUCTION

PRESSURE

SUCTION

Fig 1c

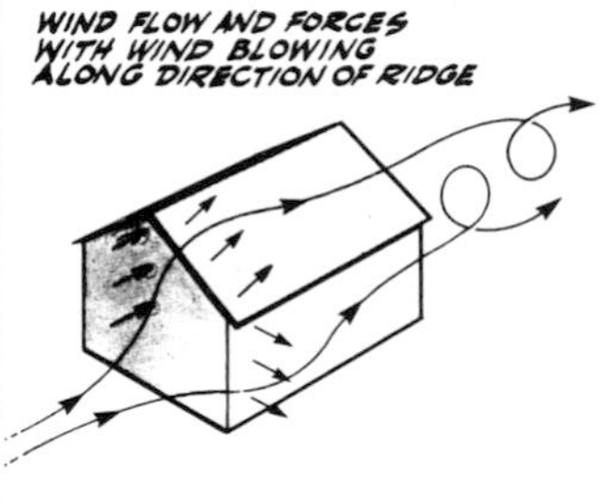

WIND FLOW AND FORCES
WITH WIND BLOWING
ALONG DIRECTION OF RIDGE

Fig 2

GENERATION OF PRESSURES AND SUCTIONS

When the wind blows more or less square-on to a building, it is slowed down against the front face with a consequent build-up of pressure against that face. At the same time it is deflected and accelerated around the end walls and over the roof with a consequent reduction of pressure (ie suction) exerted on these areas. These effects are shown in Figure 1. The greater the speed of the wind, the greater will be the suction.

The sides of a building can experience severe suction, and it is greatest near the windward edge. Access openings through and under large slab-like blocks are usually subjected to high wind speeds because of the pressure difference between the front and rear faces of the building. The facings of such openings are particularly prone to high suction which may damage the glazing and cladding.

Channelling of the wind between two buildings causes some additional suction effects on the sides facing the gap between them.

The wake behind the building is a low pressure region which exerts a suction on the rear face. This is of lower intensity than that on the sides of the building.

In the wake of a building the flow has reduced momentum providing substantial shelter from the mean flow to other buildings downwind, although the peak pressures and suctions are not reduced to the same extent. This shelter is not exploited in BS 6399:Part 2 or in other current Codes of practice.

Effect of roof pitch

On the windward slope of a roof, the pressure is dependent on the pitch. When the roof angle is below about 30°, the flow separates and the windward slope can be subjected to severe suction. Roofs steeper than about 35° generally present a sufficient obstruction to the wind for the flow to remain attached and a positive pressure to be developed on their windward slopes. Even with such roofs there is a zone near the ridge where suction is developed and insecure roof coverings may be dislodged. Leeward slopes are always subject to suction. Gabled roofs of all pitches are affected by suction along their windward edges when the wind blows along the direction of the ridge — see Figure 2. This does not occur with hipped roofs.

The suction over a roof, particularly a low-pitched one, is often the most severe wind load experienced by any part of a building. Under strong winds the uplift on the roof may be far in excess of its self weight, requiring firm positive anchorage to prevent the roof from being lifted and torn from the building.

Variation of pressure over a surface

The distribution of pressure or suction over a wall or roof surface is generally far from uniform. Pressure tends to be greatest near the centre of a windward wall and falls off towards the upwind edges. The most severe suction is generated at the corners and along the edges of walls and roofs; careful attention must be paid to the fixings at these locations.

Any projecting feature, such as a chimney stack, dormer window or tank room, will generate eddies in the air flow causing local loads on the feature as well as modifying the loads on the roof in their vicinity. The roof cladding around projections needs special attention. Roof overhangs are subject to an upward pressure on the underside which must be taken into account in assessing the total roof uplift. These effects are described in more detail in Part 6.

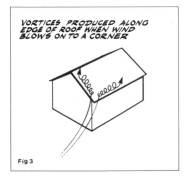

Fig 3

Fig 4

Vortex action on roofs

When the wind blows obliquely on to a building it is deflected round and over the building. The pressures on the walls are generally less severe than where it blows square-on, but strong vortices are generated as the wind rolls up and over the edges of the roof — see Figure 3. These give rise to very high suctions on the edges of the roof which must be resisted by especially firm fixing of the roof structure and covering. Since at most sites the wind can blow from any direction, *all edges and corners need special attention* — see Figure 4. Most wind damage to roofs is caused by this effect.

THE RESPONSE OF BUILDINGS TO THE WIND

The response of buildings to wind effects is strongly dependent on the characteristics of the building itself. The principal features are the natural frequency of the first few modes of vibration and the size of the building. A small structure will be completely loaded by quite small gusts but as the size increases the smaller gusts will not act simultaneously and will tend to cancel each other. Only the longer period gusts, of lower overall intensity, are significant.

A stiff building will have a high natural frequency of vibration and will tend to follow any fluctuations of load without magnification. The only design parameter to be considered is the maximum load likely to be experienced in the building's intended lifetime. Such a building is described as *static* for wind loading design purposes.

Conversely, a flexible building will have a low natural frequency: only those components of load at frequencies below the natural frequency of the structure will not be modified. Load fluctuations above the natural frequency will be attenuated in the response; the response at fluctuations near the natural frequency will be amplified, such that it may be greater than the static component. Such a structure is described as *dynamic*.

When a structure becomes very flexible the deflection may interact with the aerodynamic loads to produce various types of instability. Such structures are described as *aeroelastic*.

The majority of buildings constructed are static and can clearly be recognised as such. Only a very few are potentially aeroelastic; they are usually specialised structures. In between, there is a wide range of structures, completely static to highly dynamic. If the dynamic magnification of the response is small, this can be handled by the application of a magnification factor to the static response such that the general procedures used for the design of 'static' structures can be used. For more dynamically sensitive structures, full dynamic response procedures are necessary.

If a structure is predicted to be aeroelastic, its design has to be modified so that the interaction of the structural response with the wind is reduced. The instabilities can then be avoided and the structure is then considered as dynamic.

Methods have been developed to enable the designer to categorise his structure simply so that it can be designed by the appropriate method. This simplified procedure is described in Part 2.

WIND CLIMATE
The nature of the wind
Within the area of a wind storm, many local influences modify the
general wind flow. There is a convection causing mixing of the air
masses, and a mechanical stirring caused by the friction of the air
over the ground. The scale of the turbulence varies over wide
limits. Some of the major eddies may be several thousand metres in
extent and give rise to squalls lasting several minutes. At the other
end of the scale, small (though possibly severe) eddies may be due
to the passage of the wind past a building or other minor
obstruction. They may last only a fraction of a second. Usually the
pattern is complex with small eddies superimposed on larger ones,
so that wind speeds vary greatly from place to place and from
moment to moment. The result of any measurement of wind speed
will depend on the duration over which the sample is taken. A long
averaging time allows the inclusion of a large eddy, while a brief
averaging time may cover only a small superimposed eddy, but this
may have a higher speed.

Wind speeds in the United Kingdom
The United Kingdom lies in the range of latitudes where the climate
is characterised by the eastward passage of large weather systems.
As the UK is in the southern part of this latitude range, most
depressions pass over or to the north of Scotland resulting in a
marked gradient in the increasing severity of strong winds in the
UK from south-east to north-west.

The greatest risk is from the direction of the prevailing winds: from
the south-west. As the UK is small compared with the depressions
which cause strong winds, directional characteristics show no
significant variation with location after correction for the site
exposure. This is also true of the seasonal variations: January is the
windiest month and the least windy period is between June and
August. These effects are described in detail in Part 3 where design
values are given.

Wind speeds in other countries
The climate of the UK is dominated by prevailing westerly winds
caused by large frontal depressions. Other storm mechanisms, such
as squalls, thunderstorms or tornadoes, either produce less strong
winds or have very low probabilities of occurrence and can be
ignored in design. This is not the general case in other countries.
Tropical regions are subject to hurricanes, cyclones or typhoons
with intensities and likelihood of occurrence higher than winds
from general frontal depressions. In such cases, the estimation of
wind speeds is more difficult than for the UK and the mixed
climate data have to be separated before analysis can be
undertaken. Owing to the likely greater dispersion of these wind
speeds, higher safety factors are frequently required to achieve the
same level of reliability as that adopted for the UK.

There are regions, including parts of the UK, where there is a risk
of tornadoes or other local intense storms for which the wind
speeds cannot be predicted by the method in this Digest. If these
need to be considered in design, recourse has to be made to local
records which include such phenomena; usually, special local
regulations will apply. In the UK the risk of tornadoes is considered
only for high security structures, such as nuclear power plants.

DESIGN WIND SPEEDS
The wind speed to be used in design must take into account several parameters.

These are:

- the location of the site in the UK
- the altitude of the site
- the direction of the wind
- the seasonal exposure of the structure
- the terrain in which the building is sited
- the height of the building
- the dimensions of the building
- the topography, when significant (ie if the site is at, or near, the crest of a hill, ridge or escarpment).

The detailed procedures taking these parameters into account to derive the wind speed for design are set down in later parts of this Digest, but an outline description of each is given below.

Location *Part 3*
The majority of design applications are concerned with the performance of a building over many years, so the extreme wind speeds used for design purposes have been chosen to have an annual probability of exceedence of 0.02. Analysis of wind data has provided isopleth contours on a map of the UK of such extreme wind speeds. The location of the site is the first requirement in the design process, so that the appropriate wind speed can be read from the map.

Altitude *Part 3*
Wind speed increases with altitude and so the map speed (which is related to a uniform level of 10 m above sea level) must be modified for the altitude of the site.

Direction *Part 3*
The prevailing winds are from the south-west; for buildings which are wind-direction sensitive, appropriate allowance can be made for the reduced wind speeds from other directions.

Seasonal exposure *Part 3*
Allowance can be made for the fact that winter months are the most windy, summer months the least. This can be useful for temporary works during construction.

Terrain *Part 4*
As the wind blows from the sea over the land, and from rural to urban terrain, it is slowed down but made more turbulent. This is due to increased surface friction. Account must be taken of the distance of the site from the sea and whether the site is in country or rougher town terrain.

Height of building *Part 4*
Wind velocity increases with height, the variation depending on the terrain upwind of the site.

Dimensions of building *Part 4*
For static structures it is necessary to derive the appropriate size, and from that the intensity, of gust which will embrace the loaded area of the building. The appropriate wind load is then derived from that gust speed.

Topography *Part 5*
The map wind speed, corrected for altitude, takes account of the general level of the site above sea level. It does not allow for local topographic features such as hills, valleys, cliffs, escarpments or ridges. These can significantly alter the wind speed in their vicinity.

Near the summits of hills, or the crests of cliffs, escarpments or ridges, the wind speed will be accelerated. In valleys or near the foot of cliffs the flow may decelerated. In all cases, the variation of wind speed with height is modified from that appropriate to level terrain by a topography factor.

In terrain that is sensibly level (that is where the average slope of the ground does not exceed 0.05 within a 1 km radius of the site) the effect of topography is negligible.

LOAD ASSESSMENT FOR STATIC STRUCTURES
For most typical buildings, two aspects must be considered:

● the load on the structural frame taken as a whole;

● the loads on individual units, such as the walls and roof, their elements of cladding and fixings.

The appropriate gust speed for each aspect must be derived, differing due to the appropriate dimensions, and converted to a dynamic pressure to obtain the winds loads.

The dynamic pressure of the wind
If the wind is brought to rest against the windward face of an obstacle, all its kinetic energy is transferred to a pressure q, sometimes referred to as the *stagnation pressure* or *dynamic pressure*.

This is calculated:

$$q \quad = k V_{REF}^2 \;\; N/m^2$$

where $k \quad = 0.613 \; kg/m^3$

$$V_{REF} = \text{wind speed in m/s}$$

Pressure on a surface
The pressure on any surface exposed to the wind varies from point to point over the surface, depending on the direction of the wind and the pattern of flow. The pressure p at any point can be expressed in terms of q by the use of a pressure coefficient C_p. Thus:

$$p = C_p q$$

A negative value of C_p indicates that p is negative (a suction rather than a positive pressure). The load on a structure or element from the pressure or suction always acts in a direction normal to the surface.

In assessing the overall loading on a structure (for example, for the design of foundations) only overall coefficients are required, but this would not be adequate generally. In the calculation of wind load on any structure or element it is essential to take account of the pressure difference between opposite faces. For clad structures it is necessary to know the internal pressure as well as the external values, and it is convenient to use distinguishing pressure coefficients C_{p_e} and C_{p_i} to differentiate between them.

Pressure coefficients for typical rectangular buildings are given in Part 6. Other building types are covered by specific Digests (eg Digest 284 *Wind loads on canopy roofs*).

Allowances for dynamic response
As already noted, the majority of structures can be treated as static but a magnification factor can be applied to the static loading to account for any small dynamic amplification. This factor is a function of the type of building, its height, frequency of vibration and damping characteristics. It also depends on the basic wind speed for the site and, to a lesser extent, on the terrain in which the building is situated. Part 2 describes the simplifying assumptions that have been incorporated in tables from which the magnification factor can be derived for the majority of normal buildings. Where such assumptions are inappropriate, or where a more accurate derivation of the factor is required, the necessary equations are provided.

The loading for static structures is $P = p \, C_R A$

where: p is the pressure on the surface

C_R is the magnification factor

A is the reference area

Load factors
The load P derived from the above procedure has been assessed by statistical analysis of the data to be the load having an annual probability of exceedance of 0.02. The combinations of wind speed, pressure coefficients and dynamic magnification factor have been chosen such that this level of probability is provided for the loading.

The loads may thus be used with appropriate partial load and materials factors for both serviceability and ultimate loading conditions. The derivation of the partial factors has been determined separately and is not considered in these Digests.

DESIGN PROCEDURE

The stages required to derive wind loads on buildings and cladding are:

- Determine whether the structure can be treated as static and hence within the scope of the procedures covered by these Digests. The criterion is described in Part 2 and is dependent on the geometric and structural parameters of the building. From this Digest the relevant dynamic magnification factor is determined. Additional parameters are required if a more accurate value to the factor is required.

- Determine the site wind speed for each wind direction required from Part 3, dependent on the location and latitude of the site, and on the seasonal exposure of the structure.

- Determine the reference wind speed for design purposes from Part 4, using the site wind speed and the appropriate gust size dependent on the terrain of the site.

- Determine whether the reference wind speed needs to be modified for the effects of topography using Part 5.

- Determine the loading on the structure from:

    the reference wind speed;

    the dynamic magnification factor obtained from Part 2;

    the pressure coefficients from the Digest appropriate to the particular building type.

- Apply the appropriate partial load factors for the ultimate or serviceability limit state.

FURTHER READING

British Standards Institution
CP3:   Code of basic data for the design of buildings
         Chapter V:Part 2:1972 Wind loads

BS 6399: Loading for buildings
          Part2 Wind loading (*in preparation*)

Building Research Establishment
COOK, N J. The designer's guide to wind loading of building
structures. Part 1: Background, damage survey, wind data and
structural classification. BRE Report. London, Butterworths, 1985.
(Part 2: Static structures *to be published late 1989*).

COOK, N J, The assessment of design wind speed data: manual
worksheets with ready-reckoner tables. Garston, BRE, 1985.

COOK, N J; SMITH, B W and HUBAND, M V. BRE Program
STRONGBLOW : user's manual. BRE microcomputer package.
Garston, BRE, 1985.

Other BRE Digests
141 Wind environment around tall buildings
206 Ventilation requirements
210 Principles of natural ventilation
284 Wind loads on canopy roofs
295 Stability under wind load of loose-laid external roof insulation boards
302 Building overseas in warm climates
311 Wind scour of gravel ballast on roofs
346 Part 2: Classification of structures
     Part 3: Wind climate in the UK
     Part 4: Terrain and building factors and gust peak factors
     Part 5: Assessment of wind speed over topography
     Part 6: Loading coefficients for typical buildings
     Part 7: Wind speeds for serviceability and fatigue assessments

ACKNOWLEDGEMENT
The Building Research Establishment gratefully acknowledges
the assistance of Flint and Neill Partnership in the preparation
of this Digest.

Printed in the UK and published by Building Research Establishment, Department of the Environment.   *Crown copyright 1989*
Price Group 3. Also available by subscription. Current prices from:
Publications Sales Office, Building Research Establishment, Garston, Watford, WD2 7JR (Tel 0923 664444).
Full details of all recent issues of BRE publications are given in BRE News sent free to subscribers.

Printed in the UK for HMSO. Dd.8157497, 6/89, C150, 38938.   ISBN 0 85125 394 6

BRE Digest

Concise reviews of building technology

The assessment of wind loads
Part 2: Classification of structures

This is the second in a series of Digests which is compatible with the proposed British Standard BS 6399:Part 2. It deals with the methods developed to categorise structures according to their sensitivity to dynamic behaviour when subjected to wind loading.

These methods allow the majority of structures to be designed statically, as at present. Mildly dynamic structures can still be treated statically by using a dynamic magnification factor. The procedures have been simplified in the British Standard so that only basic structural and geometric parameters are used to assess the appropriate category of structure, and to define whether it can be designed statically or, in very rare cases, whether a full dynamic treatment is required.

BACKGROUND TO CLASSIFICATION

A number of methods of analysing structures for wind effects are available; they range from simple static loading to sophisticated statistical methods using power spectral techniques. Generally, the simple methods can be used with adequate accuracy for most everyday building structures. It is only in the case of wind sensitive structures, such as tall, slender towers and major bridges, where wind effects are the principal loading to be considered that the more advanced methods are needed. With these, the structure's inherent flexibility is likely to make them respond more significantly to wind effects. Between the extremes there are buildings which may exhibit some dynamic magnification, that is they may respond more severely than predicted from an equivalent static load. Up to now, the designer has had no way of knowing whether or not his structure will respond in this way.

The purpose of the classification procedures is to make this distinction quantitatively and to define the appropriate analytical procedure to be used.

Building Research Establishment
DEPARTMENT OF THE ENVIRONMENT

Technical enquiries to:
Building Research Establishment
Garston, Watford, WD2 7JR
Telex 923220 Fax 664010

FULL PROCEDURE

Static structures

Small stiff structures, including cladding panels and conventional low-rise buildings, can be assessed using static methods and are small enough for the relevant wind information to be specified as a wind speed at a single point in space. No allowance is needed for variation of wind speed over the surface of the structure nor will the structure respond to any dynamic magnification.

Larger stiff structures can also be designed statically, but account may need to be taken of the variation of wind speed over the surface of the structure, so that advantage can be taken of the reduction in wind speed averaged over the whole surface. The size of the building can be defined by a diagonal dimension for design purposes. The appropriate gust can be determined from the gust peak factor, dependent on the height of the structure and the relevant diagonal dimension. This is described in Part 4.

Dynamic structures

These structures are not stiff enough to be assessed by static methods, but remain sufficiently stiff to prevent aeroelastic instabilities, such as vortex response, galloping and flutter. Such structures are likely to respond significantly to wind effects with large deflections causing cracking of partitions etc, and motion sickness to occupants. They require a full dynamic response analysis to assess the effects of wind loading and are excluded from the scope of BS 6399:Part 2.

Classification of static and dynamic structures

To assess the response of structures, a parameter K_F was defined relating the actual displacement of the structure, in its lowest frequency mode, to the corresponding static displacement. It can then be inferred that:

- A value of $K_F = 0$ indicates the structure is small and static responding to short high-intensity gusts.

- A value of $K_F < 0$ indicates the structure is large and static responding to lower intensity gusts.

- A value of $K_F > 0$ indicates the structure responds more than from a short intensity gust and is therefore dynamic.

When K_F is between about 0.1 and 2.0, the structure will be mildly dynamic; this warrants an increase in loading above the quasi-static values, but not enough to require a full dynamic analysis. The peak deflection (and hence peak internal forces) can be obtained by applying a factor to the static deflection where the factor is the ratio of the actual to static peak deflections. This is defined as the dynamic magnification factor C_R given by:

$$C_R = \frac{1 + (S_G^2 - 1)\sqrt{1 + K_F\ K_T}}{S_G^2} \qquad (1)$$

where S_G is the gust factor appropriate to the size of the structure and terrain (see Part 4);

$$K_T = 1.33 \text{ for sea terrain}$$
$$1.00 \text{ for country terrain}$$
$$0.75 \text{ for town terrain}$$

By this mean 'mildly' dynamic structures can still be designed statically by applying C_R to the static load effects. Only those structures where K_F exceeds about 2.0 (implying $C_R > 1.4$) require a full dynamic analysis. Calibration studies have shown that this static approach, using the dynamic magnification factor, can be used with confidence up to $C_R \simeq 1.5$.

SIMPLIFIED PROCEDURE

The designer is interested only in whether he can use a static procedure and, if so, what value of C_R he should adopt. Consequently, a direct reading graph of C_R has been derived dependent only on the parameters:

H the building height

n_o the frequency of the lowest mode

ζ the damping ratio

$\overline{V}_B$ the basic hourly mean wind speed

The resulting expression for K_F is:

$$K_F = \left(\frac{S_s}{24 n_o^2 b}\right)^{2/3} \left(\frac{\overline{V}_B}{24}\right)\left(\frac{1}{\zeta}\right) \qquad (2)$$

where $S_s = S_{SC}$ for country terrain
$S_{SC} S_{CT}$ for town terrain

For preliminary classification purposes, n_o is assumed to be given approximately by: $60/\sqrt{bH}$ and $\overline{V}_B$ may be assumed to be 24 m/s

where b is the diagonal of the building given by

$$b = \sqrt{H^2 + W^2}$$

where W is the width of the building

Thus

$$K_F = \frac{K_H}{K_B} \qquad (3)$$

where

$$K_H = (S_s H)^{2/3} \qquad (4)$$

and

$$K_B = 2000\zeta \qquad (5)$$

These expressions for K_H and K_B are tabulated in BS 6399:Part 2 and shown in Tables 1 and 2 respectively using accepted values for the damping appropriate to the different forms of construction. Equation 2 can be used if more appropriate values of $\overline{V}_B$, n_o or ζ are available. The resulting graph to determine C_R from K_F is given in Fig 1.

DAMPING

Damping is a function of the material used in the structure, the form and quality of construction, the frequency and the stiffness of the structure. At low amplitudes, damping is provided primarily by the inherent damping of the material. At higher amplitudes, movement at joints provides additional damping through friction, so an all-welded steel structure will provide less damping than one of bolted construction. At large amplitudes, load transfer to cladding and internal walls contributes even more damping, so that damping for ultimate limit state design is higher than for the serviceability limit states.

The value of damping to use at the design stage is extremely difficult to quantify, and may vary significantly between two notionally identical structures. Data have been collected on a wide range of completed buildings and other structures and a reasonable set of damping values can be established for defined classes of structure. This has been incorporated in the tabulated values for K_B given in Table 2.

Table 1   Factor K_H

H (m)	Sea	Country	Town
5	3.5	3.0	2.5
10	5	4.5	4
20	9	8	7
30	12	11	10
50	18	16	14
100	29	27	25
200	50	47	43
300	67	64	60

Terrain types are defined in Part 4

Fig 1   Factor C_R

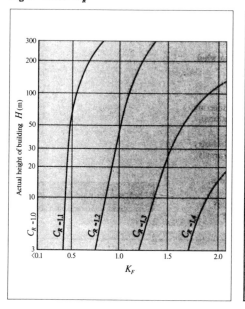

Table 2 Structural damping ratios

Type of building	Factor K_B	Implied damping ratio ζ
Welded steel unclad frames	8	.004
Bolted steel and rc unclad frames, rc chimneys (no lining)	12	.006
Lightly stiffened flexible steel or rc structures; no internal walls	20	.01
Rc core with light cladding; no internal walls	40	.02
Steel and rc frames with structural walls round lifts and stairs only, eg open-plan office buildings	60	.03
Shear wall structures; steel and rc frames with internal and/or external structural shear walls	100	.05
Buildings with structural walls round lifts and stairs and additional masonry walls subdividing floors, eg blocks of flats. All masonry and timber structures	150	.08

FREQUENCY

The calibration procedure is extremely sensitive to the natural frequency of the structure; it is therefore important to be able to assess this parameter as accurately as possible. Unfortunately, at the initial design stage this is not possible so reliance has to be made either on empirical formulae or on the analysis of similar structures.

The most common empirical formula is:

$$n_o = \frac{K}{H}$$

where K is a constant, varying from about 45 to 70
H is the height of the structure in metres.

This works reasonably well for tall structures but is not satisfactory for buildings with a lower aspect ratio. For this reason BS 6399:Part 2 has assumed the form:

$$n_o = \frac{60}{\sqrt{bH}}$$

This produces, generally, a lower bound estimate for n_o and avoids any artificial cut-offs for lower structures for which it works reasonably well.

DYNAMIC ANALYSES

Dynamic response analyses must be undertaken for structures which are extremely sensitive to dynamic effects. The classification procedure provides the limit for which static analyses, augmented by the dynamic magnification factor, are no longer applicable. This is when factor K_F exceeds 2.0, implying a maximum value of C_R of about 1.4.

Dynamic analyses require not only a modal analysis to determine frequencies and mode shapes, but a response analysis in which the wind loading spectrum is defined in terms of the scales and intensities of turbulence, and the structural and aerodynamic damping is assessed. Analytical methods for the response of dynamic structures to wind loading have been published and reference to the documents listed below should be made for guidance. Further advice can be obtained from specialists.

Analytical methods for the response of dynamic structures to wind loading are given in the following documents:

- Engineering Sciences Data, Wind Engineering Sub-series (4 volumes). London, ESDU International.

   NOTE: A comprehensive index covering all items of Engineering Sciences Data is available on request from ESDU International, 27 Corsham Street, London N1. Tel: 01 490 5151.

- Wind engineering in the eighties. London, Construction Industry Research and Information Association. 1981 CIRIA, 6 Storey's Gate, London SW1P 3AU. Tel: 01 222 8891.

- SIMIU, E and SCANLAN, RH. Wind Effects on Structures. New York, John Wiley and Sons, 1978.

- Supplement to the National Building Code of Canada, 1985. NRCC, No 23178. Ottawa, National Research Council of Canada, 1985.

> For further reading see Part 1

Printed in the UK and published by Building Research Establishment, Department of the Environment.   *Crown copyright 1989*
Price Group 3. Also available by subscription. Current prices from:
Publications Sales Office, Building Research Establishment, Garston, Watford, WD2 7JR (Tel 0923 664444).
Full details of all recent issues of BRE publications are given in *BRE News* sent free to subscribers.

Printed in the UK for HMSO. Dd.8157511, 8/89, C150, 38938.   ISBN 0 85125 400 4

BRE Digest

Concise reviews of building technology

Digest 346

August 1989

CI/SfB (J4)

The assessment of wind loads
Part 3: Wind climate in the United Kingdom

This is the third in a series of Digests which is compatible with the proposed British Standard BS 6399:Part 2. It deals with the derivation of the hourly mean wind speed for sites in the United Kingdom. This site wind speed is then used in Part 4 of the Digest as the basis for the appropriate wind speeds to be used for the structure to be designed.

BASIC WIND SPEED

The Meteorological Office records the hourly mean wind speeds and maximum gust speeds each hour at stations throughout the UK. Previous analyses only used the maximum speeds each year for which records were available, but recent analysis of these data extracts the maximum wind speed from every individual storm. This increases greatly the data available for analysis.

Recent analyses by BRE also adopted a more accurate model than that used previously to derive the required extreme wind speeds to be used for structural design. This has resulted in the wind speed map in Figure 1. It gives the basic maximum hourly mean wind speed $\overline{V}_B$ as isopleths, at 10 m above ground at sea level, adjusted for standard 'country' terrain. $\overline{V}_B$ has an annual probability of exceedance of 0.02, irrespective of direction, and was previously referred to as the 50-year return period wind speed. The notion of 'return periods' however has caused confusion with designers so this form of definition has been abandoned in favour of annual probability.

$\overline{V}_B$ must be adjusted for altitude and direction and, for structures of limited sub-annual periods of exposure, for seasonal effects. These adjustments

provide the hourly mean wind speed appropriate to the site at 10 m above standard terrain.

The hourly mean site wind speed $\overline{V}_{SITE}$ for any specific direction is given by:

$$\overline{V}_{SITE} = \overline{V}_B \times S_{ALT} \times S_{DIR} \times S_{TEM}$$

where $\overline{V}_B$ is the basic wind speed

S_{ALT} is an altitude factor

S_{DIR} is a direction factor

S_{TEM} is a seasonal building factor

The statistical factor S_3 in CP3 chapter V:Part 2 is no longer needed because adjustment for risk is made by the partial factors for temporary and permanent structures. The other S factors in CP3 are replaced by equivalent factors in Parts 3 and 4 of this Digest.

Adjustments for the actual site terrain together with the derivation of the appropriate gust speed to be used in the design of static and mildly dynamic structures (see Part 2) are then described in Part 4. Allowance for the effects of topography, if relevant, can be made using the procedures described in Part 5.

Building Research Establishment
DEPARTMENT OF THE ENVIRONMENT

Technical enquiries to:
Building Research Establishment
Garston, Watford, WD2 7JR
Telex 923220 Fax 664010

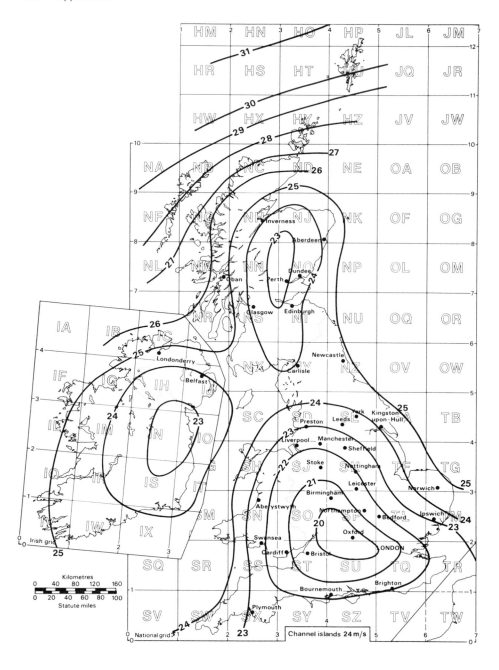

ALTITUDE FACTOR S_{ALT}

The analyses of the wind data from the Meteorological Office records show a dependence on site altitude. The analyzed data are, therefore, adjusted such that the wind speed map shown in Figure 1 is related to 10 m above the ground at sea level.

To derive the wind speed for any site at altitude A in metres above mean sea level, an adjustment of 10% per 100 m of altitude must be made to the basic wind speed from the map.

Therefore $S_{ALT} = 1 + 0.001A$

This correction accounts only for the effect of large-scale, slowly changing topography. The effects of rapid topographic changes (hills, cliffs and escarpments etc) are dealt with separately by the topography factor (see Part 5).

For guidance, a topography factor will need to be included in deriving the appropriate wind speed for design when the upwind slope is in excess of 0.05. In all other cases, the altitude factor accounts for the site level — see Figure 2.

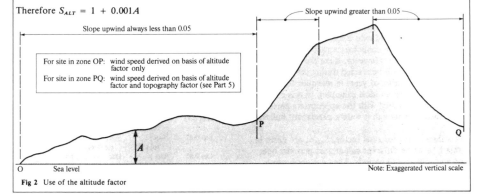

Fig 2 Use of the altitude factor

DIRECTION FACTOR S_{DIR}

The directional characteristics of extreme winds in the United Kingdom, used for design purposes, show no significant variation with location, so the directional factor is a function of the wind direction only. The highest winds come from the directions of the prevailing winds, between south-west and west. The directional extreme factor determined by direction approaches the value of the all-direction factor (strictly the value irrespective of direction) for winds from these directions. If directional factors were adopted on the basis that the annual risk in a given direction were 0.02 the overall risk from all directions would be greater owing to the contributions from other directions. Further analysis was necessary to derive the direction factors which are plotted in Figure 3; it can be seen that values greater than unity are obtained for the prevailing wind direction, but less than unity elsewhere in order to keep the overall risk at the 0.02 level. It is these factors which have been incorporated as the directional factor S_{DIR} and which are tabulated in Table 1.

The factors apply to 30° sectors; for intermediate directions values can be interpolated. Account should be taken of any uncertainty in the orientation of the building at the design stage; for those buildings or components which may be sited in any orientation a factor of 1.05 must be applied.

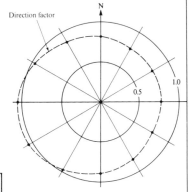

Fig 3 Direction factor

Table 1 Direction factor

Wind Direction	N			E			S			W		
°True	0	30	60	90	120	150	180	210	240	270	300	330
S_{DIR}	0.81	0.76	0.76	0.77	0.76	0.83	0.89	0.97	1.05	1.04	0.95	0.86

SEASONAL FACTOR S_{TEMP}

The highest extreme winds in the UK are expected in December and January. In the summer months of June and July winds may be expected to be only about 65% of these highest extremes. Structures which are expected to be exposed only in these more favourable conditions could be designed for a lower wind speed whilst maintaining the same risk of exceedance.

Typical of these applications are temporary structures: marquees, buildings erected solely for summer events (such as sporting fixtures) and buildings under construction. Generally, the structure will not be exposed for more than one season so that the seasonal factor needs to be used with a lower partial safety factor appropriate to temporary structures, to achieve consistent reliability. However, there are instances when a structure will be erected during the same short period over a number of years (a marquee for an annual event being an ideal example); in this case, the seasonal factor is used with the appropriate partial safety factors as though it were a permanent building.

The values of the seasonal factor, S_{TEMP}, are given in Table 2 for three different sub-annual periods: one, two and four-month periods. They are appropriate throughout the United Kingdom. To use the one-month values, it is necessary to have confidence in the building programme, or in the repeatability of the annual event. For example, a structure designed with a seasonal factor for August of 0.71 would, for the same reliability, be exposed to a 33% increase in loading if its construction or use were delayed to September. Factors for the six-month summer and

winter periods are also shown in Table 2. No advantage is given by this factor if the building is to be erected or in use at least during the two-month period of December and January.

Table 2 Values of S_{TEM}

1 month	2 months	4 months	
	Sub-annual periods		
Jan 0.98	Jan) to) 0.98	Jan) to) 0.98	
Feb 0.83	Feb) to) 0.86	Apr)	Feb) to) 0.87
Mar 0.82	Mar) to) 0.83	Mar) to) 0.83	May)
Apr 0.75	Apr) to) 0.75	Jun)	Apr) to) 0.76
May 0.69	May) to) 0.71	May) to) 0.73	Jul)
Jun 0.66	Jun) to) 0.67	Aug)	Jun) to) 0.83
Jul 0.62	Jul) to) 0.71	Jul) to) 0.86	Sep)
Aug 0.71	Aug) to) 0.82	Oct)	Aug) to) 0.90
Sep 0.82	Sep) to) 0.85	Sep) to) 0.96	Nov)
Oct 0.82	Oct) to) 0.89	Dec)	Oct) to) 1.00
Nov 0.88	Nov) to) 0.95	Nov) to) 1.00	Jan)
Dec 0.94	Dec) to) 1.00 Jan)	Feb)	Dec) to) 0.98 Mar)

The factor for the six-month winter period October to March inclusive is 1.0, and for the six-month summer period April to September inclusive is 0.84.

PROCEDURE

To derive the site hourly mean wind speed:

(1) From the location of the site determine the basic hourly mean wind speed, $\bar{V}_B$, from the map in Figure 1.

(2) From the altitude of the site (in metres) determine the altitude factor S_{ALT}.

(3) Determine the direction factor S_{DIR} for each wind direction to be considered. For a building which may be sited in any orientation S_{DIR} is taken as 1.05.

(4) For seasonal or temporary structures determine the seasonal factor S_{TEMP}. Note the sensitivity of this factor and only use values less than 1.0 if it is certain that the building will be exposed only for the specified sub-annual periods.

(5) Calculate the site hourly mean wind speed for each direction from:

$$\bar{V}_{SITE} = \bar{V}_B \times S_{ALT} \times S_{DIR} \times S_{TEMP}$$

For further reading see Part 1

Printed in the UK and published by Building Research Establishment, Department of the Environment.   *Crown copyright 1989*
Price Group 3. Also available by subscription. Current prices from:
Publications Sales Office, Building Research Establishment, Garston, Watford, WD2 7JR (Tel 0923 664444).
Full details of all recent issues of BRE publications are given in *BRE News* sent free to subscribers.

Printed in the UK for HMSO. Dd.8157511, 8 /89, C150, 38938.   ISBN 0 85125 399 7

Digest 346
August 1989

CI/SfB (J4)

Concise reviews of building technology

The assessment of wind loads
Part 4: Terrain and building factors and gust peak factors

This Digest is the fourth in a series which is compatible with the proposed British Standard BS 6399:Part 2. It uses the 'full' method of the British Standard to derive the appropriate gust wind speeds to be used for the design of 'static and mildly dynamic' structures (as defined in Part 2) from the site hourly mean wind speed (derived in Part 3). A more accurate assessment of gust speeds can be obtained from the use of the BRE computer program *STRONGBLOW*.

DERIVATION OF REFERENCE WIND SPEEDS TO BE USED FOR DESIGN

Having selected the appropriate basic speed from the wind map, and taken due account of the site's altitude and wind direction, the hourly mean site speed $\bar{V}_{SITE}$ can be derived (see Part 3). This speed must be adjusted further to account for the terrain of the site and for the height above ground for which the wind speed is required. In addition, the appropriate gust speed needs to be used for the design of static and mildly dynamic structures (see Part 2). These parameters can be accounted for by the use of further S factors so that the reference wind speed V_{REF} at any height can be derived, for sites in country terrain, from:

$$V_{REF} = \bar{V}_{SITE}\, S_{TB} \qquad (1)$$

where S_{TB} is the terrain and building factor given by:

$$S_{TB} = S_{SC}(1 + g_{GUST}\, S_{TSC} + S_{TOP}) \qquad (2)$$

This factor combines the roles of factors S_1 and S_2 in CP3 Chapter V: Part 2.

For sites in town terrain the above factors S_{SC} and S_{TSC} are modified by two further factors resulting in, for town sites,

$$S_{TB} = S_{SC}\, S_{CT}(1 + g_{GUST}\, S_{TSC}\, S_{TCT} + S_{TOP}) \qquad (3)$$

BS 6399: Part 2 includes a simplified procedure for use with common structures which allows S_{TB} to be obtained directly from tables.

The following factors are described later:

S_{SC} and S_{CT} are fetch factors which modify the hourly mean wind speed to take account of the terrain of the site.

S_{TSC} and S_{TCT} are turbulence factors which modify the turbulence effects to take account of the terrain of the site.

g_{GUST} is a gust peak factor.

S_{TOP} is the topography factor described in Part 5.

S_{TOP} is an increment to be added in equations (2) and (3) to derive the factor S_{TB}. In this respect it is different from the factor S_1 (described in CP3 Chapter V: Part 2) which was a multiplying factor to apply to the wind speed to account for topographical effects.

Building Research Establishment
DEPARTMENT OF THE ENVIRONMENT

Technical enquiries to:
Building Research Establishment
Garston, Watford, WD2 7JR
Telex 923220 Fax 664010

TERRAIN CATEGORIES

The roughness of the ground surface controls both the mean wind speed and its turbulent characteristics. The wind speed is higher near the ground over a smooth surface, such as open country, than over a rougher surface, such as a town. By defining three basic terrain categories wind speeds can be derived accounting for the influence of upstream categories different from that of the site. These three basic categories are:

Sea This applies to any offshore location and to inland lakes of at least 5 km upstream of the site. Such a category must also be defined so that the gradual deceleration of the wind speed inland from the coast can be quantified for any land-based site.

Country This covers a wide range of terrain, from the flat, open, level or nearly level country with no shelter (fens, airfields, moorland or farmland with no hedges or walls), to undulating countryside with obstructions, such as occasional buildings and windbreaks of trees, hedges or walls.

Town Town terrain includes suburban regions in which the general level of roof tops is about 5 m above ground level (all two-storey housing) provided that such buildings are at least as dense as normal suburban developments for at least 100 m upwind of the site. Whilst it is not easy to quantify, it is expected that the plan area of the buildings is at least 8% of the total area over that 100 m and within a 30° sector of the site — see Fig 1.

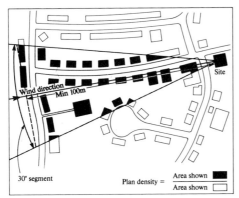

Wind direction
Min 100m

Site

30° segment   Plan density = $\dfrac{\text{Area shown} \ \blacksquare}{\text{Area shown} \ \square}$

Fig 1 Typical town site

Variation of fetch

Fetch refers to the terrain directly upwind of the site. The adjustment of wind speed characteristics as the wind flows from one terrain to another is not instantaneous. At a change from a smooth to a rougher surface the wind speed is gradually slowed near the ground.

This adjustment requires time to work up through the wind profile; at any site downwind of a change in terrain the mean speed lies between that for the smooth terrain and that for the fully developed rough terrain. This is shown diagrammatically in Fig 2.

This gradual deceleration of the mean speed is accounted for by defining the site by its distance downwind from the coast and, if it is in a town, as its distance from the edge of the town.

Shelter of a site from a upwind town has not been allowed for in the procedures in these Digests, other than if the site is in a town itself. To do so would introduce too much complexity with only a marginal saving in the resulting wind loads. The BRE computer program *STRONGBLOW* can be used to account for such effects.

It can be seen that by introducing the location of a site with respect to its distance from the sea and, if relevant, its distance from the edge of a town, different wind speeds will be obtained from different directions. For the site shown in Fig 3 a southerly wind will pass over country (AB) and town (BO) causing some deceleration of the wind. A westerly wind will be slowed down more because CD is greater than AB and DO is greater than BO.

In the example given in Fig 3, the direction factor S_{DIR} produces higher basic winds from the west ($S_{DIR} = 1.04$) than the south ($S_{DIR} = 0.89$); in this case the site wind speeds for these two directions may not be significantly different. However, an easterly wind would be markedly lower as both the direction factor and the effects of fetch changes would decrease the wind speed.

It is important, if directional effects need to be considered, to take full account of both the effects of terrain upwind of the site and the direction factor. This becomes even more significant if the effects of topography need to be considerd: the topography factor S_{TOP} will have a major influence on the value and the direction of the most critical wind speed.

Fetch and turbulence factors

Sites in country terrain

To account for the effects of terrain on the hourly mean wind speed, a set of fetch factors S_{SC} has been defined; these factor the hourly mean site wind speed $\bar{V}_{SITE}$ to obtain the hourly mean wind speed at any height above ground for a site in country terrain at various distances from the sea. This is the term outside the brackets in equation (2). However, as the designer is concerned with the appropriate gust wind speed in assessing static and mildly dynamic structures, equation (2) incorporates the appropriate factors to do this.

Fig 2 Variation of fetch

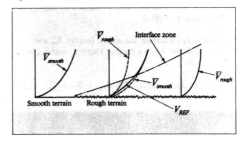

V_{rough} Interface zone

V_{smooth}

V_{smooth} V_{rough}

Smooth terrain   Rough terrain

V_{REF}

Table 1

Effective height H_{EFF} (m)	Factor	\multicolumn{7}{c}{Factors S_{SC} and S_{TSC}}

Effective height H_{EFF} (m)	Factor	\multicolumn{7}{c}{Distance from site to sea (km)}
		0.1	0.3	1.0	2.0	10	30	100 or more
2 or less	S_{SC}	.873	.840	.812	.792	.774	.761	.749
	S_{TSC}	.203	.215	.215	.215	.215	.215	.215
5	S_{SC}	1.06	1.02	.990	.966	.944	.928	.913
	S_{TSC}	.161	.179	.192	.192	.192	.192	.192
10	S_{SC}	1.21	1.17	1.13	1.10	1.07	1.06	1.04
	S_{TSC}	.137	.154	.169	.175	.178	.178	.178
20	S_{SC}	1.32	1.31	1.27	1.23	1.21	1.19	1.17
	S_{TSC}	.127	.132	.145	.157	.163	.164	.164
30	S_{SC}	1.39	1.39	1.35	1.31	1.28	1.26	1.24
	S_{TSC}	.120	.122	.132	.145	.155	.159	.159
50	S_{SC}	1.47	1.47	1.46	1.42	1.39	1.36	1.34
	S_{TSC}	.112	.113	.117	.125	.135	.145	.147
100	S_{SC}	1.59	1.59	1.59	1.57	1.54	1.51	1.48
	S_{TSC}	.097	.100	.100	.100	.110	.120	.126
200	S_{SC}	1.74	1.74	1.74	1.73	1.70	1.67	1.65
	S_{TSC}	.075	.075	.075	.078	.083	.093	.095
300	S_{SC}	1.84	1.84	1.84	1.83	1.82	1.78	1.76
	S_{TSC}	.065	.065	.065	.067	.068	.080	.081

Note: Interpolation may be used

Table 2

Effective height H_{EFF} (m)	Factor	\multicolumn{6}{c}{Adjustment Factors S_{CT} and S_{TCT} for sites in 'town' terrain}

Effective height H_{EFF} (m)	Factor	\multicolumn{6}{c}{Distance from site to edge of town (km)}
		0.1	0.3	1.0	3.0	10	30 or more
2 or less	S_{CT}	.695	.653	.619	.596	.576	.562
	S_{TCT}	1.93	1.93	1.93	1.93	1.93	1.93
5	S_{CT}	.846	.795	.754	.725	.701	.684
	S_{TCT}	1.47	1.61	1.63	1.63	1.63	1.63
10	S_{CT}	.929	.873	.828	.796	.770	.751
	S_{TCT}	1.18	1.39	1.51	1.52	1.52	1.52
20	S_{CT}	.984	.935	.886	.853	.824	.804
	S_{TCT}	1.00	1.17	1.35	1.44	1.45	1.45
30	S_{CT}	.984	.965	.915	.880	.851	.830
	S_{TCT}	1.00	1.07	1.25	1.38	1.43	1.43
50	S_{CT}	.984	.984	.947	.912	.881	.859
	S_{TCT}	1.00	1.00	1.14	1.28	1.38	1.42
100	S_{CT}	.984	.984	.984	.948	.917	.894
	S_{TCT}	1.00	1.00	1.00	1.14	1.28	1.38
200	S_{CT}	.984	.984	.984	.980	.947	.924
	S_{TCT}	1.00	1.00	1.00	1.07	1.19	1.31
300	S_{CT}	.984	.984	.984	.984	.964	.940
	S_{TCT}	1.00	1.00	1.00	1.04	1.14	1.24

Note: Interpolation may be used

The turbulence factor S_{TSC} depends on the same parameters as the fetch factor, for example effective height of the building and the site terrain. These are combined in Table 1.

The gust peak factor g_{GUST} is dependent on the size of the structure and, for practical purposes, is independent of the terrain. It can, therefore, be defined separately as described later.

Sites in town terrain
To account for the further decelerating effect of the mean wind speed for sites in towns, an adjustment fetch factor S_{CT} is used; it is always less than unity. Similarly, to account for the increased turbulence over rougher town terrain, an adjustment turbulence factor S_{TCT} is used; this is always greater than unity. These factors are shown in Table 2, related to the distance of the site from the edge of the town and the effective height of the building.

Effective height
In rough terrain, such as towns and cities, the wind tends to skip over the buildings at, or below, roof top level, leaving sheltered regions below. The height of this sheltered zone is a function of the area density of the buildings and the general height of the obstructions. The effective height of any building H_{EFF} in such terrain is the actual building height, H, less the height of the sheltered zone. An empirical formula given by:

$$H_{EFF} = H - 0.8\, H_{OBS} \text{ or } H_{EFF} = 0.4\, H_{OBS} \qquad (4)$$
whichever is the greater

has been shown to correlate well with available data, taking H_{OBS} to be 5 m where buildings are generally two to three storeys high and 15 m where buildings are at least three storeys high.

However, this provides negative or very small values of H_{EFF} for low buildings surrounded by higher buildings and so a minimum value of $0.4\, H_{OBS}$ has been proposed to account for this situation.

In towns where there is an open area upwind of the building, extending at least twice the structure's height (buildings facing open parkland), there will be minimal shelter and the effective height should be taken as the actual height, H, of the building.

Fig 3 Effects of terrain

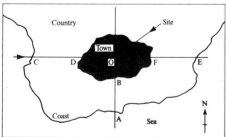

GUST PEAK FACTOR

A simplified formula for g_{GUST} given by

$$g_{GUST} = 0.42 \ln (3600/t) \dots\dots\dots\dots (5)$$

where t is the gust duration time in seconds has been shown to be within a few percent of more complex formulations. For the purposes of these procedures, the simplified formula was considered quite adequate. However, it is a factor dependent on the gust duration, t, which is not of direct interest to the designer. His concern is to choose, for static structures, the appropriate gust speed which will envelop his structure or component to produce the maximum loading.

Fortunately for bluff type structures, such as buildings, which can be designed statically, there is a simple empirical relationship between the duration, t, and the size of the structure or element, t, given by:

$$t = \frac{4.5b}{\overline{V}} \dots\dots\dots\dots\dots\dots\dots (6)$$

where $\overline{V}$ is the relevant mean wind speed given by S_{SC} $\overline{V}_{SITE}$ for country and S_{SC} S_{CT} $\overline{V}_{SITE}$ for town and where b is the diagonal dimension of the loaded area under consideration. This may be the whole building, a single cladding element or any intermediate part.

By combining equations 5 and 6, a graph can be plotted of height against size to give values of the gust peak factor g_{GUST} — see Fig 4. For design purposes it is likely that $\overline{V}_{SITE}$ will lie within the range 20 to 30 m/s so that for a typical $b = 20$ m, g_{GUST} varies from about 2.86 to 3.0. This variation of g_{GUST} makes only about a 20% difference in the resulting gust speeds. Consequently, for these purposes the values of g_{GUST} adopted have been based on a $\overline{V}_{SITE}$ fixed at 25 m/s. The resulting values of size, b, are then shown as the abscissa on the graph of Fig 4 which enables g_{GUST} to be read directly for given heights and sizes.

Factor g_{GUST} is given in BS 6399:Part 2 in a table for various heights and sizes of loaded area.

CLASSIFICATION PROCEDURE

The conventional gust factor S_G which is required for the full classification method is the ratio of the gust and the mean wind speeds given by:

$$S_G = \frac{1 + g_{GUST} \, S_{TSC} \, S_{TCT} + S_{TOP}}{1 + S_{TOP}} \quad (7)$$

This reduces to:

$$S_G = 1 + g_{GUST} \, S_{TSC} \, S_{TCT}$$

where topography is insignificant.

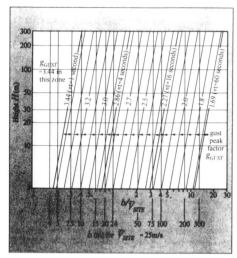

Fig 4 Gust peak factor

For further reading see Part 1

PROCEDURE
To derive the reference wind speed:

(1) From Part 3, determine the site hourly mean wind speed for each wind direction to be considered.

(2) From Table 1, determine factors S_{SC} and S_{TSC} for the site according to its distance from the sea in the upwind direction and building height.

(3) For sites in towns determine factors S_{CT} and S_{TCT} according to the upwind distance of the site from the edge of the town and the effective height, H_{e}.

(4) For sites affected by topography, from Part 5 calculate S_{TOP} appropriate to the wind direction being considered.

(5) Determine S_{TP} from equation (2) for sites in country and Equation (3) for sites in town.

(6) Determine V_{REF} from equation (1).

Printed in the UK and published by Building Research Establishment, Department of the Environment.   *Crown copyright 1989*
Price Group 3. Also available by subscription. Current prices from:
Publications Sales Office, Building Research Establishment, Garston, Watford, WD2 7JR (Tel 0923 664444).
Full details of all recent issues of BRE publications are given in *BRE News* sent free to subscribers.

Printed in the UK for HMSO. Dd.8157511, 8 /89, C150, 38938.   ISBN 0 85125 398 9

BRE Digest

Concise reviews of building technology

Digest 346
November 1989

CI/SfB (J4)

The assessment of wind loads
Part 5: Assessment of wind speed over topography

This Digest is the fifth in a series which is compatible with the proposed British Standard BS 6399:Part 2. It deals with the assessment of wind speeds over topographic features such as hills, ridges, escarpments and cliffs for wind loading calculations. The information here is generally similar to that in Digest 283, but now the new topography factor S_{TOP} is *added* to the wind speed for flat terrain. The proposals for the topography factor are soundly based on theory and agree in value to better than 10 per cent for typical UK topography with the more empirical rules incorporated in BS 8100:Part 1: *Lattice towers and masts*.

This Digest supersedes Digest 283 which is now withdrawn.

bre **Building Research Establishment**
DEPARTMENT OF THE ENVIRONMENT

Technical enquiries to:
Building Research Establishment
Garston, Watford, WD2 7JR
Telex 923220 Fax 664010

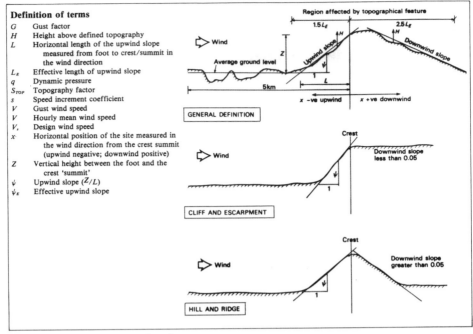

Definition of terms

G	Gust factor
H	Height above defined topography
L	Horizontal length of the upwind slope measured from foot to crest/summit in the wind direction
L_E	Effective length of upwind slope
q	Dynamic pressure
S_{TOP}	Topography factor
s	Speed increment coefficient
V	Gust wind speed
V	Hourly mean wind speed
V_s	Design wind speed
x	Horizontal position of the site measured in the wind direction from the crest summit (upwind negative; downwind positive)
Z	Vertical height between the foot and the crest 'summit'
ψ	Upwind slope (Z/L)
ψ_E	Effective upwind slope

Fig 1   Definitions of topographical dimensions

THE EFFECTS OF TOPOGRAPHY ON THE WIND

Near the summits of hills, or the crests of cliffs, escarpments or ridges, the wind speed is accelerated. In the valleys or near the foot of steep escarpments or ridges the flow may be decelerated. Topography is classified by two parameters: the upwind slope and the form or shape of the topography — see Fig 2; each parameter can be deivided into three categories:

The three categories dependent on the upwind slope are:

Gentle topography: where ψ is less than 0.05;
Shallow topography: where ψ is 0.05 to 0.3;
Steep topography: where ψ is greater than 0.3;

The three categories dependent on the form of the topography are:

Valleys: where the ground level falls then returns to the original level;
Hills and ridges: where the ground rises then returns to the original level;
Escarpments and cliffs: where the ground rises or falls and then remains at the new level.

If the downwind slope is sensibly level (slope less than 0.05) for a distance exceeding both L and $3.3Z$ the feature should be treated as an escarpment of cliff; otherwise the feature should be treated as a hill, ridge or valley. In undulating terrain it is often difficult to define the base level from which to assess the dimensions of the feature; the average level of the terrain for a distance of 5 km upwind of the site should be taken as the base level in these cases.

Gentle topography

When changes in ground level are gentle (ψ less than 0.05), the balance between the mean wind speed and the turbulence is not significantly disturbed. There is, however, a small but significant dependence on site altitude which affects equally the mean wind speed, the turbulence intensity and the gust speeds. Each of these wind speeds may be taken as increasing by one per cent per 10 metres of site altitude; for the mean speeds this is accounted for by the altitude factor defined in Part 3.

Shallow topography

When changes in ground level are not gentle (ψ greater than 0.05) but the slopes remain below a critical slope of $\psi = 0.3$ (an angle of about 17°), the balance between the mean wind speed and the turbulence is disturbed. The mean wind speed undergoes significant changes in value but the turbulence undergoes small distortions without significantly changing the turbulence intensity. Gust speeds, which are formed by the action of the turbulence superimposed on the mean wind speed, are therefore affected less than the mean speed.

Fig 2   Categories of topography

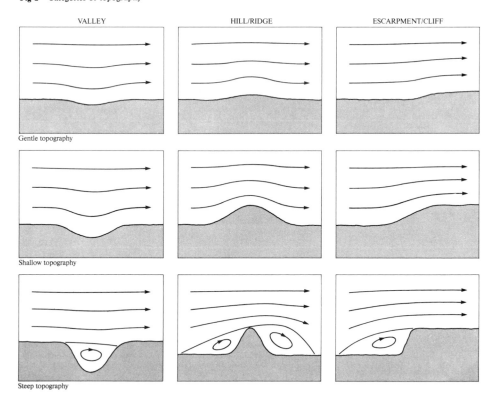

Wind blowing across a shallow valley in an otherwise flat plain decelerates down the upwind slope to a minimum at the valley bottom, then accelerates up the downwind slope back to the initial speed. As wind blowing along the axis of a valley is not significantly changed, there is no advantage to cross-axis shelter. Accordingly, the effects of this form of valley are not included in these design rules. This form of 'rift' valley is rare, the typical valley in the UK being between hills or ridges.

Wind blowing over a shallow hill, ridge or escarpment accelerates up the upwind slope to a maximum at the summit or crest. The effect varies with height and is greatest near the ground. Downwind of the summit of a hill or ridge the wind decelerates, returning to the initial speed by about $2.5L$ downwind of the summit. Downwind of the crest of an escarpment, wind decelerates more slowly, converging towards a final speed appropriate to the change in altitude for gently topography. These effects have been studied in New Zealand, comparing results from wind tunnel models with full scale[1][2]. These and other data confirm the result from theory[3] that the change in mean wind speed over shallow topography is everywhere proportional to the upwind slope.

The design rules given below were derived from theory and the New Zealand data which apply to shallow topography. When comparisons were made between these rules and BS 8100 it was found that all the major hills in the UK used in the test calculations for the British Standard (and on which towers had been constructed) were within the shallow category (0.05 to 0.3). The majority of UK hills will therefore be in this category.

Steep topography

Where the upwind or downwind slope measured in the wind direction exceeds the critical angle of about 17°, the flow of wind separates from the ground surface leaving regions of separated flow. Such situations are rare in practical sites of construction in the UK. (Topography that is steep when the wind is normal may become shallow when the wind is skewed.)

Wind blowing across a steep valley separates from the upwind edge and jumps across to the downwind edge, leaving a large sheltered region of separated flow. It is not possible to give general design rules to cater for this effect. No shelter occurs when the wind blows directly along the axis of the valley, and it is recommended that no shelter is assumed for any wind direction when making wind loading assessments.

Wind blowing over a steep hill, ridge or escarpment separates from the ground surface ahead of the upstream slope and jumps to a point just below the summit or crest. The boundary of the separated region forms an effective slope equal to the critical value of 17°, so that the flow over the summit or crest is the same as an effective upwind slope $\psi = 0.3$, irrespective of the actual upwind slope ψ: this gives an upper limit to the possible acceleration. The length of the upwind separation region becomes the corresponding effective length of the upwind slope $L_E = 3.3Z$. The flow separates from the ground downwind of the crest or summit if the downwind slope is also steep. Again, it is not possible to give general design rules for the large regions of separated flow, but the design rules given later may be used as upper-bound values.

RANGE OF APPLICABILITY

The design rules for gentle topography are always applicable. The range of applicability of the design rules for shallow and steep topography is shown in Fig 3. The design rules apply accurately only in those cases shown without shading, but these do represent the majority of typical sites affected by topography. The design rules provide upper-bound values in those cases shown by dark shading, where in reality there will be reduced acceleration or even shelter. In the cases shown by light shading the site is either unaffected or sheltered and it is recommended that no topographic corrections be applied.

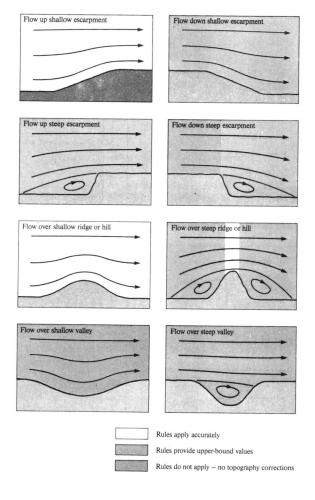

Flow up shallow escarpment

Flow down shallow escarpment

Flow up steep escarpment

Flow down steep escarpment

Flow over shallow ridge or hill

Flow over steep ridge or hill

Flow over shallow valley

Flow over steep valley

Rules apply accurately

Rules provide upper-bound values

Rules do not apply – no topography corrections

Fig 3   Range of applicability

DESIGN SPEED INCREMENT COEFFICIENT s

An important result of the theory[3] is that a good estimate of the accelerated mean wind speed near the ground at the summit or crest, and $\overline{V}_{crest}$ expressed as a ratio of the incident mean wind speed $\overline{V}$ is given by:

$$\frac{\overline{V}_{crest}}{\overline{V}} = 1 + 2\psi$$

where the topography is shallow and there is no flow separation.

As the increase in mean wind speed is everywhere proportional to the upwind slope ψ, the effect elsewhere can be quantified by a speed increment coefficient $s\{x/L, H/L\}$ (where the brackets denote functional dependence on the position from the crest x and the height above ground H in terms of the upwind slope length L). This coefficient takes values between $s = 0$ where the topography has no effect to $s = 1$ near the ground at the summit or crest. This approach extends to give upper-bound values for steep topography when the effective length L_E of the upwind slope is used in place of L.

Design values of the speed increment coefficient $s\{x/L_E, H/L_E\}$ are plotted against position x/L_E and height above ground H/L_E in terms of the effective upwind slope length in Fig 4.

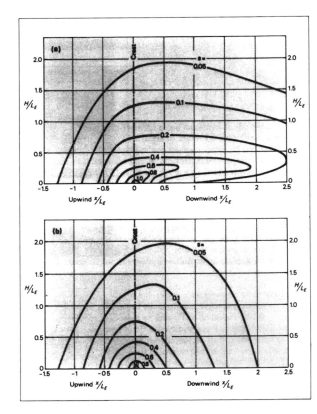

Fig 4   Factors for (a) Cliff and escarpment
(b) Ridge and hill

TOPOGRAPHY FACTOR S_{TOP}

For design purposes the effects of topography can be accounted for by the introduction of a topography factor, S_{TOP}, given by:

$$S_{TOP} = 2 \, s \, \psi \text{ for values of } \psi \text{ up to } 0.3$$

S_{TOP} represents the mean wind speed increment due to the topography and is incorporated directly in the equation of the derivation of the reference wind speed for the site V_{REF}, from the site hourly mean wind speed, $\overline{V}_{SITE}$, as shown in Part 4.

Assessment of topographic dimensions

The first stage in an assessment is to establish the relevant topographic dimensions as defined in Fig 1 for each relevant wind direction (steps of 30° are convenient). Determine from the upwind slope ψ whether the topography is gentle, shallow or steep. If gentle, the wind speeds derived for the site altitude should be used without any further correction. If shallow or steep, the wind speed already determined at the site altitude must be reduced to that at the altitude of the surrounding terrain. This is because the level of the site above mean sea level is already accounted for in the altitude factor. The topography factor, however, is based on wind speeds relative to the altitude of the general surrounding terrain. To ensure that the height correction is not applied twice, the wind speed defined for the site level must be reduced to that at the altitude of the surrounding terrain. This may be done by factoring the site wind speed by:
$(1 - 0.001h')$

where h' is the site height *above* the general level of the surrounding terrain.

This is shown diagrammatically in Fig 5.

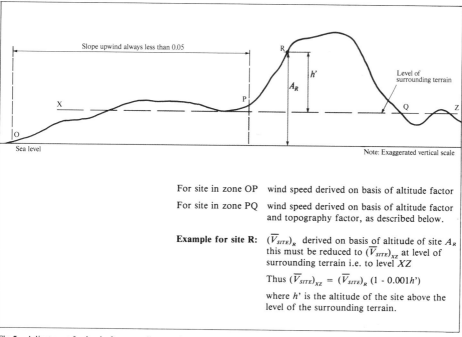

For site in zone OP   wind speed derived on basis of altitude factor

For site in zone PQ   wind speed derived on basis of altitude factor and topography factor, as described below.

Example for site R: $(\overline{V}_{SITE})_R$ derived on basis of altitude of site A_R this must be reduced to $(\overline{V}_{SITE})_{XZ}$ at level of surrounding terrain i.e. to level XZ

Thus $(\overline{V}_{SITE})_{XZ} = (\overline{V}_{SITE})_R \, (1 - 0.001h')$

where h' is the altitude of the site above the level of the surrounding terrain.

Fig 5   Adjustment for level of surrounding terrain

It is then necessary to select the appropriate values of upwind slope ψ and slope length L_E from Table 1.

Table 1 Effective parameters for shallow and steep topography

Upwind slope $\psi = (Z/L)$	
Shallow (ψ 0.05 to 0.3)	**Steep (ψ greater than 0.3)**
$L_E = L$	$L_E = Z/0.3$
$S_{TOP} = 2\,s\,\psi$	$S_{TOP} = 0.6$

Calculate the position x of the site relative to the crest or summit and the required height or heights above the ground H as ratios of L_E. If the position is in the range -1.5 to 2.5, the site is influenced by the topography and the assessment should proceed; otherwise the site is not influenced by the topography (and the Topography factor $S_{TOP} = 0.0$).

For each wind direction, select the values of speed increment coefficient $s\{x/L_E, H/L_E\}$ from Fig 4, appropriate to the position, height above ground and shape for that direction.

The topography factor S_{TOP} can then be obtained from Table 1 dependent on the parameters ψ, L_E and s.

ASSESSMENT OF MEAN WIND SPEEDS
Mean wind speeds are required as the basis for the assessment of wind loads on both static and dynamic structures (Parts 1 to 4) and for other purposes such as the assessment of natural ventilation (Digest 210). Equation (1), (2) and (3) of Part 4 may be used for this purpose taking $g_{GUST} = 0.0$ for hourly mean speeds.

For sites in country terrain $\overline{V}_{REF} = S_{SC}\,(1 + S_{TOP})\,\overline{V}_{SITE}$

For sites in town terrain $\overline{V}_{REF} = S_{SC}\,S_{CT}\,(1 + S_{TOP})\,\overline{V}_{SITE}$

where $\overline{V}_{REF}$ is the hourly mean wind speed appropriate to the building height, and the other S factors are defined in Part 4.

DESIGN WIND SPEED AND DYNAMIC PRESSURE
For the assessment of wind loads on static and mildly dynamic structures, the reference wind speed, V_{REF}, and the resulting pressure, p, should be assessed for the site in accordance with the principles given in Part 1. Values of the topography factor S_{TOP} are provided by this Digest. The other S factors given in the formula below are described in Parts 1 to 4.

The reference wind speed V_{REF} for country terrain is then calculated from the formula:

$$\overline{V}_{REF} = V_{SITE}\,S_{SC}\,[1 + (g_{GUST}\,S_{TSC}) + S_{TOP}]$$

and the design dynamic pressure from the formula:

$$q = k\,V^2{}_{REF}$$

where $k = 0.613$ kg/m^3 as described in Part 1.

REFERENCES

1 BOWEN, A J. Some effects of escarpments on the atmospheric boundary layer. (PhD Thesis) Christchurch, New Zealand, University of Canterbury, Department of Mechanical Engineering, June 1979.

2 PEARSE, J R. The influence of two-dimensional hills on simulated atmospheric boundary layers. (PhD Thesis) Christchurch, New Zealand University of Canterbury, Department of Mechanical Engineering, August 1979.

3 JACKSON, P S and HUNT, J C R. Turbulent wind flow over a low hill. Quart. J. R. Met. Soc 101, 929-955.

4 ESDU. Strong winds in the atmospheric boundary layer: Part 1: mean-hourly wind speeds. Data item B2026. London, ESDU International 1982. 27 Corsham Street, London N1.

FURTHER READING
Building Research Establishment
Digest 210 Principles of natural ventilation.

British Standards Institution
BS 8100: Lattice towers and masts
    Part 1: 1986 Code of practice for loading
    Part 2: 1986 Guide to the background and use of
                Part 1 'Code of practice for loading'

See also Further reading *in Part 1*

Printed in the UK and published by Building Research Establishment, Department of the Environment.   *Crown copyright 1989*
Price Group 3. Also available by subscription. Current prices from:
Publications Sales Office, Building Research Establishment, Garston, Watford, WD2 7JR (Tel 0923 664444).
Full details of all recent issues of BRE publications are given in *BRE News* sent free to subscribers.

Printed in the UK for HMSO. Dd.8245005, 10/89, C150, 38938   ISBN 0 85125 420 9

BRE Digest

Concise reviews of building technology

Digest 346

November 1989

CI/SfB (J4)

The assessment of wind loads
Part 6: Loading coefficients for typical buildings

This is the sixth in a series of Digests which is compatible with the proposed British Standard BS 6399: Part 2. It provides data on pressure coefficients for the walls and roofs of bluff-shaped buildings to enable loads to be derived. The procedure outlined in the Digest is limited to rectangular buildings with flat, gabled or hipped roofs but it is applicable in principle to buildings of more complex shape. The assessment of pressure coefficients for such buildings is provided in Part 2 of *The designer's guide to wind loading of building structures.*

When assessing wind loads on bluff-shaped buildings it is necessary to provide pressure distributions over the various surfaces so that loads can be derived, both for small elements, such as windows and cladding, as well as on whole faces; from this data, overall loads can be determined.

Parts 1 to 4 of this Digest outline a means of deriving dynamic pressures for the appropriate gust duration; these are used in conjunction with the pressure coefficients to determine the loads.

The design gust duration is determined by the size of the loaded area using equation (6) in Part 4.

INFLUENCE OF SLENDERNESS RATIO
The height or, more specifically, the slenderness ratio of a building affects the flow pattern of the wind around or over it.

If the height of the building is greater than about half the crosswind width, the wind tends to flow around the sides rather than over the top, except for the zone very near the top. The critical dimension, is therefore, the width of the building.

If the building is low, that is when the height is less than about half the crosswind width, the wind tends to flow over the top rather than the sides, except for zones close to the ends of the building. The critical dimension is the height of the building.

INFLUENCE OF WIND DIRECTION
The flow conditions are different between wind normal to a face and wind skew to a face. Figures 1 and 3 of Part 1 show that the zones of high suction differ in the two cases. When the wind direction factor S_{DIR} (see Part 3) is taken into account, the overall moments and shears will depend on the orientation of the structure and may not be greatest when the flow is normal.

REFERENCE DYNAMIC PRESSURE
The choice of reference height z_{ref} for the reference dynamic pressure, q, is made to minimise the variation in loading coefficient values over structures of the same shape and form but of different size.

In the case of line-like structures or lattice structures, in which the divergence of the wind flow past the structure is relatively small, the reference pressure should be calculated at the local level. With bluff structures the flow is diverted considerably, so the local dynamic pressure does not provide a set of coefficients which are invariant with the size of the building. Unfortunately, any one fixed reference height will not provide a constant 'universal' coefficient and a compromise has to be sought. In this Digest, the pressure coefficients for bluff structures whose height H is less than $4L$ are related to the reference dynamic pressure as calculated at the top of the building. Such coefficients tend to give vertical zones of constant value which enables local regions to be defined as vertical strips.

Building Research Establishment
DEPARTMENT OF THE ENVIRONMENT

Technical enquiries to:
Building Research Establishment
Garston, Watford, WD2 7JR
Telex 923220 Fax 664010

PRESSURE COEFFICIENTS
Vertical walls of rectangular buildings

On the windward face, the slenderness ratio affects the flow pattern as already described. With a slender building, the pressure contours are predominantly vertical and they scale to the crosswind breadth B in Fig 1. As buildings become more squat, the central region of relatively low pressure coefficients expands and the contours move towards the end of the face, and their distance from the edges scales to twice the height $2H$. This enables coefficients to be defined in zones dimensioned in terms of the smaller of the height or face width.

For the side face, the slenderness ratio affects the size of the local high suction region at the upwind end of the face. The width of that region will still depend on the slenderness of the corresponding upwind face, that is still to the crosswind breadth B or twice height $2H$. Loaded regions can, therefore, still be defined as vertical strips for design purposes, dimensioned in terms of the smaller of the height or the upwind face length.

For the rear leeward face the wall experiences a fairly uniform suction throughout.

The peak cladding loads experience similar effects but are modified by atmospheric and building-generated turbulence; the effects of these generally increase the high-load regions.

Pressure coefficients for walls are required for different zones of loaded areas, depending on the proportions of the building and the wind direction. The definitions of dimensions are given in Fig. 1.

The loaded zones, **A** to **D**, are defined in Fig 2 as vertical strips in terms of b where:

$b = B$ or $2H$ whichever is the smaller

Zone **A** is always at the upwind edge of the face (see Fig 1). Depending on the proportion of the face not all the zones will exist or be their full defined size.

For walls, the wind angle θ is defined from normal to the wall being considered. For a rectangular building with the flow normal to the windward wall $\theta = 0°$, for the leeward wall $\theta = 180°$ and for both side walls $\theta = 90°$.

The loaded areas should then be determined as follows:

● Determine the crosswind breadth, B, for the relevant wind direction.

● Determine the height, H, to the top of the wall. The reference dynamic pressure, q, should be determined at this height.

● Calculate b.

● Define zone A with width $b/5$ starting from the upwind edge of the face.

● Define zone **B**, extending from $0.2b$ up to b away from the upwind edge of the face. For tall buildings when $H \geq B/2$, zones A and B occupy the whole face and the procedure is complete. If not:

● Define zone **D** from the downwind edge with width b. If there is insufficient room for the defined size zone **D** occupies the remainder of the face after zones A and B have been defined.

● Define zone **C** as the remainder of the face having fully defined A, B and D.

In the special case of wind exactly normal to the face, both edges of that face should be treated as upwind edges and zones A and B are defined away from both these edges. Zone D does not appear.

For leeward faces the same convention applies of taking zone A from the *upwind* corner of the leeward face (see Fig 1).

Pressure coefficients for each of these zones are given in Table 1.

Fig 1   Definitions of dimensions

Crosswind width

Fixed dimensions

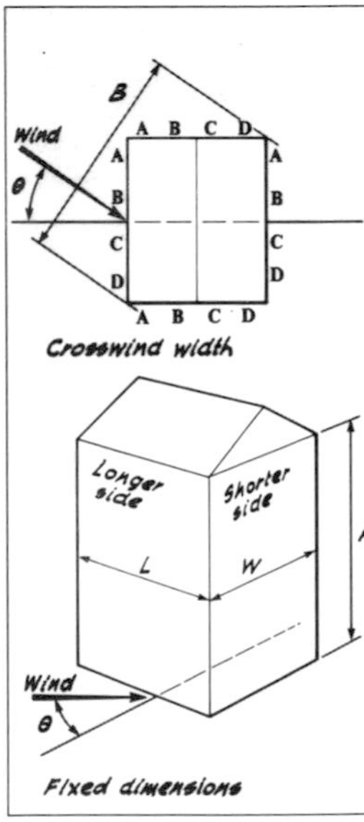

Table 1 Pressure coefficients for vertical walls

Zone	$H/B \leq 0.5$				$H/B = 1$		$H/B = 2$		$H/B \geq 4$	
	A	B	C	D	A	B	A	B	A	B
PRESSURE COEFFICIENTS										
$\theta = 0°$	0.66	0.83	0.86	0.83	0.78	0.89	0.67	0.80	0.62	0.76
30°	0.80	0.80	0.71	0.49	0.88	0.80	0.83	0.71	0.80	0.68
60°	0.24	0.51	0.40	0.26	−0.85	0.23	0.31	0.31	0.20	0.25
90°	−0.91	−0.68	−0.42	−0.12	−1.21	−0.92	−1.02	−0.82	−1.09	−0.94
120°	−0.63	−0.63	−0.46	−0.29	−0.64	−0.64	−0.54	−0.54	−0.67	−0.67
150°	−0.34	−0.34	−0.26	−0.32	−0.54	−0.54	−0.51	−0.51	−0.51	−0.51
180°	−0.34	−0.34	−0.22	−0.34	−0.24	−0.24	−0.40	−0.40	−0.54	−0.54
PEAK SUCTIONS										
90°	−1.21				−1.27		−1.29		−1.32	

Interpolation may be used. When the result of interpolating between positive and negative values is in the range ± 0.2, the coefficient should be taken as equal to ± 0.2 and both possible values used.

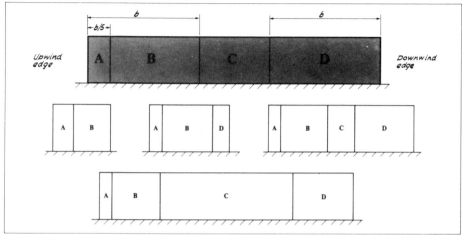

Fig 2 Key to wall pressure data

Roofs

The flow condition over flat roofs with wind normal to a face is similar to flow around the side wall: flow separates at the upwind edge and may re-attach at some distance downwind to form a separation bubble. The size of the separation bubble scales in the same manner to the smaller of the crosswind breadth, *B*, or twice the height, *2H*, as for walls. However, the corners of vertical walls remain normal to the flow for all wind angles and the separation bubbles form as cylindrical vortices. For roofs this occurs only for wind normal to the eave, at all other angles conical vortices form from the upwind corner.

The flow over monopitch roofs, that is roofs formed by one plane face at a pitch angle, is dependent on that angle. As the angle of pitch, α, increases the vortices formed at the upwind edges decrease in strength and size while the overall pressure rises, such that when $\alpha = 30°$ the vortices have disappeared and the overall pressure becomes positive. At 45° pitch angle, the coefficients exceed unity when the reference pressure is taken at the eave as the wind speed at the high downwind eave is in excess of the chosen reference value. When α is negative, the overall pressure starts to fall; when α reaches $-30°$ the pressure distribution is nearly uniform, with an overall uplift.

Duopitch roofs are formed by two plane faces joined along a common edge to form either a high ridge (positive α) typical of most houses, or a low trough (negative α). The upwind face behaves similarly to a monopitch roof; the downwind face is influenced significantly by the upwind face but is less onerously loaded. Little data are available on unequal pitch duopitch roofs and specialist advice should be sought.

Conventional hipped roofs are formed from duopitch roofs by replacing the gable ends with triangular pitched roofs or 'hips'. Vortices form along each of the ridges which might suggest that loadings on such roofs are more severe. This is not the case because the verge vortices are less severe than for a duopitch roof resulting in the loading on hipped roofs being much less severe.

Definitions

The various forms and parts of roofs are defined as follows — see Fig 3.

Eave   the horizontal edge of the roof — taken as the longer edges for flat roofs and hipped roofs

Verge   the non-horizontal edge of the roof, such as the gable edge of monopitch and duopitch roofs — taken as the shorter edges for flat roofs and hipped roofs

Hip   triangular pitched face at each end of the main faces of a hipped roof

Ridge   the highest horizontal line formed where the two faces of a duopitch roof meet. On hipped roofs, the horizontal line is the main ridge; this distinguishes it from the hip ridges which are at the junction of the main face and hip faces

Trough   the lowest horizontal line formed where the two faces of troughed duopitch roofs meet.

Fig 3 Definitions of roof types and parts

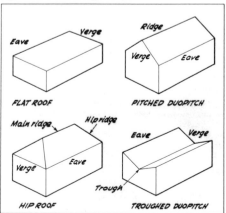

FLAT ROOFS — *below 5° pitch*

The roof should be subdivided into zones behind each upwind eave/verge. The general notation and definition of the loaded areas is given in Fig 4; this shows the zones on a rectangular roof when edges 1 and 2 are both upwind.

E_1 and E_2 are the lengths of the upwind eaves/verges measured from upwind external corner to downwind external corners

H is the height to the eaves. The reference dynamic pressure, q, should be determined at this height.

θ_n is the wind angle expressed as a local wind angle from normal to the eave/verge, n, ie θ_1 and θ_2 in Fig 4.

Loaded zones

Zones of constant pressure coefficients are defined in strips parallel to the eave/verge. These are each further divided downwind from the upwind corner.

The zones at the upwind corner are divided from those along the adjacent eave/verge by the line through the corner in the direction of the wind. This allows zones to be defined for any corner angle.

Once the zones have been defined behind each upwind eave/verge, pressure coefficients for all zones over the whole roof can be obtained.

Pressure coefficients for flat roofs with sharp eaves

Pressure coefficients for each zone for flat roofs with sharp eaves are given in Table 2 in terms of the local wind angle each side from normal to the eaves. These coefficients apply both to pressure for overall loading and for cladding loads.

Table 2 Pressure coefficients for flat roofs with sharp eaves

Local wind direction θ_n	Zone						
	A	**B**	**C**	**D**	**E**	**F**	**G**
0	−1.47	−1.25	−1.15	−1.15	−0.69	−0.71	±0.20
± 30	−2.00	−1.70	−1.38	−1.03	−0.66	−0.67	±0.20
± 60	−1.70	−1.24	−1.10	−0.64	−0.61	−0.42	±0.20
± 90	−1.20	−0.75	−0.52	−0.24	−0.62	±0.20	±0.20

Interpolation may be used. Where both positive and negative values are given both values should be considered.

Curved or chamfered eaves and parapets generally give lower values; see *The designer's guide to wind loading of building structures Part 2.*

Extent of zones

The extent of the zones and their types should be determined as follows:

● Determine the length E_n of the eave being considered

● Determine the height H_n of the corresponding wall (ie to the ground for simple cuboidal buildings)

● Calculate e as:
  $e = E$ or $2H$ whichever is the smaller

● Draw the boundary line through the upwind corner in the wind direction

● Mark out the depth of the edge zones parallel to the eave $e/10$ behind the eave and define zones A to D such that:
  Zone A extends $e/10$ from upwind corner
  Zone B extends from $e/4$ from upwind corner
  Zone C extends from $e/4$ to $e/2$ from upwind corner
  Zone D extends from $e/2$ to downwind corner

  When the wind is exactly normal to an eave/verge define the zones inwards from both corners.

● Mark out the depth of the central region parallel to the eave from $e/10$ to $e/2$ behind the eave. Define zones E and F such that:
  Zone E extends $e/2$ from upwind corner
  Zone F extends from $e/2$ from upwind corner to the downwind corner

● All the remainder of the roof downwind of zones E and F is zone G

● Repeat the above for the adjacent upwind eave (2 for the case considered in Fig 4).

Fig 4 Key for flat roof pressure data

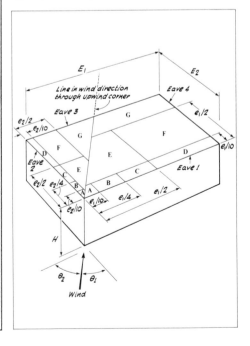

MONOPITCH AND DUOPITCH ROOFS

Figures 5 and 6 show the notation and zones for monopitch and duopitch roofs, in which:

L is the length of the upwind eaves

W is the width of the upwind verge. For duopitch roofs W is the total width of both verges (see Fig 6).

α is the pitch angle of the roof defined from normal to the upwind eave. For monopitch roofs α is taken as positive with the low eave upwind and negative with the high eave upwind.

For duopitch roofs α is taken as positive when the roof has a central ridge and negative when the roof has a central trough.

H is the height to the upwind eave for monopitch roofs and to the upwind eave for each face of duopitch roofs (see Fig 6).

θ is the wind angle from normal to the horizontal eave or ridge.

The loaded zones, **A** to **J**, are defined in Fig 5 as strips parallel to the eave and verge in terms of the widths e and v where:

$e = L$ or $2H$ whichever is the smaller
$v = W$ or $2H$ whichever is the smaller

The reference dynamic pressure, q, is taken at the effective height of each face appropriate to the height of the upwind corner (see Fig 6). Conservatively a single value appropriate to the highest point of the roof could be used.

Fig 5 Key for monopitch and duopitch roofs

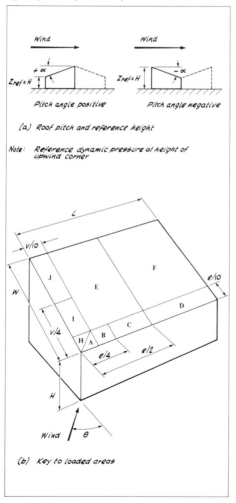

Fig 6 Key for duopitch roofs

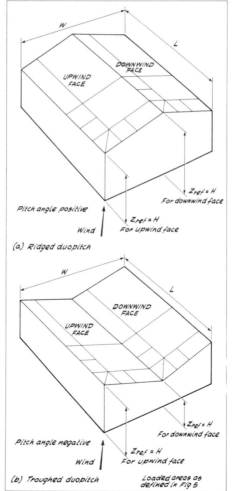

Loaded zones

The zones, over which the pressure coefficients are taken as constant, are defined from the upwind corner of each face, and should be determined from Fig 5 appropriate to the values of e and v. These zones are constant for all wind directions for which the corner is upwind.

Monopitch roofs

Pressure coefficients for each of the zones are given in Table 3.

Table 3 Pressure coefficients for monopitch roofs and upwind face of duopitch roofs

Pitch angle α	Wind direction	Zone								
		A	B	C	D	E	F	H	I	J
-45°	0°	-0.61	-0.58	-0.56	-0.41	-0.76	-0.78	-0.62	-0.79	-0.94
	30	-0.53	-0.50	-0.49	-0.55	-0.55	-0.81	-0.52	-0.58	-0.58
	60	-1.11	-1.29	-1.36	-0.96	-0.97	-0.91	-1.05	-0.97	-1.17
	90	-1.25	-0.81	-0.62	-0.42	-0.77	±0.20	-1.48	-1.05	-0.97
-30°	0°	-0.76	-0.68	-0.60	-0.50	-0.76	-0.63	-0.76	-0.85	-0.90
	30	-1.13	-1.02	-1.89	-0.79	-0.84	-0.76	-1.17	-0.87	-0.73
	60	-2.06	-2.33	-2.17	-1.22	-1.03	-0.80	-1.69	-1.18	-1.21
	90	-1.28	-0.94	-0.70	-0.37	-0.70	±0.20	-1.54	-1.10	-1.01
-15°	0°	-1.08	-1.05	-0.97	-0.92	-0.88	-0.82	-1.10	-0.96	-0.83
	30	-2.64	-2.37	-1.71	-1.00	-0.93	-0.85	-2.75	-1.66	-1.11
	60	-2.25	-2.15	-1.85	-1.02	-0.76	-0.72	-2.44	-1.60	-1.07
	90	-1.22	-0.79	-0.58	-0.31	-0.60	±0.20	-1.51	-1.12	-1.13
-5°	0°	-1.49	-1.34	-1.19	-1.12	-0.83	-0.82	-1.47	-0.91	-0.67
	30	-2.36	-2.21	-1.63	-1.04	-0.82	-0.77	-2.24	-1.30	-0.91
	60	-1.85	-1.57	-1.28	-0.77	-0.65	-0.54	-2.10	-1.67	-1.09
	90	-1.30	-0.79	-0.58	-0.27	-0.59	±0.20	-1.65	-1.13	-1.20
+5°	0°	-1.39	-1.24	-1.11	-1.19	-0.56	-0.59	-1.39	-0.69	-0.43
	30	-1.78	-1.64	-1.34	-1.09	-0.62	-0.60	-1.75	-1.02	-0.76
	60	-1.67	-1.33	-1.12	-0.71	-0.64	-0.42	-2.05	-1.51	-1.05
	90	-1.21	-0.83	-0.55	-0.25	-0.61	±0.20	-1.48	-1.12	-1.30
+15°	0°	-0.91	-0.83	-0.78	-0.81	-0.21	-0.31	-0.90	-0.36	-0.30
	30	-0.84	-0.88	-0.82	-0.83	-0.21	-0.37	-0.63	±0.20	-0.32
	60	-1.27	-0.86	-0.70	-0.61	-0.54	-0.33	-1.57	-1.21	-0.93
	90	-1.20	-0.84	-0.58	-0.27	-0.64	±0.20	-1.42	-1.10	-1.30
+30°	0°	0.38	0.69	0.79	0.77	0.39	0.40	±0.20	±0.20	±0.20
	30	0.75	0.74	0.22	0.59	0.41	0.26	0.78	0.69	0.47
	60	-0.14	0.43	0.39	0.33	±0.20	±0.20	-0.80	-0.89	-0.83
	90	-1.13	-0.94	-0.77	-0.19	-0.78	±0.20	-1.25	-1.06	-1.36
+45°	0°	0.82	1.02	1.11	1.13	0.75	0.74	0.69	0.56	0.43
	30	1.11	1.09	1.03	0.88	0.77	0.55	1.12	1.00	0.85
	60	0.79	0.69	0.62	0.46	0.38	0.21	0.84	0.82	0.54
	90	-1.17	-0.96	-0.86	-0.33	-0.88	-0.28	-1.25	-1.08	-1.36

Interpolation may be used. When the result of interpolating between positive and negative values is in the range of ± 0.2, the coefficient should be taken as equal to ± 0.2 and both positive values used.

Duopitch roofs

Pressure coefficients for each zone for the upwind face are given in Table 3. Pressure coefficients for each zone of the downwind face can be obtained from Table 4.

These coefficients are appropriate to duopitch faces of equal pitch but may be used without modification provided the upwind and downwind pitch angles are within ± 5° of each other. For duopitch roofs of greater difference in pitch angles, see *The designer's guide to wind loading of building structures Part 2.*

Table 4 Pressure coefficients for the downwind face of duopitch roofs

Pitch angle α	Wind direction	Zone								
		A	B	C	D	E	F	H	I	J
−45°	0°		−0.92		−0.75	−0.75	−0.75		−0.63	
	30		−1.12		−0.52	±0.20	−0.52		−0.32	
	60		−1.04		−0.24	−0.73	−0.24		−1.05	
	90	−1.17	−0.96	−0.86	−0.33	−0.88	−0.28	−1.25	−1.08	−1.36
−30°	0°		−0.78		−0.66	−0.66	−0.47		−0.40	
	30		−0.44		−0.52	−0.52	±0.20		±0.20	
	60		−0.74		−0.27	−0.27	−0.62		−1.01	
	90	−1.13	−0.94	−0.77	−0.19	−0.78	±0.20	−1.25	−1.06	−1.36
−15°	0°		−0.69		−0.52	−0.52	−0.26		−0.21	
	30		±0.20		±0.20	±0.20	±0.20		−0.55	
	60		−0.67		±0.20	±0.20	−0.65		−1.03	
	90	−1.20	−0.84	−0.58	−0.27	−0.64	±0.20	−1.42	−1.10	−1.30
−5°	0°		−0.34		−0.25	−0.25	−0.25		−0.28	
	30		±0.20		±0.20	±0.20	−0.26		−0.48	
	60		−0.69		±0.20	±0.20	−0.66		−0.88	
	90	−1.21	−0.83	−0.55	−0.25	−0.61	±0.20	−1.48	−1.12	−1.30
+5°	0°	−0.32	−0.27	−0.28	−0.28	±0.20	±0.20	−0.36	−0.30	−0.24
	30	−0.70	−0.46	−0.30	−0.23	−0.31	±0.20	−0.71	−0.59	−0.46
	60	−1.04	−0.90	−0.52	±0.20	−0.56	±0.20	−0.97	−0.83	−0.73
	90	−0.90	−0.83	−0.58	±0.20	−0.60	±0.20	−0.89	−0.89	−1.09
+15°	0°	−0.83	−0.81	−0.80	−0.78	−0.39	−0.40	−0.85	−0.55	−0.39
	30	−1.32	−1.14	−1.11	−0.88	−0.46	−0.34	−1.47	1.25	−0.81
	60	−1.31	−0.92	−0.72	−0.58	−0.57	−0.23	−1.45	−1.08	−0.75
	90	−0.81	−0.74	−0.54	±0.20	−0.58	±0.20	−0.83	−0.77	−0.92
+30°	0°	−0.29	−0.26	−0.25	−0.30	−0.30	−0.30	−0.31	−0.32	−0.33
	30	−0.74	−0.63	−0.52	−0.43	−0.39	−0.43	−0.76	−0.51	−0.40
	60	−1.04	−1.05	−0.98	−0.64	−0.58	−0.47	−1.02	−0.67	−0.64
	90	−0.66	−0.61	−0.49	−0.21	−0.49	±0.20	−0.67	−0.58	−0.69
+45°	0°	−0.21	−0.21	−0.21	−0.20	−0.23	−0.23	−0.21	−0.24	−0.26
	30	−0.21	−0.20	−0.20	−0.27	−0.23	−0.26	−0.20	−0.21	−0.22
	60	−0.54	−0.54	−0.51	−0.41	−0.44	−0.38	−0.55	−0.47	−0.50
	90	−0.55	−0.46	−0.38	−0.20	−0.40	±0.20	−0.60	−0.45	−0.47

Interpolation may be used.

Hipped roofs

The following provisions apply to conventional hipped roofs on cuboidal-plan buildings, where all faces of the roof have the same pitch angle in the range $5° < \alpha < 45°$.

The data are valid both for the trapezoidal main faces and the triangular hip faces, with the wind angle, θ_n, expressed as a local wind angle from normal to the eave/verge, ie θ_1 and θ_2 in Fig 7.

Loaded zones

Loaded zones, over which the pressure coefficients may be taken to be constant, are determined fron Fig 7 in which:

L is the length of the main face eave

W is the length of the hip face eave

H is the height of the ridge.

   For the main face, e should be taken as $2H$ or L whichever is the smaller.

   For the hip face, e should be taken as $e = 2H$ or W whichever is the smaller.

θ_n is the local wind direction normal to the eave/verge of the face.

The reference dynamic pressure, q, should be determined at height H, the height of the ridge, or, for those zones immediately adjacent to the eaves, and zones E and F on the upwind faces, it may be taken as the height of the eaves.

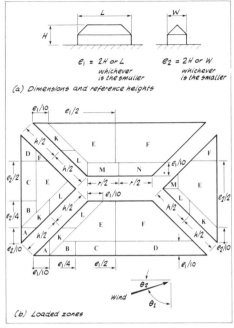

Fig 7   Key for hipped roofs

Pressure coefficients

The pressure coefficients for the upwind faces for zones **A** to **F** may be obtained for the corresponding pitch angle from Table 3 (for positive values of α only). Pressure coefficients for zones **K** and **L** are given in Table 5.

The pressure coefficients for the downwind faces for zones **E** and **F** may be obtained for the corresponding pitch angle from Table 4 (for positive values of α only). Pressure coefficients for zones **K** to **N** are given in Table 5.

Table 5 Pressure coefficients for hipped roofs

Pitch angle α	Wind direction	Upwind face zone K	Upwind face zone L	Downwind face zone K	Downwind face zone L	Downwind face zone M	Downwind face zone N
	0°	−0.56	−0.56	−0.31	−0.45	−0.58	−0.58
5°	30	−0.62	−0.62	−0.60	−0.46	−0.47	−0.54
	60	−1.13	−0.63	−0.76	−0.51	−0.38	−0.36
	90	−1.19	−0.76	−0.89	−0.50	−0.61	±0.20
	0°	−0.31	−0.31	−0.44	−0.83	−1.17	−1.17
15°	30	−0.37	−0.37	−1.00	−0.99	−1.31	−1.13
	60	−0.94	−0.52	−1.43	−0.71	−0.78	−0.80
	90	−1.09	−0.77	−0.97	−0.59	−0.64	±0.20
	0°	0.40	0.40	−0.53	−0.33	−0.28	−0.28
30°	30	0.26	0.26	−0.74	−0.55	−0.51	−0.50
	60	−0.99	−0.47	−1.25	−0.82	−0.77	−0.49
	90	−1.10	−1.01	−1.40	−0.62	−0.78	±0.20
	0°	0.74	0.74	−0.65	−0.24	−0.20	−0.20
45°	30	0.55	0.55	−0.52	−0.22	−0.22	−0.28
	60	−1.11	−0.33	−0.67	−0.35	−0.32	−0.41
	90	−1.22	−0.71	−1.35	−0.43	−0.88	−0.28

For further reading see Part 1

Interpolation may be used. When the result of interpolating between positive and negative values is in the range ± 0.20 the coefficient should be taken as equal to ± 0.20 and both possible values used.

Printed in the UK and published by Building Research Establishment, Department of the Environment.   *Crown copyright 1989*
Price Group 3. Also available by subscription. Current prices from:
Publications Sales Office, Building Research Establishment, Garston, Watford, WD2 7JR (Tel 0923 664444).
Full details of all recent issues of BRE publications are given in *BRE News* sent free to subscribers.

ISBN 0 85125 422 5

BRE Digest

Concise reviews of building technology

Digest 346
November 1989

CI/SfB (J4)

The assessment of wind loads

Part 7: Wind speeds for serviceability and fatigue assessments

This is the seventh in a series of Digests which is compatible with the proposed new British Standard BS 6399: Part 2. It deals with the assessment of more frequent parent wind speeds in the United Kingdom from the extreme wind speeds given by Part 3. Two procedures are given:

● for estimating the values of wind speeds occurring for between one and one hundred hours per year, for making serviceability assessments

● for estimating the number of occurrences of wind speeds for making fatigue assessments.

The procedure for serviceability assessments is based on the approach used in BS 8100 *Lattice towers and masts* which has been augmented with new data and has been further refined since publication of BS 8100. The procedure for fatigue assessments is based on analysis of extreme meteorological and loading data in the UK, but gives very similar results to the procedure in
ECCS Recommendations for calculating the effects of wind on constructions.

Building Research Establishment
DEPARTMENT OF THE ENVIRONMENT

Technical enquiries to:
Building Research Establishment
Garston, Watford, WD2 7JR
Telex 923220 Fax 664010

PARENT WIND SPEEDS

The Meteorological Office records the mean wind speed, $\bar{V}$, and maximum gust speed V, in each hour at their 140 anemograph stations across the United Kingdom. The term *parent* is used to describe this complete data record. It is most usefully represented by the cumulative distribution function (CDF), denoted by P, which quantifies the probability that the wind speed is *below* any given value. The probability of *exceeding* a value will be denoted here by the symbol Q, and is given for any wind speed, V, by:

$$Q_V = 1 - P_V \qquad (1)$$

As the parent wind is continuous, the proportion of time that the wind speed is below or above the value of V is represented by P_V and Q_V respectively.

As there are 8766 hours in a year, it does not take many years to define the CDF with considerable accuracy. The simplest statistical model assumes the variations of wind speed to be Gaussian or Normally distributed in two orthogonal directions, eg north-south and east-west. This leads to the expectation that the CDF of the parent wind speed *irrespective of direction* will be described by the Rayleigh distribution:

$$P_V = 1 - \exp \frac{-\bar{V}^2}{2V'^2} \qquad (2)$$

where V' is the standard deviation of $\bar{V}$.

Figure 1(a) shows parent data for Lerwick fitted to the Rayleigh distribution.

The Rayleigh distribution is a special case of the whole family of distributions called Weibull distributions given by:

$$P_V = 1 - \exp(-(c\bar{V})^k) \qquad (3)$$

where c and k are values which define the shape of the distribution.

Comparing this equation with (2) shows that the Rayleigh distribution corresponds to a Weibull distribution with $k = 2$. The differences between the Rayleigh model equation (2) and the parent wind data in Fig 1(a) are reduced by adopting the Weibull distribution (3) and allowing the value of k to vary. The same Lerwick data are shown refitted to the Weibull distribution with $k = 1.85$ in Fig 1(b). The fit is excellent except at the lowest wind speeds, where the difference is due to friction in the bearings of the standard cup anemometer.

The parent wind speed distribution, irrespective of direction, is very well represented in the UK by Weibull distributions with k in the range 1.7 to 2.5

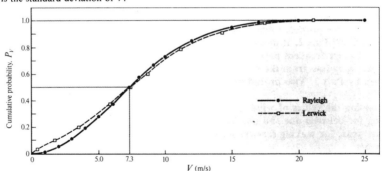

(a) Compared with a Rayleigh distribution

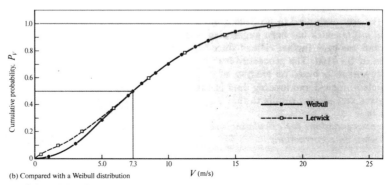

(b) Compared with a Weibull distribution

Fig 1   Parent wind speed CDF for Lerwick

EXTREME WIND SPEEDS

The requirement for structures to resist the strongest winds expected in their lifetimes means that the design calculations for the ultimate limit state are made in terms of the extreme wind speeds. An 'extreme' is defined as the maximum value occurring in a set period — usually one year to give annual maxima. If the parent values were statistically independent, the CDF of the maximum wind speed $\bar{V}_{max}$, would be given by:

$$P_{\bar{V}_{max}} = P_V^N \tag{4}$$

where N is the number of independent values in the set period.

In the case of $N = 1$, there is no change from the parent, but as N increases the CDF shifts to higher values corresponding to the upper 'tail' of the parent. Unfortunately, the parent wind data are not statistically independent because the wind speed at any hour is related by the wind climate to the speed for some hours before and after. The value of N for a one year period is not 8766, but is in the range $100 \sim 300$.

Instead of predicting the extreme wind speeds from the parent, the extreme wind climate was conventionally examined directly in terms of the CDF of measured annual maxima. More recently, BRE reanalysed the UK wind climate using the maximum wind speed in every independent storm, resulting also in a direct estimate of the number of storms N. The CDF of annual maxima was then calculated using (4). This procedure is the origin of the map in Part 3 of this Digest. As storm maxima and the annual maxima are related by (4), the possibility exists to reverse the process and estimate parent wind speeds from the extreme wind data given by Parts 3, 4 and 5, at least for the upper tail of the parent CDF in the range $1 \le N \le 100$.

HOURS OF WIND

The parent CDFs were formed from the meteorological data for the same sites and over the same period as were used to derive the wind speed map in Part 3. When the parent wind speed was expressed as a fraction of the basic extreme wind speed and the probability of exceedence, Q, was expressed in terms of the hours per year of exceedances, $h = 8766 \times Q$, it was found the CDF for all sites collapsed into a band. The width of this band, which represents the uncertainty of the estimate for any individual site, corresponded to a factor of 3 on hours of exceedence, h. This demonstrated that it is possible to estimate the hours of exceedence of any given serviceability wind speed, V_s, from the reference extreme wind speed, V_{REF}, given by Parts 3, 4 and 5 to an expected accuracy of a factor of 3.

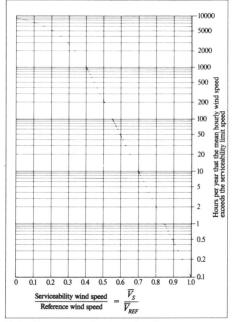

Fig 2   Hours per year that the serviceability wind speed is exceeded

Figure 2 gives a design curve for this purpose. It represents the upper boundary of the band, so gives a safe result when the effect of wind is detrimental to the design. It can be used to determine the serviceability wind speed, $\bar{V}_s$, corresponding to an acceptable number hours of exceedance per year, h, — or the number of hours of exceedence per year corresponding to an acceptable threshold wind speed. In both cases, the reference extreme wind speed, V_{REF}, must first be determined from Parts 3 & 4 and, if topography is significant, from Part 5.

Should the serviceability wind speed be beneficial to the design assessment (perhaps when calculating the concentration of pollutants) the value for the lower boundary of the band is required and the value of h from Fig 2 should be divided by 3. In making a cost-benefit analysis (perhaps when comparing the cost of providing a wind-power generator against the expected supply of electrical power) a value in the middle of the band is more appropriate and the value of h from Fig 2 should be divided by 1.5.

Figure 2 was derived for all hours of wind, irrespective of direction. As the strongest winds approach the value of $\bar{V}_{REF}$ the directional characteristics converge towards those given by the Direction Factor, S_{DIR} in Part 3. However, in light winds the wind direction is strongly affected by thermal effects, such as sea breezes near coasts. Accordingly, S_{DIR} from Part 3 should be taken to apply only for wind speeds corresponding to the range $h \le 100$.

FATIGUE LOAD CYCLES

There are two types of fatigue that may need to be considered:

- high-cycle fatigue of dynamic structures caused by oscillations of the structure by resonance at one or more of its natural frequencies, n_i, and

- low-cycle fatigue of static structures by the repeated action of gust loads

High-cycle fatigue occurs when the structure is subjected to very many thousands of load cycles at a small proportion of the ultimate capacity of the structure. The commonest source of regular oscillations of tall and line-like structures is vortex shedding which occurs when the wind speed is close to the critical wind speed (say 10% either side). There are two other sources. The first is buffeting from the turbulence of the wind or wakes of other structures. This occurs at all wind speeds, but increases as the square of wind speed so is much more important in the fewer hours of the strongest winds. The second source is galloping or flutter which occurs only above a given threshold. An estimate of the number of cycles of oscillation, N, is given from the hours for which the condition occurs, $h_1 - h_2$, and the natural frequency of the structure, n (in Hz), by:

$$N = 3600n\,(h_1 - h_2) \qquad (5)$$

where h_1 is the number of hours of exceedence of the *lower* boundary to the condition;
h_2 is the number of hours of exceedence of the *upper* boundary.

If the maximum stress in each cycle can be estimated by calculation or from observations of the motion, the fatigue life of the structure can be determined from the stress-cycle (S-N) curve for the material used. High-cycle fatigue is essentially a serviceability problem, provided the structure is properly inspected and maintained, and that fatigue cracks can be repaired or components replaced before their fatigue life is exhausted. This is common practice, for example, for the holding-down bolts of slender steel chimney stacks. If these precautions are neglected, fatigue damage can accumulate, reducing the strength until a serious structural failure or collapse occurs.

Low-cycle fatigue occurs when the structure is subjected to relatively few load cycles close to the ultimate strength of the structure, that is for a few tens to a few thousand cycles. Such high stresses tend to occur in thin metal claddings around the fixing points.

Owing to the intermittent nature of storms, low-cycle fatigue is accumulated in a few short periods corresponding to the strongest storms. In the case of severe tropical cyclones, such as cyclone Tracy at Darwin in 1974, extensive failures can occur in only a few hours of exposure. A satisfactory inspection of a structure after one severe storm may not guarantee survival of the next.

Since 1974 rules to cope with fatigue in tropical cyclones have been incorporated into Australian regulations, but these are too onerous and inappropriate for the depression-dominated climate of the UK and Europe. A table of cycles at various proportions of the ultimate design load appropriate to a 50-year design life in the UK was proposed by BRE in 1984 and is given in Table 1. This table was derived by counting the number of cycles caused by the highest, second-highest, next-highest ..., *ie* working *downwards* from the extreme. Contemporary to this, the European Convention for Constructional Steelwork (ECCS) recommended a loading sequence derived by counting *upwards* from the parent. These two independent approaches gave almost identical results, giving confidence as to their accuracy.

Table 1 Fatigue test representing typical UK service loads in 50-year exposure period

	Number of cycles	Percentage of design pressure
	1	90
	960	40
Apply sequence	60	60
five times	240	50
	5	80
	14	70
Finish with	1	100

The order of the load cycles in Table 1 is designed to represent the random sequence of loads occurring in nature as a practical loading sequence for applying proof loads to a structure or component. The main sequence of 1280 cycles from zero to between 40% and 90% of design load is repeated five times, giving 6400 cycles in all; then the structure is proof loaded by one cycle of the design load.

More recently, the development of the BRE computer-controlled test rig, BRERWULF, now enables records of real storms to be applied to structural components, obviating the need for a standard load-cycle sequence.

FURTHER READING

REDFEARN D. A test rig for proof-testing building components against wind loads. BRE Information Paper IP19/84. Garston, Building Research Establishment, 1984.

ECCS. Recommendations for calculating the effect of wind on constructions, (Second edition). Brussels, European Convention for Constructional Steelwork, 1987.

COOK N J, KEEVIL A P and STOBART R K. BRERWULF — the Big Bad Wolf. Journal of Wind Engineering and Industrial Aerodynamics, 29, (1988), 99 — 107.

See also Further Reading in *Part 1.*

Printed in the UK and published by Building Research Establishment, Department of the Environment.   *Crown copyright 1989*
Price Group 3. Also available by subscription. Current prices from:
Publications Sales Office, Building Research Establishment, Garston, Watford, WD2 7JR (Tel 0923 664444).
Full details of all recent issues of BRE publications are given in *BRE News* sent free to subscribers.

Printed in the UK for HMSO. Dd.8245006, 10/89, C150, 38938   ISBN 0 85125 421 7

Building Research Establishment Digest

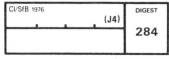

CI/SfB 1976   (J4)   DIGEST **284**

New edition 1986

Wind loads on canopy roofs

This digest deals with the assessment of wind loads on free-standing canopy roofs without walls and includes the effects of blockage caused by stacked contents. It should be used in conjunction with Digest 119. The assessment of wind loads (1984 New edition) and, if topography is significant, with Digest 283 The assessment of wind speed over topography.

The recommendations are compatible with the 1985 amendment to the British Standard Code of practice CP3: Chapter V: Part 2: 1972, but gives revised data for blocked canopies based on recent research at full and model scale.

Scope

This digest presents revised data in the same form as Table 13 of the 1985 amendment to the BS Code of practice CP3: Chapter V: Part 2: 1972. It should be used to assess the design wind loads of free-standing canopy roofs, such as dutch barns, petrol station canopies and similar shelters that do not have permanent walls. It may be used to assess the design wind loads on canopies extending from enclosed buildings, such as loading bays, provided the adjacent building is not significantly taller than the canopy. It can also be used to assess the maximum wind loads on grandstand roofs when the rear wall is solid, but more detailed estimates of design loads on grandstands, including the effect of wind direction, may be obtained by treating them as conventional mono-pitch or duo-pitch buildings with a dominant opening in the front face.

Differences from conventional buildings (Digest 119)
When the wind blows on to the face of a conventional, impermeable building it is slowed down against the front face with a consequent build-up of pressure against that face. At the same time it is deflected and accelerated around the side walls and over the roof and the consequent reduction in pressure exerts a net suction on those areas. This situation, shown in Fig 1(a) is covered by Digest 119.

An empty, free-standing canopy has no side walls to restrict the flow and the wind is free to pass above and below the canopy, shown in Fig 1(b). In this situation the principal forces on the canopy are 'lift' forces acting normal to the canopy surfaces (through pressure differences across the canopy generated in a way similar to those on a kite or aircraft wing) and 'drag' forces acting at a tangent to the canopy surfaces (through friction of the flow against both sides of the canopy surface and through pressure on any vertical fascia).

When goods are stored under a canopy, they tend to restrict the flow of wind beneath the canopy. The worst case for wind loads is generally when the canopy is completely blocked to the downwind eaves, shown in Fig 1(c); here the situation is like a conventional building with a dominant opening on the windward face. This digest can be used to assess the wind loads on any free-standing canopy roof from fully empty to completely blocked by contents. Wind loads on the contents are not assessed and it is assumed that these loads are not transferred to the structure of the canopy.

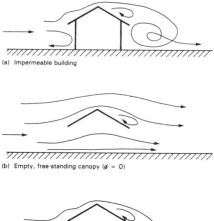

(a) Impermeable building

(b) Empty, free-standing canopy ($\phi = 0$)

(c) Canopy blocked to the downwind eaves by stored goods ($\phi = 1$)
Fig 1 Airflow over buildings

Prepared at Building Research Station, Garston, Watford, WD2 7JR
Technical enquiries arising from this Digest should be directed to Building Research Advisory Service at the above address.

Differences from Table 13 of the 1985 amendment to BS CP3: Chapter V: Part 2: 1972
During the late 1970's BRE refined methods of accurately reproducing turbulent wind loads in wind tunnels[1] and developed a method for assessing data to give the required design risk[2]; these have enabled a review of the data in the BS Code of practice CP3: Chapter V: Part 2: 1972 to be started. The data presented in the first edition of this digest and in Table 13 of the 1985 amendments to the BS Code were the first step in this review.

In a study of wind loads on mono-pitch and duo-pitch empty canopies up to 30° in pitch, the largest loads were often found at wind directions skewed to the canopy, not normal to the canopy as previously assumed. Canopy proportions were investigated in the range $1 < l/w < 3$ and height above ground in the range $0.15 < h/w < 1$ for all wind directions. Blockage by stored goods was also investigated over a range of typical conditions. In order to retain the previous format of Table 13 in the BS Code and to restrict the data to a manageable amount, only the most onerous values given by these ranges of parameters were given in the first edition.

Investigations have continued, particularly in respect of fully blocked canopies. Measurements made on full-scale dutch barns stacked with straw bales over an extensive range of stacking patterns[3] and model-scale data on grandstand designs with solid rear walls and with or without end gable walls have led to a better understanding of the effect of blockage. It is now clear that the effect of blockage on low-pitch canopies is more onerous than previously assumed, and on higher-pitched canopies blockage is less onerous and sometimes even beneficial. Changes incorporated in this edition of the digest are confined to the data for fully-blocked canopies. It is also known that the provision of deep fascias or solid gables above the eaves of duo-pitch canopies can reduce the maximum uplift at the expense of increasing the horizontal drag in the wind direction, although it is not yet possible to quantify this effect reliably.

Design wind speed and dynamic pressure
The design wind speed, V_s, and the resulting dynamic pressure, q, should be assessed for the site in accordance with the principles and data given in Digest 119 (1984 New edition). If topography at the site is significant, the Topography factor S_1 should be assessed using Digest 283. The Surface roughness and height factor S_2, the Statistical or Building life factor S_3, and the Direction factor S_4 are obtained from Digest 119. As the pressure coefficient data presented in this digest are the most onerous values irrespective of wind direction, there is little to be gained by adopting the optional Direction factor S_4 unless the site is influenced by topography. In this event, both the Topography factor S_1 and the Direction factor S_4 should be assessed by direction and the

most onerous combination used. The Direction factor S_4 will tend to offset the Topography factor S_1 unless the upwind slope of the topography is in the prevailing wind direction.

The design wind speed, V_s, is calculated from the formula:

$$V_s = V \times S_1 \times S_2 \times S_3 \times S_4$$

and the design dynamic pressure from the formula:

$$q = kV_s^2$$

where k = 0.613 in SI units (N/m² and m/s)
  k = 0.0625 in metric technical units (kgf/m² and m/s)
  k = 0.00256 in imperial units (lb/ft² and mile/h)

as described in Digest 119.

Forces normal to the canopy
The use of pressure coefficients to describe the forces normal to a solid surface is fully described in Digest 119. In brief, the pressure p at any point can be expressed in terms of q by means of a pressure coefficient C_p.

Thus:

$$p = C_p.q$$

In the calculation of wind load on any structure it is essential to take account of the pressure difference between opposite faces of a surface. For clad structures it is therefore necessary to know the internal pressure as well as the external pressure. For canopies it is convenient to combine the effect of the pressure on both sides of the canopy surface into a single overall coefficient.

Degree of blockage The degree of blockage under the canopy is described by the solidity ratio $ø$; this is the area of obstructions under the canopy divided by the gross area under the canopy, both areas normal to the wind direction ($ø = 0$ represents an empty canopy and $ø = 1$ represents the canopy fully blocked with contents to the downwind eaves). Values of C_p for intermediate solidities may be linearly interpolated between these two extremes and apply upwind of the position of maximum blockage only. Downwind of the position of maximum blockage the coefficients for $ø = 0$ may be used.

Pressure coefficients are given for single-bay duo-pitch canopies in Table 1 and for mono-pitch canopies in Table 2. The column headed *Overall coefficients* gives values to be applied to the whole canopy area when assessing overall loads for the design of the structure. For mono-pitch canopies the centre of pressure should be taken to act at 0.25 w from the windward edge. For duo-pitch canopies the centre of pressure should be taken to act at the centre of each slope. Each canopy must be able to support the Maximum (downward) and the Minimum (upward) loads, the latter depending on the degree of blockage under the canopy. In addition, a duo-pitch canopy must be able to support one slope at the Maximum or Minimum value with the other slope unloaded. Each bay of a multi-bay duo-pitch canopy may be assessed by applying the reduction factors of Table 3 appropriate to each bay to the Overall coefficients of Table 1.

The Local coefficients are used to determine loads on elements of cladding and their fixings in the area marked on the Key plans. Where these areas overlap in the corners of the canopies the more onerous of the two values should be used.

Table 2 Mono-pitch canopies

Section   Key plan

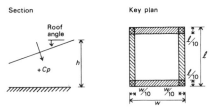

Roof angle degrees		Overall coefficients	Local coefficients □	▨	▧
0	Maximum all ϕ	+0.2	+0.5	+1.8	+1.1
	Minimum $\phi=0$	-0.5	-0.6	-1.3	-1.4
	Minimum $\phi=1$	-1.3	-1.5	-1.8	-2.2
5	Maximum all ϕ	+0.4	+0.8	+2.1	+1.3
	Minimum $\phi=0$	-0.7	-1.1	-1.7	-1.8
	Minimum $\phi=1$	-1.4	-1.6	-2.2	-2.5
10	Maximum all ϕ	+0.5	+1.2	+2.4	+1.6
	Minimum $\phi=0$	-0.9	-1.5	-2.0	-2.1
	Minimum $\phi=1$	-1.4	-2.1	-2.6	-2.7
15	Maximum all ϕ	+0.7	+1.4	+2.7	+1.8
	Minimum $\phi=0$	-1.1	-1.8	-2.4	-2.5
	Minimum $\phi=1$	-1.4	-1.6	-2.9	-3.0
20	Maximum all ϕ	+0.8	+1.7	+2.9	+2.1
	Minimum $\phi=0$	-1.3	-2.2	-2.8	-2.9
	Minimum $\phi=1$	-1.4	-1.6	-2.9	-3.0
25	Maximum all ϕ	+1.0	+2.0	+3.1	+2.3
	Minimum $\phi=0$	-1.6	-2.6	-3.2	-3.2
	Minimum $\phi=1$	-1.4	-1.5	-2.5	-2.8
30	Maximum all ϕ	+1.2	+2.2	+3.2	+2.4
	Minimum $\phi=0$	-1.8	-3.0	-3.8	-3.6
	Minimum $\phi=1$	-1.4	-1.5	-2.2	-2.7

Table 1 Single bay, duo-pitch canopies

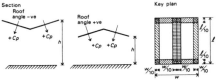

Section   Key plan

Roof angle degrees		Overall coefficients	Local coefficients □	▨	▧	▤
-20	Maximum all ϕ	+0.7	+0.8	+1.6	+0.6	+1.7
	Minimum $\phi=0$	-0.7	-0.9	-1.3	-1.6	-0.6
	Minimum $\phi=1$	-1.3	-1.5	-2.4	-2.4	-0.6
-15	Maximum all ϕ	+0.5	+0.6	+1.5	+0.7	+1.4
	Minimum $\phi=0$	-0.6	-0.8	-1.3	-1.6	-0.6
	Minimum $\phi=1$	-1.4	-1.6	-2.7	-2.6	-0.6
-10	Maximum all ϕ	+0.4	+0.6	+1.4	+0.8	+1.1
	Minimum $\phi=0$	-0.6	-0.8	-1.3	-1.5	-0.6
	Minimum $\phi=1$	-1.4	-1.6	-2.7	-2.6	-0.6
-5	Maximum all ϕ	+0.3	+0.5	+1.5	+0.8	+0.8
	Minimum $\phi=0$	-0.5	-0.7	-1.3	-1.6	-0.6
	Minimum $\phi=1$	-1.3	-1.5	-2.4	-2.4	-0.6
+5	Maximum all ϕ	+0.3	+0.6	+1.8	+1.3	+0.4
	Minimum $\phi=0$	-0.6	-0.6	-1.4	-1.4	-1.1
	Minimum $\phi=1$	-1.3	-1.3	-2.0	-1.8	-1.5
+10	Maximum all ϕ	+0.4	+0.7	+1.8	+1.4	+0.4
	Minimum $\phi=0$	-0.7	-0.7	-1.5	-1.4	-1.4
	Minimum $\phi=1$	-1.3	-1.3	-2.0	-1.8	-1.8
+15	Maximum all ϕ	+0.4	+0.9	+1.9	+1.4	+0.4
	Minimum $\phi=0$	-0.8	-0.9	-1.7	-1.4	-1.8
	Minimum $\phi=1$	-1.3	-1.3	-2.2	-1.6	-2.1
+20	Maximum all ϕ	+0.6	+1.1	+1.9	+1.5	+0.4
	Minimum $\phi=0$	-0.9	-1.2	-1.8	-1.4	-2.0
	Minimum $\phi=1$	-1.3	-1.4	-2.2	-1.6	-2.1
+25	Maximum all ϕ	+0.7	+1.2	+1.9	+1.6	+0.5
	Minimum $\phi=0$	-1.0	-1.4	-1.9	-1.4	-2.0
	Minimum $\phi=1$	-1.3	-1.4	-2.0	-1.5	-2.0
+30	Maximum all ϕ	+0.9	+1.3	+1.9	+1.6	+0.7
	Minimum $\phi=0$	-1.0	-1.4	-1.9	-1.4	-2.0
	Minimum $\phi=1$	-1.3	-1.4	-1.8	-1.4	-2.0

Table 3 Multi-bay canopies

Section

Loads on each slope of multi-bay canopies are determined by applying the following factors to the overall coefficients for isolated duo-pitch canopies.

Bay	Location	Factors for all ϕ on maximum overall coefficient	on minimum overall coefficient
1	end bay	1.00	0.81
2	second bay	0.87	0.64
3	third and subsequent bays	0.68	0.63

Horizontal forces

In addition to the pressure forces normal to the surface of the canopy, there will be horizontal loads due to the pressure of wind on any fascia at the eaves or on any gable between eaves and ridge on duo-pitch canopies, or to friction forces acting on top and bottom surfaces of the canopies. For any wind direction, only the more onerous of these two forces needs be taken into account, since a significant fascia or gable tends to shield the canopy from friction forces and, conversely, the friction over an extensive canopy tends to shield a small downwind fascia.

Fascia and gable loads should be calculated on the area of the surface facing the wind using a pressure coefficient of 1.3 on the windward fascia/gable and 0.6 on the leeward fascia/gable.

Frictional drag F is calculated from:

$$F = 0.025 \times A_s \times q$$

where A_s is either the combined area of top and bottom surfaces of an empty canopy or the area of the top surface only for a fully blocked canopy. Values of A_s for intermediate solidities may be linearly interpolated between these two extremes.

References

1. COOK, N. Simulation techniques for short test-section wind tunnels: roughness, barrier and mixing-device methods. Wind Tunnel Modelling for Civil Engineering Applications, pp. 126–136. London; Cambridge University Press, 1982.
2. COOK, N and MAYNE, J R. Design methods for Class A structures. Wind Engineering in the Eighties, pp 8.1–8.29 London; Construction Industry Research and Information Assoc; 1980.
3. ROBERTSON, A P, HOXEY, R P and MORAN, P. A full-scale study of wind loads on agricultural ridged canopy roof structures and proposals for design. Journal of Wind Engineering and Industrial Aerodynamics 21 (1985), 167–205.

Other BRE Digests
119 (New edition 1984) The assessment of wind loads
283 The assessment of wind speed over topography

Printed in the UK and published by Building Research Establishment, Department of the Environment.   © Crown Copyright 1986
Price Group 3. Also available by subscription. Current prices from:
Publications Sales Office, Building Research Station, Garston, Watford, WD2 7JR (Tel 0923 674040.)
Full details of all recent issues of BRE publications are given in BRE News sent free to subscribers

(3225/86) Dd8895382 C90 11/86 H P Ltd So'ton G3371   ISBN 0 85125 187 0

Building Research Establishment Digest

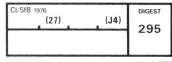

CI/SfB 1976		DIGEST
(27)	(J4)	**295**

March 1985

Stability under wind load of loose-laid external roof insulation boards

Increased costs of energy has made the addition of thermal insulation to existing buildings an economic proposition. If the building has no cavity walls or loft space, it may be necessary to apply the insulation externally; many insulation systems have been developed to fill this role. On walls and steep roofs. these systems must be bonded or fixed mechanically to the existing structure. It may be difficulty to achieve a satisfactory bond when a building is old and the finish of the external surface has deteriorated. If a fault occurs, for example if rainwater enters the structure, it may be difficult to determine whether this is a fault in the new insulation system or a pre-existing fault in the original structure. On flat or nearly flat roofs, an external insulation system may be loose-laid, relying on self-weight or ballast to provide stability against wind-induced uplift. Such systems are usually porous to rainwater and so require the existing structure to be water-tight. Because no fixings penetrate the existing structure, a loose-laid system has advantages to both client and supplier: faults can be readily ascribed to either the original structure or the insulation system and the insulation can be temporarily removed to effect repairs.

This digest gives guidance on the uplift pressures to which loose-laid external roof insulation boards can be subjected, and how they can be restrained using ballast or mechanical fixing. A worked example is given to show how to determine the spacing and adequacy of ballast.

Scope

The guidance in this digest is restricted to systems that meet the following conditions:

(a) the existing roof on which the system is laid is impermeable and is itself able to withstand the design imposed loads, namely the increased dead load due to the weight of the insulation system, the design snow load (given by Digest 290) and the design wind loads (given by Digest 119 or CP3:Chapter V:Part 2);

(b) the insulation boards are laid directly on, but not bonded to, the surface of the roof (bonded systems should be designed to withstand the full loads given by CP3:Chapter V:Pt 2 or Digest 119);

(c) the top surface of each insulation board is flush with its neighbours;

(d) the area of each individual board does not exceed 2 m²;

(e) any space remaining between the bottom of the insulation boards and the roof surface is less than 5 mm high, when averaged over the area of the board;

(f) the gap between each board and its neighbours is not less than 1 mm when averaged over the length of the joint;

(g) wind is prevented from blowing under the boards at the perimeter of the roof, or of any uninsulated area on the roof, either by means of an eaves trim or flashing, or by a parapet. The height of a parapet, measured from the top of the insulation system, should be greater than the thickness of the system and also greater than the distance between the rear face of the parapet and the edge of the first board (eg the width of any gutter).

The Code of practice on wind loads, CP3:Chapter V: Part 2, gives design values of the external pressure around buildings. If the building has a single external skin, the overall load on any face is given by the difference in external and internal pressure integrated over the face and is all borne by the single skin. However, if the building skin is composed of a number of layers of different permeability, the overall load is unequally distributed through these layers.

The case of a single permeable skin of cladding over an impermeable structural wall or roof has been studied in some detail over the last few years[1-4]; this has resulted in a complete revision of the Code of practice for slating and tiling. There are two main sources of wind loads on an element of cladding in a permeable outer skin:

(a) loads due to the pressure field around the building; these are created by the flow of wind around the building as a whole: *the global pressure field;*

(b) loads due to the flow of wind past the element: *the local velocity field.*

The effect of the global pressure field depends on the characteristics of the gaps between each cladding panel and of the void between the cladding and the im-

Prepared at Building Research Station, Garston, Watford WD2 7JR
Technical enquiries arising from this Digest should be directed to Building Research Advisory Service at the above address.

permeable structural surface. When the gaps are small and the void volume is large, the pressure in the void cannot respond to the fast changes in external pressure and tends towards the mean external pressure averaged over the response time of the void and the area of the cladding. In this case, the load on a cladding element is the difference between the local instantaneous external pressure and this void pressure. The direct load on the structural wall is the difference between the void pressure and the internal pressure of the building, plus the indirect loads transmitted by the cladding through its fixings. When the gaps are large and the void volume is small, the fluctuations of external pressure tend to equalise across each cladding element. In this case the load on the cladding is small and almost all the wind load is carried directly by the structural wall. The response of actual cladding systems lies somewhere between those two extremes.

The effect of the local velocity field depends on the external shape of the cladding. If the cladding is ribbed or has upstanding edges (like the steps formed by overlapping courses of tiles), these protrusions disturb the flow and generate local forces. In the case of roofing tiles laid on impermeable sarking or boards, these local loads are much larger than the small loads caused by the external pressure field[2] and so are the sole design criterion. If the cladding forms a smooth flush surface, the flow is not disturbed. The local velocity field contributes no normal pressure forces, but only a small component of friction parallel to the surface.

Design loading

Building Research Establishment studies on insulation systems have shown that the pressure in the void responds to changes in the external pressure in the following manner:

(a) changes in external pressure with time are followed in less than 0.1 s;

(b) changes in external pressure with position extend less than one board width.

These characteristics lead to a model for the response under wind loads in which the void pressure around the edge of each board equalises to the instantaneous local external pressure and the distribution of pressure under each board is linear, see Fig 1. This loading model has been confirmed in wind-tunnel tests[3] by the National Research Council of Canada. This model implies that:

(a) in areas of uniform external pressure or where the external pressure gradient is also linear, there will be no net load on each board;

(b) uplift of a board occurs only in repsonse to non-linear gradients of external pressure;

(c) uplift will be zero around the periphery of the board and greatest in the centre;

(d) the loading of each board is independent of its neighbours, so that adjacent boards cannot be simultaneously loaded to their maximum value.

Unfortunately, design wind loading guidance for buildings gives only the envelope of maximum absolute external pressures and suctions, and there is virtually no data on gradients of pressure. In order to use the bulk of available design guidance, it is necessary to establish a working equivalence between the net uplift force coefficient, C_F, acting on a board and the external pressure coefficient Cp_e. Full-scale measurements on an array of insulation boards on a roof at the Building Research Establishment have shown that the net uplift coefficient is a variable proportion of the measured external

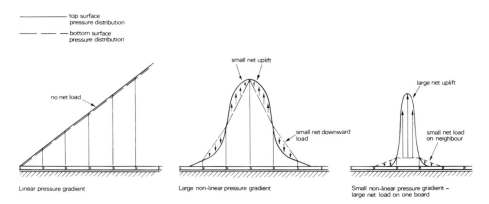

Fig 1 Effects of different pressure gradients

pressure coefficient, but that the net uplift coefficient remains near or below one third of the design external pressure coefficient values given in the wind loading code.

It is recommended that the uplift force coefficient on a single insulation board, C_f, is taken as one third of the external pressure coefficient, Cp_e, applicable to the area in which the insulation board lies.

Design restraint
Without interlock

If each board is restrained solely by deadweight, then the self-weight should exceed the design uplift. Otherwise, additional ballast should be provided until self-weight plus ballast weight exceeds the design uplift for each individual board.

With interlock

An interlock between boards is often provided in the form of a tongue-and-groove joint along the long edges of each rectangular board; this provides additional restraint normal to the long axis of the boards. Additional restraint parallel to the long axis of the boards can be provided by staggering each row of boards by half the board length, so that each board interlocks with two of its neighbours on each side.

upon in both horizontal dimensions. It is feasible to postulate a two-dimensional gust which produces a loading pattern in the shape of a wave which passes over the array. The most severe loading occurs when the wave size equals the size of the board, Fig 2. (If a wave is large enough to envelope two boards, the pressure at the joint between the boards equalises to the pressure at the centre of the wave, reducing the load on both boards.) The most severe alignment of the wave is parallel to the long axis of the boards. In this alignment, the neighbouring board along each short side is also loaded, so that the continuity of restraint is of no benefit. In this event, the restraint of an individual board comes from the neighbouring boards on either long side (half their self-weight each) and the self-weight of the board, giving a minimum net restraint of twice the self-weight. It is recognised that this interpretation of the restraint mechanism contains several major assumptions, but the possible errors tend to cancel out. The wind speed at which interlocked model insulation arrays begin to fail in wind tunnel tests[3] matches these assumptions quite well.

It is recommended that the restraint of an interlocked insulation board is taken as twice its self-weight within the array, and equal to its self-weight along any free edge (unless mechanically restrained).

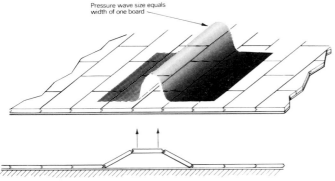

Pressure wave size equals width of one board

Fig 2 Wave size equals width of one board

If the independence of loading could be totally relied upon, each of the neighbouring boards would contribute to the restraint. In a staggered array, lifting a board would cause the two boards interlocked along each long edge to rotate about the interlock with the next row of boards (thus each contributing half their self-weight); the staggered interlock would also fully lift the next board at each short edge (thus each contributing their full self-weight). The continuity of the interlock through the staggered array would spread this effect along the direction of the long axis of the board. Owing to the rotation possible at each tongue-and-groove joint, there is not continuity of restraint in the direction normal to the long axis of each board.

However, the independence of loading cannot be relied

Ballast

If gravel is used as ballast, the designer should ensure that the critical speed at which the gravel blows off the roof is above the design wind speed, using a source of design guidance such as Ref [5].

If paving stones are used as ballast, the weight of the stones should be added to the self-weight of the boards when considering their stability. The stability of the paving stones, acting alone, should also be considered.

In general, it is recommended that the self-weight of the insulation boards should be significant to ensure stability outside the areas of high local loads. Ballast should normally be distributed evenly over the areas of high local loads. However, for structural reasons, the ballast may also be concentrated around the perimeter of the roof provided it can be demonstrated by mechanical testing that the interlock is sufficient to restrain the un-ballasted boards in the local high area farthest from the ballast.

The weight of any ballast must be included with the weight of the insulation system in the design imposed load for the roof.

Mechanical restraint

An alternative to perimeter ballast is some form of mechanical restraint. This is most useful when there is no parapet and some form of eaves trim is necessary to prevent wind from blowing under the perimeter boards. In this case it should be proved by mechanical testing that the eaves restraint and the interlock between boards is sufficient to restrain the un-ballasted boards in the local high load area.

References and further reading
1 KRAMER C, GERHARDT H J and KUSTER H-W. On the windload mechanism of roofing elements. *Journal of Wind Engineering and Industrial Aerodynamics*, 4 (1979), 415-427.
2 HAZELWOOD R A. The interaction of the two principal wind forces on roof tiles. *Journal of Wind Engineering and Industrial Aerodynamics*, 8 (1981), 39-48.
3 KIND R J and WARDLAW R L. Failure mechanisms for loose-laid roof insulation systems. *Journal of Wind Engineering and Industrial Aerodynamics*, 9 (1982) 325-341.
4 GERHARDT H J and KRAMER C. Wind loads on wind-permeable building facades. Proceedings of the 5th Colloquim on Industrial Aerodynamics, Aachen, June 1982. Aachen, Fachhohschule Aachen, 1982.
5 KIND R J and WARDLAW R L. Design of rooftops against gravel blow-off. NRC Report No 15544. Ottawa, National Research Council of Canada, 1976.

British Standards Institution
CP3:Chapter V:Part 2:1972. Wind loads
BS 5534: Code of practice for slating and tiling.
Part 1: 1978 Design

Other BRE Digests
119 Assessment of wind loads
283 The assessment of wind speed over topography
284 Wind loads on canopy roofs
290 Loads on roofs from snow drifting against vertical obstructions and in valleys

Example calculation
The following is an example design calculation for a flat-roofed industrial building located in the suburbs of Manchester.

Building dimensions:
length $l = 60$ m, width $w = 30$ m, height $h = 20$ m; roof pitch less than 5°.
From CP3:Ch V:Pt 2:
Basic (map) wind speed; $V = 45$ m/s
Topography factor: $S_1 = 1.0$
Ground roughness, building size and height above ground factor: $S_2 = 0.95$ (Class A for cladding)
Statistical factor: $S_3 = 1.0$
giving:
Design wind speed:
$V_s = 45 \times 1.0 \times 0.95 \times 1.0 = 42.8$ m/s
Design dynamic pressure: $q = 0.613\ Vs^2 = 1123$ Pa
General external pressure coefficient: $Cp_e = -0.8$
Local external pressure coefficient: $Cp_e = -2.0$
Width of local area around perimeter: $y = 0.15w = 4.5$ m
giving:
Uplift on boards in general area: $p = q\ Cp_e/3 = 299$ Pa
Uplift on boards in local area: $p = q\ Cp_e/3 = 749$ Pa
From system data:
Self-weight of boards = 176 Pa; size 0.5 m by 1.0 m with tongue-and-groove interlock, edge restraint by ballast.
Self-weight of paving stones = 1000 Pa; size 0.5 m square.
giving:
Restraint in un-ballasted area = (self-weight) $\times 2 = 352$ Pa
Restraint in ballasted area = (self-weight + ballast) $\times 2$, thus ballast required in local area = 749/2 − 176 = 199 Pa.
Restraint at ballasted edge = self-weight + ballast, thus ballast required at edge = 749 − 176 = 573 Pa.

Note: loads expressed in Pascals (1 Pa = 1 N/m²) indicate equivalent u.d.l. or pressure.

Option 1 Distribute paving stones over 4.5 m wide strip around perimeter corresponding to local high load area.
Problem: determine spacing of paving stones and whether general area is stable.

(1) Restraint in un-ballasted area (352 Pa) exceeds general uplift (299 Pa), therefore general area further than 4.5 m from edge is stable.

(2) Single paving stone weighs $1000 \times 0.5 \times 0.5 = 250$ N.

(3) Additional ballast required at edge of boards is equivalent to 573 Pa. Row of paving stones (0.5 m wide) along edge spaced at 800 mm centres (300 mm gap between stones) is equivalent to $250/(0.8 \times 0.5) = 625$ Pa, and is sufficient to ensure that edge is stable.

(4) Additional ballast required in remaining 4 m-wide strip of local high load area is equivalent to 199 Pa. Paving stones spaced at 1 m centres in both horizontal directions (500 mm gap between stones) is equivalent to 250 Pa, and is sufficient to ensure the local high load area is stable.

Option 2 Single row of paving stones around perimeter.
Problem: determine if this is adequate.

(1) Restraint in un-ballasted area (352 Pa) exceeds general uplift (299 Pa), therefore general area further than 4.5 m from edge is stable.

(2) Ballast at edge (1000 Pa) exceeds required value (573 Pa), therefore perimeter boards are stable.

(3) Remaining 4 m strip of local area requires 199 Pa ballast for dead-weight stability:
(i) Demonstrate that the interlock between boards in this area can withstand an uplift of 749 Pa by applying $749 \times 0.5 \times 1.0 = 375$ N upwards force to a single board 4 m from edge, or
(ii) apply additional ballast as in Option 1.

Printed in the UK and published by Building Research Establishment, Department of the Environment.   *Crown copyright 1985*
Price Group 3. Also available by subscription. Current prices from:
Publications Sales Office, Building Research Station, Garston, Watford, WD2 7JR (Tel 0923 674040).
Full details of all recent issues of BRE publications are given in *BRE News* sent free to subscribers.

Dd8839084 2/85 10170 (516)   ISBN 0 85125 091 2

Building Research Establishment Digest

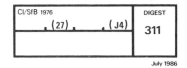

CI/SfB 1976		DIGEST
, (27) ,	, (J4)	**311**

July 1986

Wind scour of gravel ballast on roofs

The wind speed required to blow gravel off a flat roof depends on three primary factors: stone size, aerodynamic force exerted by the wind on the stones and the height or distance that the stones must travel to leave the roof top. The aerodynamic force is not simply related to nominal wind speed because it depends on the detailed structure of the airflow over the stones; this in turn depends on the nature of the terrain upwind of the building, on the orientation and geometry of the building, and on the effects of nearby buildings.

This digest gives a method of estimating the threshold wind speeds for scour; it is based on published information from wind-tunnel tests conducted for this specific purpose. It should be used in conjunction with Digest 119; it should also be used with Digest 283 if topography is significant, and with Digest 295 if paving stones are used on the roof.

A layer of gravel or crushed stone is often laid loose on flat roofs as a decorative finish, to protect bitumen-felt from ultra violet degradation, as ballast to restrain un-bonded insulation elements or impermeable membranes in inverted roof systems, or other practical and aesthetic reasons. Experience has shown that loose gravel will move under the scouring action of the wind in storms and may be blown off the roof, causing damage to glazing[1] or injury to passers-by. When the gravel is ballast essential to the stability of the roof cladding systems, wind scour may lead to structural damage. This digest can be used to predict the threshold wind speeds for scouring of loose-laid gravel or crushed stone on flat-roofed buildings with or without parapets. The objective is to ensure that these threshold wind speeds are below the design wind speed for the building. When the method indicates that gravel scour will occur, the size of gravel may be increased or gravel in the vulnerable areas of the roof may be replaced by paving stones.

Photo by courtesy of SECO, Brussels

Prepared at Building Research Station, Garston, Watford WD2 7JR.
Technical enquiries arising from this Digest should be directed to Building Research Advisory Service at the above address.

Basis of method

The method, adapted from References 2 to 5, predicts the critical or threshold 3-s gust wind speed for scour, V_c, using the formula:

$$V_c = V_{ref} \times F_S \times F_P \times F_\alpha \qquad \ldots (1)$$

where:

V_{ref} is a reference wind speed for scour;

F_S is a factor to adjust for gravel size;

F_P is a factor for parapet size, building shape and size of the area vulnerable to scour;

F_α is a factor for wind direction (the use of F_α is optional: direction α is measured from normal to the building face; $F_\alpha = 1$ at $\alpha = 45°$ is the most onerous case).

Figure 1 gives the principal dimensions of the roof. The L-shaped shaded area is the vulnerable area in which scour may occur.

The shape of the building is classifed as follows:

if $2.5\ (H + 3p) \leq W$ the building is *Low-rise*

if $H \geq (W + L)$ and $L < 1.5\ W$ the building is a *Tower*;

if $H \geq (W + L)$ and $L > 1.5\ W$ the building is a *Slab*;

otherwise the building is *Intermediate*.

Three critical wind speeds of increasing severity may be estimated:

V_{c1}: the 3-s gust speed at which gravel is first moved;

V_{c2}: the 3-s gust speed at which scouring will clear gravel from the vulnerable area; and

V_{c3}: the 3-s gust speed at which gravel is blown off the roof (over the upwind edge).

A key to the corresponding reference wind speed and factors is given in Table 1.

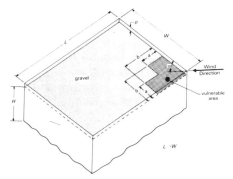

Fig 1 Principal dimensions

For Intermediate-shaped buildings, the value of F_P should be interpolated between that for Low-rise and that for Tower or Slab (or the larger value should be taken). Figures 3, 4, and 5 give more than one curve on each graph: the heavy line marked $a = b = 0$ represents the value of the factor for no scour, the light lines represent scour within the vulnerable area defined by the corresponding values of a and b. (Note the different horizontal scales: p/H in Fig 3 and p/W in Figs 4 and 5).

Site design wind speed

The site design 3-s gust wind speed is predicted by using Digest 119 and, if topography is significant at the site, Digest 283. The site design wind speed, V_s, is the 3-s gust speed at the height of the roof, H, given by:

$$V_s = V \times S_1 \times S_2 \times S_3 \times S_4 \qquad \ldots (2)$$

using the data for V, S_2 and S_3 in Digest 119 and the data for S_1 in Digest 283. S_1 is required only when topography is significant at the site. Select S_2 for the height of the roof and terrain roughness category (from the '3-s gust' columns only of Table 2 in Digest 119). Choose the value of S_3 for a 'building life' appropriate to the consequences of gravel removal. If the gravel is ballast essential for the stability of the roof or to retain insulation panels (see Digest 295), the 'building life' should be 50 years. If the gravel is merely decorative, a shorter 'life' may be acceptable. Use of the direction factor S_4 is optional: direction factor S_4 is measured relative to North, so correspondence with F_α depends on building orientation; $S_4 = 1$ is the most onerous case.

Table 1 Reference wind speeds and factors

	V_{ref} m/s			F_S	F_P			F_α
	Low-rise	Tower	Slab		Low-rise	Tower	Slab	
V_{c1}	28	28	24	Fig 2	Fig 3a	Fig 4a	Fig 5a	Fig 6
V_{c2}	35	34	30	Fig 2	Fig 3a	Fig 4a	Fig 5a	Fig 6
V_{c3}	38	36	30	Fig 2	Fig 3b	Fig 4b	Fig 5b	Fig 6
						Intermediate		

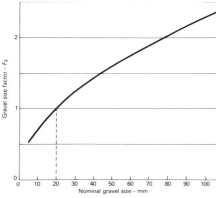

The gravel size factor accounts for the effects of using a gravel size different from 20mm, the reference size; graph applies for V_{C1}, V_{C2} and V_{C3} only.

Fig 2 Gravel size factor

Prediction

1 Check for scour

Calculate the critical 3-s gust wind speeds for scour, V_{C1}, V_{C2} and V_{C3} from Equation 1 and the data of this digest. Calculate the design 3-s gust wind speed, V_s, from Equation 2 and the data of Digests 119 and 283. (Check the most onerous case: $F_\alpha = 1$, $S_4 = 1$, before making detailed directional assessment). If $V_s < V_{C1}$, the design is satisfactory. If $V_{C1} < V_s < V_{C2}$, minor scour may occur which would require maintenance. If $V_s > V_{C2}$, serious scour is likely occur and the design is not satisfactory. If $V_s > V_{C3}$, gravel is likely to be blown from the roof, endangering passers-by and the glazing of neighbouring buildings.

2 Frequency of scour

The frequency or risk of occurrence is given by the 'building life factor', S_3, of Digest 119. Solve Equation 2 for the value of S_3 by substituting the critical wind speed, V_c, for the design wind speed, V_s, . Thus:

$$S_3 = V_c/(V \times S_1 \times S_2 \times S_4) \qquad \ldots (3)$$

Look up the 'design life' corresponding to this value of S_3 on the solid line (marked 'Probability level 0.63') on Fig 6 of Digest 119. Scour is expected to occur, on average, at least once in this period. (Values of $S_3 < 0.77$ indicate scour is expected more frequently than every two years).

3 Gravel size for no scour

The value of gravel size factor, F_s, to ensure that no scour occurs is obtained by substituting the design wind speed, V_s, for the first critical wind speed, V_{C1}, in Equation 1 and solving for the value of F_s. Thus:

$$F_s = V_s/(V_{ref} \times F_P \times F_\alpha) \qquad \ldots (4)$$

Look up the corresponding gravel size in Fig 2.

4 Size of scoured area

If a reasonable gravel size is not obtained by the previous procedure, the gravel in the vulnerable area can be replaced by paving stones. If so, it will be necessary to know the size of the vulnerable area; this will be largest at $\alpha = 45°$ where $F_\alpha = 1$. Substitute the design wind speed, V_s, for the first critical wind speed, V_{C1}, in Equation 1 and solve for the value of F_P using $F_\alpha = 1$. Thus:

$$F_P = V_s/(V_{ref} \times F_s) \qquad \ldots (5)$$

Look up the corresponding position of F_P on Fig 3a, 4a or 5a, as appropriate. This should lie above the heavy line marked 'a = b = 0'. Interpolate between the lighter lines to give the values of a and b corresponding to the size of the vulnerable area. In any range of wind direction, the vulnerable area is always in the upwind corner. The stability of paving stones should be checked using Digest 295.

5 Parapet height for no scour

Alternatively, the parapet height could be increased until no scour is predicted, but this is unlikely to be an economic solution. Calculate F_P by Equation 5 above and look up the required value of parapet height on the line 'a = b = 0' of Figs 3a, 4a or 5a as appropriate.

Fig 3 Parapet height/paving stone array factor for low-rise buildings

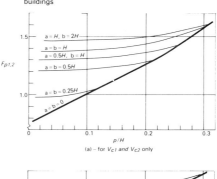

(a) – for V_{C1} and V_{C2} only

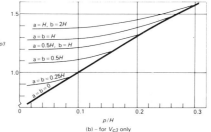

(b) – for V_{C3} only

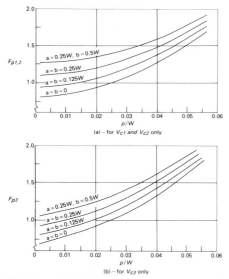

(a) – for V_{C1} and V_{C2} only

(b) – for V_{C3} only

Fig 4 Parapet height/paving stone array factor for tower buildings

Fig 5 Parapet height/paving stone array factor for slab buildings

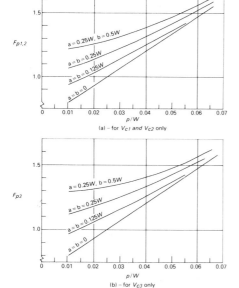

(a) – for V_{C1} and V_{C2} only

(b) – for V_{C3} only

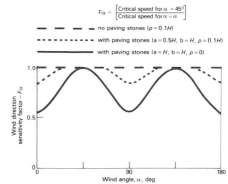

$$F_\alpha = \left[\frac{\text{Critical speed for } \alpha = 45^\circ}{\text{Critical speed for } \alpha = \alpha} \right]$$

– – – – no paving stones ($p = 0.1H$)

········ with paving stones ($a = 0.5H$, $b = H$, $p = 0.1H$)

──── with paving stones ($a = H$, $b = H$, $p = 0$)

Fig 6 Wind direction sensitivity factor

Example

Design a roof top so that gravel will not be blown off by winds with a 30-year return period.

Building dimensions: $H = 4.5$ m, $W = 23$m, $L = 30$ m.

Calculate 3-s gust speed from Digest 119 and, if necessary, Digest 283. For this calculation, assume this to be 25 m/s.

Provisionally select a gravel size of 20 mm, a 75 mm parapet height and no paving stones.

From Fig 2, $F_s = 1.0$

2.5 $(H + 3p) < W$, so building is low-rise; from Fig 3(a), $F_{P1,2} = 0.8$ ($p/H = 0.017$) and from Fig 3(b) $F_{P3} = 0.75$.
$V_{C1} = 28 \times 10 \times 0.8 \quad = 22$ m/s
$V_{C2} = 35 \times 1.0 \times 0.8 \quad = 28$ m/s
$V_{C3} = 38 \times 1.0 \times 0.75 = 29$ m/s

The design will therefore be acceptable because V_{C3} is greater than the design wind speed (25 m/s). Since V_{C2} is also greater than 25 m/s, winds with a 30-year return period should not cause significant scouring.

References
1 BEASON W L et al. Recent window glass breakage in Houston. Proc. 5th U.S. National Conference on Wind Engineering. Lubbock, Texas Tech University, November 1985.
2 KIND R J and WARDLAW R L. Design of roof tops against gravel blow-off. NRC Report No. 15544. Ottawa, National Research Council of Canada, 1976.
3 KIND R J. Estimation of critical wind speeds for scouring of gravel or crushed stone on rooftops. NRC, NAE LTR-LA-142, January 1974. Ottawa, National Research Council of Canada, 1974.
4 KIND R J. Wind tunnel tests on some building models to measure wind speeds at which gravel is blown off rooftops. NRC, NAE LTR-LA-162, June 1974. Ottawa, National Research Council of Canada, 1974.
5 KIND R J. Further wind tunnel tests on building models to measure wind speeds at which gravel is blown off rooftops. NRC, NAE LTR-LA-189, 1976. Ottawa, National Research Council of Canada, 1976.

Other BRE Digests
119   The assessment of wind loads
283   The assessment of wind speed over topography
284   Wind loads on canopy roofs
295   Stability under wind load of loose-laid external roof insulation boards

Appendix L   Mean overall loading coefficients for cuboidal buildings

L.1 Scope

These data are taken directly from the study of Akins and Peterka [277] which gives a self-consistent set of coefficients referenced against the average wind speed over the height of the building, as defined below. The data are valid for tall cuboidal buildings in the range of proportions:

$$0.5 \leqslant H/L \leqslant 4 \qquad 0.5 \leqslant H/W \leqslant 4 \qquad 1 \leqslant L/W \leqslant 4$$

where the dimensions H, L and W are defined below.

The data may be used with the peak gust dynamic pressure, $\hat{q}$, to estimate overall peak base shear forces and moments through the quasi-steady model. They may also be used with the complimentary quasi-steady models for dynamic response implemented in Part 3 of the Guide. (Data for the overall vertical force, effectively the load on the roof, is ony included for completeness, since the flat roof data in Chapter 20 are more comprehensive.)

These data are currently unique in that a standard deviation is given for each coefficient which gives the variability of the value and so describes the reliablity of the result. Normally, with standard values of safety factors, the value of the loading coefficient is used without modification. However, without these safety factors, the standard deviation can be used to deduce the required partial factor or the reliability of this part of the design assessment. The other elements of the design assesment, the reference dynamic pressure, the material strengths and the calculation process itself, also have inherent variability and corresponding standard deviations which must be included in any reliability assessment.

L.2 Definitions

Building dimensions are defined as:

H the height;
L the longer horizontal dimension;
W the shorter horizontal dimension.

The axes convention is defined in Figures 12.7 and 12.9, with the origin at the centre of the base, the x-axis normal to the longer face, the y-axis normal to the shorter face and the z-axis pointing vertically upwards. The base shear forces, F_x,

F_y and vertical force F_z, act along the respective axes (Figure 12.7) and the base moments, M_x, M_y and M_z clockwise around the respective axes (Figure 12.9) in the standard convention.

The reference dynamic pressure, q_{ref}, whether a peak or mean value, is formed from the average wind speed over the height of the building, V_{ave}. Thus:

$$V_{ave} = \frac{1}{H} \int_0^H V\{z\}\, dz \tag{L.1}$$

$$q_{ref} = \tfrac{1}{2}\, \rho\, V_{ave}^2 \tag{L.2}$$

The overall force coefficients are defined as:

$$\overline{C}_{F_x} = \overline{F}_x / (\overline{q}_{ref}\, L\, H) \tag{L.3}$$

$$\overline{C}_{F_y} = \overline{F}_y / (\overline{q}_{ref}\, W\, H) \tag{L.4}$$

$$\overline{C}_{F_z} = \overline{F}_z / (\overline{q}_{ref}\, L\, W) \tag{L.5}$$

The overall moment coefficients are defined as:

$$\overline{C}_{M_x} = \overline{M}_x / (\overline{q}_{ref}\, W\, H^2) \tag{L.6}$$

$$\overline{C}_{M_y} = \overline{M}_y / (\overline{q}_{ref}\, L\, H^2) \tag{L.7}$$

$$\overline{C}_{M_z} = \overline{M}_z / (\overline{q}_{ref}\, L\, W\, H) \tag{L.8}$$

Note that these definitions correspond to the Guide convention of loaded area as the reference area (§12.3.6).

The x-axis moment is caused by the y-axis force acting on the moment arm through the centre of force. The y-axis moment is similarly caused by the x-axis force. Instead of the moment coefficients, the moments can be defined in terms of the forces and the moment arm as shown in Figure 12.9. Hence we may have:

z_{F_x} the height of the centre of x-force producing **positive** y-moment; and

z_{F_y} the height of the centre of y-force producing **negative** x-moment.

L.3 Loading coefficients

Values of the mean force and moment coefficients are given in Table L.1 below. These are empirical values that have not been smoothed or otherwise manipulated except for coordinate transformation into the standard Guide convention defined above.

553

Table L.1 Mean overall loading coefficients for cuboidal buildings

Ref $\hat{q}$ from average wind speed over height of building

Wind angle from normal to longer face, θ (degrees)

Shape	L/W	0	20	40	50	55	60	65	70	75	80	90
$\bar{C}_{F_x}$	1	+1.47 ±0.08	+1.54 ±0.08	+1.40 ±0.03	+1.20 ±0.11	+1.17 ±0.12	+0.98 ±0.18	+0.86 ±0.18	+0.59 ±0.25	+0.47 ±0.22	+0.15 ±0.28	0.00 ±0.00
	2	+1.78 ±0.24	+1.76 ±0.14	+1.64 ±0.08	+1.50 ±0.11	+1.38 ±0.12	+1.26 ±0.11	+1.11 ±0.11	+0.94 ±0.10	+0.72 ±0.09	+0.50 ±0.11	0.00 ±0.00
	4	+1.90 ±0.19	+1.86 ±0.18	+1.66 ±0.18	+1.53 ±0.16	+1.43 ±0.16	+1.31 ±0.16	+1.19 ±0.14	+1.03 ±0.14	+0.76 ±0.28	+0.64 ±0.13	0.00 ±0.00
$\bar{C}_{F_y}$	1	0.00 ±0.00	−0.65 ±0.21	−1.28 ±0.6	−1.36 ±0.08	−1.44 ±0.09	−1.49 ±0.09	−1.51 ±0.08	−1.52 ±0.09	−1.53 ±0.08	−1.53 ±0.07	−1.50 ±0.07
	2	0.00 ±0.00	−0.16 ±0.36	−1.02 ±0.06	−1.31 ±0.09	−1.43 ±0.10	−1.51 ±0.11	−1.56 ±0.11	−1.58 ±0.11	−1.60 ±0.12	−1.51 ±0.10	−1.41 ±0.06
	4	0.00 ±0.00	+0.29 ±0.18	−0.74 ±0.06	−1.14 ±0.09	−1.29 ±0.11	−1.37 ±0.10	−1.41 ±0.10	−1.42 ±0.11	−1.39 ±0.10	−1.35 ±0.09	−1.34 ±0.07
$\bar{C}_{F_z}$	1	−1.16 ±0.12	−1.06 ±0.09	−0.88 ±0.05	−0.83 ±0.08	−0.88 ±0.06	−0.89 ±0.07	−0.94 ±0.09	−0.99 ±0.09	−1.05 ±0.10	−1.10 ±0.14	−1.18 ±0.13
	2	−1.35 ±0.08	−1.29 ±0.06	−1.12 ±0.11	−1.01 ±0.12	−0.98 ±0.11	−0.92 ±0.12	−0.89 ±0.11	−0.84 ±0.12	−0.84 ±0.13	−0.85 ±0.11	−0.79 ±0.14
	4	−1.28 ±0.36	−1.26 ±0.31	−1.30 ±0.18	−1.26 ±0.12	−1.23 ±0.13	−1.15 ±0.14	−1.02 ±0.13	−0.87 ±0.13	−0.75 ±0.11	−0.63 ±0.10	−0.47 ±0.06
$\bar{C}_{M_x}$	1	0.00 ±0.00	+0.34 ±0.12	+0.67 ±0.01	+0.72 ±0.05	+0.76 ±0.06	+0.79 ±0.06	+0.80 ±0.05	+0.83 ±0.05	+0.84 ±0.04	+0.84 ±0.04	+0.86 ±0.02
	2	0.00 ±0.00	+0.05 ±0.16	+0.55 ±0.07	+0.73 ±0.06	+0.80 ±0.05	+0.86 ±0.07	+0.92 ±0.07	+0.96 ±0.10	+1.03 ±0.13	+0.99 ±0.14	+0.93 ±0.18
	4	0.00 ±0.00	−0.08 ±0.26	+0.45 ±0.07	+0.69 ±0.10	+0.76 ±0.10	+0.83 ±0.10	+0.87 ±0.14	+0.89 ±0.19	+0.89 ±0.22	+0.83 ±0.14	+0.88 ±0.10

Table L.1 *continued*

Shape		Ref $\hat{q}$ from average wind speed over height of building										
		Wind angle from normal to longer face, θ (degrees)										
	L/W	0	20	40	50	55	60	65	70	75	80	90
$\bar{C}_{M_y}$:	1	+0.85 ±0.03	+0.87 ±0.06	+0.77 ±0.01	+0.63 ±0.06	+0.60 ±0.06	+0.49 ±0.09	+0.43 ±0.08	+0.28 ±0.13	+0.21 ±0.11	+0.06 ±0.15	0.00 ±0.00
	2	+1.02 ±0.08	+0.98 ±0.05	+0.89 ±0.05	+0.80 ±0.06	+0.73 ±0.07	+0.65 ±0.06	+0.55 ±0.06	+0.46 ±0.06	+0.34 ±0.05	+0.24 ±0.05	0.00 ±0.00
	4	+1.02 ±0.10	+1.05 ±0.18	+0.88 ±0.11	+0.84 ±0.08	+0.76 ±0.12	+0.68 ±0.10	+0.61 ±0.08	+0.53 ±0.07	+0.46 ±0.08	+0.32 ±0.08	0.00 ±0.00
$\bar{C}_{M_z}$:	1	0.00 ±0.00	+0.11 ±0.02	+0.02 ±0.00	−0.02 ±0.00	−0.05 ±0.00	−0.07 ±0.01	−0.09 ±0.02	−0.11 ±0.02	−0.12 ±0.03	−0.11 ±0.03	0.00 ±0.00
	2	0.00 ±0.00	+0.17 ±0.03	+0.18 ±0.03	+0.16 ±0.05	+0.12 ±0.05	+0.10 ±0.06	+0.04 ±0.06	+0.01 ±0.09	−0.04 ±0.10	−0.07 ±0.09	0.00 ±0.00
	4	0.00 ±0.00	+0.28 ±0.07	+0.41 ±0.13	+0.41 ±0.18	+0.38 ±0.20	+0.35 ±0.20	+0.29 ±0.16	+0.26 ±0.14	+0.26 ±0.11	+0.22 ±0.08	0.00 ±0.00
z_{F_x}/H:	1	0.58	0.57	0.55	0.53	0.52	0.50	0.50	0.48	0.45	0.37	
	2	0.57	0.56	0.54	0.53	0.53	0.51	0.50	0.49	0.47	0.48	
	4	0.54	0.56	0.53	0.55	0.53	0.52	0.51	0.51	0.60	0.50	
z_{F_y}/H:	1		0.52	0.52	0.53	0.53	0.53	0.53	0.55	0.55	0.55	0.57
	2		0.32	0.53	0.55	0.56	0.57	0.59	0.60	0.64	0.65	0.66
	4		0.27	0.60	0.61	0.59	0.61	0.62	0.63	0.64	0.62	0.65

Appendix M A semi-empirical model for pressures on flat roofs

M.1 Introduction

Maruta's model for wind speeds around the base of tall buildings described in §19.2.3.2.2 is an empirical 'mapping' method, in that it allows regions to be mapped as contours of given wind speeds. While working at BRE, Maruta applied the same technique to pressures on the flat roofs of the wedge-shaped models used to deduce the effect of corner angle (§17.3.2.4) and produced a prototype empirical model which allowed the position of the contour of a given pressure to be mapped onto the surface of the roof. This is useful for compiler of design guidance who wishes to draft a design chart, but less useful to the building designer who wishes to reverse the process and obtain pressures in terms of position and thus requires a 'predictive' method.

The ideal situation would be to be able to frame an equation for pressure coefficient, c_p, in terms of position from the corner, thus:

$$c_p = \mathbf{f}_1\{r, a\} \tag{M.1}$$

where r is the radial distance from the corner and a is the angle from the eave, to give a 'predictive' model, but which could also be solved for either positional coordinate:

$$r = \mathbf{f}_2\{c_p, a\} \tag{M.2}$$

$$a = \mathbf{f}_3\{c_p, r\} \tag{M.3}$$

There is a reason for preferring polar coordinates instead of the more usual cartesian coordinates, x and y. In the conical 'growth' region the range, r, becomes the coordinate along the vortex describing its growth and the angular position, a, becomes the coordinate across the vortex describing its size.

It would be even more convenient if the two components of the unknown functions $\mathbf{f}$ were independent, i.e. if:

$$\mathbf{f}_1\{r, a\} = \mathbf{g}_1\{r\}\, \mathbf{h}_1\{a\} \tag{M.4}$$

and could be derived from theory to make an exact analytical model, or at least from theoretical considerations to make a semi-empirical model (§1.2).

M.2 Candidate predictive functions

M.2.1 Angular position function

A good approximation to the ideal can be obtained for the function $\mathbf{h}_1\{a\}$, since it is expected to take a form similar to the Rankine vortex (§2.2.8.3) for which the limiting forms for large and small radii (here angular radius a) are given by Eqns 2.17 and 2.20. A form of equation is needed that is transitional between these limits, and a good candidate is the form:

$$\mathbf{h}_1\{a\} = 1 / (1 + A^2) \tag{M.5}$$

where A is the non-dimensional position through the vortex. For conical vortices, i.e. in the 'growth' region, this is the non-dimensional angle defined by:

$$A = (a - a_o) / a_c \tag{M.6}$$

where a_o represents the angular position of the centre of the vortex and a_c represents the angular radius of its core (§2.2.8.3). Similarly, in cylindrical vortices, i.e. in the 'mature' region, this is the non-dimensional distance normal to the eave defined by:

$$A = (y - y_o) / y_c \tag{M.7}$$

where y_o represents the position of the centre of the vortex and y_c represents the radius of its core.

Equation M.5 satisfies the three necessary criteria:

1 to decay towards zero as $1/A^2$ for large A, since $1 + A^2 \rightarrow A^2$,
2 to decay from unity as A^2 for small A, since the first two terms of the binomial expansion of $(1 + A^2)^{-1} = 1 - A^2$ and all higher powers of A become negligible as $A \rightarrow 0$, and
3 to take the value 0.5 at $A = 1$, corresponding to the core radius of the Rankine vortex.

Equation M.5 should not be taken as exact, since the vortex will be distorted by the flow field around the building, the separating shear layer and the solid surface of the roof. In any case, the vortex is above the roof surface and the pressure distribution over the roof will not be the pressure distribution through the middle of an ideal Rankine vortex.

M.2.2 Range function

The function $g_1\{r\}$ can only be determined empirically. In the conical 'growth' region the suction increases towards the corner and plotting c_p against $\log(r)$ for any given angular position produces a straight line and forms a family of parallel straight lines at other angular positions. This suggest that the form:

$$g_1\{r\} = I - S \log(r) \quad \text{for } x \leqslant x_{max} \tag{M.8}$$

where S is the slope common to the family of lines, I is the individual intercept value at $r = 1$ for the given angular position and x_{max} is the position along the eave of the boundary between the 'growth' and 'mature' regions.

In the cylindrical 'mature' region the suctions remain constant parallel to the eave, giving $g_1\{x\}$ = constant. As the conical 'growth' region and cylindrical 'mature' region must merge at their common boundary:

$$g_1\{x\} = I - S\log(x_{max}) \quad \text{for } x > x_{max} \tag{M.9}$$

M.3 Prototype model

M.3.1 Predictive form

Combining the candidate functions $\mathbf{g}_1$ and $\mathbf{h}_1$ through Eqn M.4 on the assumption of independence gives an equation of the form:

$$c_p = c_{p_0} + (c_{p_1} - c_{p_0})(1 - S\log[R_{eff}]) / (1 + A_{eff}^2) \tag{M.10}$$

where R_{eff} is the effective non-dimensional range (position along the eave in terms of the scaling length b) given by:

$$\begin{aligned} R_{eff} = R &= (x^2 + y^2)^{1/2} / b & \text{for } R \leqslant R_{max} \\ &= R_{max} = (x_{max}^2 + y^2)^{1/2} / b & \text{for } R > R_{max} \end{aligned} \tag{M.11}$$

A_{eff} is the effective non-dimensional angular position (position normal to the eave) given by Eqn M.6 with the effective angle, a_{eff}, given by:

$$\begin{aligned} a_{eff} &= \arctan(y / x) & \text{for } x/b \leqslant R_{max} \\ &= \arctan(y / x_{max}) & \text{for } x/b > R_{max} \end{aligned} \tag{M.12}$$

These equations are framed in polar coordinates for the conical 'growth' region. The effective range is restricted to $0 \leqslant R_{eff} \leqslant R_{max}$. The effective angular position, A_{eff}, in the 'mature' region is the value of the actual position, A, projected onto the line $x = R_{max}$, parallel to the eave. Hence the model defined by Eqns M.10 to M.12 applies to both 'growth' and 'mature' regions but, strictly, not to the 'decay' region. However, as the presures of the 'decay' region are less onerous than the 'mature' region values, extending the 'mature' region to the downwind corner produces a safe model.

The other parameters in Eqns M.10 to M.12 are model parameters for which values must be found by fitting measured data. These are

R_{max} — the position of the 'growth/mature' region boundary;

c_{p_0} — the value of pressure coefficient at $R_{eff} = 1$ and $A_{eff} = \infty$, representing a datum maximum (most positive) value;

c_{p_1} — the value of pressure coefficient at $R_{eff} = 1$ and $A_{eff} = 0$, representing a datum minimum (most negative) value;

Note: neither c_{p_0} or c_{p_1} are 'real' values in that they do not correspond to actual values on the roof, since $R_{eff} < 1$ because real values of $x_{max} < 1$ and the angular position is limited by the wind angle to $a \leqslant 90° - \theta$.

S — the slope of the logarithmic decay with position from the corner in the 'growth' region (in Eqn M.8);

a_o — the angular position of the centre of the conical vortex in the 'growth' region;

a_c — the angular radius of the core of the conical vortex in the 'growth' region.

M.3.2 Mapping form

The form of the predictive equation, Eqn M.10, for pressure coefficient in terms of position is soluble for either positional coordinate in terms of pressure coefficient and the other coordinate. Thus:

$$\log[R_{\text{eff}}] = 1 - (c_p - c_{p_0})\,(1 + A_{\text{eff}}^2)\,/\,(c_{p_1} - c_{p_0}) \tag{M.13}$$

$$A_{\text{eff}} = ([c_{p_0} + (c_{p_1} - c_{p_0})\,(1 - S\log[R_{\text{eff}}])] - 1)^{1/2} \tag{M.14}$$

M.4 Application and accuracy of prototype model

Table M.1 shows the values obtained for the parameters of Eqns M.10–M.12 when the model was fitted measured data for $\tilde{c}_p\{t = 4\,\text{s}\}$ on the roof of a squat cuboid with $H/L = 1/3$ and $L/W = 1$, together with the rms standard error of the fit. Values are given for a range of wind angles, θ, and consistent trends are evident. The standard error is low, typically $\varepsilon\{c_p\} \simeq 0.1$, indicating that the data fit the semi-empirical model well.

Table M.1   Model parameters for $3 \times 3 \times 1$ cuboid

for $\tilde{c}_p\{t = 4\,\text{s}\}$:	$H/L = 1/3$,	$L/W = 1$,	$\hat{q}_{\text{ref}}$ at $z = H$				
Wind angle	θ	0°	15°	30°	45°	60°	75°
'Growth/mature' boundary	x_{max}/b	0.24	0.48	0.60	0.75	0.90	1.11
Maximum datum pressure	$\tilde{c}_{p_0}\{t = 4\,\text{s}\}$	0.42	0.99	1.016	1.22	0.613	0.72
Minimum datum pressure	$\tilde{c}_{p_1}\{t = 4\,\text{s}\}$	− 0.60	− 0.56	− 0.59	− 0.48	− 0.33	− 0.28
Slope of logarithmic decay	S	1.01	1.01	1.02	1.02	1.41	1.09
Vortex centre angle	a_0	17.8°	5.0°	6.8°	5.5°	20.3°	17.8°
Vortex core radius angle	a_c	41.9°	59.7°	52.5°	61.8°	68.1°	65.2°
Standard error of model	$\varepsilon\{c_p\}$	0.09	0.08	0.12	0.13	0.10	0.07

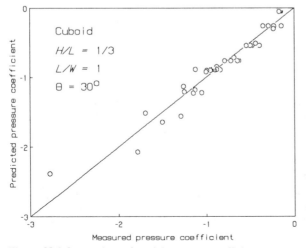

Figure M.1 Comparison of model pressure coefficients

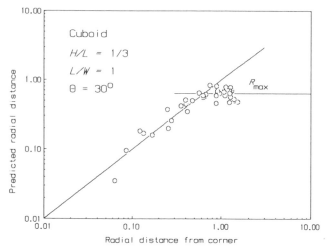

Figure M.2 Comparison of model radial distance

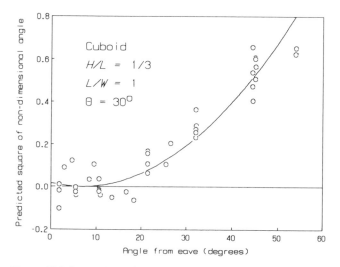

Figure M.3 Comparison of model angular position

Taking the wind angle $\theta = 30°$ as an example, Figure M.1 indicates the scatter for the predictive form of the full model in Eqn M.10 by plotting the measured data directly against the corresponding predicted values. This compares very favourably with the wind tunnel model to model comparisons in Figure 13.45 and the assessment method comparisons in Figures 15.7 and 15.8. The two component functions of the model, **g** and **h**, can be assessed separately. Figure M.2 compares the actual radial distance, R, against the model values from the mapping form of Eqn M.13, confirming the logarithmic decay in the 'growth' region and the limit at R_{max} in the 'mature' region. Similarly, Figure M.3 compares the actual effective angular position, a_{eff}, against the square of the model non-dimensional values, A_{eff}^2, showing the expected parabolic form.

M.5 Discussion

This semi-empirical model is still at an early stage of development, but the comparisons shown here and in §17.3.3.2.5 indicate that the model gives a good representation of the measured data for flat roofs. Initial studies show that the effects of roof pitch can be included and that the model will work for the eave/verge vortex pair on monopitch roofs and the upwind face of duopitch roofs and also for the ridge/verge vortex pair on the downwind face of duopitch roofs. The expectation is that other parameters, such as parapet height, curved eaves, etc. can also be accomodated by variations in value of the model parameters so that the semi-empirical model may prove to be a useful design tool.

To be able to use the model in design, the sensitivity of the model parameters to the critical flow and structural parameters must be determined over the typical ranges of value. Consistent trends with wind direction θ are evident in Table M.1. At $\theta = 0°$ the 'growth/mature' boundary is very close to the corner, so that the majority of the vortex is cylindrical as expected for the normal flow case, but as the flow becomes more skewed the conical 'growth' region extends further from the corner. The slope of the logarithmic decay, S, is consistently near 1.0, except at $\theta = 60°$ which may be experimental error.

As the semi-empirical model parameterises the problem in a logical way, study of the way the parameters change with with building proportions, Jensen number, etc. may lead to a better understanding of the physical processes involved. The semi-empirical model may therefore also prove to be a valuable research tool.

References

1 **ISO.** *Performance standards in building – Definitions and means of performance of a whole-building.* Draft International Standard ISO/DIS 7164. Geneva, International Organisation for Standardisation, 1985.

2 **Armer G S T and Mayne J R.** Modern structural design codes – the case for a more rational format. *CIB Journal of Building Research and Practice,* 1986, **14** 212–217.

3 **Armer G S T, Fewell A R and Mayne J R.** *Papers from Symposium on 'The use of partial factors in structural design'.* Garston, Building Research Establishment, 1987.

4 **BSI.** *Code of basic data for the design of buildings: Chapter v Loading, Part 2 Wind loads.* CP3 ChV Pt2 1972 (and amendments 1–3, 1986 and 4, 1988). London, British Standards Institution, 1972.

5 **Bearman P W.** Some measurements of the distortion of turbulence approaching a two-dimensional bluff body. *Journal of Fluid Mechanics,* 1972, **53** 451–467.

6 **Holmes J D.** Pressure fluctuations on a large building and along–wind structural loading. *Journal of Industrial Aerodynamics,* 1975/6, **1** 249–278.

7 **Lawson T V.** *Wind effects on buildings, Volume 1: Design applications.* London, Applied Science Publishers, 1980.

8 **ESDU.** *Wind engineering subseries, Volume 2, Mean loads on structures.* London, ESDU International.

9 **Rice S O.** Mathematical analysis of random noise, Pt III. *Bell System Technical Journal,* 1945, **19** 46–156.

10 **Davenport A G.** The application of statistical concepts to the wind loading of structures. *Proceedings of the Institution of Civil Engineers,* 1961, **19** 449–472.

11 **Davenport A G.** Note on the distribution of the largest value of a random function with application to gust loading. *Proceedings of the Institution of Civil Engineers,* 1964, **28** 187–196.

12 **Davenport A G.** Gust loading factors. *Proceedings of the American Society of Civil Engineers, Structural Division,* 1967, **93** 11–34.

13 **Dalgliesh W A.** Statistical treatment of peak gusts on cladding. *Proceedings of the American Society of Civil Engineers, Structural Division,* 1971, **97** 2173–2187.

14 **Dalgliesh W A, Templin J T and Cooper K R.** Comparisons of wind tunnel and full–scale building surface pressures with emphasis on peaks. *Wind engineering: Proceedings of the fifth international conference, Fort Collins, Colorado, USA, July 1979.* (Ed: Cermak J E), pp. 553–565. Oxford, Pergamon Press, 1980.

15 **Mayne J R and Walker G R.** *The response of glazing to wind pressure.* Current Paper CP44/76. Garston, Building Research Establishment, 1976.

16 **Eaton K J, Mayne J R and Cook N J.** Wind loads on low-rise buildings – effects of roof geometry. *Proceedings of the fourth international conference on wind effects on buildings and structures, Heathrow, 1975.* (Ed: Eaton KJ), pp. 95–110. Cambridge, Cambridge University Press, 1977.

17 **Cook N J.** Calibration of the quasi-static and peak-factor approaches to the assessment of wind loads against the method of Cook and Mayne. *Journal of Wind Engineering and Industrial Aerodynamics,* 1982, **10** 315–341.

18 **Matsui G, Suda K and Higuchi K.** Full-scale measurement of wind pressures acting on a high–rise building of rectangular plan. *Journal of Wind Engineering and Industrial Aerodynamics,* 1982, **10** 267–286.

19 **Lawson T V.** The measurement of short term average pressure in a wind tunnel investigation. *Journal of Wind Engineering and Industrial Aerodynamics,* 1975/76, **1** 233–238.

20 **Peterka J A and Cermak J E.** Wind pressures on buildings – Probability densities. *Proceedings of the American Society of Civil Engineers*, 1975, **101** 1255–1267.

21 **Mayne J R and Cook N J.** *On design procedures for wind loading.* Current Paper CP 25/78. Garston, Building Research Establishment, 1978.

22 **Holmes J D.** Non-gaussian characteristics of wind pressure fluctuations. *Journal of Wind Engineering and Industrial Aerodynamics*, 1981, **7** 103–108.

23 **Fisher R A and Tippett L H C.** Limiting forms of the frequency distribution of the largest or smallest member of a sample. *Proceedings of the Cambridge Philosophical Society*, 1928, **24** 180–190.

24 **Gumbel E J.** *Statistics of Extremes.* New York, Columbia University Press, 1958.

25 **Stathopoulos T, Davenport A G and Surry D.** The assessment of effective wind loads acting on flat roofs. *Proceedings of the third colloquium on industrial aerodynamics, Aachen, 1978.* (Eds: Kramer C and Gerhardt H-J), **1** 225–239. Aachen, Fachhochschule Aachen, 1978.

26 **Peterka J A.** Selection of local peak pressure coefficients for wind tunnel studies of buildings. *Journal of Wind Engineering and Industrial Aerodynamics*, 1983, **13** 477–488.

27 **Lieblein J.** *Efficient methods of extreme-value methodology.* Report NBSIR 74–602. Washington, National Bureau of Standards, 1974.

28 **Cook N J and Mayne J R.** Design methods for Class A structures. *Wind engineering in the eighties.* London, Construction Industry Research and Information Association, 1981.

29 **Cook N J and Mayne J R.** A novel working approach to the assessment of wind loads for equivalent static design. *Journal of Industrial Aerodynamics*, 1979, **4** 149–164.

30 **Sockel H.** Local pressure fluctuations. *Wind Engineering: Proceedings of the fifth international conference, Fort Collins, Colorado, 1979.* (Ed: Cermak J E), pp. 509–518. Oxford, Pergamon Press, 1980.

31 **Peterka J A.** Predicting peak pressures *vs* direct measurement. *Wind tunnel modeling for civil engineering applications.* (Ed: Reinhold T A), pp. 313–319. Cambridge, Cambridge University Press, 1982.

32 **Baker B.** The Forth Bridge. *Engineering*, 1884, **38** 213–215 and 223–225.

33 **Irminger J O V.** Nogle forsøg over trykforholderne paa planter og legemer paavirkede af luftstrømninger. *Ingeniøren*, 1894, **17** and **18**.

34 **Eiffel G.** Travaux scientifiques executés á la tour de trois cents mètres de 1889–1900. *Maretheux*, 1900.

35 **Dryden H L and Hill G C** Wind pressures on circular cylinders and chimneys. *Journal of Research of the National Bureau of Standards*, 1930, **5** 653–693.

36 **Bailey A.** *Wind pressures on buildings.* Selected Papers No.139. London, Institution of Civil Engineers, 1933.

37 **Rathbun J C.** Wind forces on a tall building. *Transactions of the American Society of Civil Engineers*, 1940, **105** 1–41.

38 **Davenport A G.** Perspectives on the full-scale measurement of wind effects. *Journal of Industrial Aerodynamics*, 1975, **1** 23–54.

39 Survey of full–scale investigations of wind effects on structures for symposium on full-scale measurements of wind effects on tall buildings and other structures, June 23–29, 1974. *Journal of Industrial Aerodynamics*, 1975, **1** 126–137.

40 **Irminger J O V and Nøkkentved C.** *Wind pressures on buildings. Experimental researches (first series).* Copenhagen, Danmarks Naturvidenskabelige Samfund, 1930.

41 **Flachsbart O.** Model experiments on wind loadings on lattice girder structures. Part 2: Spatial lattice girder structures. *Die Bautechnik*, 1934.

42 **Flachsbart O.** Model experiments on wind loadings on latticed girder structures. Part 1: Separate plane lattice girders. *Die Bautechnik*, 1934, **7** (10) 73–79.

43 **Flachsbart O.** Model experiments on wind loadings on latticed girder structures. Part 1: Single lattice beams in one plane. *Die Bautechnik*, 1934, **7** (9) 65–69.

44 **Flachsbart O and Winter H.** Model experiments of wind loadings on latticed girder structures. Part 2: Spatial lattice girder structures. *Stahlbau*, 1935, **8** 57–63, **9** 65–69 and **10** 73–77.

45 **Irminger J O V and Nøkkentved C.** *Wind pressures on buildings. Experimental researches (second series).* Copenhagen, Danmarks Naturvidenskabelige Samfund, 1936.

46 **Bailey A and Vincent N D G.** Wind-pressure on buildings including the effect of adjacent buildings. *Journal of the Institution of Civil Engineers*, 1943, **20** 243–275.

47 **Jensen M.** Some lessons learned in building aerodynamics research. *Proceedings of the international research seminar on wind effects on buildings and structures, Ottawa, 1967*, pp. 1–18. Toronto, University of Toronto Press, 1968.

48 **Jensen M.** The model law for phenomena in natural wind. *Ingeniøren (international edition)*, 1958, **2** 121–128.

49 **Chien N, Feng Y, Wang H J and Siao T T.** *Wind-tunnel studies of pressure distribution on elementary building forms.* Iowa City, Iowa Institute of Hydraulic Research, 1951.

50 **Pris M R.** Etudes aérodynamiques VII, Determination des pressions dues a l'action du vent sur les toitures des bâtiments rectangulaires en plan en contact avec le sol. *Annales de l'Institut Technique de Bâtiment et des Travaux Publics,* 1963, **186** 590–620.

51 **Leutheusser H J.** *The effects of wall parapets on the roof pressure coefficients of block–type and cylindrical structures.* Department of Mechanical Engineering, TP 6404. Toronto, University of Toronto, 1964.

52 **Lythe G and Surry D.** Wind loading of flat roofs with and without parapets. *Journal of Wind Engineering and Industrial Aerodynamics,* 1983, **11** 75–94.

53 **Baines W D.** Effects of velocity distribution on wind loads and flow patterns on buildings. *Wind effects on buildings and structures: proceedings of the conference held at the National Physical Laboratory, June 1963,* pp. 198–223. London, HMSO, 1965.

54 **Hellers B G and Lundgren S.** *Vindbelasting på huskroppar av allmän form – modellprov.* Byggforskningen Rapport R22:1974. Stockholm, Statens Institut för Byggnadsforskning, 1974.

55 **Vickery B J.** Fluctuating lift and drag on a long cylinder of square cross–section in a smooth and in a turbulent stream. *Journal of Fluid Mechanics,* 1966, **25** 481–494.

56 **Bearman P W.** An investigation of the forces on flat plates normal to a turbulent flow. *Journal of Fluid Mechanics,* 1971, **346** 177–198.

57 **R Pris.** Préparation des essais sur maquettes de bâtiments au laboratoire aérodynamique et applications à la vraie grandeur. *Wind effects on buildings and structures: proceedings of the conference held at the National Physical Laboratory, June 1963,* pp. 227–253. London, HMSO, 1965.

58 **Franck N.** Model law and experimental technique for determination of wind loads on buildings. *Wind effects on buildings and structures: proceedings of the conference held at the National Physical Laboratory, June 1963,* pp. 182–196. London, HMSO, 1965.

59 **Colin P E and d'Havé R.** Exécution en tunnel aérodynamique d'essais sur maquettes de bâtiments en rapport avec mesures faites sur constructions réelles. *Wind effects on buildings and structures: proceedings of the conference held at the National Physical Laboratory, June 1963,* pp. 256–281. London, HMSO, 1965.

60 **Gill G C.** The helicoid anemometer. *Atmosphere,* 1973, **11** 145–155.

61 **Gill G C.** Development and use of the Gill UVW anemometer. *Boundary-Layer Meteorology,* 1975, **8** 475–495.

62 **Hicks B B.** Propellor anemometers as sensors of atmospheric turbulence. *Boundary-Layer Meteorology,* 1972, **3** 214–228.

63 **Drinkrow R.** A solution to the paired Gill-anemometer response function. *Journal of Applied Meteorology,* 1972, **11** 76–80.

64 **Bowen A J and Teunissen H W.** *Correction factors for the directional response of propellor anemometers.* Research Report MSRB-84-2. Toronto, Atmospheric Environment Service, 1984.

65 **Dayoub A H.** A new method for correcting turbulence data caused by angle deviation from the nominal 45° X-wire probes. *Dantec Information,* 1986, **2** 12–13.

66 **Guenkel A A, Patel R P and Weber M E.** On the development of a hot-wire probe for highly turbulent and rapidly reversing flows. *Industrial and Engineering Chemistry Fundamentals,* 1971, **10** 627–629.

67 **Cook N J and Redfearn D.** Calibration and use of a hot-wire probe for highly turbulent and reversing flows. *Journal of Industrial Aerodynamics,* 1975/76, **1** 221–231.

68 **Bradbury L J S and Castro I P.** A pulsed-wire technique for turbulence measurements. *Journal of Fluid Mechanics,* 1971, **49** 657–691.

69 **Wood C J.** On the use of static tubes in architectural aerodynamics. *Journal of Industrial Aerodynamics,* 1978, **3** 374–378.

70 **Huey L J.** A yaw-insensitive static pressure probe. *Transactions of the American Society of Mechanical Engineers, Journal of Fluids Engineering,* 1978, **100** 229–231.

71 **Blackmore P A.** Pressure probes for use in turbulent three–dimensional flows. *Journal of Wind Engineering and Industrial Aerodynamics,* 1987, **25**, 207–218.

72 **Mayne J R.** A wind-pressure transducer. *Journal of Physics E: Scientific Instruments,* 1970, **3** 248–250.

73 **Lam L C H.** Investigations on the dynamic behaviour of a wind pressure measuring system for full-scale measurement. *Journal of Wind Engineering and Industrial Aerodynamics,* 1981, **7** 129–134.

74 **Waldek J L.** Effects of the reference-pressure system on the dynamic response of pressure transducers in full-scale experiments. *Journal of Wind Engineering and Industrial Aerodynamics,* 1986, **23** 37–50.

75 **Cook N J.** *Manufacture and calibration of restrictor and averaging manifolds for the measurement of fluctuating pressures.* BRE Note 54/80 (unpublished). Garston, Building Research Establishment, 1980.

76 **Holmes J D.** Effect of frequency response on peak pressure measurements. *Journal of Wind Engineering and Industrial Aerodynamics,* 1984, **17** 1–9.

77 **Bergh H and Tijdeman H.** *Theoretical and experimental results for the dynamic response of pressure measuring systems.* Report NLR-TR F238. National Aerospace Laboratory (The Netherlands), 1965.

78 **Irwin H P A H, Cooper K R and Girard R.** Correction of distortion effects caused by tubing systems in measurements of fluctuating pressures. *Journal of Industrial Aerodynamics,* 1979, **5** 93–107.

79 **Irwin P A.** Pressure model techniques for cladding loads. *Journal of Wind Engineering and Industrial Aerodynamics,* 1988, **29** 69–78.

80 **Durgin F H.** Instrumentation requirements for aerodynamic pressure and force measurements on buildings and structures. *Wind tunnel modeling for civil engineering applications.* (Ed: Reinhold TA), pp. 329–348. Cambridge, Cambridge University Press, 1982.

81 **Gumley S J.** Tubing systems for pneumatic averaging of fluctuating pressures. *Journal of Wind Engineering and Industrial Aerodynamics,* 1983, **12** 189–228.

82 **Gumley S J.** A detailed design method for pneumatic tubing systems. *Journal of Wind Engineering and Industrial Aerodynamics,* 1983, **13** 441–452.

83 **Holmes J D and Lewis R E.** *Optimisation of dynamic–pressure–measurement systems.* DBR Internal Paper No.86/12 (unpublished). Melbourne, Commonwealth Scientific and Industrial Research Organisation, 1986.

84 **Holmes J D and Best R J.** An approach to the determination of wind load effects on low-rise buildings. *Journal of Wind Engineering and Industrial Aerodynamics,* 1981, **7** 273–287.

85 **Surry D and Stathopoulos T.** An experimental approach to the economical measurement of spatially-averaged wind loads. *Journal of Industrial Aerodynamics,* 1977/1978, **2** 385–397.

86 **Surry D, Stathopoulos T and Davenport A G.** Wind loading of low rise buildings. *Proceedings of the Canadian structural engineering conference 1978.* Toronto, Canadian Steel Industry Construction Council, 1978.

87 **Surry D, Stathopoulos T and Davenport A G.** Simple measurement techniques for area wind loads. *Proceedings of the American Society of Civil Engineers, Structural Division,* 1983, **109** 1058–1071.

88 **Cook N J.** A sensitive 6-component high-frequency-range balance for building aerodynamics. *Journal of Physics E: Scientific Instruments,* 1983, **6** 390–393.

89 **Isyumov N and Poole M.** Wind induced torque on square and rectangular building shapes. *Journal of Wind Engineering and Industrial Aerodynamics,* 1983, **13** 183–196.

90 **Bardowicks H.** A new six-component balance and applications on wind tunnel models of slender structures. *Journal of Wind Engineering and Industrial Aerodynamics,* 1984, **16** 341–349.

91 **Tschanz T.** Measurement of total dynamic loads using elastic models with high natural frequencies. *Wind tunnel modeling for civil engineering applications.* (Ed: Reinhold TA), pp. 296–312. Cambridge, Cambridge University Press, 1982.

92 **Tschanz T and Davenport A G.** The base balance technique for the determination of dynamic wind loads. *Journal of Wind Engineering and Industrial Aerodynamics,* 1983, **13** 429–439.

93 **Hansen S O.** Static wind load measurements with a six-component high frequency balance based on piezoelectric crystals. *Journal of Wind Engineering and Industrial Aerodynamics,* 1986, **24** 87–91.

94 **Whitbread R E.** The measurement of non-steady wind forces on small-scale building models. *Proceedings of the fourth international conference on wind effects on buildings and structures, Heathrow 1975.* (Ed: Eaton K J), pp. 567–574. Cambridge, Cambridge University Press, 1977.

95 **Roy R J and Holmes J D.** Total force and moment measurement on wind tunnel models of low rise buildings. *Designing with the wind,* (Eds: Bietry J, Duchene-Marullaz P and Gandemer J), paper IX-2. Nantes, Centre Scientifique et Technique du Bâtiment, 1981.

96 **Roy R J.** Wind tunnel measurements of total loads on a mobile home. *Journal of Wind Engineering and Industrial Aerodynamics,* 1983, **13** 327–338.

97 **Newberry C W, Eaton K J and Mayne J R.** *Wind pressure and strain measurements at the Post Office Tower.* Current Paper CP 30/73. Garston, Building Research Establishment, 1973.

98 **Eaton K J and Mayne J R.** The measurement of wind pressures on two-storey houses at Aylesbury. *Journal of Industrial Aerodynamics,* 1975, **1** 67–109.

99 **Leicester R H and Hawkins B T.** Large models of low rise buildings loaded by the natural wind. *Journal of Wind Engineering and Industrial Aerodynamics,* 1983, **13** 289–300.

100 *Wind effects on buildings and structures. Proceedings of the conference held at the National Physical Laboratory, Teddington, June 1963.* London, HMSO, 1965.

101 *Proceedings of the international research seminar on wind effects on buildings and structures, Ottawa, Canada, 11–15 September 1967.* Toronto, University of Toronto Press, 1968.

102 *Proceedings of the third international conference on wind effects on buildings and structures, Tokyo, 1971.* Tokyo, Saikon Co Ltd, 1971.

103 **Eaton K J** (Ed). *Proceedings of the fourth international conference on wind effects on buildings and structures, Heathrow, 1975.* Cambridge, Cambridge University Press, 1977.

104 **Cermak J E** (Ed). *Wind engineering: proceedings of the fifth international conference, Fort Collins, Colorado, 1979.* Oxford, Pergamon Press, 1980.

105 **Cook N J.** Jensen number; a proposal. *Journal of Wind Engineering and Industrial Aerodynamics,* 1986, **22** 95–96.

106 **Leutheusser H J and Baines W D.** Similitude problems in building aerodynamics. *Proceedings of the American Society of Civil Engineers, Hydraulics Division,* 1967, **93** 35–49.

107 **Cermak J E.** Wind tunnel testing of structures. *Proceedings of the American Society of Civil Engineers, Engineering Mechanics Division,* 1977, **103** 1125–1140.

108 **Tieleman H.** Similitude criteria based on meteorological or theoretical considerations. *Wind tunnel modeling for civil engineering applications.* (Ed: Reinhold TA), pp. 73–96. Cambridge, Cambridge University Press, 1982.

109 **Stathopoulos T.** Wind loads on low-rise buildings: a review of the state of the art. *Engineering Structures,* 1984, **6** 119–135.

110 **Lawson T V.** *Wind effects on buildings, Volume 2: Statistics and meteorology.* London, Applied Science Publishers, 1980.

111 **Simiu E and Scanlan R H.** *Wind effects on structures,* (first edition). New York, John Wiley & Sons, 1978.

112 **Simiu E and Scanlan R H.** *Wind effects on structures,* (second edition). New York, John Wiley & Sons, 1986.

113 **Surry D.** Consequences of distortions in the flow including mismatching scales and intensities of turbulence. *Wind tunnel modeling for civil engineering applications.* (Ed: Reinhold T A), pp. 137–185. Cambridge, Cambridge University Press, 1982.

114 **Reinhold T A** (Ed). *Wind tunnel modeling for civil engineering applications.* Cambridge, Cambridge University Press, 1982.

115 **Davenport A G and Isyumov N.** The application of the boundary layer wind tunnel to the prediction of wind loading. *Wind effects on buildings and structures,* pp. 201–230. Toronto, University of Toronto Press, 1968.

116 **Barrett R V.** A versatile compact wind tunnel for industrial aerodynamics. *Atmospheric Environment,* 1972, **6** 491–498.

117 **Cook N J.** A boundary layer wind tunnel for building aerodynamics. *Journal of Industrial Aerodynamics,* 1975, **1** 3–12.

118 **Sykes D M.** A new wind tunnel for industrial aerodynamics. *Journal of Industrial Aerodynamics,* 1977, **2** 65–78.

119 **Wood C J.** The Oxford University 4 m × 2 m industrial aerodynamics wind tunnel. *Journal of Industrial Aerodynamics,* 1979, **4** 43–70.

120 **Blessman J.** The boundary layer TV-2 wind tunnel of the UFRGS. *Journal of Wind Engineering and Industrial Aerodynamics,* 1982, **10** 231–248.

121 **Hertig J-A.** A stratified boundary-layer wind tunnel designed for wind engineering and diffusion studies. *Journal of Wind Engineering and Industrial Aerodynamics,* 1984, **16** 265–278.

122 **Parkinson G V.** A tolerant wind tunnel for industrial aerodynamics. *Journal of Wind Engineering and Industrial Aerodynamics,* 1984, **16** 293–300.

123 **Stathopoulos T.** Design and fabrication of a wind tunnel for building aerodynamics. *Journal of Wind Engineering and Industrial Aerodynamics,* 1984, **16** 361–376.

124 **Kramer C, Gerhardt H J and Regenscheit B.** Wind tunnels for industrial aerodynamics. *Journal of Wind Engineering and Industrial Aerodynamics,* 1984, **16** 225–264.

125 **Marshal R D.** Wind tunnels applied to wind engineering in Japan. *Proceedings of the American Society of Civil Engineers, Structural Division,* 1984, **110** 1203–1221.

126 **Hansen S O and Sørensen E G.** A new boundary-layer wind tunnel at the Danish Maritime Institute. *Journal of Wind Engineering and Industrial Aerodynamics,* 1985, **18** 213–224.

127 **Owen P R and Zienkiewicz H K.** The production of uniform shear flow in a wind tunnel. *Journal of Fluid Mechanics,* 1957, **2** 521–531.

128 **Elder J W.** Steady flow through non-uniform gauzes of arbitrary shape. *Journal of Fluid Mechanics,* 1957, **5** 355–368.

129 **Lawson T V.** Methods of producing velocity profiles in wind tunnels. *Atmospheric Environment,* 1968, **2** 73–76.

130 **Armitt J and Counihan J.** The simulation of the atmospheric boundary layer in a wind tunnel. *Atmospheric Environment,* 1968, **2** 49–71.

131 Counihan J. An improved method of simulating the atmospheric boundary layer in a wind tunnel. *Atmospheric Environment,* 1969, **3** 197–214.

132 Counihan J. Simulation of an adiabatic urban boundary layer in a wind tunnel. *Atmospheric Environment,* 1973, **7** 673–689.

133 Robins A G. The development and structure of simulated neutrally stable atmospheric boundary layers. *Journal of Industrial Aerodynamics,* 1979, **4** 71–100.

134 Cook N J. On simulating the lower third of the urban adiabatic boundary layer in a wind tunnel. *Atmospheric Environment,* 1973, **7** 691–705.

135 Tieleman H W, Reinhold T A and Marshall R D. On the wind-tunnel simulation of the atmospheric surface layer for the study of wind loads on low-rise buildings. *Journal of Industrial Aerodynamics,* 1978, **3** 21–38.

136 Cook N J. Wind-tunnel simulation of the adiabatic atmospheric boundary layer by roughness, barrier and mixing-device methods. *Journal of Industrial Aerodynamics,* 1978, **3** 157–176.

137 Cermak J E. Physical modeling of the atmospheric boundary layer in long boundary-layer wind tunnels. *Wind Tunnel Modeling for Civil Engineering Applications.* (Ed: Reinhold T A), pp. 97–125. Cambridge, Cambridge University Press, 1982.

138 Cook N J. Simulation techniques for short test-section wind tunnels; roughness, barrier and mixing-device methods. *Wind Tunnel Modeling for Civil Engineering Applications.* (Ed: Reinhold T A), pp. 126–136. Cambridge, Cambridge University Press, 1982.

139 Standen N M. *A spire array for generating thick turbulent shear layers for natural wind simulation in wind tunnels.* Laboratory Technical Report LA-94. Ottawa, National Aeronautical Establishment, 1972.

140 Schon J P and Mery P. A preliminary study of the simulation of neutral atmospheric boundary-layer using air injection in a wind tunnel. *Atmospheric Environment,* 1971, **5** 299–311.

141 Nagib H M, Morkovin M V, Yung J T and Tan-Atichat J. On modelling of the atmospheric surface layers by the counter-jet technique. *American Institute of Aeronautics and Astronautics Journal,* 1976, **14** 185–190.

142 Blessman J. The use of cross-jets to simulate wind characteristics. *Journal of Industrial Aerodynamics,* 1977, **2** 37–47.

143 Cook N J. Determination of the model scale factor in wind–tunnel simulations of the adiabatic atmospheric boundary layer. *Journal of Industrial Aerodynamics,* 1978, **2** 311–321.

144 Bienkiewicz B, Cermak J E, Peterka J A and Scanlan R H. Active modeling of large-scale turbulence. *Journal of Wind Engineering and Industrial Aerodynamics,* 1983, **13** 465–475.

145 Zunz G J, Glover M J and Fitzpatrick A J. The structure of the new headquarters for the Hongkong Shanghai Banking Corporation, Hong Kong. *The Structural Engineer,* 1985, **63A** 255–284.

146 Lythe G, Surry D and Davenport A G. *Wind profiles over Hong Kong: results of experiments with a 1:2500 and a 1:500 scale model of the area.* Engineering Science Research Report BLWT-SS15. London (Ontario), University of Western Ontario, 1981.

147 Croft D D. Cladding wind pressures on the Exchange Square project, Hong Kong. *Engineering Structures,* 1984, **6** 297–306.

148 Cook N J. On applying a general atmospheric simulation method to a particular urban site for ad-hoc wind loading or wind environment studies. *Atmospheric Environment,* 1974, **8** 85–87.

149 Pasquill F. Some aspects of boundary layer description. *Quarterly Journal of the Royal Meteorological Society,* 1972, **98** 469–494.

150 Beddingfield P S and Lapraik R D. *Boundary-layers in an urban environment.* Department of Aeronautical Engineering, BSc Thesis 136. Bristol, University of Bristol, 1970.

151 Newberry P S and Obee R W. *Flow in building complexes.* Department of Aeronautical Engineering, BSc Thesis 146. Bristol, University of Bristol, 1971.

152 McKeon R J and Melbourne W H. Wind tunnel blockage effects and drag on bluff bodies in a rough wall boundary layer. *Proceedings of the third international conference on wind effects on buildings and structures, Tokyo, 1971.* pp. 263–272. Tokyo, Saikon Co Ltd, 1971.

153 Ranga Raju K G and Singh V. Blockage effects on drag of sharp-edged bodies. *Journal of Industrial Aerodynamics,* 1975/76, **1** 301–309.

154 Maskell E C. A theory for the blockage effects on bluff bodies and stalled wings in a closed wind tunnel. *Aeronautical Research Council, Reports and Memoranda No.3400.* London, HMSO, 1963.

155 Cowdrey C F. *The application of Maskell's theory of wind tunnel blockage to very large scale models.* Aero Report 1268. Teddington, National Physical Laboratory, 1968.

156 Hackett J E, Wilsden D J and Lilley D E. *Estimation of tunnel blockage from wall pressure signatures.* NASA CR15224. Washington, National Aeronautics and Space Administration, 1979.

157 Burton G R. *Tests of a semi-automated blockage correction procedure using the wall pressure signature method.* Department of Aeronautical Engineering, Report GRB/1/83. Bristol, University of Bristol, 1983.

158 **ESDU.** *Blockage corrections for bluff bodies in confined flows.* Data Item 80024. London, ESDU International, 1980.

159 **Kirrane P P and Steward S J.** *The effect of blockage on shear flow in the wind tunnel.* Department of Aeronautical Engineering, BSc Thesis 221. Bristol, University of Bristol, 1978.

160 **Hunt A.** Wind-tunnel measurements of surface pressures on cubic building models at several scales. *Journal of Wind Engineering and Industrial Aerodynamics,* 1982, **10** 137–163.

161 **Lawson T V.** The use of roughness to produce high Reynolds number flows around circular cylinders at lower Reynolds numbers. *Journal of Wind Engineering and Industrial Aerodynamics,* 1982, **10** 381–387.

162 **ESDU.** *Mean forces, pressures and flow velocities for circular cylindrical structures: single cylinders with two-dimensional flow.* Data Item 81017. London, ESDU International, 1981.

163 **Dalgliesh W A.** Comparison of model/full-scale wind pressures on a high-rise building. *Journal of Industrial Aerodynamics,* 1975, **1** 55–66.

164 **Dalgliesh W A.** Comparison of model and full–scale tests of the Commerce Court building in Toronto. *Wind Tunnel Modeling for Civil Engineering Applications.* (Ed: Reinhold TA), pp. 575–589. Cambridge, Cambridge University Press, 1982.

165 **Marshall R D.** *The measurement of wind loads on a full-scale mobile home.* Report NBSIR-77-1289. Washington, National Bureau of Standards, 1977.

166 **Holmes J D.** Comparison of model and full-scale tests of the Aylesbury house. *Wind Tunnel Modeling for Civil Engineering Applications.* (Ed: Reinhold T A), pp. 605–618. Cambridge, Cambridge University Press, 1982.

167 **Tieleman H W, Atkins R E and Sparks P R.** Model/model and full-scale/model comparisons of wind pressures on low-rise structures. *Designing with the wind.* (Eds: Bietry J, Duchenne-Marullaz P and Gandemer J), paper IV-5. Nantes, Centre Scientifique et Technique du Bâtiment, 1981.

168 **Stathopoulos T and Surry D.** Scale effects in wind tunnel testing of low buildings. *Journal of Wind Engineering and Industrial Aerodynamics,* 1983, **13** 313–326.

169 **Holmes J D.** Comparative measurements of wind pressure on a model of the full-scale experimental house at Aylesbury, England. *Journal of Wind Engineering and Industrial Aerodynamics,* 1980, **6** 181–182.

170 **Melbourne W H.** Comparison of measurements on the CAARC standard tall building model in simulated model wind flows. *Journal of Wind Engineering and Industrial Aerodynamics,* 1980, **6** 73–88.

171 **Vickery P J.** *Wind loads on the Aylesbury experimental house: a comparison between full scale and two different scale models.* Faculty of Engineering Science, MESc Thesis. London, Ontario, University of Western Ontario, 1984.

172 **Vickery P J and Surry D.** The Aylesbury experiments revisited – Further wind tunnel tests and comparisons. *Journal of Wind Engineering and Industrial Aerodynamics,* 1983, **11** 39–62.

173 **HMSO.** *Manual to the Building Regulations 1985.* London, HMSO, 1985.

174 **HMSO.** *Approved Documents, Part A Structure.* London, HMSO, 1985.

175 **Teknisk Forlag.** *Norm for last pa konstruktioner, NP-157-N.* Dansk Standard DS 410. Copenhagen, Teknisk Forlag, 1982.

176 **Teknisk Forlag.** *Loads for the design of structures, NP-157-N.* Danish Standard DS410. (English translation edition). Copenhagen, Teknisk Forlag, 1983.

177 **ANSI.** *Minimum design loads for buildings and other structures.* ANSI A58.1-1982. New York, American National Standards Institute, 1982.

178 **NRCC.** *National Building Code of Canada 1985.* NRCC No. 223174. Ottawa, National Research Council of Canada, 1985.

179 **NRCC.** *Supplement to the National Building Code of Canada 1985.* NRCC No. 23178. Ottawa, National Research Council of Canada, 1985.

180 **Stathopoulos T, Surry D and Davenport A G.** A simplified model of wind pressure coefficients for low-rise buildings. *Proceedings of the Fourth Colloquium on Industrial Aerodynamics, Aachen, June 1980.* (Eds: Kramer C and Gerhardt H J), **1** 17–31. Aachen, Fachhochschule Aachen, 1980.

181 **Mehta K C.** Wind load provisions of the new ANSI standard. *Journal of Wind Engineering and Industrial Aerodynamics,* 1983, **14** 37–48.

182 **SIA.** *Actions sur les structures.* Norme Suisse, SIA 160. Zurich, Société Suisse des Ingénieurs et des Architectes, 1988.

183 **Hertig J A.** Peak pressure coefficient distribution around low-rise buildings. *Journal of Wind Engineering and Industrial Aerodynamics,* 1986, **23** 211–222.

184 **BSI.** *British Standard Code of Practice for design of buildings and structures for agriculture. Part 1, General considerations: Section 1.2, Design, construction and loading.* BS5502:Section 1:2:1980. London, British Standards Institution, 1980.

185 **BSI.** *British Standard Code of Practice for glazing of buildings.* BS6262:1982. London., British Standards Institution, 1982.

186 **BSI.** *British Standard code of practice for slating and tiling: Part 1 – Design.* BS5534:1978. London, British Standards Institution, 1978.

187 **Hazelwood R A.** Principles of wind loading on tiled roofs and their application in the British Standard BS5534. *Journal of Wind Engineering and Industrial Aerodynamics,* 1980, **6** 113–124.

188 **Hazelwood R A.** The interaction of the two principal wind forces on roof tiles. *Journal of Wind Engineering and Industrial Aerodynamics,* 1981, **8** 39–48.

189 **BSI.** *British Standard: Lattice towers and masts: Part 1, Code of Practice for loading.* BS8100:Part1:1986. London, British Standards Institution, 1986.

190 **BSI.** *British Standard: Lattice towers and masts: Part 2, Guide to the background and use of Part 1 'Code of practice for loading'.* BS8100: Part2:1986. London, British Standards Institution, 1986.

191 **NMI.** *Wind loading on lattice towers, Part II – A method for the prediction of wind loading on lattice towers.* Report on Project No. P/352003 (unpublished). Feltham, National Maritime Institute, 1977.

192 **NMI.** *Wind loading on lattice towers, Part I – Wind-tunnel measurements on models of typical tower sections.* Report on Project No. P/352003 (unpublished). Feltham, National Maritime Institute, 1977.

193 **Scruton C and Gimpel G.** *Memorandum on wind forces and pressures exerted on structures and buildings.* NPL Aero Report 391. Teddington, National Physical Laboratory, 1959.

194 **Cowdrey C E.** *Aerodynamic forces and moments on two sections of a Forth Crossing tower.* NPL Aero Special Report 027. Teddington, National Physical Laboratory, 1969.

195 **NMI.** *Wind-load measurements on a range of configurations for a model of a typical lattice tower.* Report on Project No. P/352020 (unpublished). Feltham, National Maritime Institute, 1978.

196 **Smith B W.** *Comparison of the Draft Code of Practice for lattice towers with wind load measurements on a complete lattice tower.* Occasional Paper. Garston, Building Research Establishment, 1985.

197 **Cook N J and Mayne J R.** A new approach to the assessment of wind loads for the design of buildings. *Proceedings of the third colloquium of industrial aerodynamics, Aachen, 1978.* (Eds: Kramer C and Gerhardt HJ), **1** 294–306. Aachen, Fachhochschule Aachen, 1978.

198 **Cook N J and Mayne J R.** A refined working approach to the assessment of wind loads for equivalent static design. *Journal of Wind Engineering and Industrial Aerodynamics,* 1980, **6** 125–137.

199 **Cook N J.** Further development of a working approach to the assessment of wind loads for equivalent static design. *Journal of Wind Engineering and Industrial Aerodynamics,* 1982, **9** 389–392.

200 **Holmes J D.** Discussion of 'A refined working approach to the assessment of wind loads for equivalent static design'. *Journal of Wind Engineering and Industrial Aerodynamics,* 1981, **8** 295–297.

201 **Harris R I.** An improved method for the prediction of extreme values of wind effects on simple buildings and structures. *Journal of Wind Engineering and Industrial Aerodynamics,* 1982, **9** 343–379.

202 **Gumley S J and Wood C J.** A discussion of extreme wind-loading probabilities. *Journal of Wind Engineering and Industrial Aerodynamics,* 1982, **10** 31–45.

203 **Mayne J R and Cook N J.** Acquisition, analysis and application of wind loading data. *Wind engineering: proceedings of the fifth international conference on wind engineering, Fort Collins, Colorado, 1979.* (Ed: Cermak J E), pp. 1339–1355. Oxford, Pergamon Press, 1980.

204 **Everett T W and Lawson T V.** *Estimation of the design pressure by the quantile method in a wind tunnel investigation and its comparison with the estimated value using the Cook-Mayne method.* Department of Aeronautical Engineering, Report No. TVL/8209. Bristol, University of Bristol, 1982.

205 **Roshko A.** *On the drag and shedding frequency of two–dimensional bluff bodies.* NACA TN 3169. Washington, National Advisory Committee on Aeronautics, 1954.

206 **Hoerner S F.** *Fluid-dynamic drag.* New Jersey, S F Hoerner, 1958.

207 **Nakaguchi H, Hashimoto K and Muto S.** An experimental study on aerodynamic drag of rectangular cylinders. *Journal of the Japan Society of Aeronautical and Space Science,* 1968, **16** 1–5.

208 **ESDU.** Mean fluid forces and moments on rectangular prisms: surface-mounted structures in turbulent shear flow. Data Item 80003. London, ESDU International, 1986.

209 **Cook N J.** *The effect of turbulence scale on the flow around high-rise building models.* Department of Aeronautical Engineering, PhD Thesis. Bristol, University of Bristol, 1971.

210 **Fage A and Warsap J H.** The effects of turbulence and surface roughness on the drag of a circular cylinder. *Aeronautical Research Council, Reports and Memoranda No.1281.* London, HMSO, 1929.

211 **Delany N K and Sorensen N E.** *Low-speed drag of cylinders of various shapes.* NACA TN 3038. Washington, National Advisory Committee on Aeronautics, 1953.

212 **Basu R I.** Aerodynamic forces on structures of circular cross-section. Part 2. The influence of turbulence and three-dimensional effects. *Journal of Wind Engineering and Industrial Aerodynamics,* 1986, **24** 33–59.

213 **Gould R E, Raymer W G and Ponsford P J.** Wind tunnel tests on chimneys of circular section at high Reynolds numbers. *Proceedings of a Symposium on Wind Effects on Buildings and Structures, Loughborough University of Technology, April 1968.* (Eds: Johns D J, Scruton C and Ballantyne A M), paper 10. Loughborough, Loughborough University of Technology, 1968.

214 **Okamoto T and Yagita M.** The experimental investigation on the flow past a circular cylinder of finite length placed normal to the plane surface in a uniform stream. *Bulletin of the Japanese Society of Mechanical Engineers,* 1973, **16** 805.

215 **Farivar D.** Turbulent uniform flow around cylinders of finite length. *American Institute of Aeronautics and Astronautics Journal,* 1981, **19** 275.

216 **Osborne C H.** *Wind loading on chimneys.* BESc Thesis, Faculty of Engineering Science, University of Western Ontario. London, Ontario, University of Western Ontario, 1981.

217 **Simpson A and Lawson T V.** Oscillations of twin power transmission lines. *Proceedings of a symposium on wind effects on buildings and structures, Loughborough University of Technology, April 1968.* (Eds: Johns D J, Scruton C and Ballantyne A M), paper 25. Loughborough, Loughborough University of Technology, 1968.

218 **Bearman P W and Zdravkovich M M.** Flow around a circular cylinder near a plane boundary. *Journal of Fluid Mechanics,* 1978, **89** 33–47.

219 **ESDU.** *Mean forces, pressures and flow field velocities for circular cylindrical structures: single cylinder with two-dimensional flow.* Data Item 80025. London, ESDU International, 1981.

220 **Alridge T R, Piper B S and Hunt J C R.** The drag coefficient of finite-aspect-ratio perforated circular cylinders. *Journal of Industrial Aerodynamics,* 1978, **3** 251–257.

221 **Wong H Y.** A means of controlling bluff body flow separation. *Journal of Industrial Aerodynamics,* 1979, **4** 183–201.

222 **Wong H Y and Kokkalis A.** A comparative study of three aerodynamic devices for suppressing vortex-induced oscillation. *Journal of Wind Engineering and Industrial Aerodynamics,* 1982, **10** 21–29.

223 **Zdravkovich M M.** Review and classification of various aerodynamic and hydrodynamic means for suppressing vortex shedding. *Journal of Wind Engineering and Industrial Aerodynamics,* 1981, **7** 145–189.

224 **Galbraith R A McD.** Flow pattern around a shrouded cylinder at Re $= 5 \times 10^3$. *Journal of Wind Engineering and Industrial Aerodynamics,* 1980, **6** 227–242.

225 **Modi V J and Dikshit A K.** Mean aerodynamics of stationary elliptic cylinders in subcritical flow. *Proceedings of the third international conference on wind effects on buildings and structures, Tokyo, 1971.* (Ed: Hirai A), pp. 345–355. Tokyo, Saikon Co Ltd., 1971.

226 **Wiland E.** *Unsteady aerodynamics of stationary elliptic cylinders in subcritical flow.* University of British Columbia MASc Thesis. Vancouver, University of British Columbia, 1968.

227 **Fage A and Johansen F C.** On the flow of air behind an inclined flat plate of infinite span. *Proceedings of the Royal Society of London, Series A,* 1927, **116** 170–197.

228 **ESDU.** *Fluid forces and moments on flat plates.* Data Item 70015. London, ESDU International, 1972.

229 **Laneville A and Williams C D.** The effect of intensity and large-scale turbulence on the mean pressure and drag coefficients of 2D rectangular cylinders. *Wind engineering: proceedings of the fifth international conference, Fort Collins, Colorado, USA, July 1979.* (Ed: Cermak J E), pp. 397–404. Oxford, Pergamon Press, 1980.

230 **ESDU.** *Fluid forces, pressures and moments on rectangular blocks.* Data Item 71016. London, ESDU International, 1978.

231 **Lindsey W F.** *Drag of cylinders of simple shapes.* NACA Report 289. Washington, National Advisory Committee on Aeronautics, 1938.

232 **Prandtl L and Betz A.** Messungen von profiltragern. *Ergebnisse Aerodynamischen Versuchs., Gottingen,* 1927, III 151–156.

233 **Modi V J and Slater J E.** Unsteady aerodynamics and vortex–induced aeroelastic instability of a structural angle section. *Journal of Wind Engineering and Industrial Aerodynamics,* 1983, **11** 321–334.

234 **Raymer W G and Nixon H L.** *Drag and cross wind force of square and angle sections.* NPL Aero Report 281. Teddington, National Physical Laboratory, 1955.

235 Modi V J and Slater J E. Unsteady aerodynamics of an angle section during stationary and vortex excited oscillatory conditions. *Proceedings of the third colloquium on industrial aerodynamics, Aachen, June 1978.* (Eds: Kramer C and Gerhardt H J), **2** 79–90. Aachen, Fachhochschule Aachen, 1978.

236 Grant I and Barnes F H. The vortex shedding and drag associated with structural angles. *Proceedings of the fourth colloquium on industrial aerodynamics, Aachen, June 1980.* (Eds: Kramer C and Gerhardt H J), **2** 47–57. Aachen, Fachhochschule Aachen, 1980.

237 Maher F J and Wittig L E. Aerodynamic response of long H-sections. *Proceedings of the Americal Society of Civil Engineers, Structural Division,* 1980, **106** 183–198.

238 Hayne F. *Drag measurements on open angular rolled sections (standard sections).* Rep 1-NK-68-40 (English translation) Porz-Wahn (W Germany), Institute Angewandte Gasdynamik, 1968.

239 Maseyer M M and Khudyakov G Ye. Aerodynamic characteristics of construction channel iron. *Certain problems of experimental aerodynamics,* NASA TT-F-16565, pp. 73–76. Washington, National Aeronautics and Space Administration, 1975.

240 ESDU. *Structural members: mean fluid forces on members of various cross sections.* Data Item 82007. London, ESDU International, 1982.

241 Robertson A P. Design wind loads for ridged canopy roof structures. *Journal of Wind Engineering and Industrial Aerodynamics,* 1986, **24** 185–192.

242 Wood C J. The effect of base bleed on a periodic wake. *Journal of the Royal Aeronautical Society,* 1964, **68** 477–482.

243 Igarashi T. Flow characteristics around a circular cylinder with a slit. *Bulletin of the Japanese Society of Mechanical Engineers,* 1978, **21** 656–664.

244 Ramberg S E. The effects of yaw and finite length upon the vortex wakes of stationary and vibrating circular cylinders. *Journal of Fluid Mechanics,* 1983, **128** 81–107.

245 Mair W A and Stewart A J. The flow past yawed slender bodies, with and without ground effects. *Journal of Wind Engineering and Industrial Aerodynamics,* 1985, **18** 301–328.

246 Allen H J and Perkins E W. *A study of the effects of viscosity on flow over slender inclined bodies of revolution.* NACA Report 1048. Washington, National Advisory Committee on Aeronautics, 1951.

247 Lamont P J and Hunt B L. Pressure and force distribution on a sharp-nosed circular cylinder at large angles of inclination to a uniform subsonic stream. *Journal of Fluid Mechanics,* 1976, **76** 519–559.

248 Moll R and Thiele F. Windkanalversuche am Modell eines Stahlskelett-Hochregals zur Bestimmung des Widerstandsbeiwertes c nach DIN 1055 für ein filigranartiges Bauwerk. *Der Stahlbau,* 1972, **3** 65–72.

249 Taylor G I and Davies R M. The aerodynamics of porous sheets. *Aeronautical Research Council, Reports and Memoranda No. 2237.* London, HMSO, 1944.

250 Taylor G I. Air resistance of a flat plate of very porous material. *Aeronautical Research Council, Reports and Memoranda No. 2236.* London, HMSO, 1944.

251 Glauert H, Hirst D M and Hartshorn A S. The induced flow through a partially choked pipe with axis along the wind stream. *Aeronautical Research Council, Reports and Memoranda No. 1469.* London, HMSO, 1932.

252 Georgiou P N. *A study of the wind loads on building frames.* University of Western Ontario, MESc Thesis. London (Ontario), University of Western Ontario, 1979.

253 BSI. *British Standard: Rules for the design of cranes. Part 1 – Specification for classification, stress calculations and design criteria for structures.* BS2573:Part1:1983. London, British Standards Institution, 1983.

254 Walker H B. *Wind forces on unclad tubular structures.* Croydon, CONSTRADO, 1975.

255 Gould R W F and Raymer W G. *Measurements over a wide range of Reynolds numbers of the wind forces on models of lattice frameworks with tubular members.* NPL Maritime Science Report No 5-72. Teddington, National Physical Laboratory, 1972.

256 Bayar D C. Drag coefficients of latticed towers. *Proceedings of the American Society of Civil Engineers, Structural Division,* 1986, **112** 417–430.

257 ESDU. *Lattice structures. Part 2: mean fluid forces on tower-like space frames.* Data Item 81028. London, ESDU International, 1982.

258 Smith B W. *BRE Wind load measurements on a model of a complete lattice tower – Comparison with the Draft Code of Practice for lattice towers.* Report No. 592/1/1 (unpublished). London, Flint and Neill Partnership, 1979.

259 Eden J F, Butler A J and Patient J. A new approach to the calculation of wind forces on vertical and horizontal latticed structures. *The Structural Engineer,* 1985, **63A** 179–188.

260 Eden J F, Butler A J and Patient J. Wind tunnel tests on model crane structures. *Engineering Structures,* 1983, **5** 289–298.

261 **Eden J F, Butler A J and Patient J.** Discussion of 'A new approach to the calculation of wind forces on vertical and horizontal latticed structures'. *The Structural Engineer*, 1985, **63A** 384–387.

262 **Georgiou P N and Vickery B J.** Wind loads on building frames. *Wind engineering: proceedings of the fifth international conference, Fort Collins, Colorado, July 1979*. (Ed: Cermak J E), pp. 421–433. Oxford, Pergamon Press, 1980.

263 **Whitbread R E.** The influence of shielding on the wind forces experienced by arrays of lattice frames. *Wind engineering: proceedings of the fifth international conference, Fort Collins, Colorado, 1979*. (Ed: Cermak J E), pp. 405–420. Oxford, Pergamon Press, 1980.

264 **Sykes D M.** Lattice frames in turbulent airflow. *Journal of Wind Engineering and Industrial Aerodynamics*, 1981, **7** 203–214.

265 **Good M C and Joubert P N.** The form drag of two-dimensional bluff plates immersed in turbulent boundary layers. *Journal of Fluid Mechanics*, 1968, **31** 547–582.

266 **Ranga Raju K G, Loeser J and Plate E J.** Velocity profile and fence drag for a turbulent boundary-layer along smooth and rough flat plates. *Journal of Fluid Mechanics*, 1976, **76** 383–399.

267 **BRE.** *Wind loads on canopy roofs*. BRE Digest 284 (first edition). Garston, Building Research Establishment, 1984.

268 **Robertson A P, Hoxey R P and Moran P.** A full-scale study of wind loads on agricultural ridged canopy roof structures and proposals for design. *Journal of Wind Engineering and Industrial Aerodynamics*, 1985, **21** 167–205.

269 **Robertson A P and Moran P.** Comparisons of full-scale and wind-tunnel measurements of wind loads on a free-standing canopy roof structure. *Journal of Wind Engineering and Industrial Aerodynamics*, 1986, **23** 113–125.

270 **Blessman J.** *Efeitos do vento em edifícios e cúpulas*. Porto Alegre, Universidade Federale do Rio Grande do Sul, 1971.

271 **Yoshida M and Hongo T.** Wind tunnel study of wind forces on flat roofs with long span. *Annual Report of Kajima Institute of Construction Technology*, 1981, **29** 117–124.

272 **Yoshida M and Hongo T.** Wind tunnel study of wind forces on flat-round roofs with long span. *Annual Report of Kajima Institute of Construction Technology*, 1983, **31** 119–126.

273 **Holmes J D and Rains G J.** Wind loads on flat and curved roof low-rise buildings – Application of the covariance integration approach. *Designing with the wind*. (Eds: Bietry J, Duchene-Marullaz P and Gandemer J), paper V-1. Nantes, Centre Scientifique et Technique du Bâtiment, 1981.

274 **Greenway M E.** *Peak wind loads on cladding panels and their dependence upon panel size*. OUEL Report No.1289/79. Oxford, University of Oxford, 1979.

275 **Reinhold T A, Sparks P R, Tieleman H W and Maher F J.** The effect of wind direction on the static and dynamic wind loads on a square-section tall building. *Proceedings of the third colloquium on industrial aerodynamics, Aachen, June 1978*. (Eds: Kramer C and Gerhardt H-J), pp. 263–279. Aachen, Fachhochschule Aachen, 1978.

276 **Corke T C and Nagib H M.** Wind loads on a building model in a family of surface layers. *Journal of Industrial Aerodynamics*, 1979, **5** 159–177.

277 **Akins R E, Peterka J A and Cermak J E.** Mean force and moment coefficients for buildings in turbulent boundary layers. *Journal of Industrial Aerodynamics*, 1977, **2** 195–209.

278 **Klemin A, Schaefer E B and Beerer J G.** Aerodynamics of the perisphere and trylon at World's Fair. *Transactions of the American Society of Civil Engineers*, 1939, 1449–1472.

279 **Taniguchi S, Sakamoto H, Kiya M and Arie M.** Time-averaged aerodynamic forces acting on a hemisphere immersed in a turbulent boundary. *Journal of Wind Engineering and Industrial Aerodynamics*, 1982, **9** 257–273.

280 **Toy N, Moss W D and Savory E.** Wind tunnel studies on a dome in turbulent boundary layers. *Journal of Wind Engineering and Industrial Aerodynamics*, 1983, **11** 201–212.

281 **Blessman J.** *Influéncia do perfil do vento e do número de Reynolds nas pressõoes estáticas em modelos de cúpulas*. Porto Alegre, Universidade Federale do Rio Grande do Sul, 1971.

282 **Newman B G, Ganguli U and Shrivastava S C.** Flow over spherical inflated buildings. *Journal of Wind Engineering and Industrial Aerodynamics*, 1984, **17** 305–327.

283 **Macdonald P A, Kwok K C S and Holmes J D.** Wind loads on circular storage bins, silos and tanks: 1 Point pressure measurements on isolated structures. *Journal of Wind Engineering and Industrial Aerodynamics*, 1988, **31** 165–188.

284 **Briassoulis D and Pecknold D A.** Anchorage requirements for wind–loaded empty silos. *Proceedings of the American Society of Civil Engineers, Structural Division*, 1986, **112 308–325.**

285 **Maher F J.** Wind loads on dome-cylinder and dome-cone shapes. *Proceedings of the American Society of Civil Engineers, Structural Division*, 1966, 79–95.

286 **Holroyd R J.** On the behaviour of open-topped oil storage tanks in high winds. Part 1. Aerodynamic aspects. *Journal of Wind Engineering and Industrial Aerodynamics*, 1983, **12** 329–352.

287 **Sabransky I J and Melbourne W H.** Design pressure distribution on circular silos with conical roofs. *Journal of Wind Engineering and Industrial Aerodynamics*, 1987, **26** 65 – 84.

288 **Davenport A G and Surry D.** The pressures on low rise structures in turbulent wind. *Proceedings of the Canadian Structural Engineering Conference – 1974.* Ontario, Canadian Steel Industry Construction Council, 1974.

289 **Richardson G M.** *Full-scale wind load measurements on a single-span film plastic clad livestock building.* Divisional Note DN.1390. Silsoe, AFRC Institute of Engineering Research, 1987.

290 **Hoxey R P and Richardson G M.** *Full-scale measurements of the wind loads on film plastic clad greenhouses: Four span 26.8 m × 29.4 m greenhouse.* NIAE Divisional Note DN1119. Silsoe, National Institute for Agricultural Engineering, 1982.

291 **Gandemer J.** *Etude de la simulation des structures gonflables: la similitude aérodynamique.* Rapport Interne ADYM:11 – 74 (unpublished). Nantes, Centre Scientifique et Technique du Bâtiment, 1974.

292 **Toy N and Tahouri B.** Pressure distributions on semi-cylindrical structures of different geometrical cross-sections. Proceedings of the seventh IAWE conference, Aachen, July 1987. *Journal of Wind Engineering and Industrial Aerodynamics*, 1988, **29** 263 – 272.

293 **Holmes J D.** Determination of wind loads for an arch roof. *Institution of Engineers, Australia, Civil Engineering Transactions*, 1985, **CE26** 247 – 253.

294 **Davenport A G and Surry D.** Turbulent wind forces on a large span roof and their representation by equivalent static loads. *Canadian Journal of Civil Engineering*, 1984, **11** 955 – 966.

295 **Dalgliesh W A.** *Wind pressure measurements on the Post Office Building, Confederation Heights, Ottawa.* Division of Building Research, Internal Report 342 (unpublished). Ottawa, National Research Council of Canada, 1967.

296 **Peterka J A, Meroney R N and Kothari K M.** Wind flow patterns about buildings. *Journal of Wind Engineering and Industrial Aerodynamics*, 1985, **21** 21 – 38.

297 **Hongo T, Yoshida M, Sanada S and Nakamura O.** Experimental study of wind forces on tall buildings (Part 1) Aerostatic forces. *Annual Report of Kajima Institute of Construction Technology*, 1979, **27** 209 – 216.

298 **Surry D and Djakovich D.** Extreme suctions on tall building models. *Proceedings of the seventh IAWE conference, Aachen, July 1987.* (Eds: Kramer C and Gerhardt H-J), **3** 49 – 60. Aachen, Fachhochschule Aachen, 1987.

299 **Stathopoulos T.** Wind environmental conditions around tall buildings with chamfered corners. *Journal of Wind Engineering and Industrial Aerodynamics*, 1985, **21** 71 – 87.

300 **Holmes J D and Best R J.** *Wind pressures on an isolated high-set house.* Wind Engineering Report 1/78. Townsville, James Cook University, 1978.

301 **Holmes J D and Best R J.** *Model study of wind pressures on an isolated single-storey house.* Wind Engineering Report 3/78. Townsville, James Cook University, 1978.

302 **Holmes J D.** *Wind pressures and forces on tropical houses.* Final Report of Project No. 17 of the Australian Housing Research Council. Townsville, James Cook University, 1980.

303 **Melbourne W H.** The relevance of codification to design. *Proceedings of the fourth international conference on wind effects on buildings and structures, Heathrow, 1975.* (Ed: Eaton KJ), pp. 785 – 790. Cambridge, Cambridge University Press, 1977.

304 **Melbourne W H.** Turbulence effects on maximum surface pressures – a mechanism and possibility of reduction. *Wind Engineering: Proceedings of the fifth international conference, Fort Collins, Colorado, July 1979.* (Ed: Cermak JE), pp. 541 – 551. Oxford, Pergamon Press, 1980.

305 **Burton G R.** *Details of a mechanism by which high peripheral loads are produced on the leading edge of a flat roofed building.* Department of Aeronautical Engineering Report No. GRB/1/84. Bristol, University of Bristol, 1984.

306 **Saathoff P J and Melbourne W H.** The generation of peak pressures in separated/reattaching flows. *Proceedings of the seventh IAWE conference, Aachen 1987.* (Eds: Kramer C and Gerhardt H-J), **4** 11 – 20. Aachen, Fachhochschule Aachen, 1987.

307 **Stathopoulos T.** Wind pressures on flat roof edges and corners. *Proceedings of the seventh IAWE conference, Aachen 1987.* (Eds: Kramer C and Gerhardt H-J), **3** 39 – 48. Aachen, Fachhochschule Aachen, 1987.

308 **Castro I P and Dianat M.** Surface flow patterns on rectangular bodies in thick boundary layers. *Journal of Wind Engineering and Industrial Aerodynamics*, 1983, **11** 107 – 119.

309 **Stathopoulos T, Surry D and Davenport A G.** Effective wind loads on flat roofs. *Proceedings of the American Society of Civil Engineers, Structural Division*, 1981, **107** 281 – 298.

310 **Maruta E.** *The study of high wind regions around tall buildings.* PhD Thesis. Tokyo, Nihon University, 1984.

311 **Maruta E.** *A tentative prediction method for high wind regions around tall buildings.* BRE Translation (Unpublished). Garston, Building Research Establishment, 1984.

312 **Stathopoulos T.** Wind pressures on flat roofs with parapets. *Proceedings of the American Society of Civil Engineers, Structural Division,* 1987, **113** 2166–2180.

313 **Sakamoto H and Arie M.** Flow around a cubic body immersed in a turbulent boundary layer. *Journal of Wind Engineering and Industrial Aerodynamics,* 1982, **9** 275–293.

314 **Stathopoulos T and Mohammadian A R.** Wind loads on low buildings with mono-sloped roofs. *Journal of Wind Engineering and Industrial Aerodynamics,* 1986, **23** 81–97.

315 **Blessmann J.** *Ação do vento em coberturas curvas, 1a Parte.* Caderno Tecnico CT-86. Porto Alegre, Universidade Federale do Rio Grande do Sul, 1987.

316 **Blessman J.** *Vento em coberturas curvas – pavilhões vizinhos.* Caderno Tecnico CT-88. Porto Alegre, Universidade Federale do Rio Grande do Sul, 1987.

317 **Holmes J D.** *Wind loading of saw-tooth roof buildings, I – Point pressures.* Internal Report No. 83/17. Melbourne, Commonwealth Scientific and Industrial Research Organisation, 1983.

318 **Holmes J D.** *Wind loading of saw-tooth roof buildings, II – Panel pressures.* Internal Report No. 84/1. Melbourne, Commonwealth Scientific and Industrial Research Organisation, 1984.

319 **Holmes J D.** *Wind loading of single- and multi-span industrial buildings.* Internal Report No. 84/14. Melbourne, Commonwealth Scientific and Industrial Research Organisation, 1984.

320 **Moran P and Westgate G R.** *Wind loads on farm buildings: (9) Full-scale measurements on a 36.2 m long twin 20.6 m span ridged roof building of 4.8 m eaves height and 16° roof pitch.* NIAE Divisional Note DN1127. Silsoe, National Institute for Agricultural Engineering, 1982.

321 **Blessman J.** *Pressões causadas por vento turbulento e deslizante em telhados a duas águas com calha central.* Caderno Tecnico CT-26/81. Porto Alegre, Universidade Federale do Rio Grande do Sul, 1981.

322 **Blessman J.** Wind pressures on roofs with negative pitch. *Journal of Wind Engineering and Industrial Aerodynamics,* 1982, **10** 213–230.

323 **Kind R J.** *Estimation of critical wind speeds for scouring of gravel or crushed stone on rooftops.* LTR-LA-142. Ottawa, National Research Council of Canada, 1974.

324 **Kind R J.** *Wind tunnel tests on some building models to measure wind speeds at which gravel is blown off rooftops.* LTR-LA-162. Ottawa, National Research Council of Canada, 1974.

325 **Kind R J.** *Further wind tunnel tests on building models to measure wind speeds at which gravel is blown off rooftops.* LTR-LA-189. Ottawa, National Research Council of Canada, 1977.

326 **Maruta E.** On blowoff of rooftop gravel and concrete blocks. *Journal of Wind Engineering (Japan),* 1982, **11** 19–33.

327 **Kind R J and Wardlaw R L.** *Design of rooftops against gravel blow-off.* NRC No. 15544. Ottawa, National Research Council of Canada, 1976.

328 **Kind R J and Wardlaw R L.** The development of a procedure for the design of rooftops against gravel blow-off and scour in high winds. *Proceedings of the Symposium on Roofing Technology, September 21–23, 1977.* pp. 112–123. Washington, National Bureau of Standards, 1977.

329 **Vermeulen P E J and Visser G Th.** Determination of similarity criteria for wind-tunnel model testing of wind flow patterns close to building facades. *Journal of Wind Engineering and Industrial Aerodynamics,* 1980, **6** 243–259.

330 **Vermeulen P E J.** Wind-flow patterns close to building facades. *Proceedings of the fourth colloquium on industrial aerodynamics, Aachen, June 1980.* (Eds: Kramer C, Gerhardt H-J, Ruscheweyh H and Hirsch H), **1** 187–197. Aachen, Fachhochshule Aachen, 1980.

331 *Research project 'Determination of the aerodynamic factors for scaffolding'.* Berlin, Institut für Bautechnik, 1981.

332 **Yoshida M, Sanada S, Hongo T and Nakamura O.** Wind tunnel test on zone of high wind velocity around tall buildings. *Annual report of Kajima Institute of Construction Technology,* 1976, **23** 253–270.

333 **Jancauskas E D and Eddleston J D.** Wind loads on canopies at the base of tall buildings. *Proceedings of the Seventh IAWE Conference, Aachen 1987.* (Eds: Kramer C and Gerhardt H-J), **3** 1–10. Aachen, Fachhochshule Aachen, 1987.

334 **Stathopoulos T and Zhu X.** Wind pressures on buildings with various surface roughnesses and appurtenances. *Proceedings of the fourth international conference on tall buildings, Hong Kong and Shanghai, 1988.* (Eds: Cheung Y K and Lee P K K), pp. 438–444. Hong Kong, Organising Committee, 1988.

335 **Tieleman H W, Atkins R E and Sparks P R.** *An investigation of wind loads on solar collectors.* Blacksburg, Virginia, Virginia Polytechnic Institute, 1980.

336 **Machado C G and Blessman J.** *Pressões causadas por vento turbulento em telhados a duas águas de proporções em planta 4 × 1, com e sem lanternim fechado.* Caderno Tecnico CT-80. Porto Alegre, Universidade Federale do Rio Grande do Sul, 1985.

337 **Silveira N I B and Blessman J.** *Pressões causadas por vento turbulento em telhados a duas águas com e sem lanternim fechado.* Caderno Tecnico. CT-A-63. Porto Alegre, Universidade Federale do Rio Grande do Sul, 1984.

338 **Lee Y, Tanaka H and Shaw C Y.** Distribution of wind-and temperature-induced pressure differences across the walls of a twenty-storey compartmentalised building. *Journal of Wind Engineering and Industrial Aerodynamics,* 1982, **10** 287–301.

339 **Tanaka H and Lee Y.** Stack effect and building internal pressure. Proceedings of the seventh IAWE conference, Aachen, July 1987. *Journal of Wind Engineering and Industrial Aerodynamics,* 1988, **29** 293–302.

340 **Warren P R and Webb B C.** Ventilation measurements in housing. *Proceedings of the CIBS symposium on natural ventilation by design, BRE, Garston, 1980.* Garston, Building Research Establishment, 1980.

341 **Shaw C Y, Sander D M and Tamura G T.** Air leakage measurements of the exterior walls of tall buildings. *American Society of Heating, Refrigerating and Air-conditioning Engineers, Transactions,* 1973, **79**.

342 **Tamura G T and Shaw C Y.** Studies on exterior wall air tightness and air infiltration of buildings. *American Society of Heating, Refrigerating and Air-conditioning Engineers, Transactions,* 1976, **82**.

343 **Persily A K and Grot R A.** Pressurisation testing of federal buildings. ASTM STP 904. *Measured air leakage of buildings.* (Eds: Trenchsel H R and Lagus P L). Philadelphia, American Society of Testing Materials, 1986.

344 **Phaff J C and de Gids W F.** *Ventilation in small utility buildings: Measurements on air-leaks in inside walls.* Institute for Environmental Hygiene and Health Technology, Report C522. Delft, TNO (Netherlands Organisation for Applied Scientific Research), 1983.

345 **Rayleigh** (Lord). *Theory of sound.* London, Macmillan, 1896.

346 **Holmes J D.** Mean and fluctuating pressure induced by wind. *Wind engineering: proceedings of the fifth international conference, Fort Collins, Colorado, 1979.* (Ed: Cermak J E), pp. 435–450. Oxford, Pergamon Press, 1980.

347 **Liu H and Saathoff P J.** Building internal pressure: sudden change. *Proceedings of the American Society of Civil Engineers, Engineering Mechanics Division,* 1981, 309–321.

348 **Liu H and Rhee K H.** Helmholtz oscillation in building models. *Journal of Wind Engineering and Industrial Aerodynamics,* 1986, **24** 95–115.

349 **Harris R I.** The propagation of internal pressures in buildings. *Journal of Wind Engineering and Industrial Aerodynamics,* (in press).

350 **Vickery B J.** Gust factors for internal pressures in low rise buildings. *Journal of Wind Engineering and Industrial Aerodynamics,* 1986, **23** 259–271.

351 **Holdø A E, Houghton E L and Bhinder F S.** Effects of permeability on wind loads on pitched-roof buildings. *Journal of Wind Engineering and Industrial Aerodynamics,* 1983, **12** 255–279.

352 **Cook N J.** Reduction of wind loads on a grandstand roof. *Journal of Wind Engineering and Industrial Aerodynamics,* 1982, **10** 373–380.

353 **Withey M A.** *A study of internal pressure distributions in a multi-cellular building and related wind engineering.* Final year project report, BTech(Hons) in Building Technology. London, Brunel University.

354 **Cook N J, Keevil A P and Stobart R K.** BRERWULF – The big bad wolf. Proceedings of the seventh IAWE conference, Aachen 1987. *Journal of Wind Engineering and Industrial Aerodynamics,* 1988, **29** 99–107.

355 **Kind R J and Wardlaw R L.** Failure mechanisms of loose-laid roof-insulation systems. *Journal of Wind Engineering and Industrial Aerodynamics,* 1982, **9** 325–341.

356 **Kind R J, Savage M G and Wardlaw R L.** Prediction of wind-induced failure of loose laid roof cladding systems. Proceedings of the seventh IAWE conference, Aachen, July 1987. *Journal of Wind Engineering and Industrial Aerodynamics,* 1988, **29** 29–37.

357 **Amano T, Fujii K and Tazaki S.** Wind loads on permeable roof-blocks in roof insulation systems. Proceedings of the seventh IAWE conference, Aachen, July 1987. *Journal of Wind Engineering and Industrial Aerodynamics,* 1988, **29** 39–48.

358 **Ganguli U and Dalgliesh W A.** Wind pressures on open rain screen walls: Place Air Canada. *Proceedings of the American Society of Civil Engineers, Structural Division,* 1988, **114** 642–656.

359 **Gandemer J.** The aerodynamic characteristics of windbreaks, resulting in empirical design rules. *Journal of Wind Engineering and Industrial Aerodynamics,* 1981, **7** 15–36.

360 **Lee B E and Soliman B F.** An investigation of the forces on three dimensional bluff bodies in rough wall turbulent boundary layers. *Transactions of the American Society of Mechanical Engineers, Journal of Fluids Engineering,* 1977, **99** 503–510.

361 **Hussain M and Lee B E.** A wind tunnel study of the mean pressure forces acting on large groups of low-rise buildings. *Journal of Wind Engineering and Industrial Aerodynamics,* 1980, **6** 207–225.

362 **Soliman B F.** *Effect of building group geometry on wind pressure and properties of flow.* Department of Building Science, Report BS29. Sheffield, University of Sheffield, 1976.

363 **Hussain M and Lee B E.** *An investigation of wind forces on three dimensional roughness elements in a simulated atmospheric boundary layer, Part 1: Flow over isolated roughness elements and the influence of upstream fetch.* Department of Building Science, Report BS55. Sheffield, University of Sheffield, 1980.

364 **Hussain M and Lee B E.** *An investigation of wind forces on three dimensional roughness elements in a simulated atmospheric boundary layer flow, Part II: Flow over large arrays of identical roughness elements and the effect of frontal and side aspect ratio variations.* Department of Building Science, Report BS56. Sheffield, University of Sheffield, 1980.

365 **Hussain M and Lee B E.** *An investigation of wind forces on three dimensional roughness elements in a simulated atmospheric boundary layer flow. Part III: The effect of central model height relative to the surrounding roughness arrays.* Department of Building Science, Report BS57. Sheffield, University of Sheffield., 1980.

366 **Holmes J D and Best R J.** *A wind tunnel study of wind pressures on grouped tropical houses.* Wind Engineering Report 5/79. Townsville, James Cook University, 1979.

367 **Wiren B G.** Effects of surrounding buildings on wind pressure distributions and ventilative heat losses for a single-family house. *Journal of Wind Engineering and Industrial Aerodynamics,* 1983, **15** 15–26.

368 **English E C and Durgin F H.** A wind tunnel study of shielding effects on rectangular prismatic structures. *Proceedings of the fourth US national conference on wind engineering research, Seattle, Washington, August 1981.*

369 **English E C.** Shielding factors from wind-tunnel studies of mid-rise and high-rise structures. *Proceedings of the fifth US national conference on wind engineering, Texas Tech University, Lubbock, Texas, November 1985.* (Eds: Mehta K C and Dillingham R A), **4A** 49–56. Lubbock, Texas Tech University, 1985.

370 **Kamei I and Maruta E.** Study on wind environmental problems caused around buildings in Japan. *Journal of Industrial Aerodynamics,* 1979, **4** 307–331.

371 **ECCS.** *Recommendations for calculating the effects of wind on constructions.* Technical Committee 12 – Wind, Report No. 52, Second Edition. Brussels, European Convention for Constructional Steelwork, 1987.

372 **Wise A F E.** Effects due to groups of buildings. *Philosophical Transactions of the Royal Society of London A,* 1971, **269** 469–485.

373 **Britter R E and Hunt J C R.** Velocity measurements and order of magnitude estimates of the flow between two buildings in a simulated atmospheric boundary layer. *Journal of Industrial Aerodynamics,* 1979, **4** 165–182.

374 **Hamilton G F.** *Effect of velocity distribution on wind loads on walls and low buildings.* Toronto, University of Toronto, 1962.

375 **Blessman J and Riera J D.** Interaction effects in neighbouring tall buildings. *Wind Engineering: Proceedings of the fifth international conference, Fort Collins, Colorado, 1979.* (Ed: Cermak J E), pp. 381–395. Oxford, Pergamon Press, 1980.

376 **Blessman J.** *Efeitos do vento em edificios alteados vizinhos.* Caderno Tecnico CT-39/83. Porto Alegre, Universidade Federale do Rio Grande do Sul, 1983.

377 **Blessman J and Riera J D.** Wind excitation of neighbouring tall buildings. *Journal of Wind Engineering and Industrial Aerodynamics,* 1985, **18** 91–103.

378 **Ponsford P.** *Some wind-pressure measurements on a model of a group of closely-spaced cylindrical silos.* NPL Aero Note 1088. Teddington, National Physical Laboratory, 1970.

379 **ESDU.** *Cylinder groups: mean forces on pairs of long circular cylinders.* Data Item 84015. London, ESDU International, 1984.

380 **Saffir H S.** Discussion of 'Hurricane related window glass damage in Houston'. *Proceedings of the American Society of Civil Engineers, Structural Division,* 1986, 204–206.

381 **Buller P S J.** *Gale damage to buildings in the UK – an illustrated review.* Garston, Building Research Establishment, 1986.

382 **Buller P S J.** *The October gale of 1987: damage to buildings and structures in the south-east of England.* Garston, Building Research Establishment, 1988.

383 **Walker G R.** *Report on cyclone Tracy – Effect on buildings – December 1974.* Melbourne, Dept of Housing and Construction, 1975.

384 **SAA.** *Design and installation of self-supporting metal roofing without transverse laps.* Australian Standard AS 1562–1973. Sydney, Standards Association of Australia, 1973.

385 Trollope D H. *Cyclone Althea: Part 1 – Buildings.* Townsville, James Cook University, 1972.
386 Beck V R. *Fatigue and static loading of shear panels.* Report MRL 38/6. Melbourne, BHP Melbourne Research Laboratory, 1974.
387 Morgan J W and Beck V R. Failure of sheet-metal roofing under repeated wind loading. *Civil Engineering Transactions, Institution of Engineers, Australia,* 1977, **CE19** 1–5.
388 Melbourne W H. Loading cycles for simulation of wind loading. *Proceedings of the Workshop on Guidelines for Cyclone Product Testing and Evaluation.* Sydney, Experimental Building Station, 1977.
389 Beck V R and Stevens L K. Wind loading failures of corrugated roof cladding. *Civil Engineering Transactions, Institution of Engineers, Australia,* 1979, **CE21** 45–56.
390 Beck V R. Loads – the basis of repeated loading and impact criteria. *Design for tropical cyclones,* paper L. Townsville, James Cook University, 1978.
391 Beck V R and Morgan J W. *Appraisal of metal roofing under repeated wind loading – Cyclone Tracy, Darwin 1974.* Technical Report No. 1., Australian Department of Housing and Construction, 1975.
392 SAA. *SAA loading code, Part 2 – Wind forces.* Australian Standard AS 1170, Pt2 – 1975. Sydney, Standards Association of Australia, 1975.
393 DRC. *Darwin area building manual.* Darwin, Darwin Reconstruction Commission, 1975.
394 Walker G R. A simplified wind loading code for small buildings in tropical cyclone prone areas. Proceedings of the seventh IAWE conference, Aachen 1987. *Journal of Wind Engineering and Industrial Aerodynamics,* 1988, **30** 163–171.
395 Holmes J D. Wind action on Glass and Brown's Integral. *Engineering Structures,* 1985, **7** 226–230.
396 Davenport A G. *The estimation of load repetitions on structures with application to wind induced fatigue and overload.* London, Ontario, University of Western Ontario, 1966.
397 Lynn B A and Stathopoulos T. Wind-induced fatigue on low metal buildings. *Proceedings of the American Society of Civil Engineers, Structural Division,* 1985, **111** 826–839.
398 Redfearn D. *A test rig for proof-testing building components against wind loads.* BRE Information Paper IP19/84. Garston, Building Research Establishment, 1984.
399 ESDU. *Mean fluid forces and moments on cylindrical structures: polygonal sections with rounded corners including elliptical shapes.* Data Item 79026. London, ESDU International, 1980.
400 SIA. *Normen für die belastungsannahmen, die inbetriebnahme ünd die überwachungsbauten. Schweizer Norm SN.160.* Zurich, Schweizerischer Ingenieur-ünd Architekten-Verien, 1956.
401 SAA. *Minimum design load on structures, Part 2 – wind forces.* (Draft Australian Standard for comment, AS 1170 Part 2.) Sydney, Standards Association of Australia, 1987.

Amendments to Part 1

Page 109: replace Eqn 5.24 with –

$$P\{M,N\} = \frac{N!}{M!\,(N-M)!}\, P^{N-M}\, Q^M \qquad (5.24)$$

Page 219: line 13 of §9.3.2.3.2: Replace value '1.24' with –

1.12

Page 227: Eqn (9.20): Replace second line of equation with –

$$= \left[1 - \frac{\ln(z_{o,b}/z_{o,a})}{0.42 + \ln m_o} \right] \frac{\ln(10/z_{o,a})}{\ln(10/z_{o,b})} \frac{S_{E,b}}{S_{E,a}} \qquad (9.20)$$

Page 292: Figure 10.8: Replace caption on horizontal axis with –

Structural size parameter, $l = \sqrt{h^2 + b^2}$ *(m)*

Page 294: lines 11 to 13 of §10.8.2.1: Replace (deleting Eqn 10.12) with –

over flat terrain (§9.4.3.2)), calculated for a gust duration $t = 1\text{s}$ which represents quasi-steady response, the base from which R, is calculated.

Index